Ecological Studies, Vol. 218

Analysis and Synthesis

Ecological Studies

Further volumes can be found at springer.com

Volume 201
The Everglades Experiments: Lessons for Ecosystem Restoration (2008)
C. Richardson (Ed.)

Volume 202
Ecosystem Organization of a Complex Landscape: Long-Term Research in the Bornhöved Lake District, Germany (2008)
O. Fränzle, L. Kappen, H.-P. Blume, and K. Dierssen (Eds.)

Volume 203
The Continental-Scale Greenhouse Gas Balance of Europe (2008)
H. Dolman, R.Valentini, and A. Freibauer (Eds.)

Volume 204
Biological Invasions in Marine Ecosystems: Ecological, Management, and Geographic Perspectives (2009)
G. Rilov and J.A. Crooks (Eds.)

Volume 205
Coral Bleaching: Patterns, Processes, Causes and Consequences
M.J.H van Oppen and J.M. Lough (Eds.)

Volume 206
Marine Hard Bottom Communities: Patterns, Dynamics, Diversity, and Change (2009)
M. Wahl (Ed.)

Volume 207
Old-Growth Forests: Function, Fate and Value (2009)
C. Wirth, G. Gleixner, and M. Heimann (Eds.)

Volume 208
Functioning and Management of European Beech Ecosystems (2009)
R. Brumme and P.K. Khanna (Eds.)

Volume 209
Permafrost Ecosystems: Siberian Larch Forests (2010)
A. Osawa, O.A. Zyryanova, Y. Matsuura, T. Kajimoto, R.W. Wein (Eds.)

Volume 210
Amazonian Floodplain Forests: Ecophysiology, Biodiversity and Sustainable Management (2010)
W.J. Junk, M.T.F. Piedade, F. Wittmann, J. Schöngart, P. Parolin (Eds.)

Volume 211
Mangrove Dynamics and Management in North Brazil (2010)
U. Saint-Paul and H. Schneider (Eds.)

Volume 212
Forest Management and the Water Cycle: An Ecosystem-Based Approach (2011)
M. Bredemeier, S. Cohen, D.L. Godbold, E. Lode, V. Pichler, P. Schleppi (Eds.)

Volume 213
The Landscape Ecology of Fire (2011)
D. McKenzie, C.S. Miller, D.A. Donald (Eds.)

Volume 214
Human Population: Its Influences on Biological Diversity (2011)
R.P. Cincotta and L.J. Gorenflo (Eds.)

Volume 215
Plant Desiccation Tolerance (2011)
U. Lüttge, E. Beck, D. Bartels (Eds.)

Volume 216
Forest Hydrology and Biogeochemistry: Synthesis of Past Research and Future Directions (2011)
D.F. Levia, D. Carlyle-Moses, T. Tanaka (Eds.)

Volume 217
Peatlands of the Western Guayana Highlands, Venezuela: Properties and Paleogeographic Significance of Peats (2011)
J.A. Zinck and O. Huber (Eds.)

Volume 218
Aralkum - a Man-Made Desert: The Desiccated Floor of the Aral Sea (Central Asia) (2012)
S.-W. Breckle, W. Wucherer, L.A. Dimeyeva, N.P. Ogar (Eds.)

Siegmar-W. Breckle • Walter Wucherer • Liliya A. Dimeyeva • Nathalia P. Ogar
Editors

Aralkum - a Man-Made Desert

The Desiccated Floor of the Aral Sea (Central Asia)

Springer

Editors
Professor Dr. Siegmar-W. Breckle
University of Bielefeld
24-26, Wasserfuhr
33619 Bielefeld, Germany
sbreckle@gmx.de

Dr. Walter Wucherer
Michael Succow Foundation
University of Greifswald
Grimmer Str. 88
17487 Greifswald, Germany
walter.wucherer@succow-stiftung.de

Dr. Liliya A. Dimeyeva
Institute of Botany and Phytointroduction
Timiryazeva Str. 36-D
050040 Almaty, Kazakhstan
l.dimeyeva@mail.ru

Professor Dr. Nathalia P. Ogar
CRS@GIS „Terra"
83 Zhabaeva Str.
Microrayon Kok-Tobe
050010 Almaty, Kazakhstan
ogar@gis-terra.kz

ISSN 0070-8356
ISBN 978-3-642-21116-4 e-ISBN 978-3-642-21117-1
DOI 10.1007/978-3-642-21117-1
Springer Heidelberg Dordrecht London New York

Library of Congress Control Number: 2011935933

Printed on acid-free paper

Springer is part of Springer Science+Business Media (www.springer.com)

Preface

This book gives an updated picture of the various and dynamic aspects of the Aralkum. This new desert on the desiccated seafloor of the Aral Sea is a unique phenomenon. It started only in 1960, when the Aral seashore began to retreat. The hydrological equilibrium of this endorheic basin became disturbed. The desiccation of the Aral Sea has changed the environmental conditions in the whole Aral Sea basin. The increasing areas with newly formed salt deserts and their powdery salt crusts led to a serious situation in the increasing frequency and intensity of salt-dust storms. The present huge salt desert area of the desiccated seafloor, we call it the Aralkum, is now comparable with the Iranian salt desert or the Great Basin salt desert in Utah (USA). But the fundamental difference is the timescale. The latter deserts were formed by a geological process, whereas the Aralkum has been formed within a few decades, meaning a more than 100-fold speed of development. The area of the Aralkum is now about 60,000 km, and the Aralkum has reached its hydrological dynamic equilibrium (state in 2009).

Research in the area of the Aralkum was done only from a few sides. During 1977–1985 under the leadership of the Geographical Institute in Moscow, the consequences of the diversion of the irrigation water from the Amu Darya and the Syr Darya in the framework of a complex research programme were studied. Within this project, also floristic and phytocoenological investigations started. The area of the Aralkum was then about 20,000 km^2 and salt deserts had only just started to develop; the recovery of the Aral Sea seemed to be realistic.

After the dissolution of the Soviet Union and the formation of independent states in Central Asia, some international organisations started small research programmes (UNESCO).

Since 1998 a BMBF project has been active; it is coordinated by the Department of Ecology at the University of Bielefeld. When this project started, the area of the Aralkum started to exceed the area of the remnant water body of the Aral Sea. The environmental situation became catastrophic for the people in the villages and towns in the whole Aral basin. It became clear that the development of further desiccation could not be stopped. In 1994 the ever-increasing desert was termed "Aralkum" (Breckle and Agachanjanz 1994). The geographical situation changed totally, maps were no longer correct and nobody knew about the ongoing "natural" dynamics of this new huge desert area.

The first BMBF project ("Succession processes on the desiccated sea floor of the Aral Sea and perspectives of land-use", 1998–2001) was basically oriented to the new Aralkum, less to the important delta areas of the two main tributaries and their suffering tugai forests. The aims of the project were to study the vegetation dynamics, the spontaneous invasion of species and the possible future uses. The primary succession on the new land surface was very fast and very dynamic, and the temporal and spatial changes from year to year were tremendous and not predictable.

The results of the first project were presented on various occasions, e.g. BMBF-UNESCO Meeting (May 2000, Tashkent, Uzbekistan), German Association for Ecology (GfÖ) Congress (September 2000, Kiel, Germany; September 2001, Basel, Switzerland), Annual meeting of "Desert*Net" (January 2000, Bonn, Germany, February 2002, Bonn, Germany), Desert ecology working group (November 2000, Leipzig, Germany), Annual Meeting of the Reinhold-Tüxen-Association (February 2002, Hannover, Germany), Vegetation Symposium annual meeting (August, 2001, Weihenstephan, Germany), Annual Meeting of British Ecology (December 2001, Birmingham, UK), and Annual Meeting of the British Association for Ecology (August 2001, Warwick, UK), 4th Conference of the Parties (UNCCD, December 2000, Bonn, Germany), 5th Conference of the Parties (UNCCD, October 2001, Geneva, Switzerland).

The BMBF workshop "Ecological problems of sustainable land-use in deserts" (5–9 May 1999, AZK = Arbeitnehmer-Zentrum Königswinter) was also organised by the BMBF project. Results were published in the monograph *Sustainable Land Use in Deserts* (Breckle et al. 2001).

The second project ("Combating desertification and rehabilitation of salt deserts in the Aralkum", 2002–2005) had a more applied concept. The ongoing dynamics of the flora and vegetation was the basis for applying plantation experiments on the desiccated seafloor in order to optimize phytomeliorative effects on salty substrates. But in parallel the development of a practicable strategy for nature conservation for a larger area of Northern Aralkum in order to enhance the unique halophytic biodiversity was envisaged. Both were only possible by including measures of capacity building in the region and resource management on the local and regional levels. This was a starting point for improving information exchange between all stakeholders and to improve living conditions of local people. Cooperation took place, e.g., with the Center for Developmental Studies (ZEF, Bonn, Germany), with the GTZ-CCD project (Bonn, Germany), with other universities and with several organisations in Kazakhstan and Uzbekistan, and especially with some villages (Bugen, Karateren) in the affected region.

The results were presented at several meetings and symposia: International Association for Vegetation Science (July 2004, Kona, Hawaii, USA; July 2005, Lisbon, Portugal), International Association for Great Lakes Research (June 2003, Chicago, USA), International Congress of Succulent Plants (June 2004, Hamburg, Germany), International Botanical Congress (July 2005, Vienna, Austria), GfÖ (September 2004, Gießen, Germany; September 2005, Regensburg, Germany), Desert Ecology working group of the GfÖ (November 2004, Rauischholzhausen, Germany), Ecosystems Research working group of the GfÖ (May 2004, Duderstadt, Germany), Annual Meeting of Desert*Net (December 2004,

Bonn, Germany), Regional Environmental Politics as a Contribution for Sustainable Development in Central Asia (November 2004, Berlin, Germany), Symposium Research in Central Asia (April 2005, Gießen, Germany), Biodiversity – Gradients in Drylands international symposium (September 2005, Bielefeld, Germany), International Conference on Ecological Restoration in East Asia (June 2006, Osaka, Japan), Fifth Ecological Conference on Ecological Restoration (August 2006, Greifswald, Germany), LTER – All Scientists' Meeting (September 2006, Estes Park, Colorado, USA), and ECOST meeting Putting Halophytes to Work – from Genes to Ecosystems" (March 2010, Naples, Italy). Additionally, results were presented in Central Asia during several meetings and symposia, e.g. the workshop "Combating desertification and resource management in the Aral Sea region" (March 2004, Almaty, Kazakhstan) together with the GTZ-CCD international workshop "North-eastern Aral Sea - perspectives of sustainable development and nature conservation at Barsa-Kelmes" (September 2004, Aralsk, Kazakhstan). In subsequent years, several talks took place at various universities in Germany.

Both projects were done in close cooperation with the Institute of Botany and Phytointroduction of the Ministry of Education and Science, Almaty, Kazakhstan.

Activity on the second project was furthermore realized with help of Barsa-Kelmes Nature Reserve staff. Joint work has initiated conservation activity on a new level and searching for funds for future projects. The international workshop "Possibilities for the establishment of biosphere reserves in the north-eastern Aral region on the basis of the Barsakelmes Nature Reserve" was organized under the support of the International Aral Rehabilitation Fund (September 2006, Aralsk, Kazakhstan). Several activities have been done in the framework of this project: zoning of the future biosphere territory, sociological study and interviews with local people, popularization of the idea about a biosphere reserve in local communities through booklets, posters, ecological education lessons, etc.

This book gives not only a synthesis of both projects, but also includes other details from new projects and sources; it is a comprehensive interdisciplinary treatise of the knowledge on the Aralkum.

For cooperation and technical help we thank Uta Breckle, Diana Wucherer, UFZ Centre for Environmental Research Leipzig-Halle (Germany), Michael Succow Foundation for Nature Protection University of Greifswald (Germany), The Hebrew University of Jerusalem (Israel), Institute of Botany, Academy of Science, Almaty (Kazakhstan), Institute of Forest Economy, Schutschinsk (Kazakhstan), Institute of Agroecology, Kzyl-Orda (Kazakhstan), Barsa-Kelmes Nature Reserve (Kazakhstan), and Syrboi Ormany Gardening Co., Kzyl-Orda (Kazakhstan)

Great help for the first project by Dr. Joachim Kutscher (BMBF project coordinator, Jülich, Germany) and Dr. Vefa Moustafaev (UNESCO, Paris, France) is greatly acknowledged. Great help for the second project by Dr. Ingo Fitting (BMBF project coordinator, Jülich and Berlin, Germany) and by Günter Winckler and Dr. Anneke Trux (GTZ-CCD project, Bonn, Germany) as well as by DAAD, is greatly acknowledged too.

The editors gratefully acknowledge the great help and advice of Prof. Otto-Ludwig Lange. They acknowledge Dr. Andrea Schlitzberger at Springer-Verlag in

Heidelberg for her continued interest, her patience and her help during preparation of this book, Dr. Dieter Czeschlik for his steady support, as well as Mr. Ejaz Ahmad and his team (Chennai, India) for his excellent cooperation during proof-reading and processing of the book.

Bielefeld	Siegmar-W. Breckle
Greifswald	Walter Wucherer
Almaty	Liliya Dimeyeva
May 2011	Nathalia P. Ogar

Contents

Part C Means of Present and Future Development

Contributors

Zauresh Alimbetova Nature Reserve, Aralsk, Kyzyl-Orda Oblast, Kazakhstan

Madina Batyrbayeva Institute of Space Research, Joint-Stock Company National Centre of Space Research and Technology of the Republic of Kazakhstan, Almaty, Kazakhstan

Oleg Belyalov Kazakhstan Bird Conservation Union, Almaty, Kazakhstan, Belyalov

Siegmar-W. Breckle Department of EcologyUniversity of Bielefeld, Bielefeld, Germany

Olaf Bubenzer Institute of Geography, Heidelberg University, Heidelberg, Germany

Allan Buras Institute of Botany and Landscape Ecology, University of Greifswald, Greifswald, Germany

Yuriy Chikin Institute of Zoology, Uzbek Academy of Sciences, Tashkent, Uzbekistan

Liliya A. Dimeyeva Institute of Botany and Phytointroduction, Academy of Science, Almaty, Kazakhstan

Tatjana Dujsebajeva Institute of Zoology, Ministry of Education and Sciences, Almaty, Kazakhstan

Boris V. Gel'dyev CRS@GIS "Terra", Almaty, Kazakhstan

Galina V. Gel'dyeva Institute of Geography, Academy of Science, Almaty, Kazakhstan

Y.A. Grachev Institute of Zoology, Ministry of Education and Sciences, Almaty, Kazakhstan

Jürgen Groeneveld UFZ-Umweltforschungszentrum, Ökologische Systemanalyse, Leipzig, Germany

Daniela Guicking Morphology and Systematics of Plants, University of Kassel, Germany

Florian Jeltsch Vegetation Ecology and Nature Conservation, University of Potsdam, Potsdam, Germany

Ulrich Joger Naturhistorisches Museum Niedersachsen, Braunschweig, Germany

Rustem Kadyrbekov Institute of Zoology, Ministry of Education and Sciences, Almaty, Kazakhstan

Vladimir S. Kaverin Institute for Forest Economy, Schuchinsk, Kazakhstan

Eva Kleinn (deceased), formerly Deutsche Gesellschaft für Technische Zusammenarbeit GmbH, CCD-Projekt, Bonn, Germany

Zh. V. Kuz'mina Water Problems Institute of the Russian Academy of Sciences, Moscow, Russia

Fabian Löw Faculty of Geography, University of Heidelberg, Germany

Claude Miaud Laboratory of Alpine Ecology, Université de Savoie, UMR CNRS 5553 LECA-GPB, Le Bourget du Lac Cedex, France

Peter Navratil RSS - Remote Sensing Solutions GmbH, Munich, Baierbrunn, Germany

Zinoviy B. Novitskiy Scientific Production Center for Decorative Gardening and Forestry of the Republic of Uzbekistan, Darkhan Village, Taschkent Region, Uzbekistan

Nathalia P. Ogar Institute of Botany and Phytointroduction, Academy of Science, Almaty, Kazakhstan

Leah Orlovsky The Jacob Blaustein Institute for Desert Research, Ben-Gurion University of the Negev, Beer Sheva, Israel

Oleg E. Semenov Kazakh Research Institute for Ecology and Climate, KAZNIIEK, Almaty, Kazakhstan

Lev Spivak Institute of Space Research, Joint-Stock Company National Centre of Space Research and Technology of the Republic of Kazakhstan, Academy of Science, Almaty, Kazakhstan

Alexei Terechov Institute of Space Research, Joint-Stock Company National Centre of Space Research and Technology of the Republic of Kazakhstan, Almaty, Kazakhstan

Sergey Y. Treshkin Institute of Bioecology, Uzbek Academy of Sciences, Nukus, Uzbekistan

Irina Vitkovskaya Institute of Space Research, Joint-Stock Company National Centre of Space Research and Technology of the Republic of Kazakhstan, Academy of Science, Almaty, Kazakhstan

Günter Winckler Deutsche Gesellschaft für Technische Zusammenarbeit GmbH, CCD-Projekt, Tulpenfeld 2, Bonn, Germany

Christian Wissel UFZ-Umweltforschungszentrum, Ökologische Systemanalyse, Leipzig, Germany

Walter Wucherer Foundation for Nature Protection, Institute of Botany and Landscape Ecology, University of Greifswald, Greifswald, Germany

Khasii Zhamantikov Institute for Ecology and Agriculture, Kyzyl-Orda, Kazakhstan

Abbreviations

°C	Degree Celsius
ADB	Asian Development Bank
AF	Adaptation Fund
asl	Above sea level
AVHRR	Advanced Very High Resolution Radiometer
BMBF	German Federal Ministry of Education and Research
BMZ	German Federal Ministry for Economic Cooperation and Development
BP	Before present
ca.	Circa
C	Central
CA	Central Asia
CAC	Central Asian Countries
CACILM	Central Asian Countries Initiative for Land Management
CD	Combating desertification
CEC	Cation exchange capacity
CGIAR	Consultative Group for International Agricultural Research
COP	Conference of the parties (of UN-Convention member countries)
DAAD	Deutscher Akademischer Austauschdienst (German Academic Exchange Service)
DLR	Deutsches Zentrum für Luft- und Raumfahrt
dm	Dry matter
DNA	Desoxi-ribonucleic acid
DSF70	Desiccated sea floor from the 1970s
E	East
EC	Electric conductivity
EC	European Community
ET	Evapotranspiration
FAO	Food and Agriculture Organization of the United Nations
fw	Fresh weight
GDP	Gross domestic product

GEF	Global Environment Facility
GHCN	Global historical climate network
GIS	Geographical information system
GPS	Global Positioning System
GTZ	Gesellschaft für Technische Zusammenarbeit, now called GIZ
ICARDA	International Centre for Agricultural Research in Dry Areas
IFAS	International Fund for Saving the Aral Sea
IFPRI	International Food Policy Research Institute (Washington)
IPCC	Intergovernmental Panel on Climate Change
ISSR	Inter-simple sequence repeat
K	Kelvin
ka	Thousand years
KKGM	Hydrometeorological Center of Karakalpakstan
LADA	Land Degradation Assessment in Drylands
LAI	Leaf area index (m^2 m^{-2})
LCCS	Land Cover Classification System
m	Metre
Ma	Million years ago
MODIS	Moderate Resolution Imaging Spectroradiometer
MPa	Megapascal
N	North
NAP	National action programme
NDVI	Normalized differentiated vegetation index
NGO	Nongovernmental organization
NOAA	National Oceanic and Atmospheric Administration
N_t	Total nitrogen
p	Precipitation
PDE	Partial differential equation
PET	Potential evapotranspiration
PFT	Plant functional type
R	Rainfall, precipitation
RH	Relative humidity of the air
S	South
SRAP	Subregional action programme
T	Temperature
t	Tonne
UNCBD	United Nations Convention on Biological Diversity
UNCCD	United Nations Convention to Combat Desertification
UNDP	United Nations Development Programme
UNEP	United Nations Environment Programme
UNFCCC	United Nations Framework Convention on Climate Change
UNESCO	United Nations Educational, Scientific and Cultural Organization (Paris)

UzGM	Hydrometeorological Centre of Uzbekistan
vpd	Vapour pressure deficit
W	West
WMO	World Meteorological Organization

Prefixes for Units and Appropriate Factors

10^1	Ten	deca	da	10^{-1}	deci	d	tenth
10^2	Hundred	hecto	h	10^{-2}	centi	c	hundreth
10^3	Thousand	kilo	k	10^{-3}	milli	m	thousandth
10^6	Million	mega	M	10^{-6}	micro	μ	millionth
10^9	Billion[a]	giga	G	10^{-9}	nano	n	
10^{12}	Trillion[a]	tera	T	10^{-12}	pico	p	
10^{15}	Quadrillion	peta	P	10^{-15}	femto	f	
10^{18}	Quintillion	exa	E	10^{-18}	atto	a	

[a] in German: 10^9 is Milliarde, 10^{12} is Billion, 10^{15} is Billiarde, 10^{18} is Trillion etc.

Chapter 1
The Aralkum, a Man-Made Desert on the Desiccated Floor of the Aral Sea (Central Asia): General Introduction and Aims of the Book

S-W. Breckle and W. Wucherer

What is the meaning of "Aral"? The name Aral Sea is not easy to explain. Berg (1908) gave some explanations. He cited several papers where the Aral Sea is called the Sea of the Islands, e.g. by the Russian topographer Rychkov (1762). This interpretation was also taken by von Humboldt (1844) in his book *Zentralasien*, as well as by Berg (1901) in his first publication on the Aral Sea.

The khan of Khiva Abulgazi (1603–1663) in his history book writes, "...1644 I became the khan in the country *Aral*, which is located at the inflow of the Amu to the sea...". The region of the Amu Darya delta was apparently called "Aral". The old Russian geographers called the Uzbeks from the Amu Darya delta the "Aralthsy", meaning the "Aral settlers". Since the delta area is a complex region with many islands and changing water courses, the original word "Aral" was apparently used for this area of islands; later it was used for the whole Aral Sea.

In the sixteenth century the Aral Sea was also named the Blue Sea (*Sinee More*) by Russian cartographers and travellers, as e.g. in the *Kniga Bolshogo Chertezha* (1647). The map by Remesov of 1697 (after Berg 1908) used "Aral Sea" (*Aralsko More*). The Russian czar Peter the Great in an official declaration in 1697 also used the term *Aralsko More* as a synonym for Khiva Sea. In 1716, Peter the Great sent an expedition under the command of Prince Bekovich-Cherkassky to Khiva with a diplomatic mission and to investigate the Uzboi River and the Amu Darya and its relation to the Aral Sea; at that time this term was already in common use. The expedition, however, was destroyed by the khan of Khiva.

In *Kniga Bolshogo Chertezha* (1647) there is a description of the northern coast of the Blue Sea (Berg 1908). Here, the sandy desert in the northeast is called the Kara-kum ("the black sand"; in Uzbek and Turkish *kara* means "black"), which we today call the Priaralski Karakum. In 1664–1665, the Dutchman N. Witsen visited Moscow and translated several texts from the *Kniga Bolshogo Chertezha* (1647). His book (Witsen 1672) used the term "Arakum" (in Dutch *Arakoem*) instead of Karakum for the sandy areas along the northeastern coast. According to the Russians, he used the term "Blue Sea".

S.-W. Breckle et al. (eds.), *Aralkum - a Man-Made Desert: The Desiccated Floor of the Aral Sea (Central Asia)*, Ecological Studies 218,
DOI 10.1007/978-3-642-21117-1_1, © Springer-Verlag Berlin Heidelberg 2012

In fact the term “Aralkum”, which we used (Agachanjanz and Breckle 1993), is from 1672. The term “Kyzylkum” is derived from its red sands (in Uzbek and Turkish *kyzyl* means “red”).

The Aral Sea was the fourth largest lake on the globe 50 years ago. Today, the Aral Sea no longer exists.

The Aral Sea basin is located in a vast-stretching tectonic depression. This Aralo-Caspian basin in Central Asia exhibits a dry desert climate. Today, only four more or less saline lakes and some salt swamps exist. The former Aral Sea is (after the dissolution of the Soviet Union) part of Uzbekistan and Kazakhstan (Fig. 1.1). But also Turkmenistan, Kyrgyzstan and Tajikistan and even Afghanistan have territories which are part of the Aralo-Caspian basin.

The Aral Sea no longer exists. Only the Small Aral Sea now has a constant water level. The huge desiccated seafloor (Walter and Breckle 1994; Breckle et al. 1998, 2001) is the source of salt-dust storms and sandstorms. These are affecting large proportions of the villages and agricultural systems in the whole region, as well as the livelihood and health of the people. In the coming years these saline areas will increase even more (Breckle and Wucherer 2006). Within about 50 years a huge

Fig. 1.1 Geographical Map of Middle Asia, indicating countries and main cities around former Aral Sea

new desert has developed (Fig. 2.7). This desert is caused by human activities and thus is an artificial desert, but all the ongoing processes follow natural laws and are most interesting for science. The area can be called the largest primary succession experiment of mankind (Wucherer and Breckle 2001), or depending on one's viewpoint, one of the biggest ecological catastrophes (Letolle and Mainguet 1996).

There are many plans for tackling and combating the disastrous situation caused, e.g., by the dangerous salt-dust storms (Chaps. 5–7). Public awareness of this giant catastrophe has grown considerably in the last few decades, especially in the 1980s and in the 1990s; this is also documented by increasing scientific interest (Zonn et al. 2009). On the other hand, the exploration of the area in general started relatively late; there have been successful expeditions only since $<$200 years ago, e.g. in 1848–1849 by Butakov (Uhrig 2008).

The present tragic situation may be improved if the recommendations for conservation and restoration of the sea are applied. Those concepts envisage transformation of development strategies in the region, modernization of the irrigation systems, afforestation and phytoreclamation in the area (Levintanus 1992), as is shown in later chapters.

Within the last 20 years many new developmental projects have been performed. However, most of them need basic knowledge which can only be provided by scientific expertise. So it is necessary to bring together not only interdisciplinary and international scientists in the fields of ecology and geography (with remote sensing, using geographical information systems), but also climatologists, soil scientists, social scientists, economists, nature conservation organizations, developmental and health agencies, other main stakeholders in the area, administrations, decision makers and regional and national politicians.

There were few attempts to study the steadily increasing desiccated seafloor until 1995. Then the BMBF (German Federal Ministry of Education, Research and Technology) financed two research projects of the Department of Ecology at the University of Bielefeld. The first was to study "Succession processes on the desiccated seafloor of the Aral Sea and perspectives of land-use" (1998–2001); the second was to study "Combating desertification and rehabilitation of salt deserts in the Aralkum" (2002–2005). Many open questions needed answers. What ecosystems develop on the desiccated seafloor? What plants may invade those new salt deserts and what vegetation may develop? Can the natural succession be accelerated by seedings or plantings?

Both projects revealed not only interesting and new scientific data and results but they were also able to contribute to better socioeconomic conditions in some villages, and to increase the participatory involvement of stakeholders at the grass-roots level by capacity building. The cooperation with scientific institutions in Kazakhstan and Uzbekistan and with the relevant political administration on various administrative levels gave good opportunities for information exchange and increased mutual understanding in future applications of phytomelioration measures and nature conservation topics.

Sandstorms affect settlements and villages, since their close surroundings are overgrazed and all woody vegetation is cut down. Sand dunes started to develop and to move. This is a local problem in many areas. It should be solved by local wind shelter programmes and rational grazing management, involving activities of all inhabitants.

Salt-dust storms originate mostly from the younger seafloor surface covered by puffy-crusty solonchaks, and from abandoned salinized fields with salt crusts. This is a regional problem. It can only be solved by huge plantings of heat- and frost-resistant as well as drought- and salt-resistant woody plant species to minimize the effects of wind-speed and deflation processes on the soil surface of the fields and especially on the dry seafloor. Huge experimental plantings (Chaps. 15–17) with plots up to 250 ha have shown that only very few species are suitable for this purpose: *Haloxylon aphyllum* and *Halocnemum strobilaceum*. Some other species still need to be tested accordingly (*Halostachys caspica* and *Tamarix*, *Ammodendron*, *Suaeda* and *Salsola* species). Furthermore, recent evaluation of various experimental sets revealed that special techniques to plant saplings have to be applied and have to be adjusted to the relevant soil situation.

A variety of desert types have been described in the literature. On the basis of surface properties, the following main types have been mentioned: rocky and blocks surfaces (hamada); gravely surfaces (serir); stone pavement surfaces (reg); sandy surfaces (erg), clay surfaces (takyr) and saline surfaces (sabkha, playa or salina). In deserts all these types occur often rather mixed or along catenas. However, about only 15% of deserts is covered with sand fields (erg) (Breckle et al. 2008). The main deserts of the globe are listed in Table 1.1, as are the percentages of sand cover areas within them.

In the Aralkum, about 20% (about 12,000 km^2) of the area is sand and sandy-loamy deserts, mainly the first parts of the seafloor which became dry in the 1960s and 1970s (Fig. 2.7). The majority is salt desert (70%, about 42,000 km^2), and this situation is rather different from that in other deserts, where sometimes no salt desert may exist (e.g. in the Negev, the Sonoran Desert or the Chihuahua; Table 1.1). The remaining area (10%) is wetlands and remnants of tugai scrub, as well as transformed landscapes that are under the influence of the Amu Darya and the Syr Darya.

From an ecological point of view, sand deserts offer more favourable conditions than other desert types for plant cover and species diversity. This is due to the special characteristics of the sand, namely low water absorption, rapid infiltration rate and especially low evaporation losses (since capillary threads reach only 10 or 20 cm). Thus, all sandy areas represent, even in dry arid deserts, water-storing bodies. The only limitation for plant establishment and plant cover is a high frequency of extreme wind speeds that limit surface stability. In areas where the frequency of extreme wind speeds is low, the sand surface is relatively stable. Under such conditions the establishment of plants takes place spontaneously, leading to increased surface stability by further reducing wind speed and sand mobility, often favoured by biotic crust formation as the primary process (Breckle et al. 2008). This is an important fact for phytoreclamation programmes.

The changes in the Aral Sea region within the last 50 years can be briefly characterized as follows (data for 2009):

Table 1.1 Major drylands/deserts of the World

Name, type of desert	Surface area (10^6 km^2)	Sand desert area (10^3 km^2)	Salt desert area (10^3 km^2)	Location	Desert types, in brackets: (approximate percentage of sand desert), [approximate percentage of salt desert]
Sahara, subtropical ZBIII (including the Egyptian desert east of the Nile)	9.25	2750	220	Northern Africa	70% gravel, rock plains. Contrary to popular belief, the desert is only <30% sand (several erg fields) [2%]
Arabian, subtropical ZBIII	2.59	620	25	Arabian Peninsula	Gravel plains, rocky highlands; one quarter is the Rub al-Khali ("Empty Quarter"), the world's largest expanse of unbroken sand (25%) [1%]
Australian deserts, subtropical ZBIII, ZEII-III (Great Victoria, Great Sandy, Gibson, Simpson Sturt, Stewart)	1.38	400	60	Australia	Sandhills, gravel, rocks, grassland, Simpson parallel sand dunes are the longest in the world: up to 200 km (30%) [4%]
Gobi, cold winter ZBVIIa, ZBVII(rIII)	1.33	200	30	China, Mongolia	Stony, sandy soil (15%), steppes and dry grasslands [2%]
Patagonian, cold winter ZBVIIa	0.67	0	7	Argentina	Gravel plains, plateaus, basalt sheets (0%) [1%]
Kalahari, subtropical ZBIII, ZBII	0.57	400	4	South Africa, Botswana, Namibia	Sand sheets, longitudinal dunes (70%) [1%]
Great Basin, cold winter ZBVIIa	0.49	15	140	USA	Mountain ridges, valleys, sand dunes (3%) [30%]
Thar, subtropical ZEII-III	0.45	45	10	India, Pakistan	Rocky sand and sand dunes (10%) [2%]
Chihuahua, subtropical ZBIII, ZBII(rIII)	0.44	9	c.0	Mexico	Grassland, cacti savannah (2%) [0%]
Taklamakan, cold winter ZBVII(rIII)	0.36	290	4	China	Sand dunes (80%), up to 300 m high; gravel [1%]
Iranian deserts, cold winter ZBVII, ZBIII (Registan)	0.35	35	70	Iran, Afghanistan	Salt, gravel, rock, sand fields (10%) [20%]
Colorado Plateau, cold winter ZBVIIa	0.34	0	0	USA	Sedimentary rock, mesas, and plateaus – the Grand Canyon, "Painted Desert" (0%) [0%]

(continued)

Table 1.1 (continued)

Name, type of desert	Surface area (10^6 km^2)	Sand desert area (10^3 km^2)	Salt desert area (10^3 km^2)	Location	Desert types, in brackets: (approximate percentage of sand desert), [approximate percentage of salt desert]
Sonoran, subtropical ZBIII, ZEII-III	0.31	15	c.0	USA, Mexico	Cacti savannah, gravel (5%) [0%]
Kyzylkum, cold winter ZBVII(rIII)	0.30	240	10	Uzbekistan Turkmenistan, Kazakhstan	Sands, rock – name means "red sand", (80%) [3%]
Atacama, cool coastal ZBIII (Altiplano)	0.18	20	20	Chile, Peru, Bolivia	Salt basins, sand (10%), lava; world's driest desert, mountains [10%]
Mojave, subtropical ZBVIIa	0.14	15	7	USA	Mountain chains, dry alkaline lake beds, calcium carbonate dunes (12%) [5%]
Aralkum ZBVII(rIII) (new desert)	0.06	12	42	Kazakhstan, Uzbekistan	Desiccated seafloor, sand desert (20%) [70%]
Sinai (part of Sahara) ZBIII	0.060	3	c.0	Egypt	Mountains, rocks, gravel, sand (5%) [0%]
Namib, cool coastal ZBVII	0.034	15	0.5	Angola, Namibia, South Africa	Gravel plains, huge sand dunes (50%), up to 300 m [1%]
Negev ZBIII (part of Sahara)	0.013	1	c.0	Israel	Rocks, gravel, sand (5%) [0%]

According to various sources, average values. Zonobiomes (*ZB*) and Zonoecotones (*ZE*) according to Breckle (2002), after Breckle et al. (2008)

- Formation of (older) sand deserts and (younger) salt deserts
- The water level dropped by about 28 m [from 53 m above seal level (asl) to less than 25 m asl, the South Aral Sea, or "Large Aral Sea"].
- The surface area has shrunk to about one-tenth of the original area (from 67,100 to 7,000 km^2); thus, the remnant water bodies today are the North Aral Sea ("Small Aral Sea"), the West Aral Sea, the East Aral Sea, which in 2009 was almost only a huge salt swamp, but in 2010 was again a very flat sea (both forming the "Large Aral Sea") in the south, and the small Kulandy basin in the northwest.
- More than 100 km of the eastern cost has changed; the new land surface area of the dry seafloor is called the Aralkum (in 2009 about 60,000 km^2).
- The water volume decreased tremendously (from 1,100 km^3 to approximately 100 km^3), and has reached $<10\%$ of the 1960 value.
- This goes in parallel with a drastic increase in salinity (from 0.9% to 6–10% for the Large Aral Sea), which means the southern remnants of the Aral Sea are hypersaline.
- The water level of the North Aral Sea is stabilized by a dam at 42 m asl and the surface area amounts to 3,200 km^2.

The main causes for lack of water are the huge irrigation areas (for cotton, rice, etc.), which were established from 1960 during Soviet Union times. Giant canals for huge irrigation projects were constructed; water from the Syr Darya in the northeast as well as from the Amu Darya in the south was used (Breckle et al. 1998, Giese et al. 1998). These two rivers are the main tributaries for the Aral Sea. Water-saving techniques were totally ignored. Today, most irrigation systems are more or less modernized and diversified to fulfil the necessities of food production for the increasing populations of the six independent states of the region concerned. Thus, only small amounts of water are left for the remnant Aral Sea.

We will use the following terms according to geographical tradition in Russia and Germany. *Middle Asia* is mainly the Aralo-Caspian basin, including Uzbekistan, Kyrgyzstan and Turkmenistan, parts of Kazakhstan, western Tajikistan and northwestern Afghanistan. *Central Asia* is the region east of Middle Asia, mainly east of the Altai and Pamir and north of the Himalaya (Walter 1974), including most parts from Afghanistan to Xinjiang and Mongolia. *Inner Asia* is the high mountain regions of eastern Pamir, Tibet and parts of western China (Agachanjanz and Breckle 1993, Breckle 2002), often regarded as part of Central Asia. Often, however, "Central Asia" is used as general term or as a superimposed concept for all three regions.

The book is divided into four main sections. The first section provides an overview of the regional physical characteristics of the area and covers the geological, pedological, geomorphological and climatological aspects and their dynamics, especially the dust-storm dynamics.

The second section focuses on the biotic aspects. The spatial and temporal patterns of flora and vegetation, and the fauna. Especially, the halophytes play a major role in ecosystem development.

The third section covers the studies and projects dealing with attempts to improve the situation, mainly by nature conservation and by large projects of phytomelioration. Thus, this section presents a synthesis on the application of scientific results to combat desertification also regarding the issue of the sensitivity of the area to climate change. The chapters in this section go beyond ecology and touch on the socioeconomic aspects of this ecological disaster area (Erdinger et al. 2011).

The last section give the overall conclusions.

The aim of this book is to bring together the results from long-term studies on the ever-increasing desiccated seafloor of the Aral Sea and the formation of the Aralkum, as well as new developments to combat the various serious desertification processes, which almost have more than regional dimensions, and if possible, to reverse the disastrous developments by adaptive rehabilitation projects and new approaches.

Many co-authors were very distant from each other and to bring together these results took some time and patience. This was only possible by the new techniques in telecommunication, e.g., the World Wide Web and e-mail.

It is expected that this volume will be useful to biologists, plant and soil ecologists, conservation biologists, dryland foresters and agronomists, desert ecologists and geographers, geomorphologists, policy makers, site practitioners, researchers, lecturers, tutors and students, and many others with an interest in deserts and also in degradation and combating desertification.

References

Agachanjanz OE, Breckle S-W (1993) Umweltsituation in der ehemaligen Sowjetunion. Naturwiss Rdschau 47:99–106

Berg LS (1901) Ocherk phisicheskoi geographii Aralskogo morya (Physical geography of the Aral Sea). Zemlevedeniye (Earth Science), vol. 8, kn. 3–4 (Russian)

Berg LS (1908) The Aral Sea. An experience of physico-geographical monograph. Izv turkestanskogo Otd. Russk Geogr Obshchestva, St Petersburg, Vol 5, N.9. (Russian)

Breckle S-W (2002) Walter's vegetation of the earth: the ecological systems of the geo-biosphere. Springer, Heidelberg

Breckle S-W, Wucherer W (2006) Combatting desertification in the Northern Aral Sea region. In: Gao J et al (eds) Restoration and stability of ecosystems in arid and semi-arid areas. Science Press, Beijing, pp 304–316

Breckle S-W, Agachanjanz OE, Wucherer W (1998) Der Aralsee: Geoökologische Probleme. Naturwiss Rdschau 51:347–355

Breckle S-W, Veste M, Wucherer W (2001) Sustainable land use in deserts. Springer, Heidelberg

Breckle S-W, Veste M, Yair A (eds) (2008) Arid dune ecosystems: the Nizzana sands in the Negev desert. Ecology studies. Springer, Heidelberg

Erdinger L, Hollert H, Eckl P (2011) Aral Sea: An Ecological Disaster Zone with Impact on Human Health. Encyclopedia of Environmental Health, Elsevier/Burlington, p. 136–144

Giese E, Bahro G, Betke D (1998) Umweltzerstörungen in Trockengebieten Zentralasiens (West- und Ost-Turkestan). Ursachen, Auswirkungen, Maßnahmen. Vol 125. Erdkundliches Wissen, Stuttgart

Humboldt AV (1844) Zentralasien Untersuchungen zu den Gebirgsketten und zur vergleichenden Klimatologie. Das Reisewerk zur Expedition von 1829. Klemann/Berlin, Bd. 1, 2

Letolle R, Mainguet M (1996) Der Aralsee – eine ökologische Katastrophe. Springer, Berlin

Levintanus A (1992) Saving the Aral Sea. J Environ Manage 36:193–199
Rychkov PI (1762) Topographiya Orenburgskaya, to est' obstoyatelnoe opisanie Orenburgskoi gubernii, sochinennoe Petrom Rychkovym (Topography of the district Orenburg). Chast' 1–2. Sankt Peterburg. (Russian)
Uhrig M-R (ed.) (2008) Butakov Aleksej Ivanovic: Tagebuch der Aralsee-Expedition 1848/49. Edit. Buran/Zell 138 pp
Walter H (1974) Die Vegetation Osteuropas, Nord- und Zentralasiens. In: Vegetationsmonographien, vol 7. Fischer, Stuttgart
Walter H, Breckle S-W (1994) Ökologie der Erde, vol 3, 2nd edn, Spezielle Ökologie der gemäßigten und arktischen Zonen Euro-Nordasiens. Fischer, Stuttgart
Witsen N (1672) Noord-en Oost-Tartarye. Amsterdam (First edit.) [2.ed.: 1692]
Wucherer W, Breckle S-W (2001) Vegetation dynamics on the dry sea floor of the Aral Sea. In: Breckle S-W, Veste M, Wucherer W (eds) Sustainable land use in deserts. Springer, Heidelberg, pp 52–68
Zonn I, Glantz M, Kostianoy A, Kosarev A (2009) The Aral Sea encyclopedia. Springer, Berlin

Part A
Abiotic Conditions and Dynamics

Chapter 2
Dynamics of the Aral Sea in Geological and Historical Times

S-W. Breckle and G.V. Geldyeva

2.1 Introduction

The Aral Sea is a rather flat intercontinental water body within the temperate deserts of Middle Asia. The two tributaries – the Amu Darya and the Syr Darya – are the largest and most important river systems of Central Asia and Middle Asia. They are the main rivers and thus are water sources for Kazakhstan, Uzbekistan, Turkmenistan, Kyrgyzstan, Tajikistan and Afghanistan. They are all part of the huge tectonic basin forming the lowlands of Turan (Walter 1974).

The Aral Sea basin is characterised by five main geographical regions (Letolle and Mainguet 1996): (1) in the north the very flat, dry semideserts of central and northern Kazakhstan with the Barsuki Desert adjacent to the North Aral Sea, and in the south the subtropical hot deserts of Middle Asia, consisting of various sand deserts, (2) the Karakum in the southwest, (3) the Kyzylkum in the east, (4) the less sandy, more clayey plains of Ustyurt Plateau in the west, between the Caspian Sea, the Urals and the Aral Sea, and (5) finally the two rather big delta areas of the Amu Darya and the Syr Darya (Fig. 2.7).

The Aral Sea was the fourth largest lake on the globe until 1960, with a surface area of about 68,000 km^2. Mainly the huge irrigation projects in many parts of Middle Asia were responsible for the catastrophic desiccation of the Aral Sea within the last five decades. The ecological and the socioeconomic effects within the whole area have many diverse aspects (Glazovskii 1990; Glantz et al. 1993; Letolle and Mainguet 1996; Giese et al. 1998; Breckle et al. 2001; Breckle and Wucherer 2005; Micklin 2007; Micklin and Aladin 2008; Erdinger et al. 2011).

Also in former periods the Aral Sea exhibited rather drastic changes in its water budget with rather abrupt oscillations (Svitoch 2009). The huge tectonic basin is mainly fed by water from the extant mountains in the south and southeast (Boomer et al. 2009; Ni et al. 2004). The Caspian Sea west of the Aral Sea is mainly fed by water from the Volga, this means by water from northern territories with dense forest vegetation and a different climate regime. Historical changes in water balance are thus not comparable or even were often contrasting.

S.-W. Breckle et al. (eds.), *Aralkum - a Man-Made Desert: The Desiccated Floor of the Aral Sea (Central Asia)*, Ecological Studies 218,
DOI 10.1007/978-3-642-21117-1_2, © Springer-Verlag Berlin Heidelberg 2012

The present situation of the Aral Sea is characterised by the fact that only remnants of the former water body are left. Before 1960 the Small (North) Aral Sea and the Large (South) Aral Sea were one water body, connected by two water channels west and east of the island of Kokaral. The main role in water exchange between the two parts was played by the eastern connection, the "Berg Crossbar", a connection between both basins, named after the famous geographer L.S. Berg. It was about 15 km wide and 10–15 m deep. The northwestern connection between the former island of Kokaral and the Auzy-Kokaral peninsula was very shallow, only up to 1.0–1.5 m deep, and very narrow (up to 0.6 km).

The South Aral Sea depression had an area of 60,000 km^2 and a former water volume of 985 km^3. The South Aral Sea is now divided into the deep western basin and the shallow eastern basin, separated by the former island of Vozrozhdeniya, which is now part of the mainland.

The North Aral Sea is a small and shallow water body, which until 1960 was about one-ninth of the whole sea area. About 42% of the area of the North Aral Sea has a water depth of 10–20 m. The maximal depth was 29.5 m, the mean 13.3 m. The North Aral Sea has a dyke between Shevchenko and Butakov bays.

2.2 Geology and Tectonics in the Aral Basin in the Tertiary and Quaternary

The geological history of the Aral Sea is rather long. During the Mesozoic in most parts huge sedimentation within the Tethys took place. During the early Tertiary, the uplift of the old continental blocks between the Arabian plate and the drifting Indian plate caused the Tethys to disappear, and only some basins (the Mediterranean Sea, the Black Sea, the Caspian Sea and the Aral Sea) remained as water basins. The sediments are Jurassic and Cretaceous and consist of limestone, clay marls, sandstones and rather often evaporites, such as salt (NaCl), gypsum and even potassium salts, indicating various strong phases of desiccation, regressions and transgressions. The tectonic plate movements to the north caused and is still causing the formation of mountain ridges, partly consisting of those sediments. The Himalaya, the Karakorum, the Pamirs, the Hindu Kush and the Tien Shan form more or less the southern surroundings of the huge Turan basin. Also, some smaller mountain chains developed, as the Nuratau, Tamdytau, and Bukantau, now separating the upper basins of the Syr Darya and the Amu Darya.

During the Oligocene, a vast but flat sea developed, called the Sarmatian Sea, stretching from the Mediterranean Sea to the Caspian Sea and almost to the Altai mountains. The Sarmatian sediments are mainly clay, sand, limestone and conglomerates but also some evaporites, indicating a rather arid climate. During the Miocene, mainly limestone sediments developed. Some steep slopes at the former coastline are limestone chinks, others are from the later Pliocene period. During the Miocene, most parts became dry, except in the south. During the Pliocene, again most parts were inundated by a flat sea, which was less extant than the Sarmatian Sea. This sea was connected to the Caspian Sea and to the Black Sea. The connection to the Caspian Sea was mainly due to the old Palaeo-Oxus Valley south of Ustyurt Plateau. During the Pliocene, parts of the limestone layers of this river had already been eroded, forming steep chinks. This dry valley is still visible today; it is the Ungus, forming the border between Ustyurt and the Karakum.

During the Pliocene, tectonic movements continued as well as uplifts and breaking of blocks. During these slight changes, the various basins which we have today were formed, but slight movements changed the course of the rivers apparently several times. The Palaeo-Oxus discharged its water through the Uzboi, the Ungus or the Kelif Darya close to the Afghan border, but the exact facts are not known. Figure 2.1 indicates most of the old riverbeds and valleys of the last few millennia. The very dynamic hydrological activity within the last few 1,000 years in the whole Turan area is partly explained by the changing water balance of the vast input areas, but also most probably by slight tectonic movements.

In ancient times the Amu Darya was called the Oxus, the Syr Darya was called the Iaraxtes or Jakartes, and the Seravshan was called the Polytimetus. Their riverbeds changed all the time.

During the Quaternary, rather often changes of sea levels of the Caspian Sea occurred, and some of these were tremendous. There is a typical terrace formation in the Terekol basin in southern Siberia which is 60 m above the present level of the Caspian Sea. The overflow of the Caspian Sea over the Uzboi Valley to the Aral

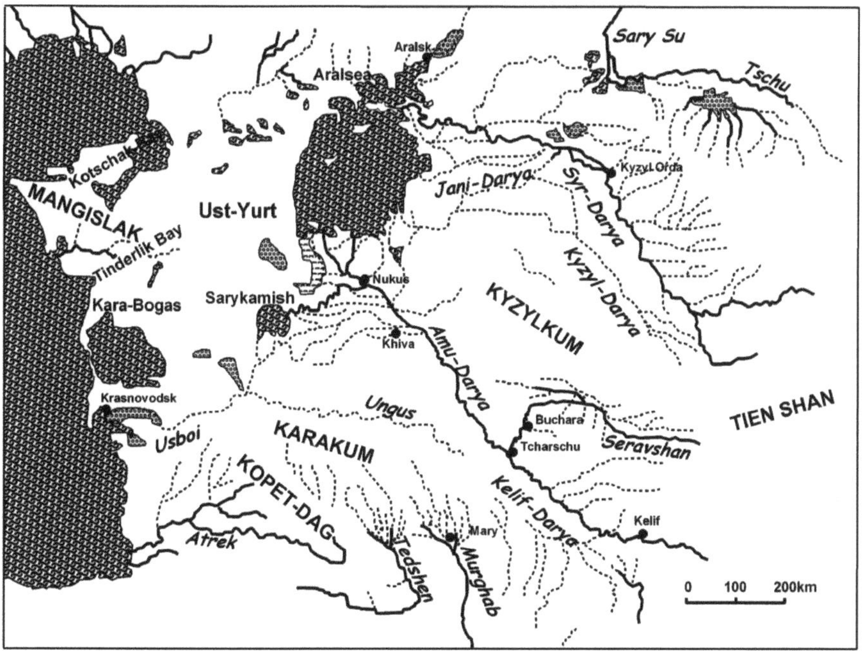

Fig. 2.1 Courses of the rivers in the Aral Sea basin within the last few millennia, indicating a strong ever-changing dynamic hydrological activity (Letolle and Mainguet 1996)

basin took place only about 5,000 years BP but most probably also between 8,000 and 15,000 years BP (Table 2.1), and then the sea level of the Caspian Sea was even higher. This was connected with the melting of the retreating ice cap in northern Siberia. The effect of this is the reason for the existence of all the organisms which developed in the Aral Sea. The more arid climate which occurred between the transgressions caused rather drastic drops of the sea levels of both the Caspian Sea and the Aral Sea. More details are given below.

2.3 Geological and Historical Dynamics of the Aral Sea Before 1960

The size and thus the coastlines of the Aral Sea have always been more or less unstable. The distinct terraces below and above the (1960) water level of 53 m above sea level (asl), the traces of old river courses on the Aral Sea floor at the eastern coast and the presence of muddy loam rich in carbonates (transgressive series) as well as soluble salts and sand (regressive series) in the central parts of the Aral Sea are evidence of the active hydrological dynamics. The Aral-Sarykamysh basin was formed tectonically during the Neogene. Additionally, deflation and karst phenomena played a role (Fedorovich 1975). The geological history of the Aral Sea exhibits three main stages:

1. The phase of sea development. This comprises the period from the middle of the Pliocene until the beginning of the Pleistocene. The Aral-Sarykamysh basin partly became inundated by the Aktshagyl and Apsheron transgressions (Table 2.2). The limits of those water bodies are still uncertain. The loamy-sandy deposits of that phase contain shells from *Dreissena*, *Adacna* and other species. The water level was about 60 m asl. The deep denudation of the Apsheron deposits indicates that at the end of this phase a long continental phase prevailed.
2. The continental phase of the Aral Sea basin lasted almost until the end of the Pleistocene. The Syr Darya discharged its water much earlier than the Amu Darya into the Aral basin. But its discharge was irregular and rather small in comparison with the overall hydrological size of the huge basin. Apparently during that period there were some smaller rivers and water bodies with almost nonsaline water in various parts of the Aral basin. They were very dynamic, with an unstable surface, and often were periodically desiccated. Thus, the stratigraphy of the sea sediments is very complex, and has not been fully elucidated.
3. The typical development of the Aral Sea started during the main melting phase of the large plateau and high mountain glaciers about 18,000–22,000 years BP. At that time the Amu Darya left the lower Karakum and started to discharge into the Khorezm and Uzboi.

Before the Amu Darya reached the Aral basin, only the Syr Darya was the main tributary to the Aral Sea. Probably also the Tschu discharged some water. Within

Table 2.1 Main events in the Aralkum region

	Geology, archaeology	Amu Darya, Uzboi	Syr Darya	Aral Sea, Karakum	Caspian Sea
Recent		Intensive irrigation projects, canals	Slight tectonic movements	Sea level drops from 53 to <30 m asl	
	Scythian	Lower Amu Darya terraces, Uzboi			Transgression 4,000 years BP
Novocaspian (Q4)	Early Bronze Age (Atlantic)	Large Lake Sarykamysh	Marine Aral Sea (*Cerastoderma edule*)	Formation of deltas from Kelif Darya, Murghab and Tedshen	Transgression 8,000 years BP
	Kelteminar, Dzheitun, Neolithic	Second terrace of Amu Darya	Darya-Lyk delta		9,000 years BP
	Mesolithic				9,500 years BP
	Kelteminar culture?		10,000 years BP		Mangyschlakian level −48 to −50 m
Chwakynian (=Kulkudukian) (Q3)		Third terrace of Amu Darya, first discharge to the Aral Sea basin, Uzboi delta		First agriculture, former deltas of Murghab and Tedshen	High sea level, about 0 m asl
	Aurignacian	Sarykamysh delta	Traces of humans (30,000 years BP)	Upper Proluvium of Kopet-Dag	Sea level about −16 to −17 m
	Later Mousterian	Traces of humans: Mousterian sites (ca. 150,000 years BP)		First formation of Aral Sea, and overflow of Caspian Sea to the Aral sea basin 130,000 years BP	
Chasarian (Q2)	Mousterian	Upper Kazakian (old alluvial land of Amu Darya)		Some former deltas of Murghab and Tedshen	Overflow of Caspian Sea to the Aral Sea
	Acheulean		Terraces of Tschu and Sary-Su, loess of Tashkent		
Aitmian (Q1) (=middle Baku)	Acheulean Chellean (Pleistocene)	Lower Kazakian (old alluvial)	Kyzylkum (0.7 million years BP)		Overflow of Caspian Sea to the Aral Sea
	Pliocene			Separation of Caspian Sea from the former Sarmatian Sea ca. +35 m (1.8 million years BP)	

Mainly after Letolle and Mainguet (1996)

Table 2.2 The main transgressive phases of the Aral Sea during the middle and late Holocene

Nomenclature	Period (years BP)	Dating	Altitude (m asl)
Aralian	50–250	Instrumental measurements and historical documents	52–53
New Aralian II	800–900	850 ± 60 (^{14}C)	>54–55
New Aralian I	2,000–3,000	2385 ± 35 (^{14}C) 2860 ± 80 (^{14}C)	>54–55
Old Aralian	5,000–6,000	5490 ± 40 (^{14}C)	>57–58

From Lvov (1959); Weinbergs and Stelle (1980); Serebryanyi and Pshenin (1980); Drenova (1985); Melnikova (1994)
asl above sea level

the Aral basin most probably three separate water bodies were present (similar to the situation in recent decades) – the depression between the island of Vozrozhdeniya and the western coast, the central part between the eastern coast and Vozrozhdeniya, and in the north the Small Aral Sea. According to Weinbergs and Stelle (1980), these water bodies had severe desiccation phases during the late Pleistocene and the beginning of the Holocene (Paskevitsh stage). The water level was most probably as low as 25 m asl. The Paskevitsh regression can be traced by mirabilite sediments. The exact dating of these salt deposits shows very variable periods.

Only during the early Holocene did the Amu Darya bend to the north. The Khorezm basin and then via the Aktsha Darya passage the Sarykamysh and the various Aral basins were filled. Aladin et al. (1996) mentioned a very huge water body, the Aral-Sarykamysh Sea, which might have had a water level of 72–73 m asl. The corresponding terraces, however, are only fragmentary and there is still some dispute about their age. They can be traced along the coastlines of the Sarykamysh and the Aral Sea. At the flat eastern coast they are almost not visible as relief lines. This transgression might have been rather short, which is unlikely, or it was a short phase earlier (before 20,000 years ago). During the early and middle Holocene, the Amu Darya discharged most of the time to the west, via the Sarykamysh basin and the Uzboi towards the Caspian Sea. The Palaeolithic and archaeological data prove the existence of a large Lake Sarykamysh and the discharge via the Uzboi during the fourth and the third millennia BC (Kes' 1969; Kes' et al. 1980). At the coastline of Lake Sarykamysh and the Uzboi, as well as in the vicinity of the Aktsha Darya delta, several traces of Neolithic hunters and fishermen and remnants of human settlements were discovered. At that time the final breakthrough of the Amu Darya to the Aral basin took place. The huge Aral-Sarykamysh Sea in the early Holocene might have had a surface area of about 95,000 km^2 (Kvasov 1980). Most scientists call this transgression the "old Aralian". Its age is under dispute, and differences in age of up to 2,000 years are common. Most scientists (Serebryanyi and Pshenin 1980; Weinbergs and Stelle 1980; Melnikova 1994; Klige et al. 1996) think that this transgression took place about 5,000 years BP. The corresponding terraces are distinctly visible in the landscapes, and are called accordingly the old Aralian terrace. After a severe regression 3,600 years BP (Mayev et al. 1983), about 3,000 years BP another transgression took place (Table 2.2). For a short time the route of the Amu Darya was along the Uzboi. The water level rose to 54–55 m asl. Again this water level can be seen geomorphologically in the relief. This transgression is called the "new Aralian I."

An increasing oasis culture occurred again during the seventh and eighth centuries AD. But only the twelfth and thirteenth centuries AD brought a prosperous and widespread oasis agriculture with sophisticated irrigation. At that time the Amu Darya completely discharged to the Aral Sea. The water level rose to about 54–55 m asl, also at Lake Sarykamysh (Kes' et al. 1980; Drenova 1985; Melnikova 1994). We call this transgression, which indicates the same water level as the new Aralian I, the "new Aralian II". The coastal relief is formed by this transgression too and is called the "new Aralian terrace".

During the last two millennia there have also been two very severe regression events at the Aral Sea, and these are also important from an ecological viewpoint (Table 2.3). About 400 AD the water level sank to only 25–30 m asl, most probably because of the deviation of the Amu Darya to Lake Sarykamysh and the Uzboi (Nikolaev 1991). From that time there are deposits of mirabilite known from the rather deep western basin, from Tschebas Bay and from the North Aral Sea. In the centre there were reed masses, which have now partly decomposed to peat. This indicates that the Syr Darya had its delta rather to the centre of the Aral Sea. Thus, huge reed stands could develop, remnants of which are found today on the desiccated seafloor along the old coastline (Fig. 2.2).

Sorrel et al. (2006, 2007) and Oberhänsli et al. (2011) observed that the runoff, resulting from warmer winter temperatures in western Central Asia and resulting in a reduction of snow cover, decreased between 100 and 300 AD, 1150 and 1250 AD, 1380 and 1450 AD, and 1580 and 1680 AD and during several low frequency events after 1800 AD. Furthermore, a negative relationship was observed between the amount of mineralization in the Aral Sea and southwestern summer monsoon intensity starting with the Little Ice Age. On the basis of these observations, it was concluded that the lake level changes during the past 2,000 years were mostly climatically controlled. Around 200 AD, 1400 AD and during the late twentieth century, human activities (namely irrigation) may also have synergistically influenced the discharge dynamics in the lower river courses.

In detail high-resolution palynological analyses, conducted on both dinoflagellate cyst assemblages and pollen grains evidenced prominent environmental

Table 2.3 The main regressive phases of the Aral Sea during the late Pleistocene and the Holocene

Nomenclature	Period (years BP)	Dating	Altitude (m asl)
Timuridic[a]	400–600	Historical documents	43–44 m
Late Antic[a]	1,600	1590 ± 140 (^{14}C)	30–35 m
Neolithic[a]	3,500	3610 ± 140 (^{14}C)	c. 30–35 m
Paskevitshic	10,000–22,000		<25 m

After (Weinbergs and Stelle 1980; Mayev et al. 1983; Nikolaev 1991)

[a]These terms are given by us, since in the literature these regression phases are not defined. The nomenclature of regression stages by Weinbergs and Stelle (1980), namely Taranglyk, Kulandy, Schomyschkol, is according to geomorphology and bathymetry not exactly defined

Fig. 2.2 Paleogeographical scheme of the Aral Sea at the time of maximal regression 1,600 years ago. *1* Shoreline of the Aral Sea before desiccation, *2* shoreline during regression, *3* sites for sampling for oxygen-isotope-studies (Nikolaev 1991), *4* the course of former rivers

changes in the Aral Sea and in the catchment area, derived from sedimentary sequences, whose total length was up to 11 m, covering approximately the past 2,000 years of the late Holocene (Sorrel 2006).

The diversity and the distribution of dinoflagellate cysts within the assemblages characterised the sequence of salinity and lake-level changes during the past 2,000 years. Owing to the strong dependence of the hydrological activity of the Aral Sea on inputs from its tributaries, the lake levels are ultimately linked to fluctuations in meltwater discharges during spring. As the amplitude of glacial meltwater inputs is largely controlled by temperature variations in the Tien Shan and Pamirs during the melting season, salinity and lake-level changes of the Aral Sea reflect temperature fluctuations in the high catchment area during the past 2,000 years (Sorrel 2006). Dinoflagellate cyst assemblages document lake lowstands and hypersaline conditions in ca. 0–425 AD, 920–1230 AD, 1500 AD, 1600–1650 AD, 1800 AD and since the 1960s, whereas oligosaline conditions and higher lake levels prevailed during the intervening periods. Besides, reworked dinoflagellate cysts from Palaeogene and Neogene deposits were a valuable proxy for extreme sheet-wash events, when precipitation is enhanced over the Aral Sea basin as from 1230 to 1450 AD.

It is proposed that the recorded environmental changes are related primarily to climate, but they may have been possibly amplified during extreme conditions by human-controlled irrigation activities or military conflicts. Additionally, salinity levels and variations in solar activity show striking similarities over the past millennium, as from 1000 to 1300 AD, from 1450 to 1550 AD and from 1600 to 1700 AD when low lake levels match well with an increase in solar activity, thus suggesting that an increase in the net radiative forcing reinforced past Aral Sea regressions.

Additionally, pollen analyses were used to quantify changes in moisture conditions in the Aral Sea basin. High-resolution reconstruction of precipitation (mean annual) and temperature (mean annual, coldest versus warmest month) parameters was performed using the "probability mutual climatic spheres" method, providing the sequence of climate change for the past 2,000 years in western Central Asia. Cold and arid conditions prevailed in ca. 0–400 AD, 900–1150 AD and 1500–1650 AD, with the extension of xeric vegetation dominated by steppe elements. Conversely, warmer and less arid conditions occurred in ca. 400–900 AD and 1150–1450 AD, where steppe vegetation was enriched in plants requiring moister conditions (Sorrel 2006).

Laminated sediments record shifts in sedimentary processes during the late Holocene that reflect pronounced changes in taphonomic dynamics. In Central Asia, the frequency of dust storms occurring during spring when the continent is heating up is mostly controlled by the intensity and the position of the Siberian High pressure system. With use of the titanium content in laminated sediments as a proxy for aeolian detrital inputs, changes in wind dynamics over Central Asia have been documented for the past 1,500 years, offering the longest reconstruction of the Siberian High variability to date. On the basis of high titanium content, stronger wind dynamics are reported from 450 to 700 AD, from 1210 to 1265 AD, from 1350 to 1750 AD and from 1800 to 1975 AD, reporting a stronger influence of the Siberian High during spring. In contrast, lower titanium contents from 1750 to 1800 AD and from 1980 to 1985 AD reflect a diminished influence of the Siberian High and a reduced atmospheric circulation. From 1180 to1210 AD and from 1265 to 1310 AD, considerably weakened atmospheric circulation is evidenced.

Sorrel (2006) concluded that as a whole, although climate dynamics controlled environmental changes and ultimately modulated changes in the climate system of western Central Asia, it is likely that changes in solar activity also had an impact by influencing to some extent the Aral Sea's hydrological balance and also regional temperature patterns in the past.

However, endogenic oscillations of the sea level by ever-increasing reed belts and subsequent die-off by lowered water levels following high transpiration rates, as well as deviations of river flows, e.g. Amu Darya to Lake Sarykamysh, and changes of land use and irrigation may cause changes in the sea level of the Aral Sea independently of climatic variations.

During the first half of the thirteenth century AD most of the oasis at the lower Amu Darya and Syr Darya was destroyed by the Mongolian conquerors. At that time additionally the Amu Darya discharged water to Lake Sarykamysh again, which caused a lower water level of about 48–50 m asl. During the second half of the thirteenth century and the fourteenth century, the Amu Darya discharged totally to the Aral Sea, and the water level rose to 52–53 m asl. This period is also characterised by another prosperous oasis culture. At the end of the fourteenth century the Timur wars brought devastation. During the fifteenth and sixteenth centuries, the water discharge of the Amu Darya was very variable, as was the sea level of the Aral Sea. Most of the time the water level was about 43–44 m asl, but it is disputed for how long this medieval regression took place. According to Nikolaev

(1991) it was 600 years, but Brodskaya (1952) mentions 400 years. Also, Bartold (1902) thought that the water level of the Aral Sea was rather low during medieval times. It is interesting that authors from the fifteenth century never mention the Aral Sea. This is also stressed by Berg (1908). This last regression is also easily indicated at many places by reed deposits. Since the seventeenth century, the Amu Darya has completely discharged to the Aral Sea, but only since the eighteenth century has the water level become stabilised at about 53 m asl. Since then this is the official mean water gauge and only slight fluctuations of the sea level of <3–4 m have occurred, caused by climatic fluctuations. Lvov (1959) and Berg (1908) reconstructed the water gauge fluctuations of the last 200 years. Three phases of high water gauges (1780–1785, 53 m asl; 1840–1845, 52.5 m asl; 1910–1961, 53 m asl, up to 53.4 m asl in 1961) and two low phases (1820–1825, 50 m asl; 1880–1885, 49–49.5 m asl) have been recorded.

In general, during the first half of the last century a slightly transgressive phase with about 53 m asl seawater gauge dominated.

The main hydrological question is "what is the reason for this active dynamics in sea level fluctuations of this large endorheic water body?". There are mainly three factors which might be responsible for such fluctuations:

1. Climatic fluctuations with fluctuations of water input in the hydrotope
2. Development and intensity of the agricultural use of irrigation water in the oasis
3. Variation of sediment transport and sedimentation, causing changes in river courses

The sediments of the Amu Darya and the Syr Darya are levelling up the seafloor at a rate of about 1 m in 2,000 years (Berg 1908). The estimations of the sedimentation figures vary according to authors. Accordingly, the rate of erosion of the river might move slightly upwards. The main riverbed is filled with mud and sediments, and the river breaks out to reach another course, forming huge meanders.

In historical times the Amu Darya rather often changed its outflow between the Sarykamysh basin and the Aral Sea basin. The Aral Sea basin is distinctly larger than the Sarykamysh basin. This could be the reason for the much shorter regressive phases of the Aral Sea in comparison with Lake Sarykamysh (Rubanov 1980; Rubanov et al. 1987). The transgressive phases of the Aral Sea, on the other hand, were much longer. This is the reason for the formation of broad and distinct terraces. Rubanov (1991) and Kes' et al. (1980) are in favour of this first hypothesis and argued that during the last few millennia the climate did not play an important role in the sea gauge fluctuations. On the other hand, Melnikova (1994) is convinced that the transgressions and regressions of the Aral Sea are mainly caused climatically. She summarised the papers of Berg (1908), Shnitnikov (1959) and Lymarev (1967). The various sea level changes of the Aral Sea and Lake Sarykamysh, as well as the present and former river courses, in comparison with the level of the Caspian Sea, are summarised in Fig. 2.3.

The proportion of melting water from glaciers for the whole discharge of the Amu Darya and the Syr Darya is not more than 20% (Melnikova 1994). The climatic dynamics, especially the precipitation regime for the Tien Shan, correlates

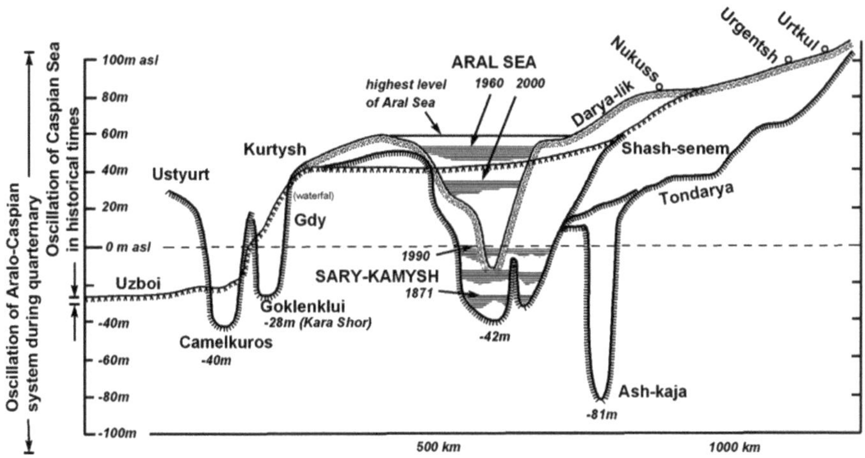

Fig. 2.3 Profiles of present and former river courses in comparison with the levels of the Aral Sea, Lake Sarykamysh and the Caspian Sea during the last few centuries along a gradient from southeast to west (Modified from Letolle and Mainguet 1996)

Table 2.4 Dynamics of the area for irrigation within the Aral Sea basin during the last 2,500 years

Period (years BP)	Area irrigated by the Amu Darya ($\times 10^6$ ha)	Area irrigated by the Syr Darya ($\times 10^6$ ha)	Total area irrigated area ($\times 10^6$ ha)
60 (1950)	1.4	1.3	2.7
200	0.5	0.25	0.75
800	0.7	0.7	1.4
2,500	0.7	1.0	1.7

After Kes' et al. (1980)

with the climatic dynamics of the Aral Sea basin for the whole of the second half of the Holocene. In other words: an increase in precipitation of the Central Asian mountains leads to an increase of discharge of the Amu Darya and the Syr Darya and thus to a transgression of the endorheic Aral Sea. The accumulation of sediments certainly plays a role in the fluctuations of the water level of the Aral Sea, at least during the geological timescale, by the causing discharge to the Caspian basin, but predominantly the climatic changes in the area are responsible for the dynamics of the last millennia. Here, we can follow Melnikova (1994).

Oasis culture with irrigation in the area has a very long tradition of about 5,000–6,000 years. In the lower Syr Darya and Amu Darya valleys, irrigation has been rather intensive and over large areas (Table 2.4). If we add the additional fields with irregular irrigation and all the areas of the middle and upper parts of both rivers in former times, then we can calculate an area of about 20,000–25,000 km^2 for the ancient area. This is comparable with the conditions during the first half of the last century, and we can conclude that irrigation of this quantity seems to have a less than negligible influence on the Aral Sea water gauge.

In general, we see that the Aral Sea as an endorheic water body in the centre of the extensive hydrotope is a very sensitive and dynamic lake. The fluctuations of his water gauge are within 40–60 m asl, except during the last severe regressions 1,600 years BP and just <3,500 years BP, and are most probably caused by climatic fluctuations within the area.

Klige et al. (1996) reconstructed a graph of the water gauge fluctuations of the Aral Sea for the last 5,000 years from geological, archaeological and historical data. They used the course of old beach lines, sediment studies and oxygen isotope analysis from sediments. However, a clear definition of the various periods during the Holocene is still lacking.

2.4 Regression of the Aral Sea After 1960 and Hydrological Conditions

2.4.1 Hydrology and Water Use

Until 1960 the Aral Sea had an area of about 68,300 km^2 together with the islands (water level of 53.0 m asl). The area of the seawater surface was 66,100 km^2. The water volume was about 1,100 km^3 (Bortnik et al. 1991; Bortnik 1996). The then established huge canals and irrigation systems diverted most of the water of the two tributaries the Syr Darya and the Amu Darya. Thus, the catastrophic desiccation of the Aral Sea started. Since 1960 more than 90% of the water volume and more than 85% of the sea surface area have been lost (Figs. 2.4 and 2.7, Table 2.5). The desiccation of the Aral Sea in a relatively short time has reached a critical phase insofar that now only remnant water bodies are left and will even become smaller in the future, except for the North Aral Sea (Small Aral Sea). The Aral Sea syndrome and the new desert, the Aralkum, are the new reality. The remnant water bodies are strongly saline (Benduhn and Renard 2004; Oberhänsli and Zavialov 2009), but similar situations can be traced in former periods (Boomer et al. 2000; Le Callonnec et al. 2005).

The first half of the twentieth century was a stable phase for the Aral Sea, with a rather constant water level of 53 m asl. The two tributaries, the Syr Darya and the Amu Darya, come from the mountains of the Pamiro-Altai, the Hindu Kush and the Inner Tien Shan in Tajikistan, Afghanistan and Kyrgyzstan. The total discharge per year was estimated as approximately 120–130 km^3 (Kirsta 1989). The discharge to the Aral Sea before the desiccation started was approximately 60–70 km^3. In the recent decades, however, it dropped to about 3–20 km^3/year. This discharge of water is still present; the water is diverted to fields, fish ponds, new water bodies, new reservoirs and dykes, and thus evaporates before reaching the former lowest point in the hydrotope, the Aral Sea.

A small proportion of water entering the Aral basin comes from groundwater discharge. The total groundwater discharge into the Aral Sea is equal to or distinctly

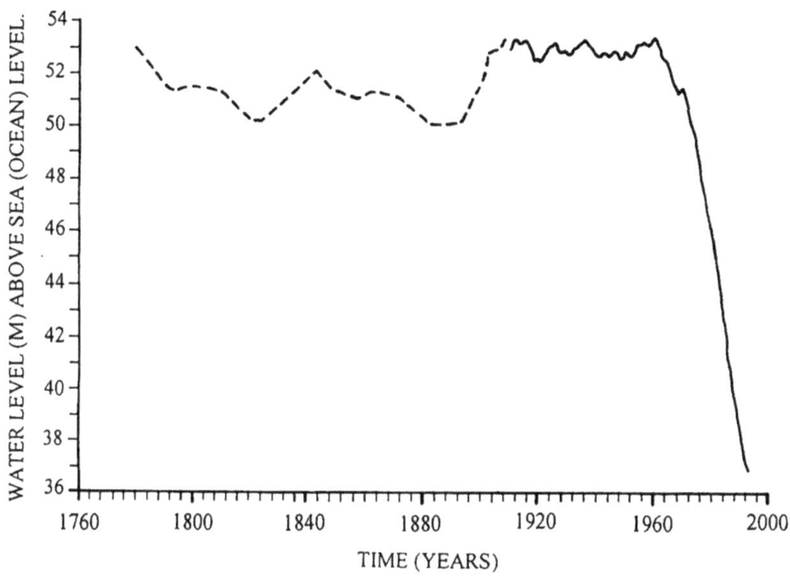

Fig. 2.4 Water-level fluctuations in the Aral Sea. *Dashed line* from reconstruction, *solid line* from instrumental data (Bortnik 1996)

Table 2.5 Dynamics of the water surface and water volume of the Aral Sea (Aralkum) within the last 50 years

Year	Water level (m asl)	Water surface, without islands (km^2)	Volume of water (km)	Salinity (% NaCl)
1960	53.4	67,100	1,080	0.9
1970	51.2	60,200	950	1.2
1980	45.4	50,800	630	1.7
1990	38.2[a]/40.5[b]	36,400	310	3.2
2000	≈ 33.5[a]/39.8[b]	26,000	160	6–8[a]/2.0[b]
2005[c]	≈ 30.7[a]/41.0[b]	19,000	110	7–15[c]/1.9[b]
2009	25[a]/42[b]	7,000	< 100	≈ 10–20[c]/1.5[b]

After (Bortnik 1996; Dukhovny et al. 2008)
[a]The South Aral Sea
[b]The North Aral Sea
[c]In 2005, the shrinking South Aral Sea was subdivided into two lakes, the western and the eastern basins, with almost no water exchange between them

greater than that in 1960 according to the increasing hydraulic gradient. Some model calculations with different relevant scenarios resulted in increased groundwater discharge of up to about 200% of the 1960 value (Jarsjö and Destouni 2004). Most likely the present (2002) discharge into the southeastern part of the Large Aral Sea is not greater than it was in 1960; the groundwater discharge into the northwestern part of the Large Aral Sea (Tschebas basin) and into the North Aral Sea, however, should have increased. Furthermore, for an unchanged absolute groundwater discharge, its relative importance for the Aral Sea(s) has increased dramatically, from being about 12% of the total river discharge in 1960 to almost 100% of total river discharge presently (Jarsjö and Destouni 2004).

The length of the Syr Darya is 2,137 km from the confluence of the Naryn and the Kara Darya to the Aral Sea, and from the Naryn spring it is about 3,020 km. The mean annual discharge is 38 km^3, bringing water from Kyrgyzstan, Tajikistan, Uzbekistan and Kazakhstan. In 1960 the irrigated area in the basin of the Syr Darya was about 1.9 million hectares; 30 years later it was about 3.4 million hectares (Ivanova 1992). Six huge retaining dams were constructed with a total volume of 37.6 km^3. This all caused a declining discharge of the Syr Darya to the Aral Sea of now only about 4–5 km^3/year.

The length of the Amu Darya is about 1,450 km from the confluence of the Pyandzh and the Vakhsh to the Aral Sea. From the Pamir springs (Zor Kul) and including the Pyandzh, the length is about 2,580 km. Nowadays, the annual discharge to the Aral Sea is only 5–15 km^3/year.

The desiccation of the Aral Sea is caused by the enormous increase of agriculture based on irrigation in the whole region for intensive production of cotton, rice, wheat and vegetables. The agriculture of Central Asia depends totally on irrigation, because of the arid desert climate with an annual mean precipitation of only 100–150 mm (mostly during the winter, see Chap. 4). Agriculture uses 87% of the whole water consumption in the area. More than 40% of the people are involved in agriculture in Uzbekistan and Turkmenistan (Klötzli 1997). The same is true for the southern districts of Kazakhstan (Shymkent, Kyzyl-Orda).

Restructuring of the economy will not be possible in the near future. Irrigation will remain the most important resource for agriculture in Central Asia, even more so when regarding population growth (see also Chap. 18).

In 1990 there was optimism that the discharge of the Amu Darya and the Syr Darya to the Aral Sea could be increased within the following 15 years to 35 km^3 and thus that the water level could be stabilised at about 40 m asl. This would have made necessary a total restructuring of the agricultural systems and a drastic reduction of irrigated areas. This was unrealistic.

2.4.2 South Aral Sea

The Aral Sea does not exist any more and will not be restored in the near or distant future. The split into smaller water basins started in 1988, when the island of Kokaral became a peninsula, after Berg Bay became dry. All three main islands, Kokaral, Barsa-Kelmes and Vozrozhdeniya, are now connected with the mainland. The development of the Large Aral Sea in the south and the Small Aral Sea in the north differs totally. The hydrological water budget of the South Aral Sea is still negative, the desiccation will continue and within these years the and since 2009 the shallow eastern basin will be a local salt lake or a system of shallow salt lakes and salt swamps between the former island of Vozrozhdeniya and the eastern coast with an area of about 4,000–6,000 km^2. The term "Large Aral Sea" is misleading now, since there now exist two almost separate basins, the very shallow eastern basin and a deep western basin, which in the future will remain relatively unchanged. Thus, the South Aral Sea is divided into two unequal and completely independent eastern and western parts. This process will be accelerated by annual fluctuations of the Amu Darya water flow together with the difference between water level rise and fall rates. Therefore, the preservation of the sea as an ecological system with reasonable water volume cannot be achieved without providing independent water inflow to the western part of the Large Aral Sea. Otherwise, the loss of up to a half of the current sea's water is inevitable just because of irrational water discharge to the eastern part alone. The preservation of the western basin at a level that is quite close to the current one requires an additional discharge of not less than 7.6 km^3/year, which may be in principle achieved by effective water conservation technologies (Salokhiddinnov and Khakimov 2004; Peneva et al. 2004).

2.4.3 North Aral Sea

Until 1960 the area of the North Aral Sea (Small Aral Sea) was about 6,000 km^2 (at a level of 53 m asl). A critical phase of desiccation started in 1990, when the water level reached <38–39 m asl and the area decreased to 2,600–3,100 km^2 (Bortnik 1996). A further sinking of the water level below 38 m asl would have led to additional particulation and thus would have caused rapid desiccation of the remnant remote western basins. A dam was constructed at Berg Bay (between the Kokaral peninsula and the eastern coast south of the Syr Darya's mouth) to stop this deleterious dynamics. The dam was constructed very fast and provisionally in July to August 1992 but was destroyed the next spring by high tides and storms. This dam was only a long line of soil, about 1 m high, across Berg Bay, to prevent direct outflow of the Syr Darya water to the southern Aral basin. But this measure caused a rise of the water level of the North Aral Sea of 1 m within 9 months, which was a successful experiment and demonstration of stabilising the water level of the North Aral Sea. In 1996–1997 another, very large dam was constructed. This dam had a height of 43 m asl, and it was about 15 km long. The relative height was between 9 and 12 m. The upper part was 4–6 m wide, and the base was more than 12 m wide. The dam was made as a "nature dam" from sand, loam and gravel without a concrete foundation or cover. It forced the Syr Darya to discharge its water only to the North Aral Sea. The hydrological budget turned very positive, which led to an increase in the water level to about 42 m asl and an area of 3,200 km^2 within 2 years. The mean discharge per year from the Syr Darya at that time was about 4–5 km^3/year, being distinctly higher than that in previous years. This higher discharge can be partly related to the declining area of irrigated paddy fields in the whole Kyzyl-Orda region (from 100,000 ha to 60,000 ha) and partly to rainy years with precipitation above average. This development led to an unexpectedly high level of the North Aral Sea and subsequently on 21 April 1999, after a strong storm, the dam broke and the water level of the North Aral Sea sank by

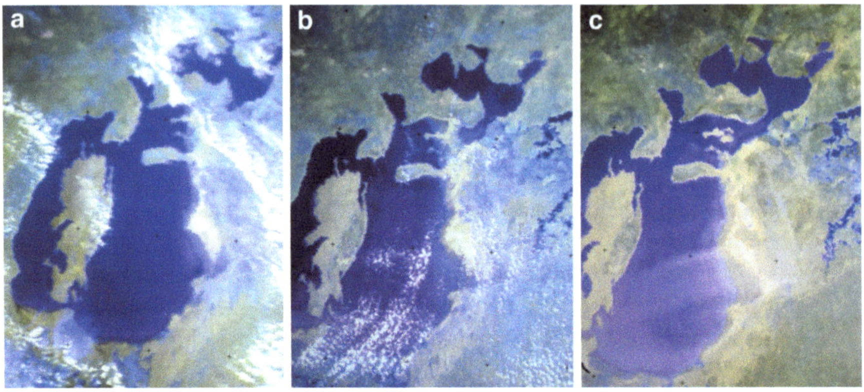

Fig. 2.5 Satellite images demonstrating the outflow from the Small Aral Sea during breakage of the dam on 21 April 1998. Situation before (**a** 19 April 1998), during (**b** 21 April 1988) and after (**c** 24 April 1998) the disaster (NASA, with compliments from B. Gelde'ev)

about 4 m (see Fig. 2.5). A huge flood was the result (Fig. 2.5b, c). The breakage of the dam was caused by strong storms and high waves. Under these conditions, the quasi-stationary sea level of the Aral Sea can be 1 m or even 2 m higher than normal. Even the mean static wave action by wind can result in a sea level 1 m higher than normal. This means the dam must be at least 2–3 m higher than the sea level. In spring 1999, the water level of the North Aral Sea had risen to more than 42 m asl. The absolute height of the dam was 43 m asl. Thus, it was certain that the next storm with high tides would damage the dam. Parts of the dam were flushed away by the breakthrough. Despite severe warnings, no direct means to save the dam were possible.

Several new canal systems have been constructed to improve the efficiency of the water budget and management in the irrigation areas and to construct a new, better dam. This dam (by a Turkish–English–Russian consortium) was ready by the end of 2006. As a result, this huge new dam keeps the water level of the North Aral Sea stable at a constant height of about 42 m asl by an overflow, mainly for the power plant producing electricity. The whole district from Kyzyl-Orda down to the delta of the Syr Darya now has a new basis for efficient and sound reliable water use and distribution for the agricultural systems as well as for fish ponds. This gave hope for an improved standard of living for the people of the whole region, and new hope for the villages around the northern Aral Sea area.

2.5 Formation of the Aralkum

The desiccation of the Aral Sea is a very complex process. Also very complex are the various changes, the hydrological budget and balance, the quality of the various water bodies, the losses of almost the total fish fauna, the formation and development of new land surfaces, the blowouts of dust and salt from the desiccated seafloor, the threat for the further existence of Barsa-Kelmes Nature Reserve, etc. The changes of the various water biota are well documented by research and many studies (Aladin 1991; Aladin and Potts 1992). The main macroecological consequences and landscape changes are:

The formation of a new desert (Aralkum)
Threat to Barsa-Kelmes Nature Reserve (see Chap. 14)
Loss of valuable coastal landscapes
Climatic changes in the Aral Sea basin (see Chap. 4)
Blowout of salt dust from the desiccated seafloor and deflation (see Chaps. 5–7)

2.5.1 Dynamics of the Aralkum

The area of the dry seafloor of the Aral Sea, called the Aralkum (Breckle and Agachanjanz 1994; Breckle et al. 2001, Breckle and Wucherer 2005; Dech and Ressl 1993), is still increasing and is now about 60,000 km^2 of seafloor which dried up inclusive of the island area. In 2009, the maximum desiccation was observed (Fig. 2.6b). In 2010, the huge, but very flat eastern basin was filled up somehow by strong winter and spring rains in the whole area; it is now in a very fluctuating hydrological dynamic equilibrium (Fig. 2.6c). It is a very unstable huge salt swamp. The dry seafloor of the Aral Sea is a new terrestrial surface (Fig. 2.7). It has developed to a new geographical object, a new desert, that has a strong environmental impact on the surroundings of the Aral Sea. The Aralkum is the largest area worldwide where a primary succession takes place. Unconsciously, humans have created a huge experiment, an experimental set, a laboratory of nature with thousands of local events. The new knowledge on vegetation dynamics in the Aralkum, which is a mosaic of sand desert and salt desert ecosystems, is very important for the understanding of the ecosystem dynamics and the vegetation mosaic in the whole Central Asian area (Walter 1974). The succession on the dry seafloor has continued for 50 years. We know the successional sequences, their age and their dynamics. This is important to identify major mechanisms that determine the rate and direction of ecosystem changes on the dry seafloor. The seafloor is normally a very loamy flat plain, surrounded by sandy beach lines and including vast saline (solonchak) plains. The geological and geomorphological structure of the desiccated seafloor is very variable and complicated. The basin of the North Aral Sea differs considerable from that of the South Aral Sea. There are also some north–south gradients. Most decisive is the colonisation by plants. This again depends on the salinity and particle size distribution of the new soils as the main factors for flora and vegetation development. From the former coastline (1960 line) to the interior to the retreating recent coastline we regularly find a zonation of ecotopes: the substrate material becomes finer and the salinity increases. Thus, distribution of soil material and vegetation changes often discontinuously and stripes of plants can be seen. The main soil types are marshy solonchaks, coastal solonchaks and degraded and sand-covered coastal solonchak and takyr soils.

During desiccation, the developing ecosystems change their properties. The particle size of the substrate becomes finer, and substrate salinity increases.

Fig. 2.6 Continued

Fig. 2.6 (Continued) Recent satellite images of the ever-increasing Aralkum (NASA): (**a**) from 2001, (**b**) from 16 August 2009 – with very low water level in the shallow eastern basin forming a salt swamp (earthobservatory.nasa.gov/images/imagerecords/39000/39944/aral_sea_20090816_lrg.jpg), (**c**) from 26 August 2010 – a very wet winter season brought some inflow and floods from the Amu Darya from the south to the shallow eastern basin (earthobservatory.nasa.gov/IOTD/view.php?id = 46685)

Accordingly, the temporal sequence of vegetation and soils does not correspond with the spatial sequence of plant associations. This is most conspicuous regarding the contrasting desiccated seafloor areas from 1960 to 1970 and from 1980 to 1990.

The desiccated seafloor is a huge open often bare surface rich in salt. This is now one of the main sources of salt-dust storms and salt blowouts, and thus of salt contamination of the adjacent cultivated lands in Kazakhstan and Uzbekistan. The extent and the effects of the salt-dust export and the effects on plants, animal and humans are only roughly known (see Chaps. 5–7).

2.6 Conclusions

The Aral Sea basin is a huge tectonic flat basin, which exhibits various shallow and deeper parts. Slight tectonic changes caused big changes in hydrological activity, in inflow and outflow and overflow to the Caspian Sea in the last several thousand years. The superimposed influence of climatic changes and sedimentation activities led to rather dynamic changes of the two main tributaries, the Amu Darya and the Syr Darya. To date there are four main separated basins: the North Aral Sea in the

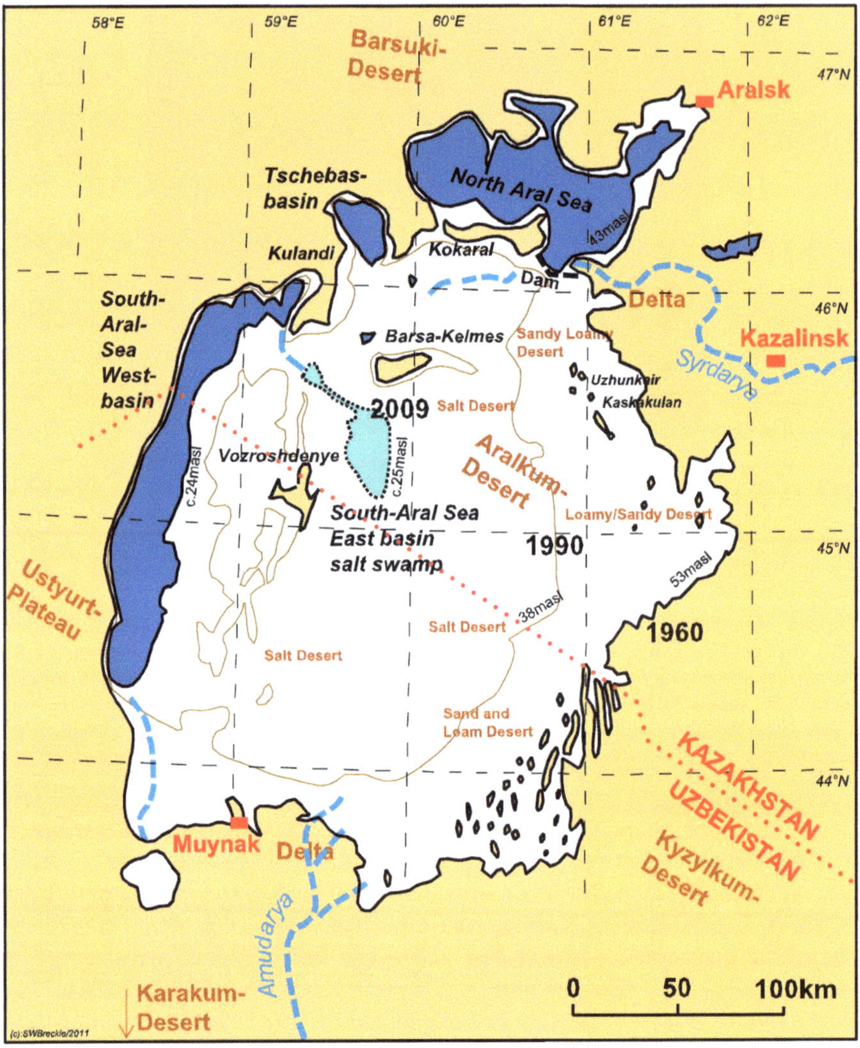

Fig. 2.7 Map of the Aralkum, indicating the old coastline (1960), the coast line from 1990 and the recent situation with the remnant isolated water bodies and indicating the adjacent desert regions

north, now a separate water body, separated by an artificial dam, the South Aral Sea, with a deep western basin, which will be the main remnant water body in the future, and a shallow eastern basin, which has became a large salt swamp, and the small Tschebas basin in the northwest. The whole area of the desiccated seafloor is now a new desert, called the Aralkum (Fig. 2.7). It consists mainly of the younger (1980–2010) salt desert (70%) and of the older (1960–1980) sand and sand-loamy deserts (20%). The salt desert is the source of salt-dust storms affecting the whole adjacent regions.

Acknowledgements Two research projects were funded by the BMBF (German Ministry for Education and Research). We thank Hans-Michael Biehl, Hans-Georg Bertram, Joachim Kutscher and Ingo Fitting from the Project Management Jülich for their support of the Aral Sea projects. The GTZ-UNCCD Project in Bonn was also involved in the research project. We thank Anneke Trux and Günther Winckler for their interest and help, which is greatly acknowledged.

References

Aladin NV (1991) Studying of the influence of water salinity in separating bays of the Aral Sea on hydrobionts. USSR Acad Sci, Proc Zool Inst, Leningrad 223: 5–18 (in Russian)

Aladin NV, Potts WTW (1992) Changes in the Aral Sea ecosystem during the period 1960-1990. Hydrobiologia 237:67–79

Aladin NV, Plotnikov IS, Orlova MI, Filippov AA, Smurov AA, Piryulin DD, Rusakova MM, Zhakova LV (1996) Changes in the form and biota of the Aral Sea over time. In: Micklin PP, Williams WD (eds) The Aral Sea basin. Springer, Berlin, pp 33–55

Bartold VV (1902) Svedeniya ob Aralskom more i nizovyakh Amudaryi s drevneishikh vremen do XVII veka (Overall view of history of the Aral Sea and Amudarya until XVII century). Scientific results of Aral Expedition. Izvestiya Turkestanskogo otdeleniya Imperatorskogo Russkogo Geograficheskogo obschestva, Sankt Peterburg, 4(2) (in Russian)

Benduhn F, Renard P (2004) A dynamic model of the Aral Sea water and salt balance. J Marine Syst 47:35–50

Berg LS (1908) The Aral Sea. Scientific results of Aral Expedition. Izvestiya Turkestanskogo otdeleniya Imperatorskogo Russkogo Geograficheskogo obschestva, Sankt Peterburg, 5(9) (in Russian)

Boomer I, Aladin N, Plotnikov I, Whatley R (2000) The palaeolimnology of the Aral Sea: a review. Quaternary Sci Rev 19:1259–1278

Boomer I, Wünnemann B, Mackay AW, Austin P, Sorrel P, Reinhardt C, Keyser D, Guichard F, Fontugne M (2009) Advances in understanding the late Holocene history of the Aral Sea region. Quaternary Int 194:79–90

Bortnik VN (1996) Changes in the water-level and hydrological balance of the Aral Sea. In: Micklin PP, Williams WD (eds) The Aral Sea basin. Springer, Berlin, pp 25–32

Bortnik VN, Kuksa VI, Tsytsarin AG (1991) Present state and possible future of the Aral Sea. Izv AN SSR, ser geogr 4:62–68 (in Russian)

Breckle S-W, Agachanjanz OE (1994) Spezielle Ökologie der gemäßigten und arktischen Zonen Euro-Nordasiens, Zonobiom VI–IX. Ökologie der Erde, vol 3, 2nd edn. Fischer, Stuttgart

Breckle S-W, Wucherer W (2005) Hat der Aralsee eine Zukunft (Has the Aral Sea a future). In: Lozan JL, Graßl H, Hupfer P, Menzel L & Schönwiese Ch-D (Hrgs.) Warnsignal Klima: Genug Wasser für alle? Wissenschaftliche Auswertungen/Hamburg, in Kooperation mit GEO, pp. 131–135 (in German)

Breckle S-W, Wucherer W, Agachanjanz O, Geldyev BV (2001) The Aral Sea crisis region. In: Breckle S-W, Veste M, Wucherer W (eds) Sustainable land use in deserts. Springer, Berlin, pp 27–37

Brodskaya NG (1952) Donnye otlozheniya i protsessy osadkoobrazovaniya v Aralskom more (Sea sediments and their dynamics in the Aral Sea). Tr Inst Geol Nauk, ser geol 57(115):106 (in Russian)

Dech SW, Ressl R (1993) Die Verlandung des Aralsees (Drying out of the Aral Sea). Geographische Rundschau 45(6):345–352 (in German)

Drenova AN (1985) Fazy obvodnennosti Sarykamysha po radiouglerodnym dannym (Flooding stadies of the Sarykamysh Sea after the radiocarbon method). Izv AN SSSR, ser geogr 4:89–93 (in Russian)

Dukhovny VA, Navratil P, Ruziev I, Stulina G, Ye R (eds) (2008) Comprehensive remote sensing and ground-based studies of the dried Aral Sea bed. SIC ICWC, Tashkent

Erdinger L, Hollert H, Eckl P (2011) Aral Sea: An Ecological Disaster Zone with Impact on Human Health. Encyclopedia of Environmental Health, Elsevier/Burlington, p. 136–144

Fedorovich BA (1975) Ravniny i gory Srednei Azii I Kazakhstana (Plains and mountains of Middle Asia and Kazakhstan). Moskva (in Russian)

Giese E, Bahro G, Betke D (1998) Umweltzerstörungen in Trockengebieten Zentralasiens (Environment destruction in dry areas of Central Asia) (West- und Ost-Turkestan). In: Ursachen, Auswirkungen, Maßnahmen (Causes, consequences, measures), vol 125. Erdkundliches Wissen, Stuttgart (in German)

Glantz MH, Rubinstein AZ, Zonn I (1993) Tragedy in the Aral Sea basin: looking back to plan ahead? Global Environ Chang 3:174–198

Glazovskii NF (1990) Aral Sea crisis. Nauka, Moscow (in Russian)

Ivanova LV (1992) Hydrological aspects of the Aral Sea problem. Vodnye Resursy 2:39–49 (in Russian)

Jarsjö J, Destouni G (2004) Groundwater discharge into the Aral Sea after 1960. J Marine Systems 47:109–120

Kes' AS (1969) Osnovnye etapy razvitiya Aralskogo morya (Main stages of the development of the Aral Sea). In: Problemy Aralskogo morya (Problems of the Aral Sea). Nauka, Moscow, pp 160–172 (in Russian)

Kes' AS, Andrianov MA, Itina MA (1980) Dinamika gidrograficheskoi seti i izmeneniya urovnya Aralskogo morya (Dynamics of the hydrological net and changes of the sea level of the Aral Sea). In: Kolebaniya uvlazhnennosti aralo-kaspiiskogo regiona v golothene (Climatic fluctuations of the Aral Sea and the Caspian Sea regions in holocene). Nauka, Moscow, pp 185–197 (in Russian)

Kirsta VT (1989) Problema Aralskogo morya i Karakumskii kanal (Problem of the Aral Sea and Karakum canal). Probl osvoeniya pustyn' 5:10–17 (in Russian)

Klige RK, Liu H, Selivanov AO (1996) Regime of the Aral Sea during historical times. Vodnye Resursy 23(4):407–412 (English translation from Russian)

Klötzli S (1997) Umweltzerstörung und Politik in Zentralasien. Eine ökoregionale Systemuntersuchung (Environment degradation and politic in Central Asia. An eco-regional system investigation). Lang, Bern (in German)

Kvasov DD (1980) Paleolimnologiya Arala (Paleolimnology of the Aral Sea). In: Kolebaniya uvlazhnennosti aralo-kaspiiskogo regiona v golothene (Climatic fluctuations of the Aral Sea and the Caspian Sea regions in holocene). Nauka, Moscow, pp 181–185 (in Russian)

Le Callonnec L, Person A, Renard M, Létolle R, Nebout N, Khelifa LB, Rubanov I (2005) Preliminary data on chemical changes in the Aral Sea during low-level periods from the last 9000 years. Compt R Geosci 337:1035–1044

Letolle R, Mainguet M (1996) Der Aralsee. Eine ökologische Katastrophe (The Aral Sea. An ecological disaster). Springer, Berlin (in German)

Lvov VP (1959) The Aral Sea level fluctuations in the last 100 years. Tr GOIN 46:80–114 (in Russian)

Lymarev VI (1967) Coasts of the Aral Sea – endorheic water basin in the arid zone. Nauka, Leningrad (in Russian)

Mayev EG, Mayeva UA, Nikolaev SD, Parunin OB (1983) Novye dannye po golotsenovoi istorii Aralskogo morya (New dates about the holocene history of the Aral Sea). In: Paleogeografiya Kaspiiskogo i Aralskogo morei v Kainizoe (Paleogeography of Caspian Sea and Aral Sea in tne cenozoic), vol 2. MGU, Moskva, pp 133–144 (in Russian)

Melnikova AP (1994) Klimaticheskii faktor v kolebaniyakh urovnya Arala (Climate impact on the fluctuations of the sea water level of the Aral Sea). Izv ANSSSR, ser geogr 4:26–36 (in Russian)

Micklin P (2007) The Aral Sea disaster. Annu Rev Earth Pl Sci 35:47–72

Micklin P, Aladin NV (2008) Reclaiming the Aral Sea. Sci Am 298:64–71

Ni A, Nurtayev B, Petrov M, Tikhanovskaya A, Tomashevskaya I (2004) The share of a glacial feeding in water balance of Aral Sea and Karakul lake. J Marine Syst 47:143–146
Nikolaev SD (1991) Razvitie Arala po izotopno-kislorodnym dannym (The Aral development according to isotope-oxygenic data). In: History of Lakes Sevan, Issyk-Kul, Balkhash, Zaisan and Aral. Leningrad, Nauka, pp. 246–250 (in Russian)
Oberhänsli H, Zavialov P (2009) Aral Sea Basin hydrological, chemical and biological dynamics today, compared with trends of the past 50 years. J Marine Syst 76:251–253
Oberhänsli H, Novotná K, Píšková A, Chabrillat S, Nourgaliev DK, Kurbaniyazov AK, Grygar TM (2011) Variability in precipitation, temperature and river runoff in W Central Asia during the past 2000 yrs. Global Planet Change 76:95–104
Peneva EL, Stanev EV, Stanychni SV, Salokhiddinov A, Stulina G (2004) The recent evolution of the Aral Sea level and water properties: analysis of satellite, gauge and hydrometeorological data. J Marine Syst 47:11–24
Rubanov IV (1980) Geologicheskie i istoricheskie svideteli kolebaniya urovnya Arala (Geological and historical facts of the fluctuation of the Aral Sea). In: Kolebaniya uvlazhnennosti aralo-kaspiiskogo regiona v golothene (Climatic fluctuations of the Aral Sea and the Caspian Sea regions in holocene). Nauka, Moscow, pp 204–209 (in Russian)
Rubanov IV (1991) Salt accumulation in the Aral hollow. In: History of Lakes Sevan, Issyk-Kul, Balkhash, Zaisan and Aral. Leningrad, Nauka, pp. 231–234 (in Russian)
Rubanov IV, Ishniyazov DP, Baskakov MA, Chistyakov PA (1987) Geology of the Aral Sea. Isd FAN, Tashkent (in Russian)
Salokhiddinnov AT, Khakimov ZM (2004) Ways the Aral Sea behaves. J Marine Syst 47:127–136
Serebryanyi LR, Pshenin GN (1980) Oledeneniye Tyanshanya i kolebaniya urovnya Arala (Glaciers dynamics in Tien Shan and fluctuations of the Aral Sea). Izv ANSSSR, ser geogr 2:52–65 (in Russian)
Shnitnikov IA (1959) The Aral Sea past and present in the context of large-scale climate rhythms. Proceedings of the 3rd Russian hydrological congress. Gidrometeoizdat, Leningrad 7:47–57 (in Russian)
Sorrel P (2006) The Aral Sea: a palaeoclimate archive. Doctoral Thesis, Universität Potsdam and Université Claude Bernard/Lyon 1, France, 109 pp. http://opus.kobv.de/ubp/volltexte/2006/780/
Sorrel P, Popescu S-M, Head MJ, Suc JP, Klotz S, Oberhänsli H (2006) Hydrographic development of the Aral Sea during the last 2000 years based on a quantitative analysis of dinoflagellate cysts. Palaeogeography Palaeoclimatology Palaeoecology 234:304–327
Sorrel P, Popescu S-M, Klotz S, Suc J-P, Oberhänsli H (2007) Climate variability in the Aral Sea basin (Central Asia) during the late Holocene based on vegetation changes. Quaternary Res 67:357–370
Svitoch AA (2009) History of the last Aral Sea. Aridnie ekosistemi 15:5–17 (in Russian)
Walter H (1974) Die Vegetation Osteuropas, Nord- und Zentralasiens. Vegetationsmonographien der einzelnen Großräume, vol 7. Fischer, Stuttgart
Weinbergs IG, Stelle VYa (1980) Pozdnechetvertichnye stadii razvitiya Aralskogo morya i ikh svyaz' s izmeneniemi klimaticheskikh uslovii etogo vremeni (Late quaternary development of the Aral Sea in connection with climatic changes). In: Kolebaniya uvlazhnennosti aralo-kaspiiskogo regiona v golothene (Climatical fluctuations of the Aral Sea and the Caspian Sea regions in holocene). Nauka, Moscow, pp 175–181

Chapter 3
Geography and Geomorphological and Lithological Characteristics of the Aralkum

G.V. Gel'dyeva, S-W. Breckle, and B.V. Gel'dyev

3.1 Introduction

The geomorphological and lithological features of the Aralkum today are characterized by the vast plains stretching on the very flat, intercontinental seafloor, which is now predominantly desiccated. The sediment layers, mainly clay, silt and sand, which came from the main tributaries within the last few millennia are rather heterogenic according to the dynamics of the river discharges. The desiccated seafloor exhibits new changes by wind erosion, causing sand and dust transport, and thus forming sand dune landscapes on the older seafloor.

3.2 Geology and Tectonic Processes

The geological and geomorphological factors play an important role in the formation and differentiation of the landscapes in the desiccated Aral Sea basin. This basin is the center of an eroded tectonic depression of the northern part of the Turanian plate (epihercynic platform). From a geomorphological point of view, the Aral Sea can be called a very flat, innercontinental lake, and the Aralkum can be called a very flat plain.

The structural basement of the relief is the tectonic elements of the northwestern Turanian platform. These are mainly the Chelkar syneclise, the Syr Darya syneclise and the Kyzylkum depression. These syneclises are large depressed structures of a continental platform crystalline basement. There are usually no distinct topographical depressions on the syneclises because they are covered by rather thick sediment layers (see Chap. 2). These elements were formed during the hercynic Ural mountain development (in the whole region north of the Aral Sea) und during the Caledonian Turgay–Kyzylkum mountain development (the region east of the Aral Sea). The basement of the platform exhibits strong dislocations and metamorphic changes. The tectonic ruptures dismember it to strata which are oriented in

S.-W. Breckle et al. (eds.), *Aralkum - a Man-Made Desert: The Desiccated Floor of the Aral Sea (Central Asia)*, Ecological Studies 218,
DOI 10.1007/978-3-642-21117-1_3, © Springer-Verlag Berlin Heidelberg 2012

submeridional directions. The west Aralian stratum corresponds mainly with the position and size of Ustyurt Plateau, including the western deep furrow of the Aral Sea with a depth of about 50–60 m. The eastern lower Syr Darya stratum forms the lower layers along the eastern coast of the Aral Sea (Bubnoff 1924; Leuchs 1935).

The Aral Sea basin with its variety of relief forms makes contact with deep faults and transregional lineament zones. Two tectonic anticlines divide the Aral Sea depression into four structural units. The first stretches from west to east (along the former island of Kokaral to the Syr Darya's mouth). This is the division between the northern, Small Aral Sea and the southern, former Large Aral Sea (now the Aralkum). The second stretches along the Kulandy peninsula, the former island of Vozrozhdeniya and the southern small islands. This is the division between the western deep furrow (western basin of the former Large Aral Sea) and the dry flat eastern basin (now the Aralkum). The submeridional deep furrow of the Aral Sea (between Vozrozhdeniya and the western coast) is the continuation of the Chelkar syneclise. The delta of the Amu Darya (the fourth structural unit) is located at the continuation of this huge deep furrow from the Palaeozoic period. Nowadays, it is filled up deeply with alluvial deposits.

For the region of the Aral Sea, the basis surface of the Aralsk layer trains of the Lower Miocene and the thickness of the layer trains of the Andasai from the Upper Oligocene/Lower Miocene was used to define the recent tectonic movements and their amplitude. For most parts of the area there have been only changes of level between 0 and 300 m since post-Miocene times. The characteristics of the tectonic structure became apparent only during the neotectonic epoch. The differentiation of the various directions of the tectonic movements is distinctly marked. In contrast, the Syr Darya syneclise and the East Aralsk depression are more characteristic. Knowledge of those tectonic movements was gained by precise nivellement measurements of the railway lines. For the Aralsk–Dzhusaly–Kyzyl-Orda section during recent decades a rather constant tectonic sinking speed of about 8 mm/year was found.

The smooth transition from the flats hills of Mugodzhary to the slight basin of Kyzyl-Orda is interrupted by the Malye Barsuki anticlinal and the sharp changes in height on the upper plain which correspond to the north Aralian lineaments. The neotectonic disjunctive disturbances without displacements or with very slight displacements only are typical for this region. They have a huge spatial size, and are clearly defined by their depth and tectonic age.

3.3 Sedimental Deposits on the Dry Desiccated Seafloor

The tectonic characteristics of the whole area are governed by the spatial distribution of the deposits from the Mesozoic and the Cenozoic periods. The lithological-stratigraphical complexes of the Aralkum region comprise deposits from the Upper Cretaceous, the Palaeogene, the Neogene and the Quaternary. These are former unique sea sediments and some continental deposits of very differing constitution and thickness. The thickness of the sediment layers is about 500 m on the anticlinal parts and 2,000–4,000 m in the synclinal basins. The basis of the platform at the northeastern coast of the Aral Sea is about 500 m deep and increases to about 3,000–4,000 m at the western coast. The sediments on the platform are mainly terrestrial sediments, which started during the Late Triassic (T_3) until almost the recent late Quaternary (Q_{IV}). The sediments from the Late Cretaceous until the Miocene (K_2-N_1) are dominated by marine sediments.

The various types of lithological basis of landscape are defined according to the genesis and composition of sediments, to mineral content of underground water and to the effective time of the exogenous processes. The oldest sediments are limestones from the Upper Cretaceous, but also marl, sandstones and reddish-brown clay layers are known from that period. They are rather common all around and are uncovered by denudation on top of the anticlinal vaults and other arches (e.g. on the former islands of Vozrozhdeniya and Lazarev, on the Kulandy and Muinak peninsulas and along the eastern edge of Ustyurt Plateau, and on the southeastern banks of the Syr Darya adjacent to the Dzhusaly vault). The thickness of the Upper Cretaceous layers is between 220 and 270 m. The groundwater is slightly to strongly brackish (0.3–13%).

Most sediments are from the Lower and Middle Palaeogene – the Saksaulski and Cheganski stratification sequence of the Eocene and Lower and Middle Oligocene. The main rocks are sandstone, clay, pebbles and gravel. Those sediments are often strongly transformed by aeolian processes. The groundwater is saline (0.5–8.5%). The Eocene–Oligocene–Pliocene sediments on the former northern and western coasts of the Aral Sea form a typical unit of the lithological landscape basis, the denudation plateaus. In some parts also the Lower Miocene sediments belong to this type (carbonatic, colored clay of the montmorillonite type, aleuritic layers, quartzite stones and limestones). These sediment layers are rather resistant against denudation; they are responsible for the formation and stability of the flat plateaus and tables in the north and west.

In the wide valleys of the ancient rivers and in some of the furrows of the northern edges of the tableland, the formation of deluvial and proluvial sediments is prominent. Those landscapes and ecosystems in these dissected reliefs are very dynamic.

During the middle Pliocene the sea retreated. All over a continental development started. The eastern Aral Sea region became strongly influenced by erosion and accumulation, which was the basis for the formation of the present lithological basis. The rocks from the Miocene to the Pleistocene are found all over except on the anticlinals and some other vaults. The middle Pliocene sediments are present as deluvial-proluvial lenses and as rock debris infiltrated with loam. Above them, the later sediments from the Pliocene are sand layers with alternate layers of clay and silt, with lenses of gravel and clayey sand. The sediments from the Pleistocene and the Holocene are mainly sands of fluviatile and aeolian origin.

The alluvial sediments of the middle Quaternary with mixed yellowish-brown quartzitic and feldspar sands, with interlayers of carbonatic clay and silt, are the lithological basis of most landscapes of the lower Syr Darya, Zhana Darya and Amu Darya (except the deltas). These layers are about 40–60 m thick. The mineralization of the groundwater is only 1–3 g/L. This Quaternary alluvium is present also on the western edge of the sands of Malye Barsuki and reaches the Aral Sea coast west of the village of Akespe. The alluvial layers in Akbidaik Bay amount to only 2 m.

The younger alluvium of the Syr Darya and Amu Darya deltas (Late Quaternary) with carbonatic clay, loam, loamy sand and sand is the lithological basis of this delta type. These sediment layers are about 3–20 m thick in the Syr Darya delta and about 30–40 m thick in the Amu Darya delta. The total load of sediments of the Aral Sea and its water exceeds 15 kg/cm^2. This coincides with the load value of the whole sediments (Mörner 1980).

The present landscape is characterized by considerable aeolian sands close to the coastal areas, mainly the Malye and Bolshie Barsuki sands, the Priaralski Karakum sands, and the Zhuvankum Kyzylkum sands.

According to the present erosion catena, the lowest parts presently are characterized by chemical sediments of the desiccated salt lakes and by solonchak soils. Here, loam, sandy loam and silt are mixed with salts or salt layers are included. Sometimes salt covers the surface. These mixed layers may be up to 8 m thick. If the groundwater is rather shallow, less than 1.5 m, the formation of solonchak soils is prominent. If it is more than 1.5 m, then crusty-puffy solonchaks are formed.

The marine deposits of the old Aralian transgression (Chap. 2) form their own type of lithological basis with fine and medium granular sands of grayish-yellow color. They are derived from the early Holocene. They exhibit intermediate layers of loam, sandy loam and clay, but often also with layers of shells. They are about 5–6 m thick. This layer is quite often also indicated by a partly or almost dissolved gypsum layer of 1.5–2 m with extraordinarily rich accumulations of shells and other marine fossils. Below this gypsum layer there is grayish and yellow sand without shells.

The youngest development started in 1960. Since then a new lithological type of silty sea sands with intermediate layers of greenish-gray clay, gypsum and sea shells has formed. This desiccated area of the seafloor since 1960 is now actively overformed by strong aeolian processes.

The desiccated seafloor sediments can be defined according to their composition. Slightly higher parts are more subjected to wind erosion than lower parts or slight depressions. In those depressions, accumulation of dust and sands takes place and solonchaks form. The edaphic processes there are very dynamic and the substrates exhibit transformations to lithological attributes similar to those of the former coastal areas.

3.4 Geomorphological Structure

The huge basin of the Aral Sea depression is part of a geological platform regime with weak tectonic movements and with the predominance of a dry climate. These are the reasons for extant accumulative plains, denudation plains and flat eroded table plains which are only very slightly dissected. The relative height differences are normally between 10 and 30 m, but in the northwest they may be up to 100–350 m. The absolute elevations of the former coastlines are between 400 m above sea level (asl) in the northwest on the plateaus and 70 m asl in the southwest in the sand deserts. The lowest point is now the dry seafloor, which is about 25–28 m asl. The elevation amplitude of the Aralkum is between 25 and 58 m asl, including the old and new Aral terraces.

Berg (1902) and Gorodezkaya and Kes (1986) studied the geomorphological features of the coasts of the Aral Sea. They distinguished four geomorphological districts: the northern (north Aralian plains), the eastern (sand deserts Priaralski Karakum and Kyzylkum and the Syr Darya delta), the southern (the Amu Darya delta and the Sarykamysh basin) and the western (Ustyurt Plateau) districts.

In the north and west the Aralkum is limited by high structural denudation plateaus. North of the Aralkum there starts a tableland plain which is dissected by some terraced steps. These steps indicate the various phases of the big Palaeogene sea, with phases of continental denudation of this territory and with the direction of erosion towards the Aral basin. The elevation of this tableland plain is about 200–300 m asl; some relict islands from a former trunk mountain reach 300–400 m asl. There is a distinct steep slope of about 30–60-m height which is the delimitation to the lower denudation plain at 100–200 m asl.

Ustyurt Plateau in the west is delimited from the Aral basin by steep slopes (chinks) from the adjacent plains and the Aral Sea. In northern Ustyurt there is a tableland plateau with incrustations armouring the surface. The elevation is about 120–170 m asl (rarely up to 220 m asl). The shallow Sam-Kosbulak basin has an elevation of 70–100 m asl.

The big basin of the Aral Sea is divided geologically into two depressions: the Small Aral Sea depression in the north and the Large Aral Sea depression in the south. The Large Aral Sea depression is situated south of the former island of Kokaral (see Chap. 2, Fig. 2.6). The delimitation of both is caused by the "River Rapid of Berg", which is a distinct submarine dike. This dike became dry in 1986–1990, and is now the southeastern border of the Small Aral Sea. Both depressions have their own undulating or delimiting slopes of the seafloor.

The Small Aral Sea is strongly dissected into four basins forming distinct bays: the Great Sarychaganak, Butakov and Shevchenko bays and the middle main part of the Small Aral Sea. The Small Aral Sea is a small and shallow water body, which until 1960 was about one ninth of the whole sea area. About 42% of the area of the Small Aral Sea has a water depth of 10–20 m. The maximal depth was 29.5 m, the mean 13.3 m. Within the submarine vaults between the basins, the depth was only 12–14 m. The basins of the Small Aral Sea exhibit an oval or roundish shape with a shallow, rather flat U-profile and a flat central part. The area of the rest of the Small Aral Sea is 3,200 km^2 and the dry seafloor in the small depression has an area of 2,400 km^2 (part of the Aralkum).

The Small Aral Sea has a dyke between the Shevchenko and Butakov bays. The Large Aral Sea is now divided into the deep western basin and the shallow eastern basin, separated by the former island of Vozrozhdeniya which is now part of the mainland.

Before 1960 the Small Aral Sea and Large Aral Sea were one water body, connected by two water channels west and east of the island of Kokaral. The main role in water exchange between the two parts was played by the eastern connection, the "Berg Crossbar", a connection between both basins, named after the famous geographer L.S. Berg. It was about 15 km wide and 10–15 m deep. The northwestern connection between the former island of Kokaral and the Auzy-Kokaral peninsula was very shallow, only up to 1.0–1.5 m deep, and very narrow (up to 0.6 km).

The Large Aral Sea had an area of 60,000 km^2. It consisted of two basins divided by the tectonic vault of the former island of Vozrozhdeniya. The eastern basin is rather shallow, almost plain to a vast extent. The western basin is rather deep with steep slopes. This western basin is a deep sea furrow originally with a surface area of 13,000 km^2 and a maximal water depth of 64 m. The eastern basin originally had a surface area of 47,000 km^2.

The eastern basin was very large but was very shallow. Before the regression the center had a depth of about 20–28 m, with a mean depth of 14.7 m. The relief of the seafloor still exhibits the traces of old riverbeds and other water discharges, and of old accumulations of sediments from sand spits, sand bars and terraces. The plain relief of the basin in some northern parts is dissected by flat vaults and small steep slopes; in the central part it is dissected by some meridionally oriented long but flat ridges or dikes (2–3 m high).

The seafloor of the Aral Sea, similar to that of other innercontinental big lakes, is divided morphologically into three parts: shelf, slopes and basin. The shelf of the western coast is very narrow and has a distinct edge where the steep slope starts. The western basin is about 150 km long from Cape Baigubekmurys to Cape Dzhidelibulak. Its width is about 15 km in the south and 5 km in the north. The eastern slope, with an absolute height of 35 m, is also characterized by a sharp edge to the shelf; this edge demarcates an old coastline. The relief of the shelf is not flat but has several island dykes and submarine bars.

The shelf of the eastern Large Aral Sea exhibits several terraces. They are 34–35, 38–39 and 41–43 m asl, which corresponds apparently with old coastlines from ancient regressions. Since the 1980s these old coastlines have been visible because of the present desiccation process.

The Small Aral Sea morphologically only exhibits a shelf with a rather shallow and flat basin and slight vaults. The eastern part also exhibits slight stepwise terraces from older regressions. The desiccated seafloor at the western coast is about 2–10 km wide. At the north coast it is only 1.5–2 km wide at the capes and 5–10 km wide within the bays.

Along most parts of the former coast at the shallow parts (53–48 m asl, corresponding to 0–5 m original water depth, isobates) the substrate sediments are well sorted by wave activity and consist of sand. The relief was and is almost plain and not dissected. Many relict relief forms such as bars, sand spits and tombolos are present there. A tombolo is a deposition landform such as a spit or bar which forms a narrow piece of land between an island and a mainland shore, or between two islands. Tombolos usually form because the island causes wave refraction, depositing sand and shingle moved by longshore drift in each direction around the island where the waves meet. Those older surface areas are practically desalinized. They are now subject to the intensive action of aeolian processes, such as deflation, sand transport and accumulation and dune formation.

Within the 5–7-m isobate line, the shelf is often somewhat steeper, indicating the transition to the submarine slope of the former Aral Sea. This is often indicated by a stepwise terrace of 0.5–0.8 m in height. The seafloor is often dissected by dykes and furrows, mostly composed of sand, which now on the desiccated part are again subject to strong aeolian processes.

Within the 7–13-m isobate, the desiccated seafloor is a slightly oblique plain at the eastern coast and the northern coast. The sediments are a mixture of sand, silt, loam and predominantly clay. Around the bigger islands (Barsa-Kelmes, Vozrozhdeniya), sand predominates. The relict relief forms with small flat hills at the eastern coast and netlike sculptured reliefs at the northern coast were caused by submarine abrasion processes. The morphometric differences of these relief forms are maintained by the differing intensities of the hydrodynamic processes and the slope angles below the water surface. The desiccated plains are now subject to halogeochemical processes (salinization, desalinazation Sydykov et al. 1983) and to wind effects (deflation, accumulation Gel'dyeva and Budnikova 1985, 1987).

Within the substrate sediments of the 13–20-m isobate area of the desiccated seafloor, we find predominately aleuritic (soft siltstone) and pelitic depositions. Nowadays, these substrates are also influenced by wind erosion and halogeochemical processes.

The newly formed land surface still exhibits the morphological structure of the seafloor with its relief form and abrasion character. Only the sandy coasts on the older desiccated seafloor are almost totally transformed by wind erosion and new sand dune landscapes have been formed.

3.5 Formation of the New Continental Plain (Dry Seafloor: Aralkum)

The strong lowering of the seawater level of the Aral Sea during the last 50 years changed the coastal dynamics. The desiccated soil sediments along the coast and the forms of the relief play an important role in the formation and development of the present landscape structure on the older seafloor. Studies on the actual landscape morphology of the former submarine coastal slopes and the dynamics of the present coastal displacement can reveal basic facts of these drastic landscape changes.

The present situation of the coasts of the Aral Sea and the changes along the fluctuating coastline are influenced by several factors and conditions. This is mainly the geological and the geomorphological activity of the Aral Sea basin as well as the adjacent mainland and the hydrodynamics (including wave intensity) of the remnant water body. The character and the influence of the waves and the related water currents depend on the wind regime along the coast and on the coastal configuration as well as on the decline of the actual coast and the type of coastal sediments. Because the Aral Sea had and has a very long and flat coastline, the combination of these factors leads to a highly diverse coastal landscape picture.

In accordance with the classification of the coast of the Aral Sea, we can distinguish the following coastal types: abrasion coast, steep abrasion coast, smooth abrasion coast, undulating coast, smooth accumulative coast, dissected accumulative coast, undulating accumulative coast, and arid denudation coast. The development of the coastal zones of the Aral Sea within the last few millennia was characterized by frequent changes of the sea level. This is again in accordance with the diversity of coastal structures seen today.

However, within the last 50 years the coasts of the Aral Sea were subject to a strong and rapid decline of the seawater level. The actual desiccated seafloor has strongly changed from a submarine coastal slope to a shallow water level and then to desiccation. Between 1960 and 1970 the annual decrease of the seawater level was about 0.09–0.50 m. Within the ever-fluctuating and ever-retreating new coastline new abrasion as well as accumulative terraces developped. They are often formed synchronously, but at different height: abrasive terraces were formed above the coastline, whereas accumulation terraces were formed below the coastline. Further retreating sea-level may cause erosion of the newly accumulated sediments again. Those complex retranslocation processes may result in a mosaic of various terraces according to changing wave-action. Then these terraces are often only shortly visible.

In general, abrasion of the lower coastal slopes is observed with transport of the sediment particles to the upper water level, which is the main process in the formation of beach lines by accumulation of submarine sediments from the coastal slopes.

The areas of the islands and of shallow underwater hills strongly increased as a result of the shrinking water level. The huge shallow ponds and plains close to the eastern coastline were enriched by accumulative sediments. Thus, underwater shelves were formed. During the 1970s (since 1972) the lowering of the sea level was about 0.6–0.8 m annually. Then accumulative relief forms of the sea came to light (underwater shelves, promontories, sand spits, tombolos), and these were partly preserved by the salt crusts, but later were changed by aeolian activity. Because of the very low slope angle at the eastern coast, huge seafloor areas became dry annually.

Since 1978, several abrasion terraces have become visible; they were formed earlier by submarine abrasion. Along the western coast as well as at the northern coast the disappearance of those terraces and cliffs was prominent. The ongoing desiccation and lowering of the water level brought those terraces to upper parts of the slopes, becoming inactive geomorphological structures. Now most of the newly formed coastal structures are accumulatively formed. Only small parts of the present retreating coastlines are still active abrasion coasts, mainly at the capes of the northwestern and western coasts.

The shrinking of the sea surface area led to a total change of the coastlines. This was the case especially along the flat eastern coast (Fig. 3.1). All the large bays such as Akkol, Bozkol, Dzhyltyrbas, Dzhyltyrbas and Muinaksky and many smaller bays became dry. All islands became part of the mainland (Vozrozhdeniya, Barsa-Kelmes, Kokaral). New islands were formed for short times where submarine bars or vaults were, but rather soon also became part of the new mainland. Also, in the Small Aral Sea (North Aral Sea) many small bays and lagoons became dry. Great Sarychaganak Bay in the northeast was completely dry. The big bays of Butakov and Shevchenko were connected with the main part of the Small Aral Sea only by small sea channels.

The recent coasts of the remnant water bodies of the divided Aral Sea are mainly of only one coastal type: they represent a typical smooth coastal border. This coastal border is characterized by a broad belt of recently desiccated flats and seafloors which are covered in the first few years by a dense halophytic, annual vegetation (Chaps. 9 and 10). The middle and upper parts of this coastal belt are regularly bare of vegetation and the loose sedimental substrates are subject to wind erosion, except at those sites where groundwater is leaking and new springs have developed (Zektzer et al. 1973). Of course, the formation and character of these coastal borders depends on the composition of the sea sediments and the actual processes forming the relief (aeolian, halogeochemical, pulverization). The seafloor

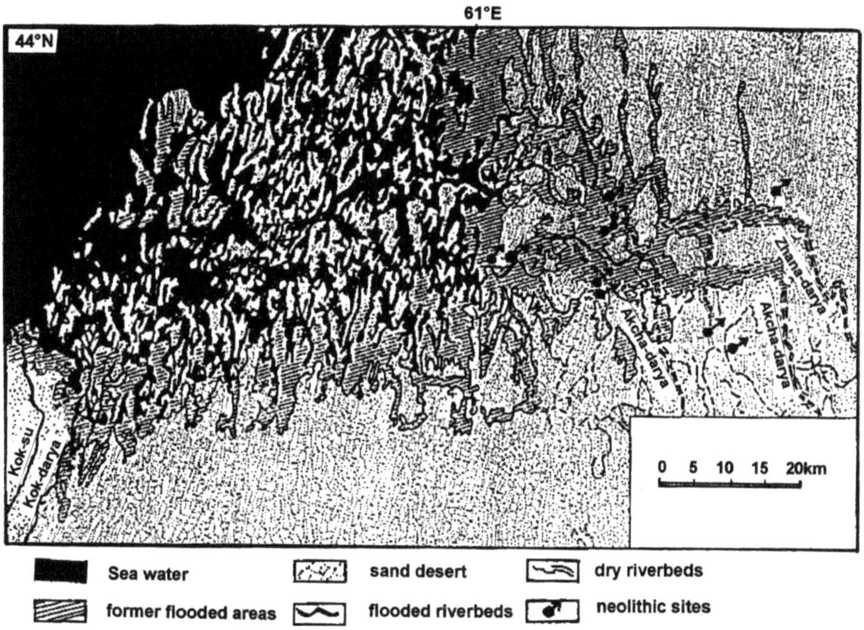

Fig. 3.1 Part of the southeastern coast of the Aral Sea around Akpekti formerly with numerous islands and bays, demonstrating the amphibiotic dynamics of a very flat coastline (Modified from Letolle and Mainguet (1996))

sediments exhibit a great variety of types: sands, aleurites, clays, shell limestones, shells, gravel, pebbles and other detritus (mainly near the abrasion capes at the northwestern and western coasts). Since 1984, mainly aulerite (rich in limestone and phosphorous), silt (at the eastern coast), clay and loam in most bay areas of the north became uncovered. Along some of the islands (Barsa-Kelmes) also sand deposits became desiccated. On all desiccated flats from the last 20–30 years, a thick crust of salts and salt powder developed on the deposits by water evaporation.

The sediments along the northern coast bays are composed of sand, silt and loam. It is remarkable that since the beginning of the 1980s a very rapid change of all the hydrodynamic and hydrochemical processes with continuing retreat of the water level has taken place and reduced the accumulative relief forms of the sea tremendously.

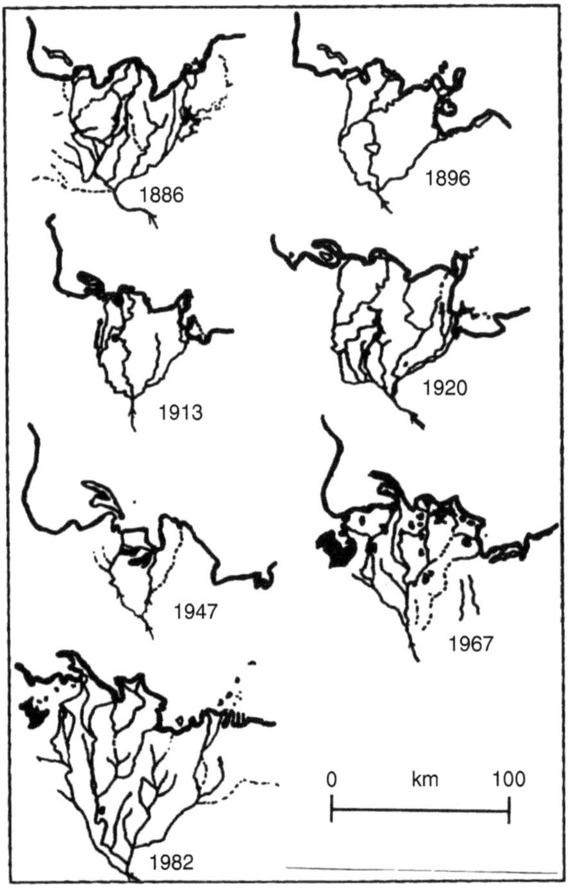

Fig. 3.2 The great variability of the delta of the Amu Darya between 1886 and 1982, indicating an increase in area caused by sedimentation but even more by the retreat of the coastline, land connections to former islands and the disappearance of delta lakes (Modified from Letolle and Mainguet (1996))

3.6 Delta Areas of the Dry Seafloor

The main tributaries to the Aral Sea, the Amu Darya and the Syr Darya, produced large delta areas on the flat plains. Their sand and silt sediment load is huge and thus those slowly growing deltas formed wide amphibiomes with always changing river courses (Fig. 3.2). During the desiccation process of the last few decades, on the one hand, the amount of water delivered to the Aral Sea was much lower than earlier and, on the other hand, the water courses became longer and longer and the old delta areas were partly channeled by river erosion. The lowering of the groundwater has led to a severe dieback of tugai vegetation. As long as water is flowing and reaching the remnants of the Aral Sea, new small deltas are formed again in lower parts. The Syr Darya has already formed a new delta in the North Aral Sea north of the new dam. Since those delta areas were formerly very important stands of extensive tugai shrub vegetation with very diverse wildlife, there is hope that the new deltas will be colonized by tugai species as well and that recovery of parts of the tugai areas may take place.

3.7 Conclusions

The geomorphological features of the new desiccated seafloor as a new continental area (Aralkum) are governed by geohalomorphic and aeolian processes. The vast-plain open sea bottom is not totally plain; it exhibits old submarine channels, hills, ridges, sand splits and tombolos, which are now overformed. Mainly the older seafloor, which became desiccated before the 1980s, is characterized by sandy substrates; the younger seafloor, which came to light after the 1980s, is sandy, silty or clayey, but is more or less strongly salinized. Ninety-seven percent of the Aralkum area is situated in the big Aral depression south of the former Kokaral peninsula (and including it) and 3% of it is situation in the small Aral depression around the Small Aral Sea.

References

Berg LS (1902) Coast morphology of the Aral Sea. In: Ezhegodnik po geologii i mineralogii Rossii, vol. 5, Ser. 6–7 (in Russian)
Bubnoff SV (1924) Der Gebirgsbau vom Osten Europas. Geol Rundschau 15:147–174
Gel'dyeva GV, Budnikova TI (1985) Eolian processes on the primary sea plains of the Priaral'e. Izv AN SSSR, ser. geogr. 5, 87–91 (in Russian)
Gel'dyeva GV, Budnikova TI (1987) Dynamics of the landscapes of the dry sea floor of the Aral Sea. Vestnik AN KasSSR 3:55–62 (in Russian)
Gorodezkaya ME, Kes AS (1986) Topography of the coasts of the Aral sea and perspectives for development. Probl Ows Pustyn 3:35–43 (in Russian)

Letolle R, Mainguet M (1996) Der Aralsee. Eine ökologische Katastrophe (The Aral Sea. An ecological desaster). Springer, Berlin (in German)
Leuchs K (1935) Der Block von Ust-Urt. Geol Rundschau 26:248–258
Mörner N-A (1980) Earth rheology, isostasy, and eustasy. Wiley, New York
Sydykov ZhS, Shapiro SM, Poryadin VI, Vinnikova TN (1983) Hydrogeological characteristics. In: Aralskoe more. Nauka, Alma–Ata, pp 5–42 (in Russian)
Zektzer IS, Ivanov VA, Meskheteli AV (1973) The problem of direct groundwater discharge to the seas. J Hydrol 20:1–36

Chapter 4
Climatic Conditions in the Aralkum

S-W. Breckle and W. Wucherer

4.1 Introduction

The Aral Sea and the Aralkum are located within the Asiatic desert belt. The northern part of the Aralkum is part of the Kazakh-Dzungarian desert, the southern part is part of the southern Turanian deserts. The climate of the surroundings of the Aralkum can be characterized by the data from eight meteorological stations in the area. It appears to be very continental with very cold winters and extremely hot summers. The consequences of the desiccation of the Aral Sea on the climatic conditions have to be estimated. Current climatic data only reveal a slight shift to even stronger continentality (see also Chap. 17). The overall water dynamics of the atmosphere does not seem to be altered in general, but intensification of water cycling by global change overlying the local and regional desiccation processes may be observed.

For ecological purposes and to characterize a distinct region, the climatic diagrams of Walter and Lieth (1967) are widely used. Meteorological stations in the whole area are scarce, and they often lack data since 1990, after the dissolution of the Soviet Union. Figure 4.1 gives an overview of the available ecological climatic diagrams for the region. It is obvious that during most the year an arid situation prevails. The temperature curve in the diagram, representing the evaporative demand (see Breckle 2002), lies above the precipitation curve (stepwise curve) most of the year. Only in winter is the evaporation low, due to the severe frosts. But even then, frozen soils, frozen stems of plants and high radiation cause severe dryness (frost drought) in plants.

S.-W. Breckle et al. (eds.), *Aralkum - a Man-Made Desert: The Desiccated Floor of the Aral Sea (Central Asia)*, Ecological Studies 218,
DOI 10.1007/978-3-642-21117-1_4, © Springer-Verlag Berlin Heidelberg 2012

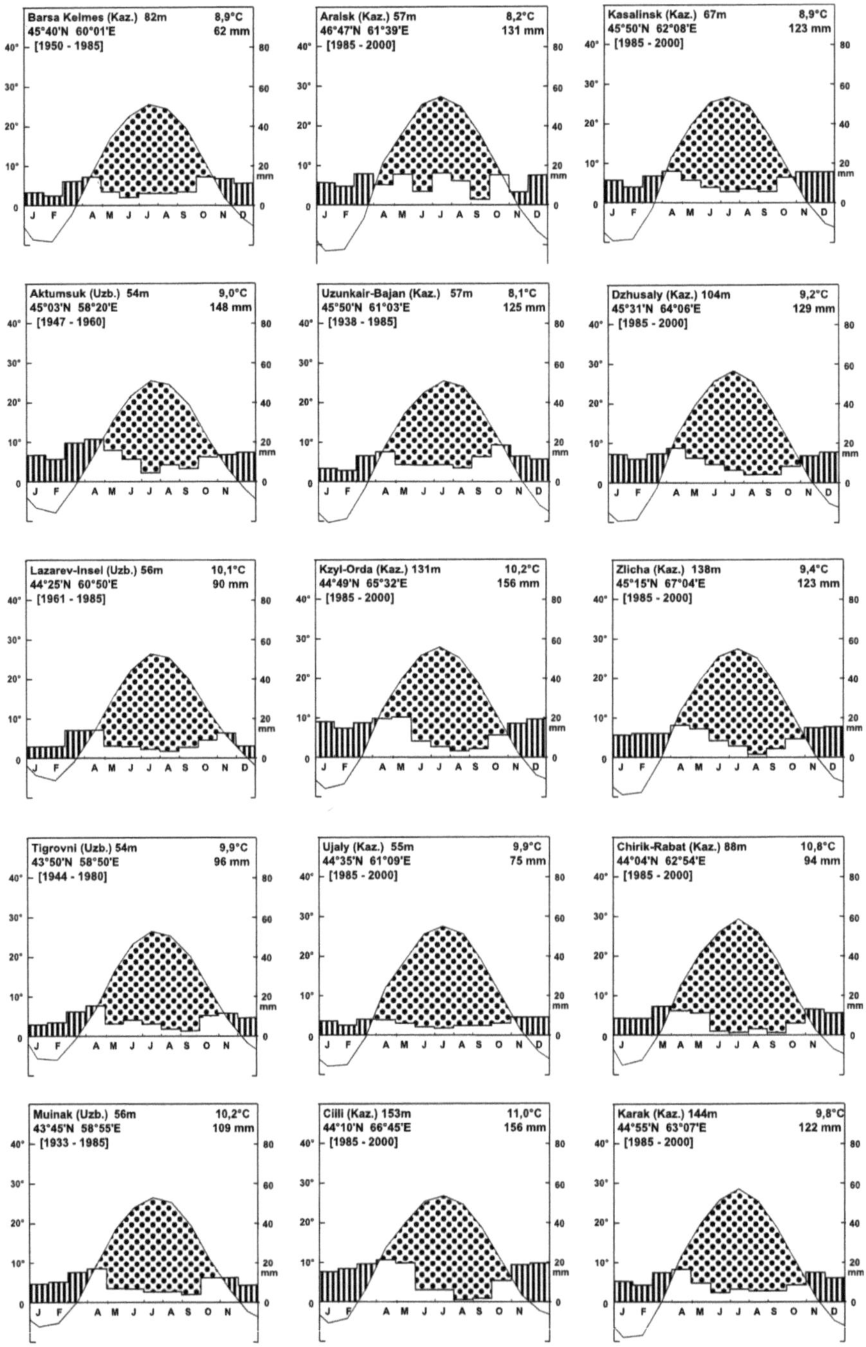

Fig. 4.1 (*Continued*)

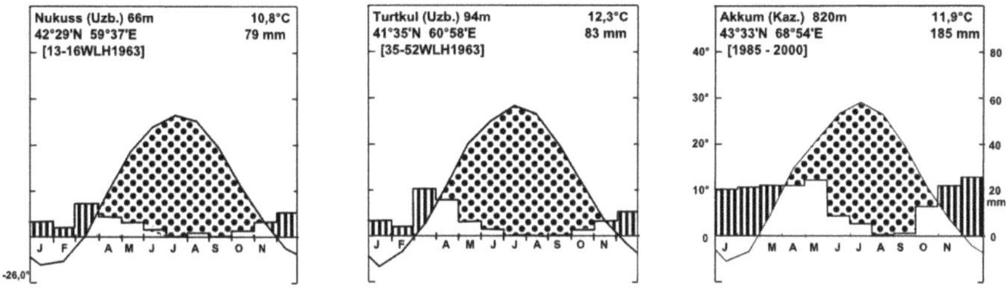

Fig. 4.1 Ecological climatic diagrams from the Aral Sea region and surroundings (Method after Walter and Lieth 1967), modified version (After Breckle (2006))

4.2 General Characteristics

The climatic conditions of the Aral Sea region are mainly governed by the relatively low elevation of the Aral basin within the center of the Asian continent (Budyko 1956, 1974; Grigoriev and Budyko 1959). The continentality is a reason for the very intensive radiation. The atmospheric circulation is not influenced by high mountains. Most of the year the region around the Aralkum is under the influence of the southwestern margin of a huge Asiatic (Siberian) anticyclone and northwesterly and northerly air intrusions (Bugaev 1957). The geographical position of the region opposite the Turanian or turgai gate plays an important role: it is the main axis for the invasion of cold air masses from the north along the Ural mountains. The low-lying basin within Middle Asia favors the development of stable cold air masses and partly considerably low temperatures in winter.

The normal synoptic processes are as follows: the stable southwestern peripheral part of the Siberian anticyclone in autumn, winter and early spring (Myachkova 1983) causes clear and dry weather and drastic temperature minima. It is a consequence of advection of air masses from continental Siberia and the Arctic, which are very low in humidity. Moist Atlantic air masses from the west normally do not reach the Aralkum region during winter, only in spring and rarely in summer. Then drastic differences in temperature between air masses can cause strong winds and storms. Those air masses from the west and south are driven by cyclones and on the back of them high humidity is brought and rains can occur. During summer, very high temperature and a clear, cloudless sky are typical weather situations. Northerly and northwesterly air masses from time to time can occur and lead to slight decreases in temperature and often to some precipitation.

The Aralkum region in general is characterized by a strong continentality, by hot and dry summers and by cold winters. The temperature amplitude during the year, expressed by the relevant monthly means can reach almost 40 K, the amplitude of the absolute temperature extremes may reach more than 85 K. Furthermore, low precipitation and rare cloudiness is typical. The number of days without clouds amounts to about 260. This together with the high radiation results in very dry and hot summer months.

4.3 Methods and Database

As in many other remote regions the use of instrumentation in meteorological stations started relatively late and often long sequence data exhibit observation gaps. Williams and Konovalov (2001) surveyed all available stations for the database of the Global Historical Climate Network. The characteristics of the database are shown in Table 4.1. More than 2% of temperature data and more than 3% of precipitation data are lacking in the continuous observation periods in Kazakhstan as well as in Uzbekistan.

The number of stations was reduced drastically during 1985–1995 to 79 for precipitation and 62 for air temperature measurements for the whole region.

Another problem for a comparable database is the change of recording methods and changing and modernization of instrumentation.

From the available data from meteorological stations, the ecological climatic diagrams have been drawn according to Walter and Lieth (1967) and Breckle (2002) to illustrate the varying primary factors temperature and precipitation as the most effective climatic parameters for vegetation (Fig. 4.1). The climatic diagrams have been modified according to Breckle (2006) by drawing the temperature curve as a curve and the precipitation data as a stepwise curve, according to the data structure. These Breckle–Walter diagrams indicate in a way similar to the former Walter diagrams the relative arid season by stippled area, the relative humid season by stripes, and the per humid months by dark shading. The location of the meteorological stations is shown in Fig. 4.15.

4.4 Climatic Parameters

4.4.1 Radiation

The radiation balance within the Aralkum region is strongly positive and reaches about 175–190 kJ cm^{-2} $year^{-1}$ (Fig. 4.2). The negative balance of radiation lasts only 1–2 months in the winter in the north of the Aralkum (Fig. 4.3). The annual

Table 4.1 Generalized characteristics of central Asia regional database with stations before 1980

Country (code-no.)	Data for mean air temperature			Data for mean precipitation		
	n	ST-y	(%) gaps	*n*	St-y	(%) gaps
Kazakhstan (211)	27	2,261	2.2	47	3,628	3.7
Kyrgyzstan (213)	45	2,404	2.3	53	3,007	2.5
Tajikistan (227)	46	2,503	1.3	40	2,167	1.3
Turkmenistan (229)	12	1,145	6.3	14	1,251	4.2
Uzbekistan (231)	64	3,708	2.2	116	6,308	3.9
Total	194	12,021		270	16,361	

According to Williams and Konovalov (2001)
n number of stations, *ST-y* sum of station years

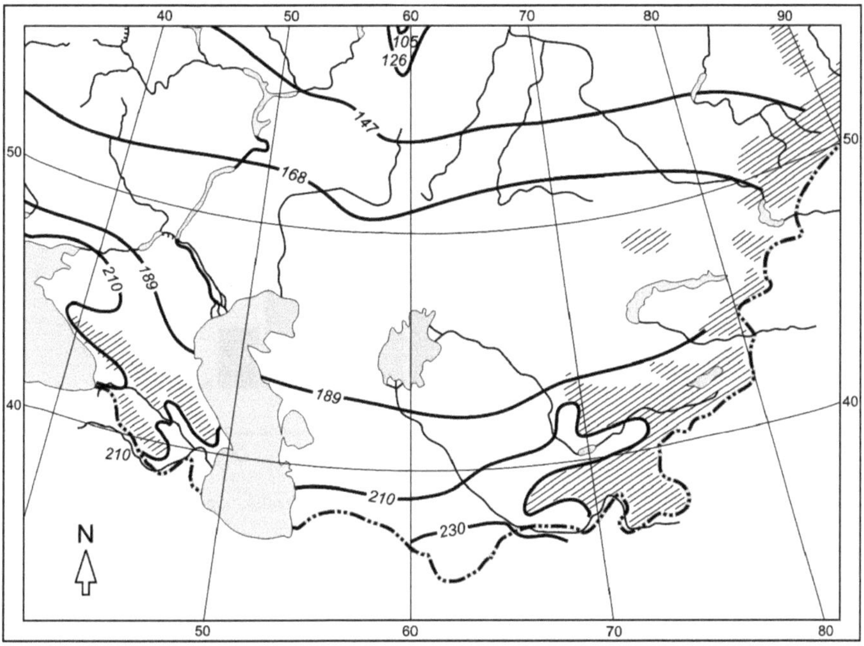

Fig. 4.2 Map of the radiation balance (kJ cm^{-2} year^{-1}) in the Middle Asia and Aralkum area (Myachkova 1983)

mean of the global radiation increases from 580 kJ cm^{-2} in the north (Aralsk, Saxaulski, Sande Barsuku) to 625 kJ cm^{-2} in the south (Fig. 4.4) (Myachkova 1983). During the long dry season, the absorbed energy is totally used for turbulent heat exchange (70–84%). Only in spring, when the soil is water-saturated, is the proportion of radiation energy used for evaporation higher (Table 4.2). On average, over the year it is only between 16% and 30%.

The Zeagly station is situated in the center of the Karakum and the Muinak station is at the southern coast of the Aral Sea. Both stations have nearly identical radiation parameters (radiation balance, heat energy for evaporation, heat energy for evaporation/radiation balance, turbulent heat flow, turbulent heat flow/radiation balance). The first three stations in Table 4.3 are characteristic for the northern Turanian deserts and the last three stations are characteristic for the southern Turanian deserts. It is obvious that the southern part of the Aralkum is situated within the climatically defined southern Turanian deserts. But Myachkova (1983) considers the Muinak station to belong to the northern deserts. The diagram of the annual radiation balance (Fig. 4.2), however, indicates the curve of 189 kJ cm^{-2} (without the data from the Muinak station) incompletely, and the monthly data for July and April (Figs. 4.5 and 4.6) show the difference between the northern and southern parts of the Aral Sea. All radiation parameters are also different in the north and in the south (Tables 4.2 and 4.3). In accordance with the high radiation and light intensity, the amount of sunshine is very high. The annual means in the

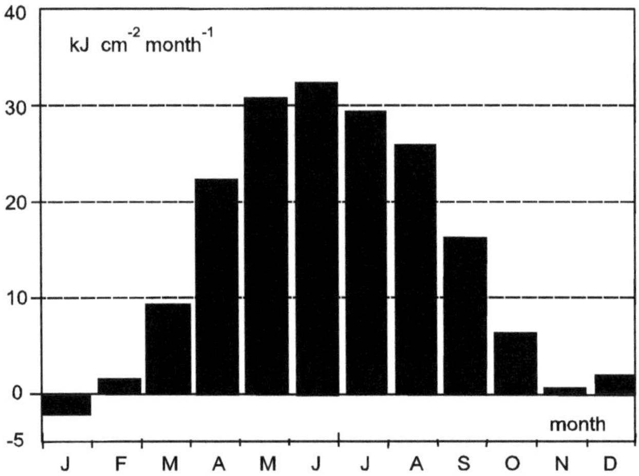

Fig. 4.3 Monthly distribution of radiation balance at the Aralsk station (kJ cm^{-2} $year^{-1}$) (Zhitomirskaya 1964)

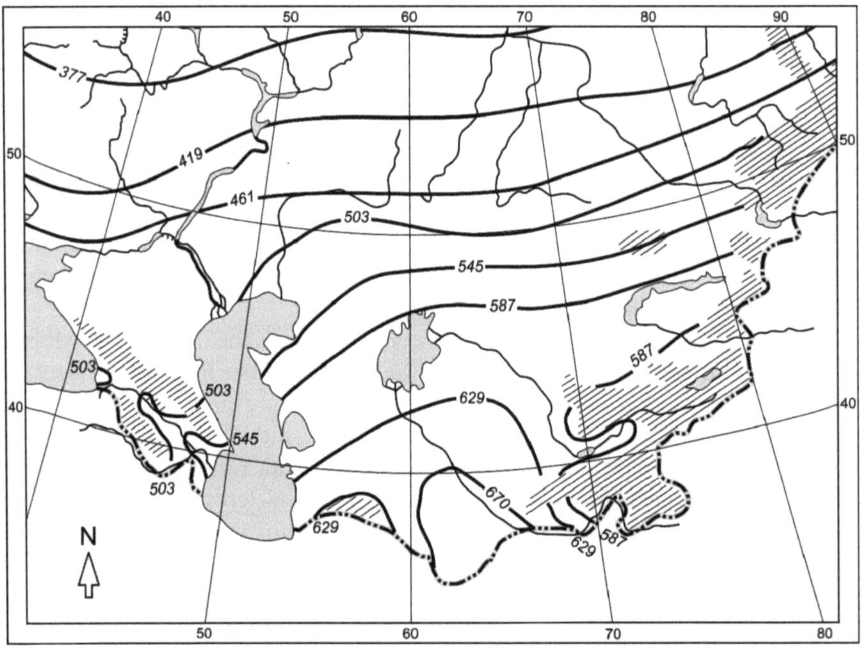

Fig. 4.4 Map of the sum of radiation (kJ cm^{-2} $year^{-1}$) in the Middle Asia and Aralkum area (Myachkova 1983)

Table 4.2 Heat balance (kJ cm^{-2}) of the earth's surface in the Middle Asian region, in spring (March to May)

	RB	HE	HE/RB (%)	TH	TH/RB (%)
Aralsk	62.8	29.3	47	23	37
Kyzyl-Orda	71.2	29.3	41	33.5	47
Burlyu-Tobe	58.7	25.1	43	25.1	43
Zeagly	67	18.9	28	37.7	56
Krasnovodsk	67	18.9	28	37.7	56
Ashgabat	69.1	25.1	36	41.9	61

According to Myachkova (1983)

RB radiation balance, *HE* heat energy for evaporation, *TH* turbulent heat flow

Table 4.3 Heat balance (kJ cm^{-2} year^{-1}) of the earth's surface in the Middle Asian region (Myachkova 1983)

	RB	HE	HE/RB (%)	TH	TH/RB (%)
Aralsk	159.2	41.9	27	117.3	73
Kyzyl-Orda	176	41.9	24	134.1	76
Burlyu-Tobe	167.6	50.3	30	117.3	70
Muinak	196.9	33.5	16	163.4	84
Zeagly	188.5	29.3	16	159.2	84
Krasnovodsk	201.1	33.5	17	167.6	83

RB radiation balance, *HE* heat energy for evaporation, *TH* turbulent heat flow

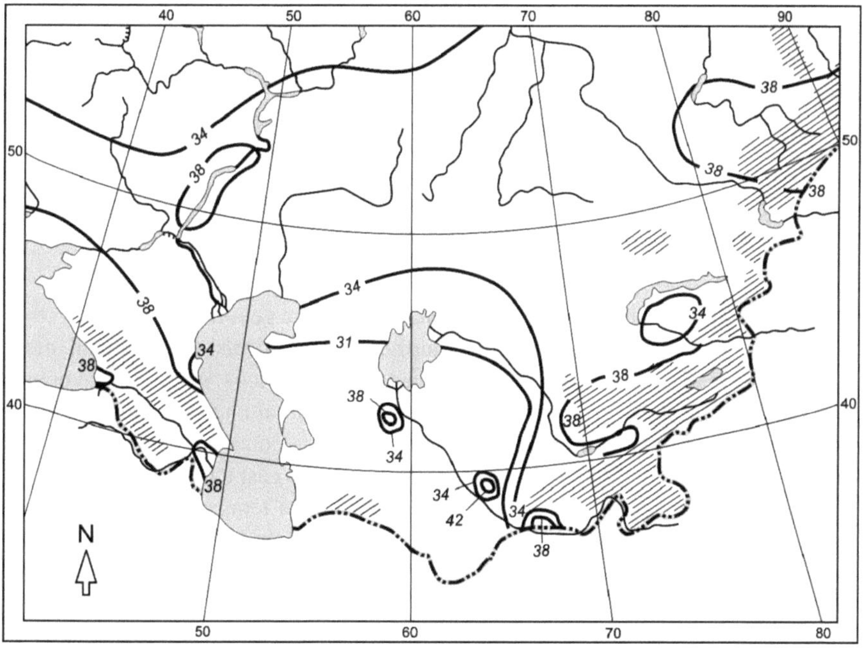

Fig. 4.5 Map of the radiation balance (kJ cm^{-2}) in the Middle Asia and Aralkum area in July (Myachkova 1983)

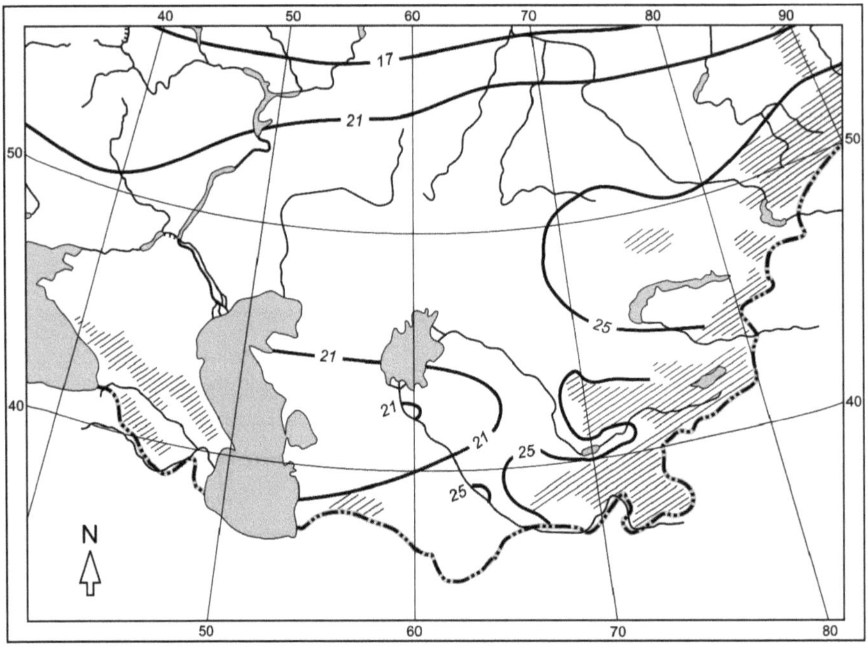

Fig. 4.6 Map of the radiation balance (kJ cm^{-2}) in the Middle Asia and Aralkum area in April (Myachkova 1983)

northern Aralkum are about 2,500–2,900 h per year and in the southern Aralkum are more than 2,900 h per year.

4.4.2 Temperature

The annual means of the air temperature change with the geographical latitude: they increase from north to south. At the northern coast (Aralsk) the annual mean temperature is 6.8°C, at the southern coast (Muinak) it is 9.8°C (Spravochnik po klimatu and Uzbekskaya 1967; Spravochnik po klimatu and Kazakskaya 1968; Spravochnik po klimatu 2004). Increasing temperature conditions can be observed with increasing distances from the Aral Sea towards the east too. In Kyzyl-Orda the annual mean temperature is 8.8°C. The monthly mean temperature in January is −13.5°C in Aralsk and −7.1°C in Muinak. This is a very clear contrast between north and south. The monthly mean temperatures in July (the hottest month) are almost the same all around the Aralkum, about 25.3–26.8°C.

The absolute maximum temperature recorded in Aralsk and Aktumsuk is 43°C, and for the other meteorological stations around the coastline of the Aral Sea the absolute maximum temperature recorded is about 40–41°C. For the Kyzyl-Orda

Table 4.4 Number of days with specific temperature limits (years before 1960, Zhitomirskaya 1964)

	Number of days per year			
	$T > -5°C$	$T > 0°C$	$T > 5°C$	Frost-free period
Aralsk	259	227	198	168
Uzun-Kair	273	238	207	198
Ujaly	294	254	217	221
Muinak	299	263	219	213

Table 4.5 Mean of monthly temperature minima of the year (°C) for some decades and the whole observation period

	1950–1960	1960–1970	1970–1980	1980–1985	19xx–1985
Aralsk	1.2	2.6	2.5	3.3	2.4 (1929–1985)
Barsa-Kelmes	5.2	6.2	5.8	6.3	5.8 (1950–1985)
Ujaly	6.5	6.6	5.4	5.8	6.2 (1942–1985)
Muinak	6.5	6.6	5.1	5.6	6.4 (1933–1985)

After Spravochnik po klimatu Kazakhstana (2004)

(46°C), Nukus (46°C) and Chirik-Rabat (47°C) stations, which are located further inland, the absolute maximum temperatures recorded are even higher.

The absolute minimum temperature recorded at the northern coast in Aralsk is −42°C and in Kyzyl-Orda is −39°C; at the southern coast the absolute minimum temperature recorded is −27°C to −33°C. The absolute annual temperature amplitude between the northern coast and inland is this more than 80 K.

The number of days per year with specific temperature limits is shown in Table 4.4.

It is obvious that the Muinak and Ujaly stations have more moderate parameters typical of stations adjacent to the sea and thus characterize the southeastern coast. During March, most parts of the coastal regions exhibit temperatures below 0°C. During April and May, a drastic rise in temperature (about 10°C) can be observed, later in the north than in the south. The main factor for the spring–summer period is a powerful warming by strong radiation, where latitudinal gradients are almost absent. Arctic intrusions together with southerly cyclones play an essential role during October to March. This specifies the differentiation of the air temperature field with latitude. In late autumn, the temperature decreases drastically: in December the monthly mean temperature goes below 0°C.

The disappearance of the Aral Sea should be an important reason for changing climatic conditions. However, the scarcity of meteorological stations and the discontinuity of observation periods of those stations within the last 100 years makes it difficult to reveal clear statistically significant trends for all meteorological parameters (Table 4.5).

Within the last 50 years, some trends are visible (Figs. 4.7 and 4.8). The annual mean temperature in the north (Aralsk station) has risen from 6.8°C to 8.2°C, and in the south (Ujaly station) it has risen from 9.0°C to 9.9°C. The mean temperatures of the warmest month, July, has risen from 26.3°C to 27.4°C in the north (Aralsk

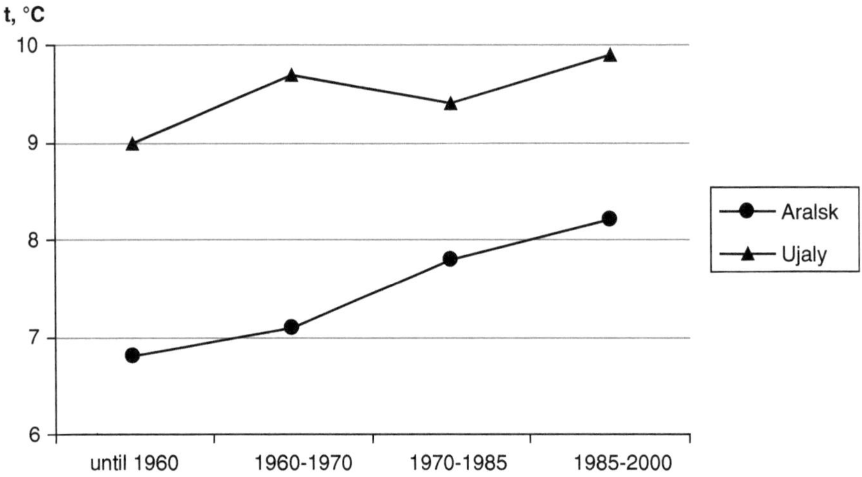

Fig. 4.7 The long-term trend of the annual mean temperature at Aralsk and Ujaly (Data from Spravochnik po klimatu Kazakhstana (2004)

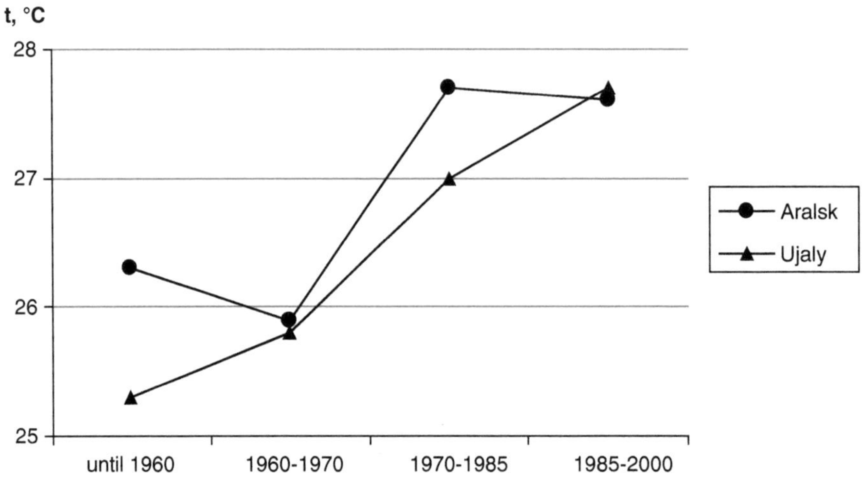

Fig. 4.8 The long-term trend of the monthly mean temperature in July at Aralsk and Ujaly (Data from Spravochnik po klimatu Kazakhstana (2004))

station) and from 25.3°C to 27.7°C (2.4°C!) in the south (Ujaly station). The mean temperature of the coldest month, January, has risen from −13.5°C to −11.5°C in the north (Aralsk station) and from −9.6°C to −7.9°C in the south (Ujaly station). This is an overall rise of almost 2 K. This is not visible with the means of the monthly minima of the whole year, the northern stations exhibiting a rise and the southern stations exhibiting a decrease in temperature (Table 4.5).

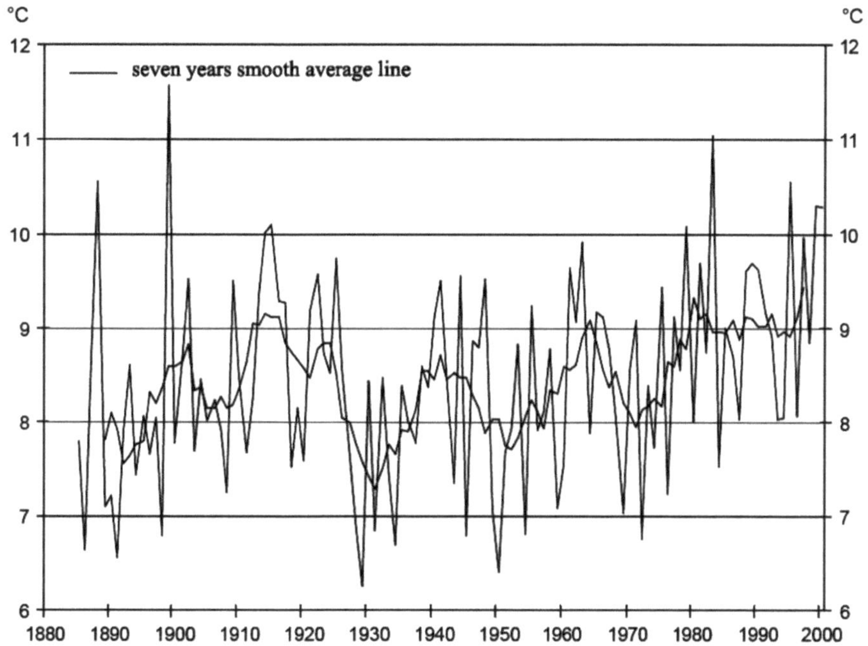

Fig. 4.9 The long-term trend of the annual mean temperature at the Kazalinsk station (Giese and Moßig 2004)

The rise of temperature is not only restricted to the coastal stations but can also be observed to a slightly lesser extent in inland stations. The Kazalinsk station (120 km east of the original coastline) and the Kyzyl-Orda station (400 km east) exhibit the same trend. In the last 50 years the annual mean temperature at the Kazalinsk station rose from 8.0°C to 8.9°C and at the Kyzyl-Orda station it rose from 8.8°C to 10.2°C. The corresponding values for the mean temperature in July are from 26.1°C to 27.3°C and from 25.9°C to 27.8°C. According to Giese and Moßig (2004), the most distinct and steady increase of the air temperatures in the plains and foothills of Central Asia was after 1970. The Kazalinsk station exhibits a statistically significant increase of 1–1.5 K within the last 70 years (Fig. 4.9). The trends in temperature at the coastal stations of the Aral Sea are parallel to those of regional and global warming within the last few decades (Ginzburg et al. 2003).

4.4.3 Precipitation

The semidesert or desert conditions are caused by the low annual precipitation in the area. Most stations receive between 60 and 140 mm precipitation per year. As can be seen from the climatic diagrams (Fig. 4.1), the precipitation maximum is rather variable between seasons and is not distinct. The fluctuations from year to

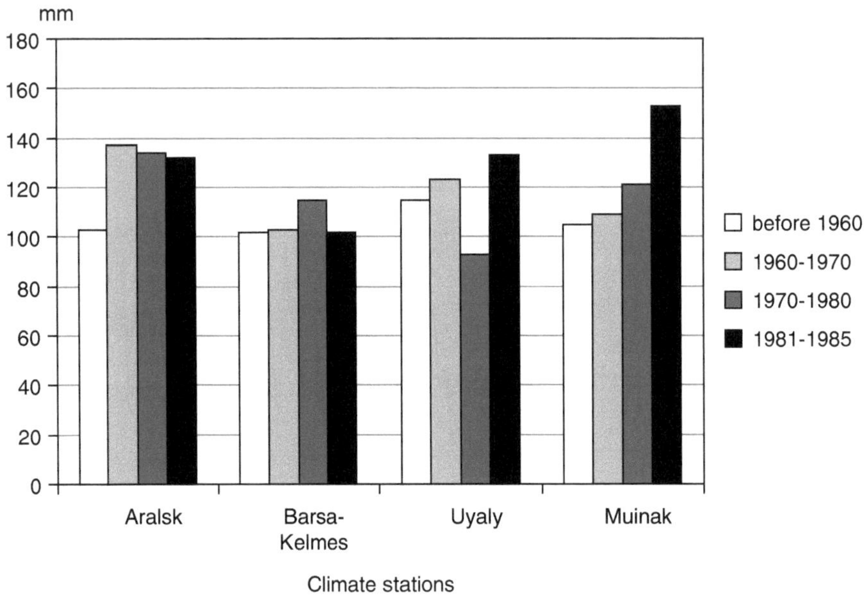

Fig. 4.10 The long-term trends of the annual precipitation (Data from Spravochnik po klimatu Kazakhstana (2004))

year are conspicuous (Nezlin et al. 2004, 2005) as is the discharge of the main rivers. Often, the necessary long-term observations are lacking (see also Chap. 17).

In autumn, there is a weak influence of the Iranian branch of the polar front, causing a slight second maximum in October/November. In summer, the dryness is caused by the strong anticyclones but is rather rarely interrupted by westerly cyclones. Thus, the precipitation on average even during the summer months is not zero. Most of the coastal areas received about 100–110 mm annual mean precipitation before 1950–1960. Only the west exhibited slightly higher values (Aktumsuk station), with about 148 mm (before 1960). In the last 50 years the picture of the precipitation dynamics is different for the different stations (Fig. 4.10). In the 1950s, the precipitation at the Aralsk station rose to 134 mm and over the next 40 years it varied around 130 mm (131–137 mm). At the Muinak, Ujaly und Kazalinsk stations the annual precipitation varied with a slight, but not significant reduction within the last 15 years (1985–2000).

As usual in arid areas, the differences from year to year in precipitation are also rather high in the Aralkum. The Ujaly climate station exhibited a maximum annual precipitation of 285 mm in 1981 and a minimum of 33 mm; the respective values for Muinak are 239 mm (1982) and 43 mm. This means an amplitude of up to a factor of 8 between years. The following years were relatively rich in precipitation: 1949, 1957–1959, 1963, 1969–1971, 1973, 1978, 1980–1982 and 1984 (see Table 4.6).

There can be precipitation maxima in three different seasons: spring, summer and autumn. The Muinak station recorded a monthly maximum for April 1978 of

Table 4.6 The precipitation maxima within a specific month (mm), and year of event (Spravochnik po klimatu Kazakhstana 2004)

Stations	January	February	March	April	May	June	July	August	September	October	November	December
Aralsk	22 1982	23 1936	43 1952	58 1949	60 1981	71 1981	81 1957	52 1949	29 1947 1948	80 1980	50 1963	34 1979
Bayan	18 1981	15 1981	24 1963	69 1973	49 1981	33 1963	36 1969	19 1983	84 1982	61 1980	60 1980	28 1971
Ujaly	21 1981	36 1981	53 1942	64 1964	34 1984	62 1981	31 1957 1970	65 1980	47 1947	57 1957	31 1978	36 1972
Muinak	31 1974	38 1960	55 1981	69 1978	34 1978	55 1963	57 1970	57 1958	77 1940	27 1971	39 1982	59 1978
Tiger	18 1974	34 1960	34 1964 1969	53 1949	20 1945	50 1963	47 1959	33 1949	11 1973	60 1946	40 1953	52 1971
Lazarev	19 1962	20 1976	44 1981	36 1969	25 1981	35 1963	19 1970	9 1979	21 1961	35 1980	40 1969	27 1979
Barsa-Kelmes	19 1974	26 1967	35 1957	38 1973	40 1981	28 1981	64 1957	65 1958	34 1982	50 1959	59 1980	48 1971

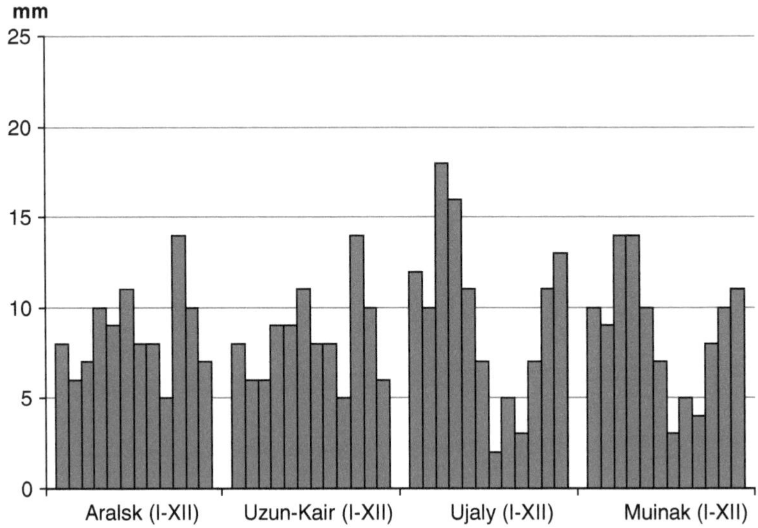

Fig. 4.11 The seasonal distribution of precipitation until 1960 (Data from Zhitomirskaya (1964))

69 mm, and the Bayan station recorded a monthly maximum for April 1973, also of 69 mm. In summer, the maximum was recorded by the Aralsk station (June 1981, 71 mm; July 1957, 81 mm). In autumn, the monthly maximum was observed at the Bayan station (September 1982, 84 mm) and at the Aralsk station (October 1980, 80 mm). There are typical synoptic events for the whole region of the Aral Sea, the adjacent deserts and the foothill regions in Middle Asia. The maximum precipitation in 1978 and in 1981, for example, was recorded by the Ujaly, Tamdy and Pskem stations. The atmospheric circulation apparently exhibits synoptic processes governed by far-reaching comparable factors.

The distribution of monthly precipitation over the year was rather uniform up to 1960 (Fig. 4.11), with slight maxima in autumn in the north (Aralsk, Uzun-Kair) and with a rather clear maximum in spring and autumn, and a minimum in summer in the south (Muinak, Ujaly). The spring and autumn precipitation maxima seem to be more pronounced after 1960 (Fig. 4.12). The precipitation minimum seems to shift from late summer to midsummer (Ujaly) and from late spring to summer (Fig. 4.13). The Aral Sea region is intermediate between the northern steppe region with summer rains and the southern desert region with Mediterranean-influenced winter rains. This intermediate position enables rather strong changes if the atmospheric circulation changes somewhat. Again, the variability from year to year allows only rough conclusions to be drawn, since the available observation periods are often too short.

A long-lasting snow cover normally starts only in mid-December (Khan and Holko 2009). On average it lasts until the second week in March. In total, in the north the snow cover lasts about 60–90 days per year and in the south it lasts only 30–50 days per year (see Table 4.7). The snow depth even in January is often not more than 5–10 cm in the north and is below 5 cm in the south. The winters of

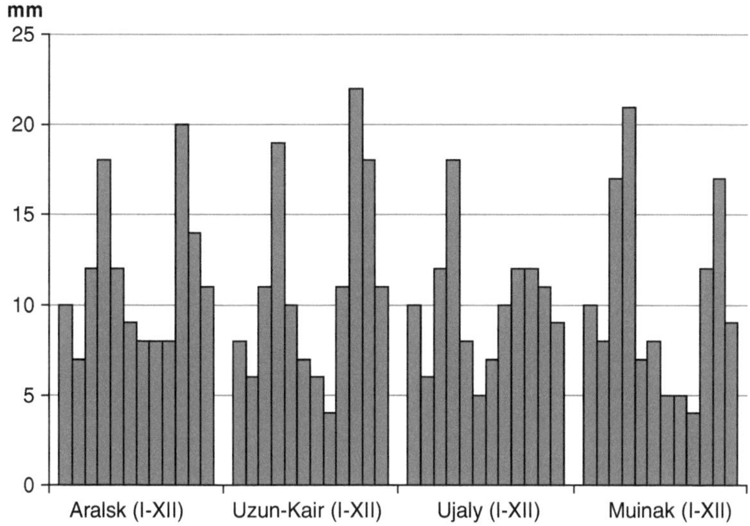

Fig. 4.12 The seasonal distribution of precipitation after 1960: 1961–1985 (Data from Spravochnik po klimatu Kazakhstana (2004))

Table 4.7 Snow cover within the former coastal region of the Aral Sea (average values) (before 1960)

Stations	Number of days per year with snowfall	Number of days per year with snow cover	Mean thickness of snow cover in January (mm)
Aralsk	24	90	10
Barsa-Kelmes	25	81	5
Ujaly	16	53	2
Muinak	18	36	2

From Zhitomirskaya (1964)

1950–1951 and 1953–1954 were rich in snow; at the northeastern coast the snow depth was 15–16 cm, on the island of Barsa-Kelmes it was 14 cm and it was 12 cm at the southern coast.

4.4.4 Humidity

The influence of the water surface of the Aral Sea on the air humidity is only important for regions adjacent to the coast. There is a drastic gradient from the coast to the desert hinterland. The highest values (Zhitomirskaya 1964) are known from stations along the coast in winter (80–89%), and there are lower values for the summer months (50–70%), except for the Aralsk station (below

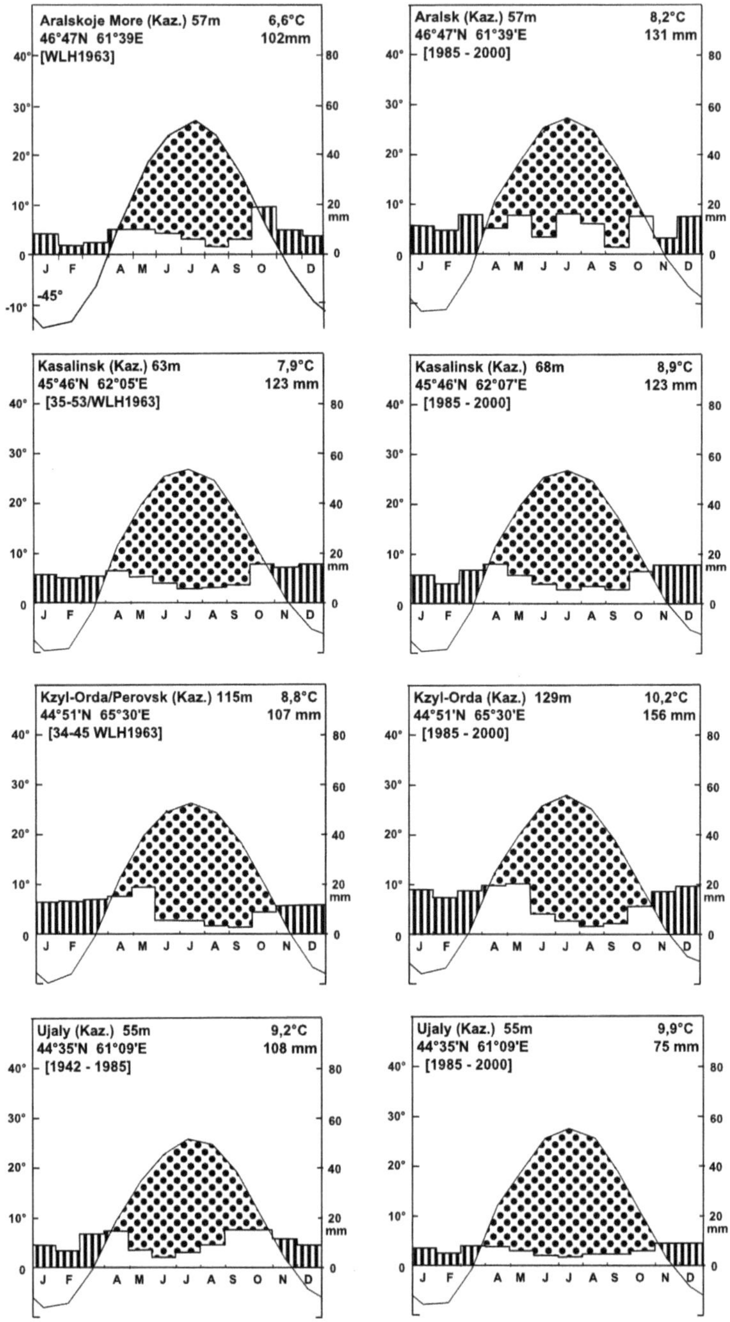

Fig. 4.13 Ecological climatic diagrams from four stations in the Aralkum area from two different periods, indicating slight changes within recent decades (WLH1963 = diagrams before 1960)

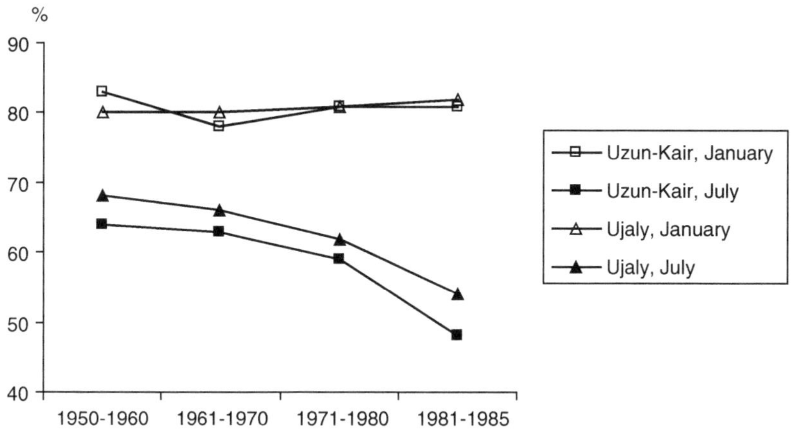

Fig. 4.14 The long-term trend in annual mean relative air humidity (%) (Data from Spravochnik po klimatu Kazakhstana (2004))

50%). Subbotina (1983) showed that the relative humidity of the coastal zone decreased during the recent decades as a consequence of the desiccation of the Aral Sea. This is especially valid for the summer months (Fig. 4.14). The humidity regime seems to be strongly influenced by the (remaining) water surface only in summer, whereas in winter the general atmospheric circulation governs the humidity conditions.

4.4.5 Evaporation

Evaporation data are scarce. According to some few data, the annual evaporation from an open water surface may range between 800 and 1,100 mm in the north and between 1,000 and 1,300 mm in the south. The temperature curve gives hints on the evaporation demand (potential evapotranspiration) and is thus used as a tool to indicate relatively arid months in the ecological climatic diagrams (see above, Fig. 4.1). It should be kept in mind that the total amount of water within the Aral Sea basin has not changed, only the active evaporation sites have switched from the sea surface to paddy and cotton fields, fish ponds, canals and water reservoirs.

4.4.6 Wind

The Aral Sea region is often and for long times in the year under the influence of the Siberian anticyclone, especially in winter. Thus, northerly winds prevail in January and February. During spring and summer, the wind regime can change: sometimes westerly and southwesterly winds can bring moister air masses. Of smaller influence seems to be a local anticyclone over Ustyurt Plateau. During autumn, northeasterly and easterly winds prevail, caused by some influence of the Siberian anticyclone. Formerly, for the coastal areas a typical local day–night breeze regime was characteristic. At the northern coast, during day there was a soft southwesterly breeze, but during the night it changed to northwesterly. At the southern coast, during the day the northwesterly breeze prevailed, changing at night to westerly. It is unknown if such a day–night breeze system has shifted to the new retreated coastline.

The wind velocities are rather equal all year round. The mean annual wind velocity for the northeasterly winds is 5–6 m s^{-1} and the proportion of these wind speeds is 20–25% (Zhitomirskaya 1964). Extremely strong cold winds can occur in winter. In spring less strong winds and more westerly directions occur. But sometimes strong storms from the north or east can be the reason for sandstorms, dust storms and salt-dust storms, which are an ecological disaster (see chapt. 5 and 7). They threaten the health of the people and are a danger because of additional salinization of the irrigation crops in the Syr Darya as well as in the Amu Darya oasis, e.g. in Karakalpakstan.

4.5 Climatic Zonation and Climatic Regions

4.5.1 General Remarks

There are contrasting opinions on the climatic situation and position of the Aralkum area. Myachkova (1983) defined this area within the desert zone as being a continental north Turanian climatic district. Alisov (1969) regarded the small area of the southeastern coast of the Aral Sea to the south Turanian climatic district as being part of a subtropical belt. According to Babushkin (1963), the northern part of the Aral Sea is located in the central Kazakhstan climatic district with a continental temperate climate and the southern part of the Aral Sea in the Turanian climatic district with a continental subtropical climate. Rachkovskaya et al. (2004) regarded most parts of the Amu Darya delta and the southern coast of the former Aral Sea as belonging to the southern Turanian province.

The Aralkum with the remnants of the Aral Sea is located in the zone of the temperate continental climate; the northern and southern parts of the Aralkum differ rather clearly. An open question is the delimitation of climatic subzones of the area. The current consensus is to delimit the area as an Aralkum district with some subdistricts, as indicated below. We will not discuss the whole Aral Sea basin area, but mainly the new desert, the Aralkum. In this respect we see the Muinak and Ujaly stations being typically southern Turanian, but most other parts as belonging to the northern Turanian deserts. From the available meteorological stations, the ecological climatic diagrams, partly for contrasting time periods, are given in

Figs. 4.1–4.8 and additional data are given in Tables 4.2–4.4 and 4.7. We distinguish two climatic subdistricts in the Aralkum area – the Aralsk subdistrict in the north and the Muinak subdistrict in the south. The limit between these two subdistricts is at the western coast from the *urochishe* (local name of the landscape) Akbulak, furthermore to the north of the former bight of Adzhibai and furthermore in the direction of the former island of Ujaly and the former bay of Kashkynsu at the eastern coast. The delimitation of both provinces is shown in Fig. 4.15.

4.5.2 Aralkum: Aralsk Subdistrict (Northern Turanian or Kazakh-Dzungarian Deserts)

This part of the Aralkum comprises the northern and western coasts of the Aral Sea (including the desiccated seafloor) and the eastern coast (to the former bay of Kashkynsu, Fig. 4.15). This is mainly the Kazakhstan part of the Aralkum. These are wide plains between 25 and 150 m asl at the northern and western coasts (the chinks belong to the Aralkum) and between 25 and 58 m asl at the eastern coast. This subdistrict is characterized by cold and long winters. The radiation balance is below 180–190 kJ cm^{-2} $year^{-1}$. The heat transfer for evaporation is about 40 kJ cm^{-2}. or slightly above 40 kJ cm^{-2}. The annual mean sunshine duration in the northern Aralkum is about 2,500–2,900 h per year.

The mean temperature in January is between −8.0 and −13.5°C. Three months of snow cover (from mid-December until mid-March) is normal. Snowstorms in winter can occur several times. The monthly mean temperature in November and in March is still about or below 0°C. The number of frost-free days per year is less than 200. The summers are hot and cloudless. The monthly mean temperature in July is about 25–28°C. The temperature sum for the days above 10°C is about 3,500–3,800 and the temperature sum for the days above 0°C is about 3,700–4,000.

The annual precipitation ranges between 90 and 150 mm. The monthly precipitation values are rather uniform all over the year with no conspicuous maxima but with slight maxima in October to November before 1960 and in October to November and April after 1960.

4.5.3 Aralkum: Muinak Subdistrict (Southern Turanian Deserts)

This part of the Aralkum comprises the southern coast of the Aral Sea (including the dry seafloor) and the southern half of the eastern coast (up to the former island of Ujaly). This is mainly the Uzbek part of the Aralkum. These are wide plains between 25 and 58 m asl.

This subdistrict is characterized by very strong radiation, more hours of sunshine and very hot summers. The radiation balance is above 185 kJ cm^{-2} $year^{-1}$. The heat

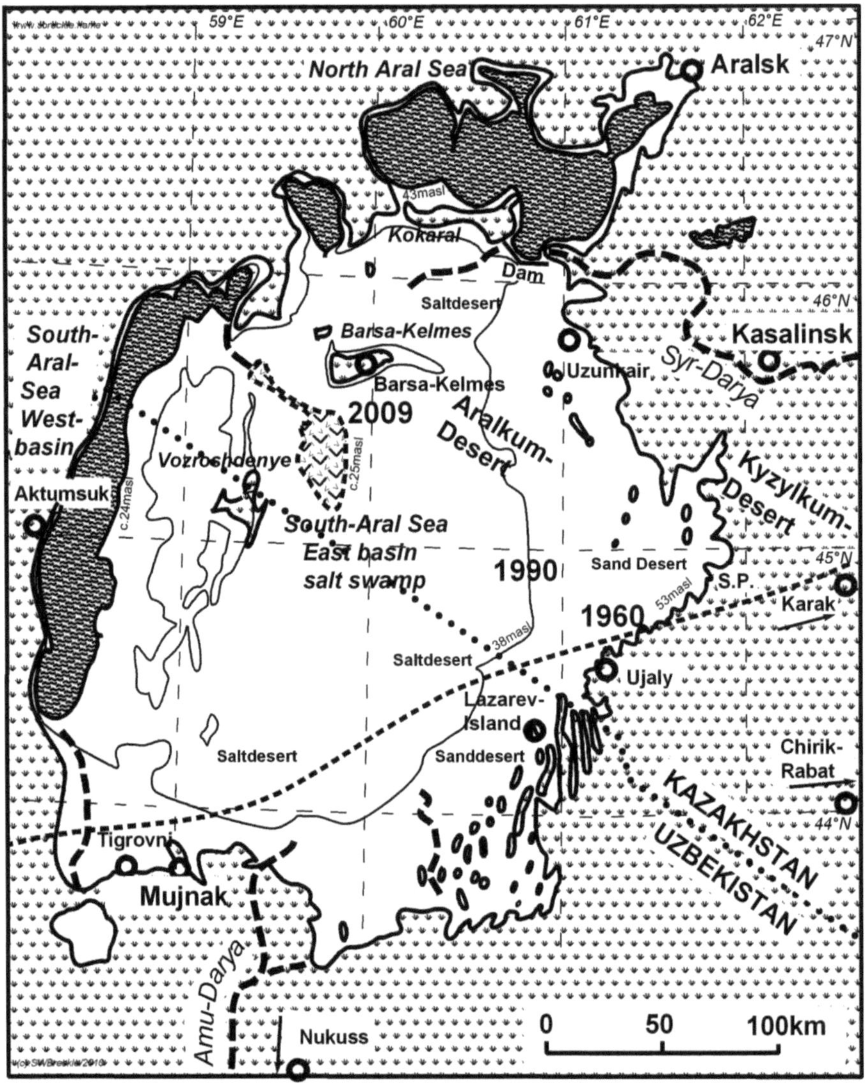

Fig. 4.15 The Aralkum, the new land between the 1960 coastline and the recent coastline. The location of meteorological stations is indicated. *S.P.* former Sandal Peninsula, *thick dashed line* limit of subdistricts (Aralsk subdistrict – northern Turanian desert, and Muinak subdistrict – southern Turanian desert)

transfer for evaporation is less than 40 kJ cm^{-2}. The annual mean sunshine duration in the southern Aralkum area is more than 2,900 h per year.

The mean temperature in January is not below $-8.0°C$. The winters are warmer than in the Aralsk subdistrict, and are about 20–25 days shorter. Snow cover lasts about 1.5–2.5 months (from end of December until end of February). The number of frost-free days per year is about or distinctly more than 200. The summers are hot and cloudless. The monthly mean temperature in July is about 25–28°C (like in the Aralsk subdistrict). The temperature sum for the days above 10°C is about 3,800–4,000 and thus is 300–500 higher than in the Aralsk subdistrict.

The annual precipitation amounts to about 90–130 mm. The seasonal distribution of precipitation is rather equal the whole year; it shows weak maxima in March and April, especially before 1960. The spring and autumn precipitation maxima are more pronounced after 1960.

4.6 Conclusions: Aridization and Climate Change?

How has the desiccation of the Aral Sea and the formation of the new desert Aralkum since 1960 influenced the climatic conditions of the whole area? Do we have a specific aridization of the climatic parameters?

Regarding the general circulation and the synoptic processes in the region, we have to distinguish two time periods: until 1960 and after 1960. Current climatic data only reveal a slight shift to even stronger continentality. The overall water dynamics of the atmosphere do not seem to be altered in general, but intensification of water cycling by global change or global climate fluctuation overlying the local and regional desiccation processes may be observed. Subbotina (1983) pointed to the fact that the desiccation of the Aral Sea occurs in parallel to an almost global change of the atmospheric circulation. Since 1960 the meridional circulation has become more and more dominant. Before 1960 there was a combination of various forms of circulation patterns with domination of zonal circulation.

Kodrau (1964) and Kuvshinova and Suzyumova (1978) showed that the influence of the Aral Sea on macroclimatic events is rather small, and only along the narrow coastal strip were there some signs of influence. The disappearance of the large water surface may cause, however, some specific changes observable in the direction of a more accentuated continentality. However, it is a fact that the loss of water-surface area is totally compensated or even overcompensated for by the development of other widespread evaporation surface areas – fish ponds, reservoirs, new canals, irrigation for paddy fields and many other crops, etc. – at other places. Thus, it cannot be expected that there is an overall influence on the water balance of the regional atmospheric processes. The Kazakh Meteorological Institute (Chichasov 1990) wanted to reveal the influence of the desiccation process on the humidity and temperature conditions of the area. During the summer months, the relative humidity is now about 20–30% lower than before 1960, and the air temperature about 1–2°C higher. According to our data, the drop in humidity is less, about 10–15%, and at the former island stations is about 10%. In general, the Aral Sea had an important, but only local influence on the climatic conditions. The global climatic trends are overlying these processes and this makes it difficult to separate them.

The meteorological data show (Muminova and Inogamova 1995) that in the whole Middle Asia region the annual precipitation has increased within the last

50 years. This is also due to the arid areas of the Turanian plain, to the high mountains at the margins of the basin (where the accumulation of glaciers plays an important role) and other areas in near the Aral Sea basin (Oberhänsli and Zavialov 2009).

Directly at the remnants of the Aral Sea, and at the Aralkum, at the stations on the former islands (Barsa-Kelmes, Lazarev, Vozrozhdeniya, Tigrovni) this trend cannot be traced significantly. However, those stations mostly have been closed since the middle or the end of the 1980s. Thus, only the last 15 or 20 years of observations would have been especially worthwhile. Only the Aralsk station clearly indicates an increase in precipitation within the 1950s, but within the last 40 years the mean precipitation seems to be stable (Fig. 4.13). The data from Aralsk from 1937 until 2005, which were checked by Kuzmina (2007), also indicate a rather distinct rise in annual precipitation (with a correlation line of $y = 1.0003x - 1{,}858$; $r = 0.40$).

According to Giese and Moßig (2004), no significant decrease or increase in the annual precipitation in Middle Asia is detectable (their only exception might be the Kazalinsk station). But Aizen et al. (2001) indicated for the past 100 years a positive trend in precipitation for western Siberia, the northern regions of Tien Shan and for the Japanese Islands in East Asia.

The climatic conditions of the area are governed by global trends. Within the last few decades (Klaus 1991) in many parts of the world (at latitudes between 35°N and 70°N) the annual temperature has been increasing. This same trend is visible in Middle Asia and the Aral Sea area (see above). Since around 1970, all statistical techniques have indicated a pronounced warming trend (Giese and Moßig 2004; Mills 2006). The global trend of rising temperatures again is to be seen in the area of the Aralkum. The annual means temperatures as well as most of the monthly mean temperatures recorded by almost all the stations around the Aral Sea have risen since 1960 by about 1–2 K. The warming trend is also indicated by Kuzmina (2007), who gave smoothed functions for the seasons for the Aralsk meteorological station, which indicate a rise in temperature of about 2–3 K between 1937 and 2002. The same is true for the hinterland stations (Kazalinsk, Kyzyl-Orda, Chilik-Rabat, Kuruk).

The warming of the climate in Central Asia has been more conspicuous than globally according to Giese and Moßig (2004). Data from the Intergovernmental Panel on Climate Change indicate that the overall global warming for the period 1950–2000 was 0.47 K for Central Asia; however, the increase of annual mean temperatures within the same period was +1.15 K up to +2.1 K. The warming of the interior of the Asian continent is thus reported to be 3–4 times more than globally. This phenomenon can be explained according to Giese and Moßig (2004) by the anthropogenic influence – mainly diversion of water resources on a very large scale in the whole of Middle Asia, and specifically by the desiccation of the Aral Sea and the formation of huge dust clouds in many parts of the Aralkum. This hypothesis seems reasonable, but to clarify the situation many more data from large-scale studies in the area and special research with remote sensing techniques are needed, since the sparse climatologic data available from meteorological stations

(Fig. 4.13b) do not indicate significant changes, or are even controversial (Lioubimtseva and Henebry (2009)).

The desiccation of the Aral Sea and the formation of the Aralkum are, however, without doubt important milestones in the formation and evolution of new Turanian geo- and bio-ecosystems and for the Middle Asian region as a whole.

Acknowledgements We thank the German Federal Ministry of Education, Science, Research and Technology (BMBF) and the project coordinator (Research Centre Jülich GmbH, PTJ Berlin) for financial support (project nos. 0339714 and 0330389), as well as the Academy of Science, Botanical Institute (Almaty) and the Geographical Institute (Almaty) for their overall cooperation. The great help of villagers and citizens in many parts of rural areas and towns is greatly acknowledged.

References

Aizen EM, Aizen VB, Melack JM, Nakamura T, Ohta T (2001) Precipitation and atmospheric circulation patterns at mid-latitudes of Asia. Int J Climatol 25(5):535–556

Alisov B (1969) Klimat SSSR. Moscow

Babushkin LN (1963) Agroklimaticheskoye raionirovanie Srednei Asii (Agricultural climatic division of Central Asia). Nauchnye trudy TashGU, Vyp. 36, Geogr. Nauki, Kn. 28:186–272

Budyko MI (1956) Heat balance of the Earth surface (*Teplpvoy balans zemnoy poverhnosti*). Hydrometeoizdat, Leningrad (Russian)

Budyko MI (1974) Change of climate (*Izmenenie klimata*). Hydrometeoizdat, Leningrad (Russian)

Bugaev VA (ed) (1957) Sinopticheskie protsessy v Srednei Asii (Synoptical processes in Central Asia). Tashkent

Breckle S-W (2002) Walter's vegetation of the earth – the ecological systems of the Geo-Biosphere. Springer, Berlin

Breckle S-W (2006) Biodiversity in deserts – is it area- or resource-related? J Arid Land Stud 16:61–74

Chichasov GN (ed) (1990) Gidrometeorologicheskie problemy Priaralya. (Hydrometeorological problems of the Pri-Aral-region). Gidrometeoizdat

Giese E, Moßig I (2004) Klimawandel in Zentralasien. ZEU Discussion Papers No. 17

Ginzburg AI, Kostianoy AG, Nickolay A, Sheremet NA (2003) Thermal regime of the Aral Sea in the modern period (1982–2000) as revealed by satellite data. J Mar Syst 43:19–30

Grigoriev AA, Budyko MI (1959) Classification of climates USSR (*Klassiphikaziya klimatov SSSR*). Informations of academy of sciences USSR. A geographical series. (*Izvestiya akademii nauk SSSR. Seriya geographicheskaya.*) Moscow. No 3. pp 3–18 (Russian)

Khan VM, Holko L (2009) Snow cover characteristics in the Aral Sea Basin from different data sources and their relation with river runoff. J Mar Syst 76:254–262

Klaus D (1991) Treibhaus "Erde". Die Expansion der Dürregebiete durch natürliche und anthropogene Klimaänderungen. In: Wüsten der Erde. Fokus, pp 145–168

Kodrau OD (1964) Vlagooborot ravninnoi i predgornoi chastei Srednei Asii. (Humidity circulation in plains of Middle Asia). Trudy GGO, 163:3–32

Kuzmina JV (2007) Long-term dynamics according to meteorological characteristics in the Prearalie Region. Izvest vizs ujebnich zawedenii, pp 73–78

Kuvshinova VK, Suzyumova GN (1978) Aralskoe more i climat Priaralya (The Aral Sea and the climate of the Aral Sea Region). Problemy kompleksnoi gidrometeorologii, Moscow, pp 67–71

Lioubimtseva E, Henebry GM (2009) Climate and environmental change in arid Central Asia: impacts, vulnerability, and adaptations. J Arid Environ 73:963–977

Mills TC (2006) Modelling current trends in N hemisphere temperatures. Int J Climatol 26(7):867–884
Muminova FA, Inogamova S (eds) (1995) Izmenchivost klimata Srednei Azii. (Climate changes in Middle Asia). SANIGMII, Tashkent
Myachkova NA (1983) Klimat SSSR. Izd. MGU, Moscow
Nezlin NP, Kostianoy AG, Lebedev SA (2004) Interannual variations of the discharge of Amu Darya and Syr Darya estimated from global atmospheric precipitation. J Mar Syst 47:67–75
Nezlin NP, Kostianoy AG, Li B-L (2005) Inter-annual variability and interaction of remote-sensed vegetation index and atmospheric precipitation in the Aral Sea region. J Arid Environ 62:677–700
Oberhänsli H, Zavialov P (2009) Aral Sea Basin hydrological, chemical and biological dynamics today, compared with trends of the past 50 years. J Mar Syst 76:251–253
Rachkovskaya EI, Safronova IN, Volkova EA (2004) Botaniko-geographicheskoe raionirovanie (Botanical-geographic regionalization). In: Rachkovskaya EI, Volkova EA, Khramtsov VN (eds) Botanical geography of Kazakhstan and Middle Asia. St Petersburg, pp 185–212
Spravochnik po klimatu SSSR, Uzbekskaya SSR (1967) Leningrad Hydrometeoizdat, vol. 19
Spravochnik po klimatu SSSR, Kazakskaya SSR (1968) Leningrad Hydrometeoizdat, vol. 18
Spravochnik po klimatu Kazakhstana (2004) Almaty RGP Kazgidromet
Subbotina OI (1983) Izmeneniya kolichestva osadkov v raznye periody usykhaniya Aralskogo morya. (Changes of precipitation in different periods of drying out of the Aral Sea). Trudy SANII Goskomgidrometa 114:16–22
Walter H, Lieth H (1967) Klimadiagramm-Weltatlas. Fischer, Stuttgart
Williams MW, Konovalov VG (2001) GHCN: Regional Data Base on Climate of Central Asia v.1.0. http://snobear.colorado.edu/Markw/Geodata/BD_INFOR_E.doc
Zhitomirskaya OM (1964) Klimaticheskoe opisaniye raiona Aralskogo morya. (Climatical description of the area of the Aral Sea). Gidrometeozdat, Leningrad

Chapter 5
Dust Storms and Sandstorms and Aerosol Long-Distance Transport

O.E. Semenov

5.1 Introduction

The general atmospheric circulation and the macrorelief is responsible for the endurance and intensity of dust storms. The average number of days with strong winds (more than 15 m s^{-1}) in most of the Aral Sea region is about 15 days, in some parts it is up to 25–29 days, and the meteorological station at Barsa-Kelmes even reported up to 44 days (Gidrometeorologicheskie problemy Priaralya 1990).

A detailed description of the dust storm situation in the Aralkum area is given in the following.

5.2 General Remarks

Dust storms are characterised by transport of huge amounts of dust particles by strong winds. Sandstorms are often combined with dust storms in areas where sand prevails (Yang et al. 2002). According to particle size, we distinguish between dust storms and sandstorms. Sand particles are between 0.1 and 1 mm in diameter and are transported close to the surface (the distance over which they are transported per day may rarely reach 1 km), normally forming sand dunes which creep 10 or 50 m. Dust particles are much smaller and are transported by wind, often to rather high parts of the atmosphere, thus achieving transport distances of hundreds of kilometres. The chemistry of dust particles is very variable. In salt deserts often a high percentage of various salts (NaCl, sulphates, carbonate) can be observed: salt dust (Orlovsky and Orlovsky 2002). Those salt-dust storms are particularly endangering human health as well as the productivity of agricultural areas by enhancing salinization. This is a prominent problem in the Aralkum region (Breckle et al. 2001).

Semenov and co-authors (Galayeva et al. 1996) have developed a physical model to evaluate the amount of aerosols transported from the desiccated seafloor

S.-W. Breckle et al. (eds.), *Aralkum - a Man-Made Desert: The Desiccated Floor of the Aral Sea (Central Asia)*, Ecological Studies 218,
DOI 10.1007/978-3-642-21117-1_5,

of the Aral Sea. The meteorological data from the available stations and the particle size distribution of the transported grains then allow the transported sand masses at the surface, the probability and direction of sand movements as well as those for the dust particles to be calculated.

5.3 Frequency of Storms

The arid climatic conditions and the open surface with fine grain sizes are favourable for the development of regular dust storms in the Aralkum area. Strong winds are often recorded: at Aktumsuk for 54 days per year and at Barsa-Kelmes for 44 days per year (Molosnova et al. 1987). The maximum wind speeds can reach 20–25 m s^{-1}. Grain sizes on average are 90–160 μm on the desiccated seafloor, smaller in young territories (desiccation 1980, 1990, 2000) and larger in older territories (1960, 1970). Adjacent desert areas (former coastal dunes, former islands, sandy deserts) have an average particle size of 170–270 μm. These are the ideal aerosol sizes for transport over a rather long distance.

Table 5.1 gives an overview of the frequency of dust storms and sand drift events between 1966 and 1992 for the 11 Kazakhstan observation stations along the Aral Sea area. The northern parts exhibit dust storms more frequently (36–84 days per year) than the eastern parts (9–23 days per year). Some islands have a frequency of only 2–5 days per year, and the southern part has a frequency of about 6–20 days per year. For Uzbekistan the frequency of dust storms seems to have been higher between 1960 and 1979 (see Table 5.2) than nowadays (Muminova and Inogamova 1995; Razakov and Kosnazarov 1996). Figure 5.1 indicates the difference in the frequency of dust storms in the north and in the south of the Aralkum and the relevant time dynamics. A similar temporal trend seems to be observable also on the Kazakhstan side (Table 5.3). However, there are three stations with increasing frequency: Ustyurt Plateau, Aralsk and Kyzyl-Orda. This might be an effect of increasing human activities and not a matter of the regional climatic conditions.

The duration of a dust storm is normally rather short. The probability of a dust storm lasting longer than 1 day is less than 5%. Table 5.4 gives some estimates of duration probabilities of dust storms for four stations and data on the longest observed dust storm.

5.4 Mass Transport by Storms

The Semenov model allows one to calculate the amount of dust which is transported through a distinct migration plain of 1 km in length by sand drift (between 10 and 30 m) and by dust transport (between 150 and 200 m).

The material transported between 0 and 10 m is often rather course sand moving by saltation, the material transported between 30 and 150 m is course dust and salty

Table 5.1 Number of days per year with prominent deflation events (sandstorms or sand drift) in the Aral Sea region (1966–1992)

	Climate station									
Year	Ujaly	Monsyr	Kazalinsk	Kyzyl-Orda	Saksaulski	Chirik-Rabat	Karak	Dzhusaly	Lazarev (I)	Barsa-Kelmes (I)	Aralsk
1966	109	5	12	55	62	82	15	30	30	0	5
1967	80	1	30	46	64	75	19	32	19	0	1
1968	86	3	21	36	70	66	26	36	31	0	0
1969	76	4	40	25	42	85	13	24	19	0	o
1970	77	35	51	30	61	80	19	16	37	0	7
1971	58	51	54	23	40	58	20	15	35	0	0
1972	99	83	50	10	41	78	29	15	13	0	0
1973	49	48	52	10	38	31	6	8	15	0	0
1974	59	27	52	22	60	52	12	6	8	0	0
1975	72	85	53	17	95	72	21	9	13	5	0
1976	79	39	27	14	47	58	9	4	16	0	3
1977	79	35	63	14	64	87	7	13	25	1	1
1978	105	5	49	3	69	35	2	2	13	1	3
1979	89	21	50	11	59	45	1	1	22	12	2
1980	59	17	39	5	55	48	2	4	18	3	0
1981	51	6	22	1	58	14	2	4	9	0	0
1982	59	1	18	0	51	42	1	2	4	3	0
1983	101	13	31	0	48	37	0	6	3	5	1
1984	115	48	39	4	72	74	0	0	6	9	1
1985	137	6	21	2	71	91	0	2	10	1	0
1986	85	12	17	1	69	43	8	1	18	10	0
1987	77	9	21	0	60	25	5	0	13	3	1
1988	93	10	43	0	111	26	0	2	55	17	0
1989	85	10	56	2	122	16	3	1	13	13	6
1990	96	18	49	0	93	7	9	4	16	14	4
1991	115	14	16	0	99	13	4	0	11	17	2
1992	69	6	0	0	110	0	0	1	7	9	5
Mean	84	23	36	12	68	50	9	9	16	5	2

I island

Table 5.2 Mean number of days per year with dust storm events in the Uzbekistan Aral Sea area (Muminova and Inogamova 1995)

	Period		
Climate station	1960–1979	1980–1990	Geographical site of the station
Karakalpakya	6	20	Ustyurt Plateau
Muinak	11	4	Amu Darya Delta
Nukus	20	9	Amu Darya Delta
Chimbai	14	11	Upper Amu Darya Delta
Tamdy	18	8	Kyzylkum
Ajakagitma	16	6	Kyzylkum

particles, and above 200 m only rather small dust particles prevail, which are often transported over very long distances. For the model calculation only the two distinct migration plains mentioned were used.

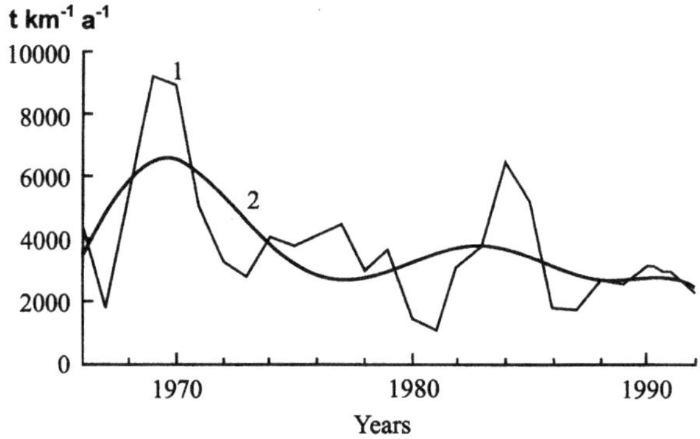

Fig. 5.1 Dynamics of sand transport at the northern coast of the Aral Sea (Aralsk station). The transported sand masses are given as tonnes per kilometre per year through a "window", a distinct plain of 1 km in length (between 10 and 30 m). *1* Annual values, *2* smoothed line

Table 5.3 Mean number of days per year with dust storm events in the Kazakhstan Aral Sea area (Muminova and Inogamova 1995)

Climate station	Period 1960–1979	1980–1990	Geographical site of the station
Aralsk	80	88	Northern coast of the Aral Sea
Ujaly	32	13	Eastern coast of the Aral Sea
Kazalinsk	13	1	Syr Darya delta
Saksaulski	65	34	Region at the northern coast
Monsyr	43	29	Region at the northern coast
Chirik-Rabat	14	3	Kyzylkum

Table 5.4 Probability of dust storm duration (%) and longest duration (h) of a dust storm between 1936 and 1972 (Muminova and Inogamova 1995)

Climate station	Integrative probability (%) of dust storm duration with a length of 1 h	3 h	6 h	9 h	12 h	Longest duration of a dust storm (h)
Aralsk	79.3	51.5	24.7	11.8	5.6	79.5
Saksaulski	67.6	38.7	16.1	5.8	2.4	37.3
Dzhusaly	82.8	55.8	29.8	11.7	4.3	83.5
Kyzyl-Orda	87.7	65.6	35.2	11.7	2.8	58.9

Table 5.5 gives calculated figures from the Semenov model for means of sand and dust masses transported through a cross-section migration plain of 1-km length in the lower atmosphere (Semenov et al. 1990). This information is very important for quantification of dust storms, which is needed for decisions on the means of protection and combating of deflation. The highest amounts of mass transport can be found at the Aral Sea region itself. The model calculation gives values of about 3,850 t km^{-1} $year^{-1}$ which are transported at Aralsk and 1,560 t km^{-1} $year^{-1}$ at Barsa-Kelmes. The lowest amounts of mass transport are registered in the Syr Darya delta (Kazalinsk station, 330 t km^{-1} $year^{-1}$). At the Aralsk station a decrease of sand transport within recent decades is registered.

Table 5.5 Statistical characteristics of transported sand and dust masses for some climatic stations in the Aral Sea region (observation period 1966–1992)

	Dust storm			Sand drift			Dust storm plus sand drift		
Climate station	M	s	C_v	M	s	C_v	M	s	C_v
Aralsk	3,189	1,503	0.47	655	904	1.38	3,845	1,979	0.51
Ujaly	1,840	2,796	1.52	1978	2,870	1.46	3,818	1,541	1.19
Monsyr	288	460	1.60	624	410	0.66	876	740	0.85
Kazalinsk	87	172	1.97	242	389	1.61	329	500	1.52
Kyzyl-Orda	475	405	0.85	701	500	0.71	1176	652	0.56
Saksaulski	3,045	390	1.28	228	158	0.69	533	513	0.96
Chirik-Rabat	492	605	1.23	2	6	3.20	494	605	1.22
Karak	669	1,027	1.53	2	8	5.10	671	1,027	1.53
Dzhusaly	1,634	4,402	2.7	30	135	4.5	1,664	4,400	2.6
Lazarev (I)	1,616	5,443	3.37	194	708	3.6	1,810	5,464	3.01
Barsa-Kelmes (I)	1,522	4,501	2.96	38	130	3.45	1,560	4,504	2.89

M (t km^{-1} year^{-1}) amount of sand and dust; s (t km^{-1} year^{-1}) mean square deviation; C_v variation coefficient (Galayeva et al. 1996)

The mass transport of sand and dust can be very variable in relation to the climatic fluctuations (Fig. 5.1) and varies considerably from time to time. For the calculation of transported sand and dust masses, vectorial dynamics was used. The vectorial roses display which masses of dust and sand are transported in the lower atmosphere by wind in which direction. The length of the vector arrow represents the amount of sand and dust transported (Fig. 5.2). This is known from changes in the shape of the newly formed barchans and other aeolian relief forms which can change their direction of movement several times according to the prevailing strong wind vectors. However, the resultant (vector addition) gives an idea of the overall long-term sand and dust movement and transport integrated over time (Fig. 5.3). In the northern parts the transportation direction is to the east or southeast; in the most central and southern parts it is to the southwest (Figs. 5.2 and 5.3). The highest transport intensity is exhibited by the island stations at Lazarev and Barsa-Kelmes and by the continental stations at Ujaly, Dzhusaly and Kyzyl-Orda, which is in contrast to the figures from the model calculation.

From the data and the model, one can extrapolate the findings to the whole amount of material transported from the desiccated seafloor. For the period 1966–1979, the average yearly amounts of dust and sand transported from the area between the former coastline and the 10-m-isobate line were 7.3×10^6 t and 70,000 t, respectively. However, the probability was only 5%. From satellite images 15×10^6–75×10^6 t was estimated (Grigorjev and Lipatov 1982). This is a rather great difference in amounts, which seems to be rather arbitrary. The upper number would almost correspond with figures of dust output from the Sahara (Semenov 1995). The aerosol output for the area from the former coastline is also very variable, but very high. Sources for dust in the Aralkum are mainly between Barsa-Kelmes and Kokaral (Figs. 5.4 and 5.5) and between Vozrozhdeniya and

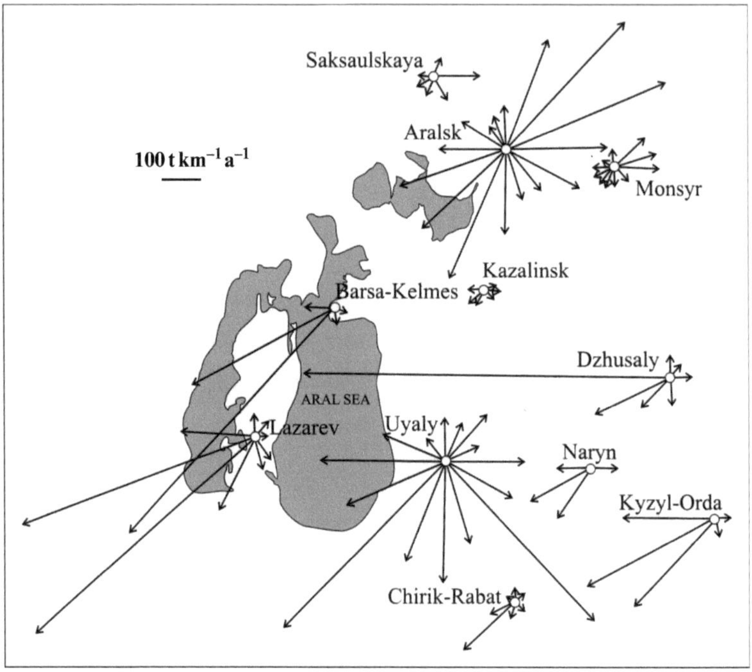

Fig. 5.2 Mean vectorial roses of sand and dust transportation in the Aral Sea region (transportation plain, see Fig. 5.1)

the former eastern coastline. The desiccation of the eastern basin will increase the salt dust source area tremendously.

The deposition at the northeastern coast is similar to that at the Amu Darya delta (70–95 t km^{-2} year^{-1}; Semenov 1998). At the eastern coast it is considerably higher (up to 145 t km^{-2} year^{-1}; Razakov and Kosnazarov 1987, 1996). Most of the transported aeolian material is deposited along the coastlines. The salt-dust clouds can be 200–400 km in length, but the finer particles are transported over more than 1,000 km (Figs 5.6 and 5.7, and some other striking examples are at

http://earthobservatory.nasa.gov/NaturalHazards/natural_hazards_v2.php3?img_id=14249;

http://earthobservatory.nasa.gov/NaturalHazards/natural_hazards_v2.php3?img_id=14253; http://earthobservatory.nasa.gov/Newsroom/NewImages/images.php3?img_id=6766;

http://earthobservatory.nasa.gov/Newsroom/NewImages/images.php3?img_id=17306).

In general, the salt-dust output from the dry seafloor cannot be explained by a simple meteorological data set and even less can it be predicted in space and time. Additional studies are needed. Good recent instrumental data of systematically taken samples of the aerosol particles are lacking. Normally only the data on the wind direction and the wind speeds during the standard times of meteorological

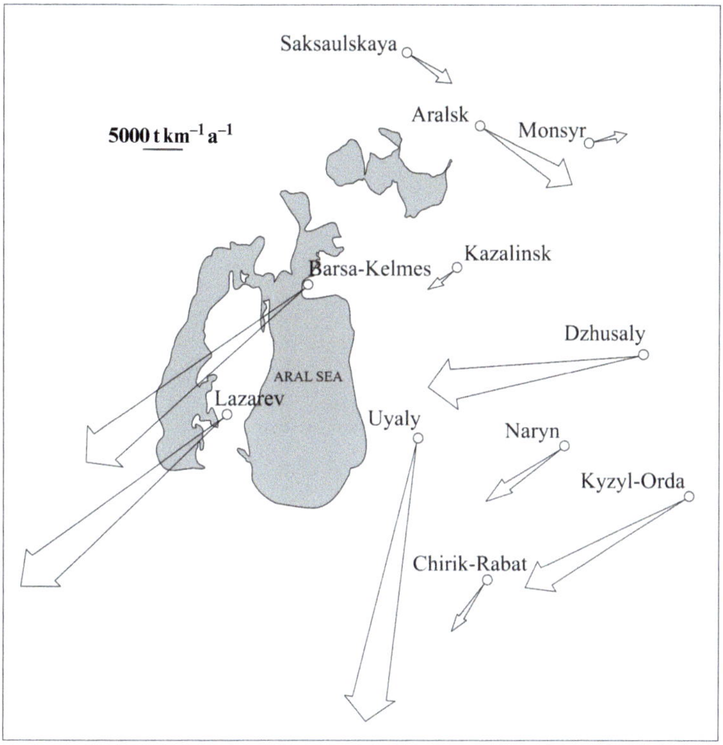

Fig. 5.3 Mean resultant vectors of sand and dust transportation in the Aral Sea region

Fig. 5.4 Aralkum plain with very scarce vegetation, a few bushes of *Halocnemum strobilaceum* on crusty solonchak, the source of salt-dust storms (photo: W. Wucherer, September 2003)

measurements are available and these data are only obtained three or four times per day. Often, only the older data from Razakov, Kosnazarov (Uzbekistan), Semenov (Kazakhstan) or Micklin (USA) are used. In Kazakhstan the monitoring of aerosol

Fig. 5.5 Typical deflation pattern caused by wind from a puffy solonchak surface with a powdery salt structure. Desiccated seafloor of the Aral Sea between Barsa-Kelmes and Kaskakulan (photo: S.-W. Breckle, 25 June 2004)

Fig. 5.6 Dust storm, shown by a satellite image from 18 April 2003 (NASA), indicating a huge dust plume stretching from northeast to southwest and indicating clearly the deflation flats of the desiccated seafloor as the main source (Image courtesy of the Image Science & Analysis Laboratory, NASA Johnson Space Center; http://earthobservatory.nasa.gov/NaturalHazards/view.php?id=11348)

Fig. 5.7 Dust storm, shown by a satellite image from the International Space Station from 16 April 2003 (NASA), indicating a long salt-dust plume stretching from north to south. North is on the left side (Image courtesy of the Image Science & Analysis Laboratory, NASA Johnson Space Center, ISS006E46653)

outputs from the desiccated seafloor was stopped in 1992. New data are only from few small research projects (see Chap. 7).

Since the environmental effects of the salt-dust storms are one of the many health impacts in the region, it is necessary to indicate the medical factors relating to the dust particles which affect the health of the people in the region, even more so, since the area of salt dust as a source of dust storms has increased tremendously in the eastern basin. It is clear that the arid climate and the structure of the desiccated seafloor with the puffy salt crusts is very favourable for the development of far-reaching salt-dust storms. A detailed study on the effectiveness of dry weather, low humidity, soil dynamics, hygroscopicity of dust, chemical compounds of crust and dust particles, on the particle size distribution and physical factors enhancing dust storms as well as the kinematics of mass transport of the lower atmosphere is strongly needed; a mesoscale model could be a good step for better predictions (Darmenova and Sokolik 2007). The cumulative, long-term effects of salt-dust inputs may cause negative effects not only on the health of people but also on ecosystems, on photosynthesis and on the productivity and growth of plants and animals.

During the further desiccation of the eastern basin in the near future, the growing huge salt flat will be an even greater source of salt-dust storms (Figs. 2.6 and 2.7). The agricultural areas to the east and south will be affected to a greater extent than now. It is therefore urgent to increase means to minimise the deflation and salt-dust output from the desiccated seafloor. One of the most important strategies is certainly to accelerate the phytomelioration projects at several starting points in order to form nuclei for spontaneous growth of vegetation cover for combating the salt desertification (Chaps. 15–17).

5.5 Conclusions

It has been observed that there have been changes in dust deposition within the last 30 years, but a detailed prognosis for the future is not possible. However, it became quite clear that an increase of the size of the desiccated seafloor with deposition of particles of even smaller size (less than 10 μm) and higher salt content, which will happen in the near future, will increase the distance over which particles are transported by wind.

We do not know exactly the chemical composition of soil nor of aeolian dust, nor the spectral behaviour of the dust clouds. We need more data on the connective air movements, on the dust devils (the whirlwinds) and on grain selection along movement trajectories; however, monitoring of such data and even basic data of aerosol output from the seafloor has ceased instead of enlarging an observation system for future countermeasures.

Acknowledgement The results are part of research which was financed by the Kazakh Academy of Science. The financial support is greatly acknowledged.

References

Breckle S-W, Wucherer W, Agachanjanz O, Geldyev B (2001) The Aral Sea crisis region. In: Breckle S-W, Veste M, Wucherer W (eds) Sustainable land-use in deserts. Springer, Heidelberg, pp 27–37

Darmenova K, Sokolik IN (2007) Assessing uncertainties in dust emission in the Aral Sea region caused by meteorological fields predicted with a mesoscale model. Glob Planet Change 56:297–310

Galayeva OS, Semenov OE, Shapov AP (1996) Ob osobennostyakh vetrovogo perenosa peska v Aralskom regione (Peculiarities of wind sand transport in the Aral Sea region). Gidrometeorologiya i ecologiya 4: 73–93 (in Russian)

Gidrometeorologicheskie problemy Priaralya (1990): (Hydrometeorological problems of the Pri-Aral-region). Gidrometeoizdat

Grigorjev AA, Lipatov VB (1982) Dinamika i ochag pylevykh bur v Priaralje (Dynamic and source of dust storms in the surroundings of the Aral Sea). Izv. ANSSSR, ser. Geogr. 5: 93–98 (in Russian)

Molosnova TI, Subbotina OI, Chanysheva SG (1987) Klimaticheskie posledstviya v zone Aralskogo morya (Climatic consequences in the surroundings of the Aral Sea). Gidrometeoizdat, Moscow (in Russian)

Muminova FA, Inogamova SI (eds) (1995) Izmenchivost klimata Srednei Azii (Climate changes in Middle Asia). Tashkent SANIGMI, Glavgidromet RUz (in Russian)

Orlovsky N, Orlovsky L (2002) White sandstorms in Central Asia. In: Yang Y, Squires V, Qi L (eds) Global Alarm: Dust and Sandstorms from the world's drylands. UN Asia Regional Coordinating Unit, Bangkok, pp 169–201

Razakov RM, Kosnazarov KA (1987) Eolovyi vynos i vypadeniye solepylevykh chastits v Priaralye (Wind transport and deposition of salt dust particles in the surroundings of the Aral Sea). In: Problemy Aralskomo morya i prirodookhrannye meropriyatiya. Tashkent SANIIRI, pp 24–31 (in Russian)

Razakov RM, Kosnazarov KA (1996) Dust and salt transfer from the exposed bed of the Aral Sea and measures to decrease its environmental impact. In: The Aral Sea Basin. NATO Asi Series. 2. Environment, vol 12: 95–102

Semenov OE (1995): Ob otsenke masshtabov vynosa massy Aralskogo aerozolya (Assessment of quantity of the aerosol transport from the Aral Sea area). Gidrometeorologiya i ecologiya, 1: 117–130 (in Russian)

Semenov OE (1998) Vypadeniya peschano-solevogo aerozolya na poverchnost Priaralya (Deposition of sand salt aerosol in the surroundings of the Aral Sea). Problemy osvoeniya pustyn 3: 37–44 (in Russian)

Semenov OE, Shapov AP, Kaipov VI (1990) Peschano-solevye buri v Priaralye (Sand salt storms in the surroundings of the Aral Sea). In: Chichasov GN (ed) Gidrometeorologicheskie problemy Priaralya (Hydrometeorological problems of the Aral Sea surroundings). Gidrometeoizdat, Leningrad, pp 132–232 (in Russian)

Yang Y, Squires V, Qi L (eds) 2002 Global Alarm: Dust and Sandstorms from the world's drylands. UN Asia Regional Coordinating Unit, Bangkok

Chapter 6
Landscape Dynamics in the Southern Aralkum: Using MODIS Time Series for Land Cover Change Analysis

F. Löw, P. Navratil, and O. Bubenzer

6.1 Introduction

With the regression of the once fourth largest sea in the world, a huge new saline desert emerged on the former seabed, which is called the 'Aralkum' (Breckle et al. 2001). It shows a wide variety of different landscape and soil types. Salt-affected areas, in particular, are now appearing after regression (e.g. a variety of solonchaks and takyrs), which are considered a major source of salt storms and dust storms in the region (Razakov and Kosnazarov 1996; Singer et al. 2003). This new terrestrial ecosystem is subjected to rapid changes. The major part of the desiccated seabed is considered to be a highly unstable landscape with a high ecological hazard degree in terms of desertification and aeolian erosion (Dukhovny et al. 2008). Diminished water flow in the Amu Darya caused dramatic loss of wetlands and the associated reed communities in the rich ecosystem of the Amu Darya delta (Sivanpillai and Latchininsky 2007).

Given the vast geographical extent of the Aralkum and its adjacent region, space-borne remote sensing, in combination with a geographical information system (GIS), offers a high potential to assess surface characteristics on a large scale, and can be utilised for monitoring and analysing changes of the land cover in a multiyear perspective.

Satellite remote sensing technology has developed rapidly over recent decades, and the increasing availability of archived and partly freely available satellite data has brought remote sensing analysis of the Aral Sea disaster into a new era (for details see Micklin 2008). The highly dynamic processes in the Aralkum result in rapid changes in land cover and surface characteristics, and thus require multiple measurements in time. Land cover change detection based on monotemporal or bitemporal image classification techniques (Jensen 2005) is dependent on the seasonal timing of data acquisition, which can considerably influence the results from change analysis when only two or a few images are used for the different time steps. In contrast, multitemporal image analysis of time series data compensates for phenological and seasonal differences in vegetation development and surface

S.-W. Breckle et al. (eds.), *Aralkum - a Man-Made Desert: The Desiccated Floor of the Aral Sea (Central Asia)*, Ecological Studies 218,
DOI 10.1007/978-3-642-21117-1_6,

characteristics, and thus offers great potential for detecting and analysing intra-annual and interannual land cover changes.

6.2 Study Objectives

The shrinking of the Aral Sea has had dramatic impact on the environment and the local population. Soil and vegetation cover dynamics and the underlying processes are key indicators of the Aral Sea disaster (Dukhovny et al. 2008), and some of the associated developments can be measured by means of remote sensing:

- The dried bottom of the Aral Sea largely became a huge salt flat (Wucherer and Breckle 2001). The area of salt desert has increased owing to the ongoing desiccation of the Aral Sea since the 1960s. This led to the formation of new soils on the dried seabed, which became a considerable source of dust storms and salt storms (Dukhovny et al. 2008). Salt and dust blown out from the Aral Sea are transported over large distances by frequent strong winds (see Chap. 5), as salt and dust were found as far away as in the Tien Shan (Giese 1997).
- The spatial distribution of reeds in the Amu Darya delta is influenced by its hydrological regime (Sivanpillai and Latchininsky 2007). Since the 1950s the amount of water flow in the Amu Darya has diminished considerably (see Chap. 3). Both reduced runoff through the delta and declining groundwater levels, caused by the desiccation of the Aral Sea, resulted in the drying up of lakes and wetlands in the Amu Darya delta (Sivanpillai and Latchininsky 2007). Thus, associated reed communities diminished significantly (Micklin 2007).
- The Aralkum is a new terrestrial surface. From a plant-ecological point of view, it is considered as the world's largest area where primary succession is taking place (Wucherer and Breckle 2001).

This study presents a method for detecting changes in land and soil cover using high-temporal-resolution time series satellite data. The key issues followed in this study are:

- Using Moderate Resolution Imaging Spectroradiometer (MODIS) time series data for a postclassification change detection between classifications of annual time series in three time slices (2000, 2004, 2008)
- Investigating the spatiotemporal dynamics of land cover classes
- Relating the findings from change analysis to ecosystem dynamics in the southern Aralkum

The results of this study are of great relevance for ecological, restoration and development projects in the region, and can be used to provide a scientific basis for further analysis.

6.3 Study Area

The study area (Fig. 6.1) is located in the southern Aralkum within the autonomous republic of Karakalpakstan (Uzbekistan), covering an area of approximately 15,000 km^2. The lower reaches of the Amu Darya delta are often referred to as southern Priaralye. They include a variety of all major (semi-)natural land cover classes of the southern Aral Sea basin, e.g. bare areas, vast sand sheets, dune fields, shrubland, reeds and various kinds of salt-affected soils (e.g. solonchaks and takyrs). About 65% of the study area is composed of former seabed sediments. Within 9 years (2000–2008), the landscape changed dramatically: whereas in 2000, the Aral Sea and smaller water bodies covered 19.34% (approximately 2,889 km^2) of the study area, the area they covered had decreased to 0.7% (approximately 105 km^2) by 2008,

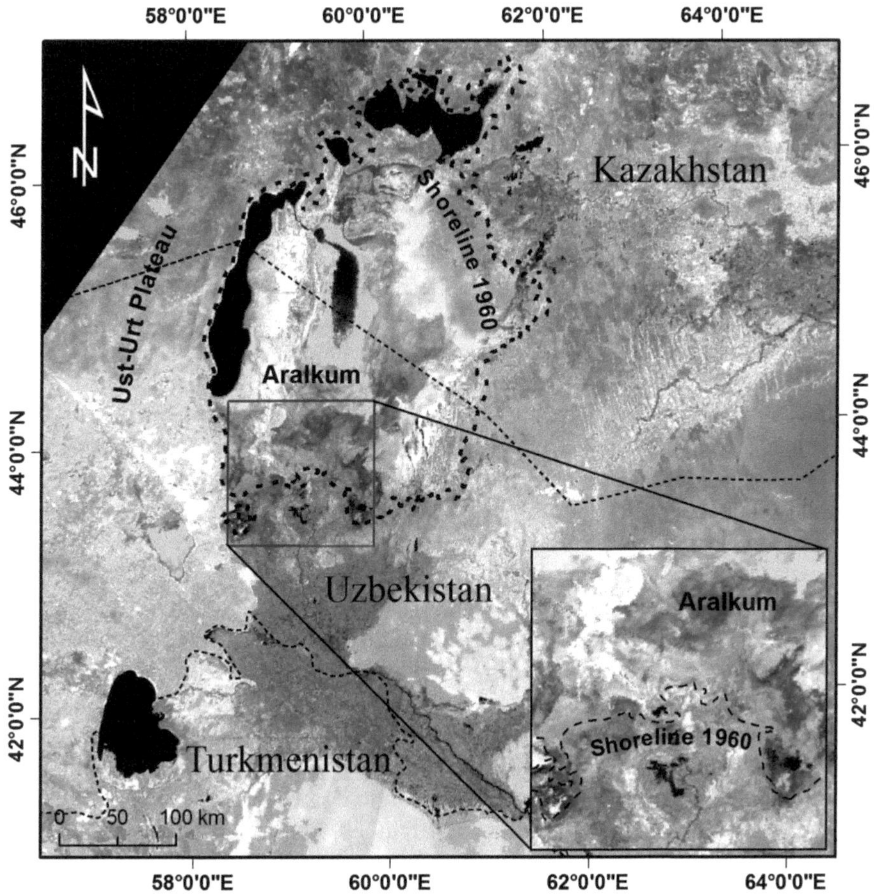

Fig. 6.1 Location of the study area in the southern Aral Sea basin. The image backdrop is a MODIS scene acquired in July 2009, displayed in false-colour band combination 7 (shortwave infrared), 2 (infrared) and 1 (red)

leaving behind only small water reservoirs, such as the lakes Ribachie, Dzhyltyrbas and Sudochie, which have been preserved from desiccation by artificial embankments. The latter dried up almost completely in May 2009.

6.4 Characterization of Land Cover Classes in the Study Area

A wide variety of land cover classes were mapped in the field during several field surveys in 2007, 2008 and 2009 (see Fig. 6.2). However, only six final classes were found suitable for classification in terms of thematic importance and spectral separability. These were finally used for analysis.

- Shrublands: sites with different halophytic shrub species composition as the dominant vegetation type in the study area. Species found in the study area are, e.g., *Haloxylon aphyllum*, *Halostachys caspica* and various *Tamarix* sp. Vegetation cover ranges between 15% and 100%. Pure shrub communities are rare; rather predominant is a mixture of shrubs, herbs and grass species (see Chap. 9).
- Reeds: reed beds in the wetlands of the Amu Darya delta, mainly composed of the common reed (*Phragmites australis)*. Pure reed communities as wells as reeds mixed with shrub vegetation were found in the study area. Pure reed stands are mostly present in flooded areas, whereas reeds mixed with vegetation are found directly adjacent to water bodies, which are not permanently flooded. Before the desiccation of the Aral Sea started, the *Phragmites australis* covered 85% of the Amu Darya delta (Sivanpillai and Latchininsky 2007).
- Bare area: sites devoid of vegetation, and sparsely vegetated areas. Sometimes scattered vegetation is present, but in general vegetation cover is below 15%. This class includes a variety of surface types such as vast sand sheets, dune fields, desert sandy soils and other unconsolidated material.
- Salt soils: highly salinized soils, e.g. solonchaks and takyrs. Surface features that allow for identification of salt accumulation in soils are puffy and sometimes cracked salt crusts as well as salt efflorescences. Capillary rise of salty groundwater through considerable evaporation leads to accumulation of salts in the top soil (Rafikov 1999; Dukhovny et al. 2008). Shallow groundwater tables can even give rise to the build-up of extensive white salt crusts (see below). Salt soils are a source of highly saline dust storms in the region (Singer et al. 2003). Salt soils can typically be identified through their high reflectance in the visible spectrum and low abundance of vegetation, resulting in very low normalised differentiated vegetation index (NDVI) values. However, low salinity and the initial stages of salinization might cause identification failures and lead to confusion between slightly saline and nonsaline soils (Metternich and Zinck 2003). Thus, low salinity levels might also be present in soils classified as 'bare area'.
- Salt crusts: white saline crusts in the study area differ from other salt-affected soils through their brightness and spectral contrast with other salt soils as well as their almost exclusive appearance in a narrow strip adjacent to the recent Aral

Fig. 6.2 Photograph samples for land cover classes mapped in the field: *1* close shrubland, *2* open shrubland, *3* sparse shrubland with vegetation cover less than 15%; *4* closed terrestrial reeds, *5* salt-affected soil, *6* white salt crust, *7* sandy soil without salt accumulation, *8* dunes

Sea (Dukhovny et al. 2008). Pure saline crusts show higher reflectance in the visible region of the spectrum when compared with other salt soils. In the latter case, salty crusts are often mixed with fine-grained material, thereby lowering overall reflectance.

- Water: the Aral Sea and shallow lakes, e.g. Lake Sudochie and Lake Dzhyltyrbas. Owing to the low spatial resolution of MODIS data, narrow rivers were not mapped.

6.5 Data and Methods

Validation data ('ground truth') for the classification of satellite images were collected during field surveys in the study area in three field campaigns in the spring to summer seasons of 2007 and 2008, as well as in May 2009. The field data were collected in accordance to the FAO's Land Cover Classification System (LCCS) (DiGregorio 2005), a globally standardised system for landscape characterization, which is highly adapted to the needs of remote-sensing-based landscape mapping. More than 200 ground-truth sampling points were collected at independent sites, representing a variety of the most important soil, vegetation and landscape types. Sampling sites were positioned by means of GPS measurements, photographs of the sampling sites and vegetation mapping by a standardised form sheet based on LCCS, tightly optimised for the study area.

Among other currently available image data, the MOD09 8-day Surface Reflectance product with 500-m spatial resolution of the MODIS sensor onboard NASA's Terra satellite was chosen as the primary data source for this study (Vermote et al. 1997). MODIS has been operational since 2000, and acquires data in 36 spectral bands (groups of wavelengths) with a daily repletion rate for the Aral Sea region. The high temporal frequency of image acquisition can be employed to form a time series, which enables temporal landscape characteristics, such as plant phenological stage and the annual hydrological cycle, to be used for the discrimination and classification of different land cover types. The data acquired by MODIS are used to generate multiple products and processing stages, ranging from raw image data to sophisticated globally available GIS products of biophysical parameters of the Earth's surface, such as snow and vegetation cover. In this study the MOD09A1 product was used, which is an estimate of the surface spectral reflectance for seven spectral bands ranging from the visible to mid-infrared electromagnetic spectrum, as would have been measured at ground level if there were no atmospheric scattering or absorption. The algorithm of this product (for details see Vermote et al. 1997) performs corrections for the effect of gaseous absorption, molecule and aerosol scattering, coupling between atmospheric and surface bidirectional reflectance functions, and the adjacency effect. MOD09A1 is a composite generated from daily images of an 8-day observation period, in order to reduce the effects of clouds and haze in the atmosphere.

Twenty-seven MODIS 8-day images ranging from Julian calendar dates 89 to 297 in each of the years 2000, 2004 and 2008 were used to generate the time series. The first seven of the 36 spectral MODIS bands, centred at 648, 858, 470, 555,

1,240, 1,640, and 2,130 nm, were processed in order to obtain the full spectral range and improve class separability.

A critical step is the generation of accurate time series and the correction of errors in the MODIS products, such as sensor artefacts, residual cloud cover, and atmospheric haze affecting data quality. The related noise introduced into the time series might lead to bad classification results, and might significantly alter the result of data analysis. Therefore, a quality assessment of the input data was performed. The TiSeG software program (Colditz et al. 2008) was used to generate high-quality MODIS time series. The software evaluates the quality at the pixel level using the quality assurance science data set provided with each MODIS image (Roy et al. 2002). Data in the time series regarded as invalid were erased and then temporally linearly interpolated.

Additional input information for the classification procedure was provided through the calculation of the NDVI from MODIS spectral bands 1 (red wavelength) and 2 (near-infrared) as well as the band ratio of MODIS bands 3 (blue wavelength) and 1, which proved valuable for discriminating salt-affected areas. The resulting data set includes nine time series (seven MODIS bands and the two indices) per year.

To reduce the high dimensionality of the data, metrics (basic statistics such as mean and standard deviation, minimum and maximum) were calculated from the time series and were further used as a baseline data set for the classification. Land surface characterization derived from time series metrics was computed in several studies dealing with land cover classification (DeFries et al. 1995, 1998; Hansen et al. 2000). Zhan et al. (2002) reported the detection of land cover changes by the use of MODIS time series.

In the scope of this study, decision trees were found to be most suitable for classifying the high-dimensional time series data, and were also successfully applied in other studies (Friedl et al. 1999; DeFries et al. 1995; Hansen et al. 2000). A regional application of a decision tree is provided by Conrad (2006) for crop classification in an irrigation system in Khorezm, Uzbekistan. The classification of the time series was conducted with the quick unbiased efficient statistical tree) (QUEST) algorithm (Loh and Shih 1997).

Validation was performed using an independent sample of validation points to assess classification accuracy. For the classified multidate MODIS image from 2008, ground verification data, collected through the field surveys, were used to assess the accuracy, whereas for 2000 and 2004, validation points were randomly generated and evaluated visually using Landsat 5 TM and AVNIR imagery as well as annual NDVI temporal signatures. The overall accuracies of the classifications for 2000, 2004 and 2008 were 82.8%, 81.5% and 79.5%, respectively. On the basis of the individual classification accuracy of the three time steps, the root mean square was calculated to assess the accuracy of the change detection maps, and was 81.95% for 2000–2004 and 80.58% for 2004–2008.

6.6 Results

6.6.1 Landscape Dynamics Between 2000 and 2008

Temporal signatures were derived from NDVI and band-ratio time series for the land cover classes. Figure 6.3 shows NDVI time profiles of the six classes. Shrubland and reeds show good separability on the basis of NDVI values. This is due to differences in annual phenology and biomass production, respectively. Reeds show high NDVI amplitude and mean values over the year and increase of NDVI values in spring, which is due to rapid biomass production. Water has permanent negative NDVI values because of high absorption in the infrared spectrum and thus clearly can be distinguished from the other classes. On the other hand, bare areas, salt soils and salt crusts produce very similar reflectance in the red and infrared parts of the spectrum during the year. However, by means of the band ratio these classes show higher reflectances in the visible spectrum and clearly can be distinguished from each other (Fig. 6.4). Water has higher reflectances but is not illustrated in Fig. 6.4.

Classification maps were generated for the 3 years under investigation (Fig. 6.5) and the individual class areas for the three years are summarised in Table 6.1. The classification results were used to perform a postclassification change-detection analysis for the time steps 2000–2004 and 2004–2008. The approach used in this study provides 'from–to' change information at the pixel level and hence the kind of landscape transformation that has occurred. To further evaluate the results of land cover conversions, change matrices of land cover change for two time steps

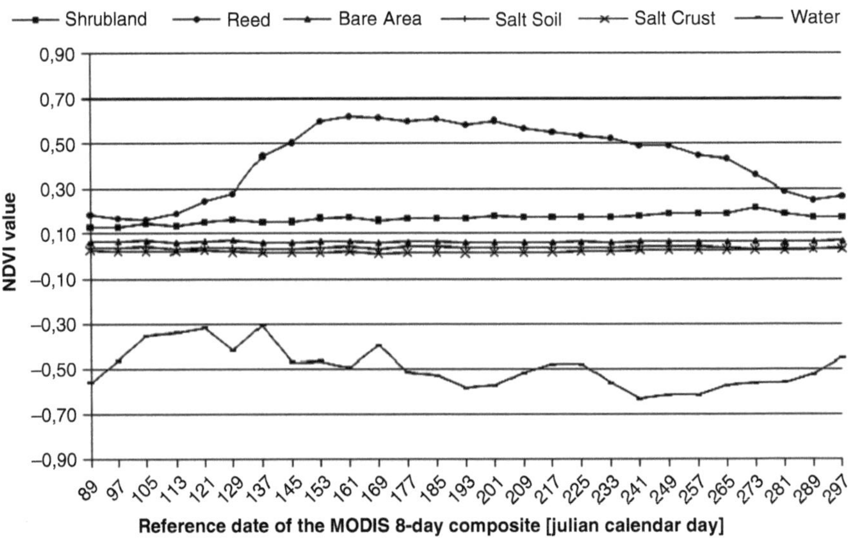

Fig. 6.3 Temporal normalised differentiated vegetation index signatures for each land cover class in 2008

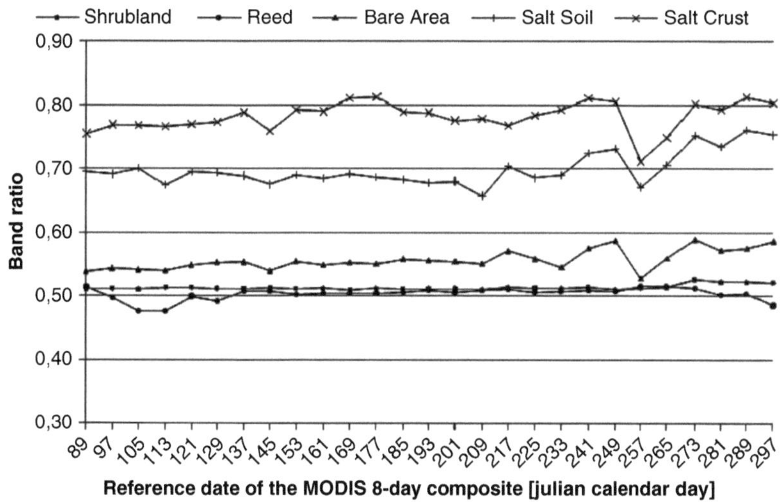

Fig. 6.4 Temporal signatures of the band ratio for each land cover class (except for water) in 2008

were created (Table 6.2). In the table, unchanged pixels are located along the major diagonal of the change matrix. However, by simply generating statistics, one cannot draw a conclusion about the question of where land cover changes are occurring. Using change maps, however, allows one to spatially disaggregate the calculated change statistics (Fig. 6.6).

The results show that the proportion of water surface in the study area diminished from 19.34% in 2000 to below 1% in 2008. Only some small natural and artificial lakes, e.g. lakes Sudochie and Ribachie, remain (Fig. 6.6). The water levels in these lakes are fluctuating as a result of the varying amount of river runoff reaching these very lowest areas of the Amu Darya delta. Simultaneously, the area of salt soils (e.g. solonchaks and takyrs) and salt crusts increased from 15.63% in 2000 to 29.61% in 2008. In summary, the results indicate that the increase of salt-affected areas is directly related to the rapid shrinking of the Aral Sea (Fig. 6.6). The total area of salt crust doubled between 2000 and 2008 (320.75 km^2 vs. 717.50 km^2, respectively). During the same time span, a major part of salt-affected area further converted into bare area. Shrubland is the most dominant land cover class, covering 30.08%, 25.62% and 37.81% of the area in 2000, 2004 and 2008, respectively. Reeds generally cover rather small areas. The results from the changed matrix (Table 6.2) indicate that the changes of the reed-covered areas are mainly the result of conversion of shrubland into reeds and vice versa. This is mostly due to changes in the inundation situation of the remaining wetland areas, as well as precipitation varying from year to year in the Amu Darya delta. In 2004, the extent of reeds reached its maximum. In this context, the changing hydrological situation of wetlands over the years is obvious (Fig. 6.5). Concerning the area, important types of conversions between 2000 and 2008 are 'water to salt soils', 'water to salt crusts' and conversions between shrubland and reeds. Relatively rare

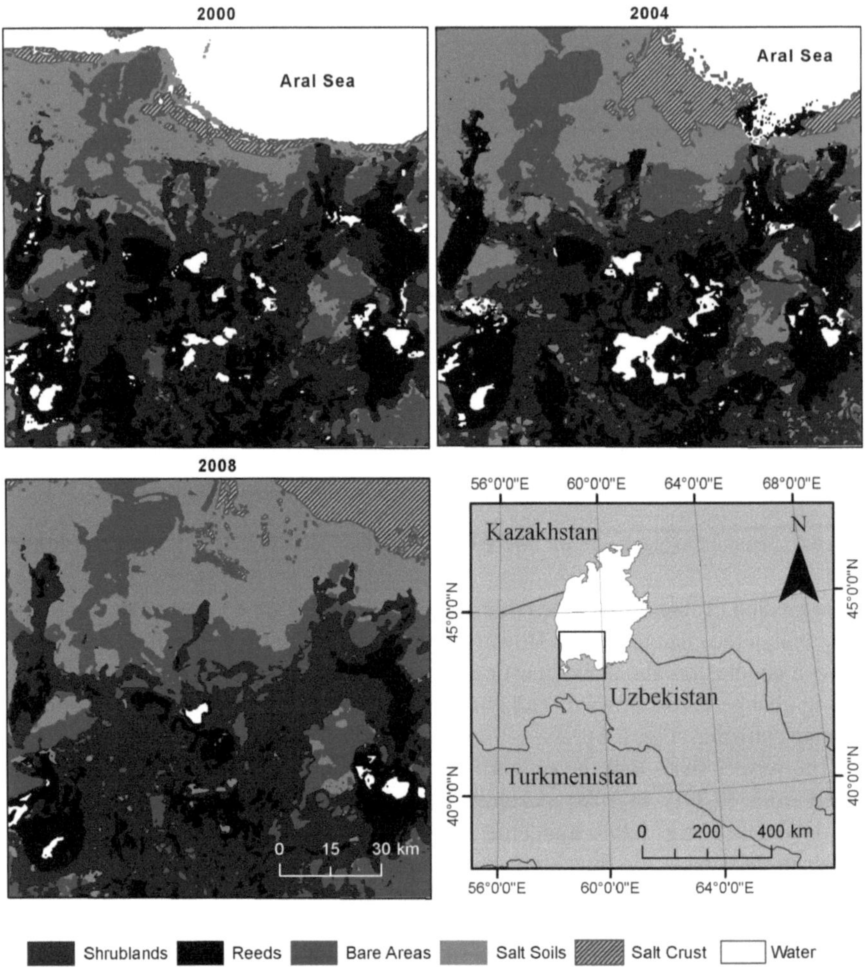

Fig. 6.5 MODIS time series classifications from 2000, 2004 and 2008

Table 6.1 Summary of MODIS time series classification area statistics for 2000, 2004 and 2008

Land cover class	2000		2004		2008	
	Area (km^2)	%	Area (km^2)	%	Area (km^2)	%
Shrublands	4,493.50	30.08	3,828.00	25.62	5,649.00	37.81
Reeds	2,961.50	19.82	3,891.25	26.05	1,737.00	11.63
Bare areas	2,261.25	15.13	1,974.25	13.21	3,024.25	20.24
Salt soils	2,014.25	13.48	3,036.75	20.33	3,707.25	24.81
Salt crusts	320.75	2.15	695.00	4.65	717.50	4.80
Water	2,888.75	19.34	1,514.75	10.14	105.00	0.70
Total	14,940.00	100	14,940.00	100	14,940.00	100

Table 6.2 Matrices of land cover and changes from 2000 to 2004 and from 2004 to 2008 (area in square kilometres)

2000–2004							
2004							2004 total
	Shrublands	Reeds	Bare areas	Salt soils	Salt crusts	Water	
Shrublands	2,804.00	465.75	467.00	75.00	0.50	15.75	3,828.00
Reeds	1,131.00	2,448.75	13.75	53.00	2.00	242.75	3,891.25
Bare areas	160.75	0.75	1,432.50	340.50	9.25	30.50	1,974.25
Salt soils	241.50	1.25	346.75	1,522.75	293.25	631.25	3,036.75
Salt crusts	5.75	1.25	0.00	14.50	15.00	658.50	695.00
Water	150.50	43.75	1.25	8.50	0.75	1,310.00	1,514.75
2000 total	4,493.50	2,961.50	2,261.25	2,014.25	320.75	2,888.75	
2004–2008							
2008							2008 total
	Shrublands	Reeds	Bare areas	Salt soils	Salt crusts	Water	
Shrublands	2,963.75	2,203.50	55.25	244.00	6.25	176.25	5,649.00
Reeds	47.50	1,553.75	0.25	10.00	1.50	124.00	1,737.00
Bare areas	799.00	6.75	1,703.75	473.00	17.75	24.00	3,024.25
Salt soils	17.75	95.50	215.00	2,295.25	572.00	511.75	3,707.25
Salt crusts	0.00	0.00	0.00	14.50	97.5	605.50	717.50
Water	0.00	31.75	0.00	0.00	0.00	73.25	105.00
2004 total	3,828.00	3,891.25	1,974.25	3,036.75	695.00	1,514.75	

types of conversions, such as bare area into reeds (0.25 km^2) between 2004 and 2008, are assumed to be largely classification errors.

6.6.2 *Ecosystem Dynamics*

The recession of the Aral Sea has resulted in a quick build-up of extensive salt crusts directly adjacent to the sea (Dukhovny et al. 2008). Almost all of these salt crusts converted into a series of different solonchak types (contained in the class salt soils) and afterwards, in some parts of the study area, into bare areas between 2000 and 2008. Lowering of the groundwater table is one reason for the transformation of hydromorphic solonchaks into automorphic solonchaks and subsequently transformation into sandy desert soils (Dukhovny et al. 2008). On elevated terrain these processes are even accelerated (Rafikov 1999). Another reason for the transformation of class 'salt soil' into class 'bare area' is overblown sand (Dukhovny et al. 2008).

A significant proportion of the emerged soil remained devoid of dense vegetation and became a salt desert. Only a small part of the salt desert in the study area, near the former Amu Darya's mouth, was converted to shrubland and reeds between 2000 and 2008 (see Fig. 6.6). Salt-affected area and salt crusts are considerable

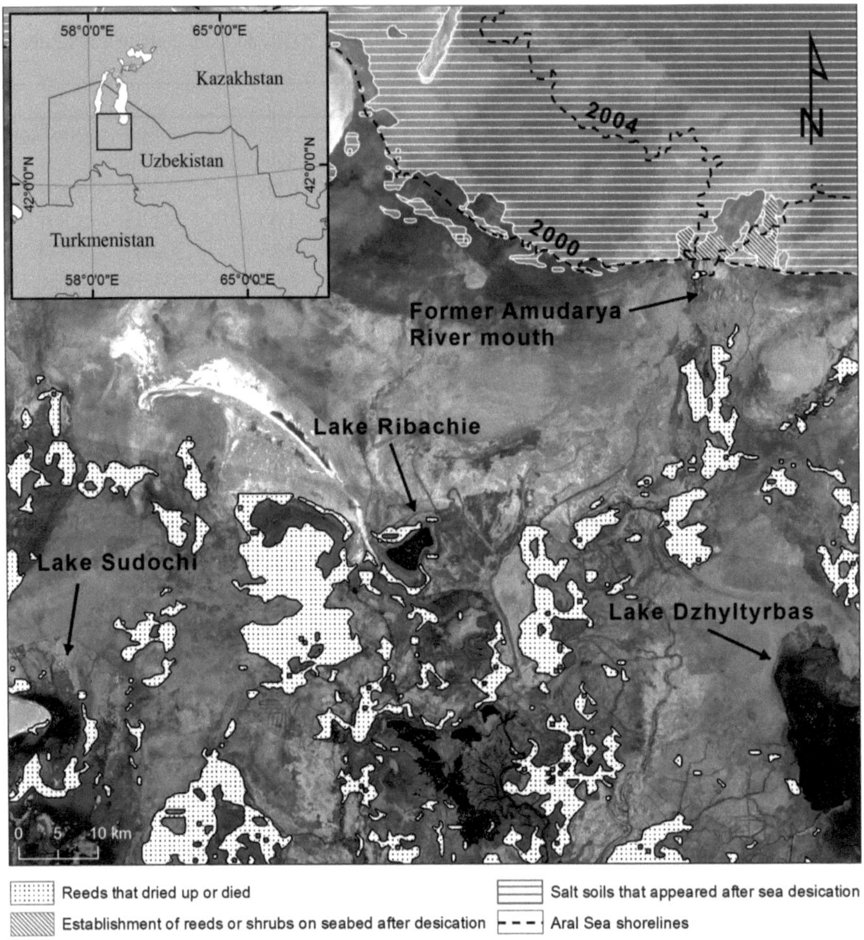

Fig. 6.6 Change map of the study area. Between 2000 and 2008 almost exclusively salt-affected soils appeared in the study area after desiccation of the Aral Sea. Within that period, shrubs and reeds established themselves on the dried seabed near the former Amu Darya's mouth. The area of reeds in the study area has decreased dramatically. The image backdrop is a Landsat 5 TM scene acquired in June 2009, displayed in false-colour band combination 7 (shortwave infrared), 5 (shortwave infrared) and 3 (red)

sources of salt storms and dust storms (Singer et al. 2003) and feature a highly unstable landscape (Dukhovny et al. 2008). Presently and in the near future, the continued desiccation of the Aral Sea will create more saline flats, and thus the potential for salt and dust storms is likely to increase (Breckle et al. 2001; Singer et al. 2003).

Shrublands and reeds show high interdependence, which can be explained by the presence of shrub–reed mixtures in the tidal areas of permanent and temporarily flooded wetlands. Wetlands, dominated by a mixture of reed species (e.g.

Phragmites sp.) and shrubs (e.g. *Haloxylon* sp.) can convert into shrubland in dry years. During wet periods, this process can reverse completely. Sivanpillai and Latchininsky (2007) stated that the area of reed bed in the Aral Sea basin shrank by 70%, whereas more than 30,000 ha of lakes and bogs have dried up completely.

6.7 Conclusions

The results demonstrate that MODIS time series classification is a valuable tool to produce accurate landscape classification, landscape change maps and statistics for large areas. For the Aralkum, the results of the postclassification change detection revealed that (1) most of the increase in salt crusts and salt soils followed from the shrinking of the Aral Sea; (2) between 2000 and 2008, no significant vegetation cover emerged on the former seabed in the study area; (3) the classes shrublands and reeds show high interdependence; and (4) the potential source area for dust storms and salt storms in the study area has increased. The results quantify the land cover change patterns and landscape dynamics in the study area. MODIS time series provide an accurate, yet cost-efficient means to map and analyse changes in land cover. This applies not only to the southern Aralkum, but also on a regional scale, covering the whole of the former Aral Sea extent and the adjacent regions in Karakalpakstan and Kazakhstan. The methodology and the resulting geo-information products are highly suitable for both integration into existing landscape information systems and monitoring surface characteristics of large areas and relatively long time periods. Monitoring land cover condition and analysing land cover change in the Aralkum is of great importance, since the ecological situation is still very dynamic and a large part of the landscape in the Aralkum is unstable.

References

Breckle S-W, Wucherer W, Agachanjanz O, Geldyev B (2001) The Aral Sea crisis region. In: Breckle S-W, Veste M, Wucherer W (eds) Sustainable land use in deserts. Springer, Berlin, pp 27–37

Colditz RR, Conrad C, Wehrmann T, Schmidt M, Dech S (2008) TiSeG: flexible software tool for time-series generation of MODIS data utilizing the quality assessment science data set. IEEE Trans Geosci Remote Sens 46(10):3296–3308

Conrad C (2006) Fernerkundungsbasierte Modellierung und hydrologische Messungen zur Analyse und Bewertung der landwirtschaftlichen Wassernutzung in der Region Khorezm (Usbekistan). Dissertation thesis, University of Würzburg, Würzburg

DeFries RS, Hansen MC, Townshend JRG (1995) Global discrimination of land cover types from metrics derived from AVHRR pathfinder data. Remote Sens Environ 54:209–222

DeFries RS, Hansen MC, Townshend JRG, Sohlberg R (1998) Global land cover classification at 8 km spatial resolution: the use of training data derived from Landsat imagery in decision tree classifiers. Int J Remote Sens 19(16):3141–3168

DiGregorio A (2005) Land cover classification system – classification concepts and user manual for Software version 2, FAO Environment and Natural Resources Service Series, No. 8, Rome
Dukhovny VA, Navratil P, Rusiev I, Stulina G, Roshenko YE (eds) (2008) Comprehensive remote sensing and ground based studies of the dried Aral Sea bed. Scientific-Information Center ICWC, Tashkent
Friedl MA, Brodley CE, Strahler AH (1999) Maximising land cover classification accuracies produced by decision trees at continental to global scales. IEEE Trans Geosci Remote Sens 37 (2):969–977
Giese E (1997) Die ökologische Krise der Aralseeregion. Geogr Rundsch 49:293–299
Hansen MC, DeFries RS, Townshend JRG, Sohlberg R (2000) Global land cover classification at 1 km spatial resolution using a classification tree approach. Int J Remote Sens 21(6–7): 1331–1364
Jensen JR (2005) Introductory digital image processing. Prentice Hall, Upper Saddle River
Loh W, Shih Y (1997) Split selection methods for classification trees. Stat Sin 7:815–840
Metternich GI, Zinck JA (2003) Remote sensing of soil salinity: potentials and constraints. Remote Sens Environ 85:1–20
Micklin PP (2007) The Aral Sea disaster. Annu Review Earth Planet Sci 35:47–72
Micklin PP (2008) Using satellite remote sensing to study and monitor the Aral Sea and adjacent zone. In: Qi J, Evered KT (eds) Environmental problems of Central Asia and their economical, social and security impacts. Springer, Dordrecht, pp 31–49
Rafikov AA (1999) Desertification in the Aral Sea region. In: Glantz MH (ed) Creeping environmental problems and sustainable development in the Aral Sea Basin. Cambridge University Press, Cambridge
Razakov RM, Kosnazarov KA (1996) Dust and salt transfer from the exposed bed of the Aral Sea and measures to decrease its environmental impact. In: Micklin PP, Williams WD (eds) The Aral Sea Basin. NATO ASI Series, Vol. 12. Springer Verlag, Berlin
Roy DP, Borak JS, Devadiga S, Wolfe RE, Zheng M, Descloitres J (2002) The MODIS land product quality assessment approach. Remote Sens Environ 83(1–2):62–76
Singer A, Zobeck T, Poberrezky L, Argaman E (2003) The PM_{10} and $PM_{2.5}$ dust generation potential of soils/sediments in the Southern Aral Sea Basin, Uzbekistan. J Arid Environ 54:705–728
Sivanpillai R, Latchininsky V (2007) Mapping locust habitats in the Amudarya river delta, Uzbekistan with multi-temporal MODIS imagery. Environ Manage 39:876–886
Vermote EF, el Saleous NZ, Justice CO, Kaufman YJ, Privette JL, Remer L, Roger JC, Tanre D (1997) Atmospheric correction of visible to middle-infrared EOS – MODIS data over land surfaces: background, operational algorithm and validation. J Geophys Res 102(D14): 17131–17141
Wucherer W, Breckle S-W (2001) Vegetation dynamics on the dry sea floor of the Aral Sea. In: Breckle S-W, Veste M, Wucherer W (eds) Sustainable land use in deserts. Springer, Berlin, pp 52–69
Zhan X, Sohlberg RA, Townshend JRG, DiMiceli C, Carroll ML, Eastman JC, Hansen MC, DeFries RS (2002) Detection of land cover changes using MODIS 250 m data. Remote Sens Environ 83:336–350

Chapter 7
Dynamics of Dust Transfer from the Desiccated Aral Sea Bottom Analysed by Remote Sensing

L. Spivak, A. Terechov, I. Vitkovskaya, M. Batyrbayeva, and L. Orlovsky

7.1 Introduction

The Aral region is an ecological disaster zone. Almost 60,000 km^2 of the Aral Sea floor is exposed. Many thousands of tons of salt and fine dust with impurities of various chemicals and poisons are blown up from dry sea bottom each year and are adversely affecting the environment (Chaps. 5 and 6). Contamination is enhanced by the fact that the Aral Sea is located within a powerful air stream from west to east which transfers these aerosols in the upper layers of the atmosphere.

7.2 Satellite Monitoring of the Aral Sea Area

Ground investigations of the Aral region, the number of weather stations and their equipment have been sharply reduced in recent years. In this regard there is an increasing need for satellite monitoring (Zakarin et al. 1999). Daily images from the National Oceanic and Atmospheric Administration (NOAA)/Advanced Very High Resolution Radiometer were used with a resolution of 1,100 m in the visible and near-infrared spectrum (1.2 channel) for the period 2000–2010 for space monitoring of the processes occurring in the region. Figure 7.1 shows the dynamics of the water area of the Aral Sea in 2000–2009 from satellite data and expressed as a percentage of the maximum value for the period. The water surface decreased within this short time by 60%. This decrease occurred mainly at the expense of the eastern part of the Aral Sea, which practically no longer receives water from rivers. Far beyond the annual fluctuations of the surface area, the eastern coastline receded by tens of kilometres. The results from satellite estimates of this tremendous disappearance of the water surface of the eastern part of the Aral Sea from May until October 2009 are documented in Fig. 7.2.

S.-W. Breckle et al. (eds.), *Aralkum - a Man-Made Desert: The Desiccated Floor of the Aral Sea (Central Asia)*, Ecological Studies 218,
DOI 10.1007/978-3-642-21117-1_7, © Springer-Verlag Berlin Heidelberg 2012

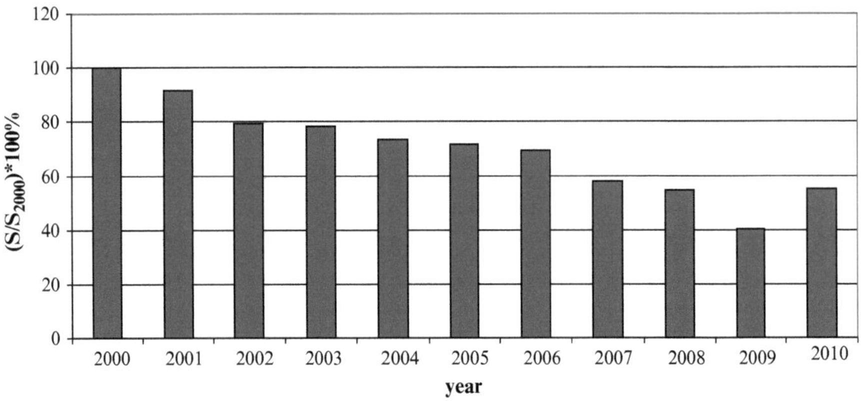

Fig. 7.1 Dynamics of water surface area of the Aral Sea in the period 2000–2010 (2000 = 100%)

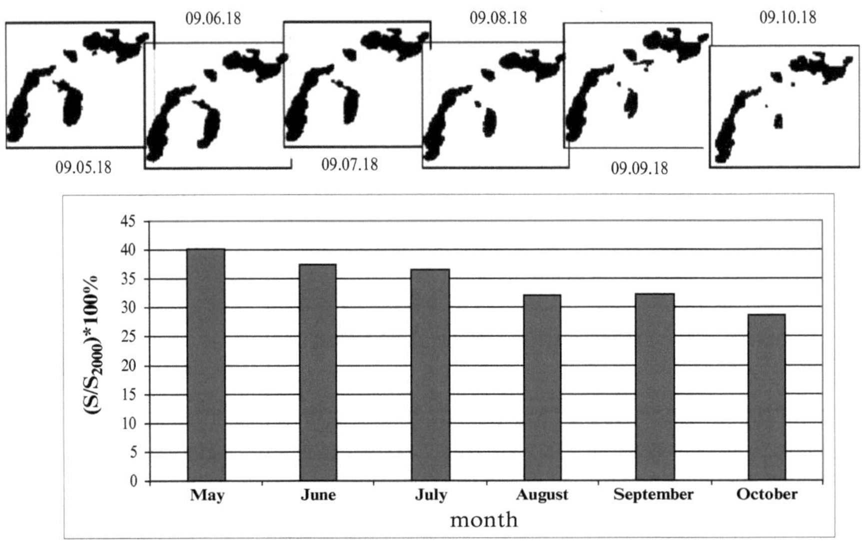

Fig. 7.2 Changes of the water surface of the Aral Sea over the period from May to October 2009

7.3 Space Monitoring of Dust Storms

The catastrophic degradation of the Aral Sea created a new phenomenon - salt-dust storms from the dry sea bottom. The dried-up surface is impregnated with salt and is subjected to erosion under conditions of strong wind. A large number of aerosol particles are generated and are carried over long distances. The likelihood of dust storms on the dry bottom of the sea is 10 times higher in comparison with the surrounding areas since the surface structure is unstable and easily subject to wind erosion (Semenov et al. 2006; see Chap. 5). The dust plumes are well recognized on satellite images especially above the water surface (Fig. 7.3).

Dust storms are most typical in spring when the soil surface dries out, getting rid of accumulated winter moisture, and vegetation is minimal. Table 7.1 shows the number of dust storm episodes for each year of observation and the direction of storm plumes obtained by space monitoring of the Aral Sea region.

Fig. 7.3 Dust storm episode from the dried bottom of Aral Sea from National Oceanic and Atmospheric Administration (NOAA) \Advanced Very High Resolution Radiometer (AVHRR) satellite data (9 May 2007)

Table 7.1 The number of dust storms recorded on National Oceanic and Atmospheric Administration satellite images for the period 2000–2009 and broken down for the direction of aerosol transfer (see also Chap. 5)

Years	Number of episodes							
	N	NE	E	SE	S	SW	W	SW
2000	6							
	–	1	4	–	–	–	1	–
2001	13							
	1	–	5	–	–	1	4	2
2002	22							
	2	1	6	2	3	5	3	–
2003	19							
	–	–	16	–	–	2	1	–
2004	7							
	–	–	5	–	–	1	1	–
2005	11							
	–	2	2	–	–	2	4	1
2006	11							
	–	–	–	–	–	5	4	2
2007	13							
	2	1	3	2	–	2	3	–
2008	15							
	–	–	–	–	–	6	9	–
2009	17							
	2	1	8	1	1	2	2	–

The maximum number of episodes was observed in 2002. This may be partly related to the large quantity of NOAA satellites in operation at this time.

7.4 Technique for Defining the Dust Aerosol Transfer Zone

The spatial referencing of salt-dust clouds to the centres of wind erosion on the dry sea bottom is a peculiarity of the Aral dust storms. Sources of removal of aerosols are formed everywhere at some distance from the water surface as a result of the dry surface caused by less filtrating water.

The mass of possible spectral characteristics of the underlying surface in the presence of atmospheric aerosol components hinders the development of automatic recognition algorithms and analysis of the characteristics of dust storms. Therefore, the basis of recognition of episodes of dust storms is expert interpretation of satellite data. The main objectives of the expert treatment are:

1. Detection of dust storm episodes, the collection of information about all the episodes in the database
2. Analysis of the dust storm episode with the release of (a) the boundaries of the aerosol cloud and (b) the boundaries of the aerosol site with a qualitative assessment of its structure in terms of the intensity of transfer

The borders of dust storms are similar to those of the usual sparse clouds and are determined by the level of brightness in contrast with that of different objects on the Earth's surface. NOAA images can be preprocessed in various ways: pseudocomposites, various indices (mathematical combination of channels available), separate channels. Then, aerosol formation and the boundaries are well identified in the satellite images.

The registration of borders of the transfer zone with the estimation of the intensity of transfer is a more difficult task.

The configuration of clouds in the Aral Sea region, especially the structure of the dust storm plume, is associated with the wind direction. Wind flow falling on the dust storm zone generates the front of the aerosol component in the atmospheric boundary layer. As result of the influence of the large-scale surface, the turbulence front splits into separate streams, which are recorded on a satellite image. This spatial feature of dust storm movement is a sign for distinguishing the windward and leeward sides of the source of transfer. A fine fraction remains on the leeward side in the air. The system of separate jets is destroyed and is mixed. More or less homogeneous dust clouds with a smooth change in optical density due to deposition processes are formed.

The existence of a characteristic shallow structure of interaction between air jets is the interpretive sign of the dust source. Multiscale turbulence generates a continuous process of recovery and deposition of relatively large aerosol particles. Only the processes of deposition and mixing of most of the fine fraction with the formation of dust storm clouds of homogeneous optical density will occur in the atmosphere behind a source of dust.

The purpose of the expert processing of the individual storm episodes on the satellite image is to construct a system of points (graph vertices) which describe the territory of the sources of dust removal and the definition of the intensity of movement. In the absence of ground-based information, the intensity of movement refers to the transparency and the degree of turbulence of dust storm clouds. Expert mapping of the intensity of the areas of dust removal by a system of points is the easiest way for subsequent rigorous mathematical operations to conduct an analysis to obtain average estimates of the configurations of the transfer zone and the internal structure.

7.5 Investigation of the Dynamics of the Zone of Aerosol Removal in the Eastern Part of the Aral Sea

The zone of dust removal and its intensity are defined using expert interpretation for a set of vertices. The contours of the aerosol zones are constructed for each year (2000–2008). A complex median filter was used to construct a regular matrix that maps the intensity of wind erosion (the higher the specific density of the vertices of the graph, the higher the expected intensity of wind erosion). For example, the results of processing of dust storm episodes occurring in 2008 are demonstrated in Fig. 7.4.

The drastic reduction of the water surface in the eastern part of the Aral Sea has tremendously increased the number of sites with an unstable surface subjected to wind erosion.

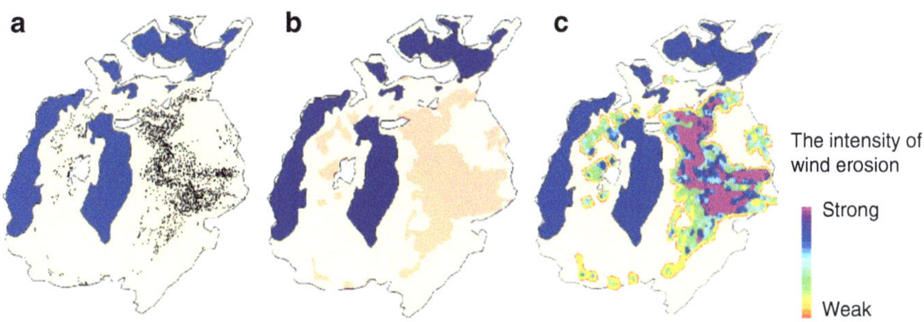

Fig. 7.4 The results of space monitoring of dust storms from the dried bottom of the Aral Sea in 2008 and stages of mathematical processing. (**a**) Result of expert estimation of the intensity of dust storms on satellite scenes for 2008. (**b**) Outlines of zones of dust removal by calculation of the contour line of dust storms (the state of the water mirror corresponds to the satellite data, May 2008). (**c**) The intensity of the process of dust removal (averaged over the yearly data by determination of the density of vertices inside the calculated area of storm removal using a complex median filter)

Potentially, the total bottom of the Aral Sea may be a source of dust storm generation. The territories with enough moisture due to seepage and groundwater are exceptions. These territories include sites adjacent to the water mirror of the Large Aral Sea and the Small Aral Sea and deltas of the Syr Darya and the Amu Darya. The deltas of rivers flowing into the remnant Aral Sea basins are more or less stable. Reduced filtration at the bottom of the eastern part of the Aral Sea is the reason for changing the zone of dust storms. This site is the most dynamic centre for dust removal (see Fig. 7.5).

There is no longer a significant water surface of the eastern part of the Aral Sea. With the almost complete disappearance of this part of the Aral Sea, which occurred during late summer 2009, we should expect a conspicuous increase of the zone of dust removal at the eastern part of the Aral Sea (Fig. 7.6e) in coming years, depending on the future low-level hydrological equilibrium from year to year (see Chap. 4).

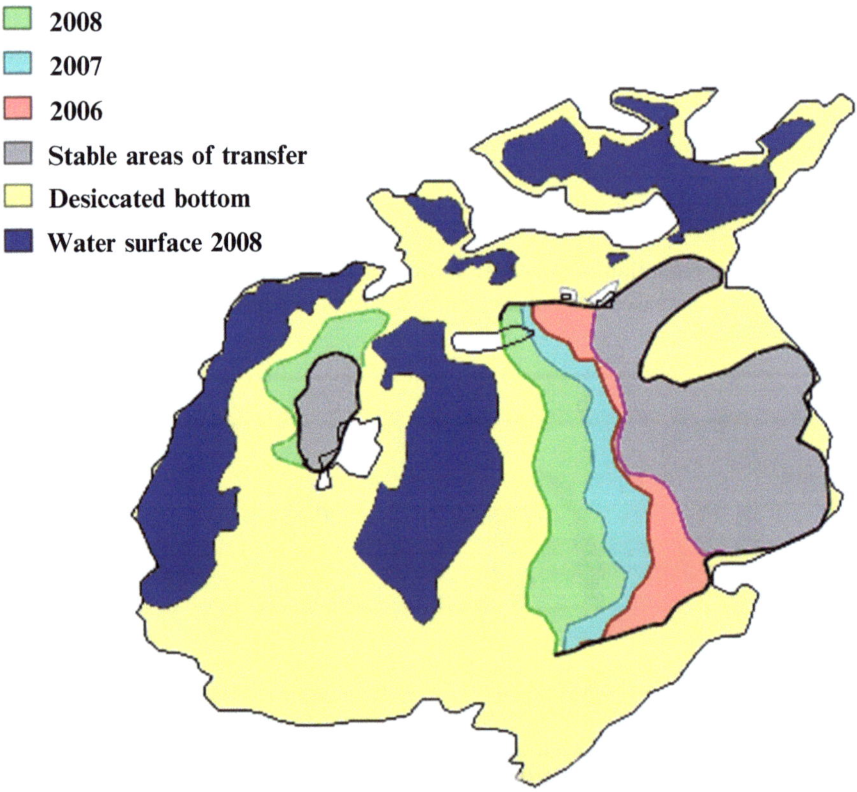

Fig. 7.5 Dynamics of the expansion of the centre of dust removal located on the eastern coastline of the Large Aral Sea (2005–2008)

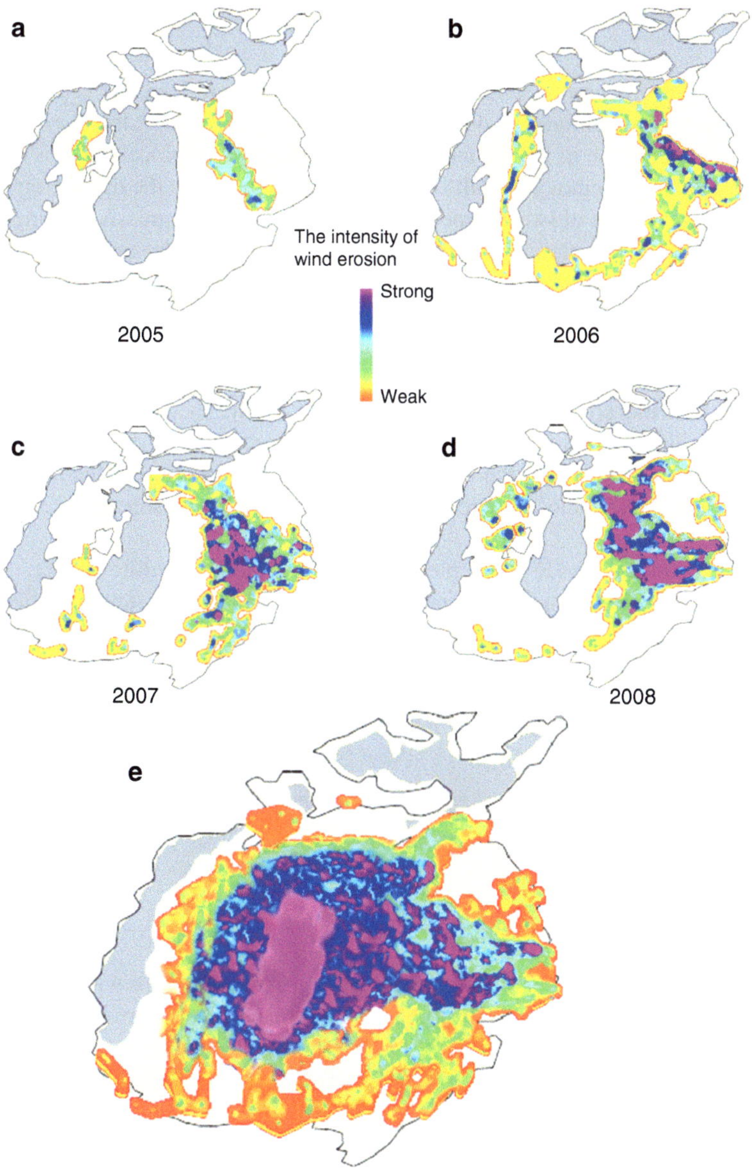

Fig. 7.6 The intensity of dust removal from the dried bottom of the Aral Sea for the period 2005–2008 (**a–d**), and the forecast of the state of this site (**e**) in the case of complete drying of the eastern part of the Large Aral Sea. Constructed using NOAA/AVHRR satellite data

7.6 Frequency of Dust Storms

The results of processing satellite information show significant differences in the frequency and magnitude of dust storms from year to year. Moreover, there is a monotonic increase in the annual activity of the transport processes. It is obvious that the process of removal of salt-dust aerosol particles from the dry bottom of the Aral Sea is the result of two independent phenomena: (1) progressive drying of the former seabed and formation of a growing area with an unstable surface not fixed by vegetation; (2) wind conditions are not stable and vary from year to year as any meteorological parameter.

Figure 7.7 shows a comparison of the annual frequency of dust storms from the dry bottom of the Aral Sea using satellite data (Fig. 7.7a) and of the average intensity of wind erosion (Fig. 7.7b) calculated by the method described above. Subjective information is largely present in the processing of satellite images in determining the contour of the zone of removal and estimation of the intensity of wind erosion. Nevertheless, the results are similar, which indicates the effectiveness of both techniques.

A dust storm episode (1 September 2006, Landsat satellite image, 240-m resolution; http://glovis.usgs.gov) is shown in Fig. 7.8. The existence of this long-lasting dust storm was confirmed by NOAA data for the period from 30 August 2006 (12:24 UTM) to 2 September 2006 (12:50 UTM). On the Landsat satellite image (Fig. 7.8a), the area of sand is clearly identified. The inkjet nature of the storm confirms the legitimacy of the use of a system of points to highlight the removal of aerosol (Fig. 7.8b).

7.7 Conclusions

The reduction of the water surface in the eastern Aral Sea basin within the period 2005–2008 was accompanied by increasing drainage of the former seabed. This process has clearly led to an increased territory with an unstable surface subjected to wind erosion.

The temporal dynamics of the zone of aerosol removal in the eastern part of the Aral Sea was studied. The results of processing remote sensing data show significant differences in the frequency and magnitude of dust storms from year to year. A monotonic increase in the annual activity of the transport processes is observed.

The results of analysis of satellite data for the period 2005–2008 show that the entire territory of the bottom of the Aral Sea is a potential area of salt-dust removal. Changes of the territory of the zone with aerosol removal occurred mainly by the reduction of infiltration of the eastern part of the Aral Sea. There is a relationship between the reduction of the water mirror of the eastern part of the Large Aral Sea and the increase in the area of the zone of salt-dust removal. Parts of the drained

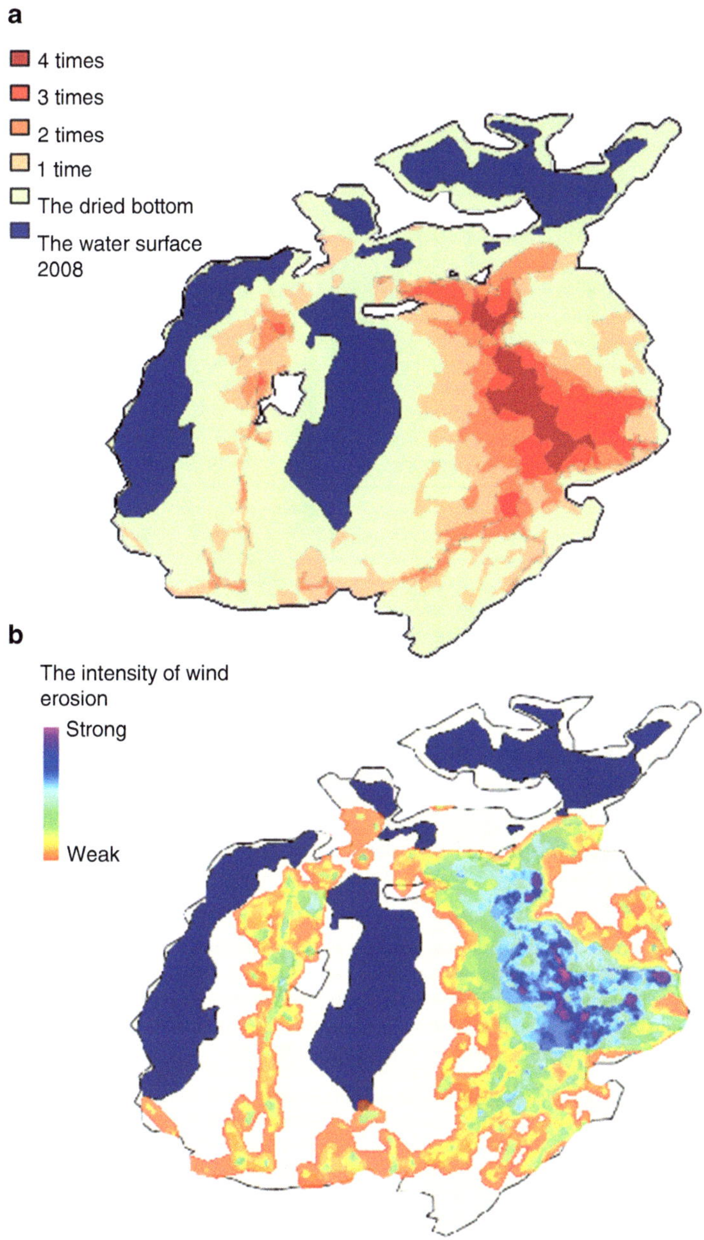

Fig. 7.7 Spatial characteristics of zones of salt-dust storms from the dried bottom of the Aral Sea (2005–2008). (**a**) Frequency of hitting the territory of the zone of dust removal in the analysis of annual samples for the period 2005–2008. (**b**) The average intensity of wind erosion for the period 2005–2008

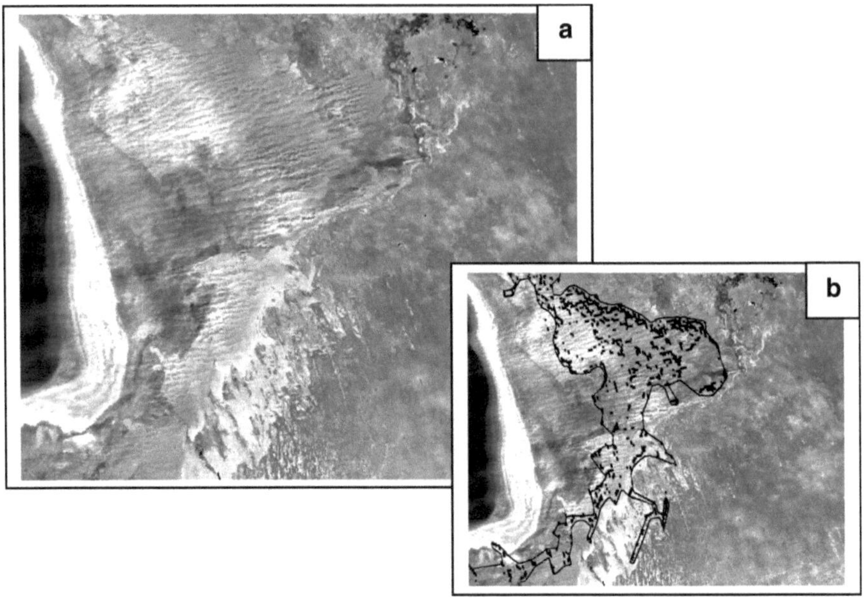

Fig. 7.8 Dust storm from the dried bottom of the Aral Sea using Landsat satellite data (1 September 2006). (**a**) Landsat satellite image. (**b**) The selection of the area of dust removal by a system of points

bottom which were hotbeds of storms in previous years also remain active in subsequent years.

The limits of the zone for the possible removal of dust are defined by the assumption of the complete disappearance of the eastern (shallow) part of the Large Aral Sea.

Acknowledgements This research was done under the project INCO Proposal PL-516721 "Long Term Ecological Research Program for Monitoring Aeolian Soil Erosion in Central Asia" (CALTER) and programmes for fundamental research "To develop mathematical methods for modeling dynamic natural and industrial processes using space technologies" funded by the Ministry of Education and Sciences of the Kazakhstan Republic.

References

http://glovis.usgs.gov

Semenov O, Shapov A, Galaeva O, Idrisova V (2006) Stem wind and sand-salt loss from the dry part of the Aral Sea/Arid areas. N 12, pp 47–58

Zakarin E, Bekmuhamedov B, Dedova T (1999) Determining areas of dust storms in the Aral Sea satellite remote sensing methods/Hydrometeorology and Ecology №4, Almaty, 1999, pp 189–196

Part B
Biotic Aspects and Ecosystems Dynamics

Chapter 8
Flora of the Aralkum

L.A. Dimeyeva, S-W. Breckle, and W. Wucherer

8.1 Introduction

The desiccated seafloor of the former Aral Sea is an area which has been growing steadily since 1960 (see Chap. 2). The area of the desiccated seafloor is a mosaic of sand and salt deserts (see Chap. 3). This area is very well defined by the former coastline of 1960, more or less equivalent to the 53 m above sea level contour line (see Fig. 4.15). This whole area is called the Aralkum (Breckle et al. 1998, 2001).

There are now 368 species of angiosperms (as of 2008). The flora of the Aralkum is rather young and has formed since 1960 as a typical immigration flora. It is certainly a flora where species have to be adapted to the rather severe ecological conditions of drought and heat in summer, to frost in winter, to moving sand conditions and to varying salinity, and also to inundation and anaerobic conditions of solonchak soils. Close to the former coastline there are mostly sandy soils which are more or less desalinized. They became dry in the 1960s. Towards the sinking seawater level they are being replaced by sandy-loamy or by loamy coastal solonchaks with some salinity. They became dry in the 1970s or later. The desiccated places which became dry from the 1980s until today are mostly very saline solonchaks, often puffy or crusty.

The desiccated seafloor area has a high percentage (24.5%) of species from the Chenopodiaceae family (approximately 90 species; Table 8.1). Along the former coastline on the sandy conditions of the Aralkum, the Polygonaceae (*Calligonum* spp.) and Fabaceae (*Astragalus* spp.) prevail. But also some psammophilous Chenopodiaceae as well as some representatives from Poaceae play a dominant role.

The new land is an area which can be spatially and historically clearly defined. It exhibits a new dynamics of vegetation complexes (see Chap. 9).

S.-W. Breckle et al. (eds.), *Aralkum - a Man-Made Desert: The Desiccated Floor of the Aral Sea (Central Asia)*, Ecological Studies 218,
DOI 10.1007/978-3-642-21117-1_8, © Springer-Verlag Berlin Heidelberg 2012

Table 8.1 The dominant plant families of the flora of the Aralkum (the desiccated seafloor of the Aral Sea, 2008)

Plant families	Number of genera	Number of species	Percentage of species
Chenopodiaceae	31	90	24.5
Asteraceae	25	48	13.0
Polygonaceae	5	39	10.6
Brassicaceae	24	34	9.2
Fabaceae	9	23	6.3
Poaceae	14	22	6.0
Boraginaceae	9	13	3.5
Tamaricaceae	1	10	2.7
Ranunculaceae	7	8	2.2
Zygophyllaceae s.l. (3)	3	7	1.9
Caryophyllaceae	3	6	1.6
Cyperaceae	3	5	1.4
Apiaceae	2	5	1.4
Additional families (28)	42	58	15.8
Total: 43	178	368	100

8.2 History of Floristic Studies

The flora and vegetation around the Aral Sea, except the delta regions of the Syr Darya and the Amu Darya, have not been completely investigated, and the new desiccated seafloor, the Aralkum, has been investigated even less. The first report on the flora and vegetation of Priaralye (direct vicinity of the Aral Sea) was by Borshchov (1865). A rather big herbarium collection was collected by Smirnov (1875) from the southeastern coast of the Aral Sea between the Syr Darya and Amu Darya deltas. On the basis of the collections of L.S. Berg (1900–1902), Litwinov (1905) published a list of 219 plant species. Sherbaev (1982) studied the flora of the southeastern islands, where he found 235 species. These papers give fragmentary ideas on the flora of the former coast of the Aral Sea. For more than 30 years from the middle of the 1960s, researchers from Herzen Pedagogical University (St. Petersburg, Russia) investigated ecosystems of the island of Barsa-Kelmes. The results of this long-term research of the insular flora were reported by Kuznetsov (1995). The floristic composition of Barsa-Kelmes consisted of 257 species (see also Chap. 14).

Between 1977 and 1985, the Institute of Geography in Moscow headed a complex programme studying the effects of irrigation projects on desiccation of the Aral Sea. The investigations within the Kazakhstan part of the Aral Sea coast resulted in some publications about flora (Kurochkina et al. 1983; Dimeyeva 1990; Dimeyeva and Kuznetsov 1999; Dimeyeva and Kurochkina 2005) and comparisons with the adjacent desert flora (Rachkovskaya and Safronova 1994). These lists of vascular plants included about 300 species. Only half of them (154 species) were recorded from the desiccated seafloor during expeditions in the 1980s.

In 1994, Wucherer studied the northern, eastern and southern coasts of the Aral Sea within the framework of a DFG project (University of Kassel). According to collected herbarium specimens, the Aralkum flora consisted of 201 species (Breckle et al. 1998). In parallel a UNESCO project revealed various desertification impacts caused by the desiccation process, mainly in the delta areas of the rivers (Geldyeva et al. 2001).

During expeditions within the first BMBF project "Succession processes on the desiccated sea floor of the Aral Sea and perspectives of land-use" (1998–2001), the list of plants was increased by 65 species (Wucherer et al. 2001). According to this publication, the Aralkum flora consisted of 266 species belonging to 34 plant families and 134 genera.

The second, BMBF–GTZ-CCD project "Combating desertification and rehabilitation of salt deserts in the Aralkum" (2002–2005) gave possibilities for a further detailed study of the northwestern and eastern coasts and the Barsa-Kelmes territory. Some additional species were found on the dry seafloor (Dimeyeva 2004; Dimeyeva and Alimbetova 2006; Dimeyeva et al. 2008). Recent information from the southern part of the Aralkum has increased the list of plants by 21 species (Sherimbetov 2008a, b, c). The number of species of vascular plants recorded from the Aralkum is now about 368.

The species concept for the Chenopodiaceae has been taken from *Flora Iranica* (Hedge et al. 1997), except for the genera *Salsola* and *Climacoptera*. Species names and plant family characteristics for other families are taken from Cherepanov (1995), except for *Calligonum* for which *Flora of Kazakhstan* was used.

8.3 Composition of the Flora

The vascular plants of the Aralkum consist of 368 species belonging to 43 families and 178 genera. The leading 13 families (Table 8.1) comprise 310 species (84.2% of the flora) belonging to 136 genera (76.4%). Species from Chenopodiaceae, Asteraceae, Polygonaceae, Brassicaceae and Poaceae families prevail. The most important genera are (Table 8.2) *Calligonum* (35 species), *Artemisia* (14 species), *Salsola* (13 species), *Atriplex* (12 species), *Astragalus* (11 species), *Tamarix* (10 species), *Suaeda* (9 species), *Climacoptera* (5 species) and *Corispermum* (5 species).

8.4 Life Form Spectrum

Life forms (Raunkiaer 1910) are represented by the following groups: therophytes, geophytes, hemicryptophytes, chamaephytes, microphanerophytes, nanophanerophytes and hydrophytes (Fig. 8.1).

Table 8.2 Species of vascular plants from the Aralkum (2008), indicating chorological type (geoelement) and life form

Species	Geoelement	Life form
Alliaceae J. Agardh		
Allium caspium (Pall.) Bieb.	Pontic–northern Turanian	GB
Allium sabulosum Stev. ex Bunge	Turanian	GB
Allium schubertii Zucc.	Irano-Turanian	GB
Amaryllidaceae J. St.-Hil.		
Ixiolirion tataricum (Pall.) Schult. & Schult. fil.	Turanian–Dzungarian	GB
Apiaceae Lindl.		
Ferula canescens (Ledeb.) Ledeb.	Pontic–northern Turanian	GR
Ferula caspica Bieb.	Pontic–northern Turanian	GR
Ferula lehmannii Boiss.	Turanian	GR
Ferula nuda Spreng.	Pontic–northern Turanian	GR
Prangos odontalgica (Pall.) Herrnst. & Heyn	Pontic–northern Turanian	GR
Asclepiadaceae R. Br.		
Cynanchum sibiricum Willd.	Eastern Mediterranean	H
Asparagaceae Juss.		
Asparagus breslerianus Schult. & Schult. fil.	Irano-Turanian	H
Asparagus inderiensis Blum ex Pasz.	Aralo-Caspian	H
Asparagus persicus Baker	Eastern Mediterranean	H
Asteraceae Dumort.		
Acroptilon repens (L.) DC.	Palaearctic	H
Amberboa turanica Iljin	Palaearctic	T
Anthemis candidissima Willd. ex Spreng.	Turanian	T
Artemisia aralensis Krasch.	Kazakhstan endemic	Ch
Artemisia austriaca Jacq.	Mediterranean	H
Artemisia diffusa Krasch.ex Poljak.	Turanian	Ch
Artemisia arenaria DC.	Mediterranean	Ch
Artemisia pauciflora Web.	Northern Turanian	Ch
Artemisia quinqueloba Trautv.	Kazakhstan endemic	Ch
Artemisia santolina Schrenk	Irano-Turanian	H
Artemisia schrenkiana Ledeb.	Eastern Mediterranean	H
Artemisia scoparia Waldst. & Kit.	Eastern Mediterranean	H
Artemisia scopiformis Ledeb.	Kazakhstan endemic	H
Artemisia semiarida (Krasch. Et Lavr.) Filat.	Kazakhstan endemic	Ch
Artemisia songarica Schrenk	Turanian–Dzungarian	Ch
Artemisia terrae-albae Krasch.	Northern Turanian–Dzungarian	Ch
Artemisia turanica Krasch.	Irano-Turanian	Ch
Chartolepis intermedia Boiss.	Eastern Mediterranean	H
Chondrilla ambigua Fisch. ex Kar. & Kir.	Turanian–Dzungarian	H
Chondrilla brevirostris Fisch. & C. A. Mey.	Northern Turanian	H
Cirsium arvense (L.) Scop.	Palaearctic	H
Cirsium ochrolepidium Juz.	Eastern Mediterranean	H
Cousinia affinis Schrenk	Turanian–Dzungarian	H
Epilasia hemilasia (Bunge) Clarke	Irano-Turanian	T
Heteracia szovitsii Fisch. & C. A. Mey.	Irano-Turanian	T
Hyalea pulchella (Ledeb.) C. Koch	Eastern Mediterranean	H

(continued)

Table 8.2 (continued)

Species	Geoelement	Life form
Inula caspica Blum ex Ledeb.	Mediterranean	H
Inula germanica L.	Pontic–northern Turanian	H
Karelinia caspia (Pall.) Less.	Eastern Mediterranean	H
Koelpinia linearis Pall.	Mediterranean	T
Koelpinia tenuissima Pavl. & Lipsch.	Irano-Turanian	T
Koelpinia turanica Vass.	Turanian	T
Lactuca serriola L.	Mediterranean	H
Lactuca tatarica (L.) C. A. Mey.	Mediterranean	H
Lactuca undulata Ledeb. Pojark	Mediterranean	T
Mausolea eriocarpa (Bunge)	Irano-Turanian	Ch
Saussurea salsa (Pall. ex Bieb.) Spreng.	Mediterranean	H
Scorzonera sericeolanata (Bunge) Krasch. & Lipsch.	Turanian	GB
Senecio noeanus Rupr.	Northern Turanian	T
Senecio subdentatus Ledeb.	Palaearctic	T
Sonchus oleraceus L.	Palaearctic	T
Taktajaniantha pusilla (Pall.) Nazarova	Mediterranean	GB
Tanacetum achilleifolium (Bieb.) Sch. Bip.	Pontic–northern Turanian	H
Taraxacum bessarabicum (Hornem.) Hand.-Mazz.	Mediterranean	H
Tragopogon marginifolius Pavl.	Turanian	GR
Tragopogon ruber S. G. Gmel.	Turanian–Dzungarian	GR
Tragopogon sabulosus Krasch. & S. Nikit.	Northern Turanian–Dzungarian	GR
Tripolium vulgare Nees	Palaearctic	T
Berberidaceae Juss.		
Leontice incerta Pall.	Turanian	GR
Boraginaceae Juss.		
Argusia sibirica (L.) Dandy	Eastern Mediterranean	H
Arnebia decumbens (Vent.) Coss. & Kral.	Palaearctic	T
Asperugo procumbens L.	Palaearctic	T
Heliotropium arguzioides Kar. & Kir.	Turanian	H
Heliotropium dasycarpum Ledeb.	Irano-Turanian	H
Heterocaryum rigidum A. DC.	Irano-Turanian	T
Heterocaryum szovitsianum (Fisch. & C. A. Mey.) A. DC.	Irano-Turanian	T
Lappula semiglabra (Ledeb.) Guerke	Eastern Mediterranean	T
Lappula spinocarpos (Forssk.) Aschers.	Irano-Turanian	T
Nonea caspica (Willd.) G. Don fil.	Palaearctic	T
Rochelia retorta (Pall.) Lipsky	Palaearctic	T
Rochelia leiocarpa Ledeb.	Mediterranean	T
Suchtelenia calycina (C. A. Mey.) A. DC.	Turanian	T
Brassicaceae Burnett		
Alyssum dasycarpum Steph.	Mediterranean	T
Alyssum turkestanicum Regel & Schmalh.	Mediterranean	T
Cardaria pubescens (C. A. Mey.) Jarm.	Eastern Mediterranean	T
Chorispora tenella (Pall.) DC.	Irano-Turanian	T
Descurainia sophia (L.) Webb ex Prantl	Holarctic	T

(continued)

Table 8.2 (continued)

Species	Geoelement	Life form
Diptychocarpus strictus (Fisch. ex Bieb.) Trautv.	Irano-Turanian	T
Draba nemorosa L.	Northern Turanian–western Siberian	T
Erysimum sisymbrioides C. A. Mey.	Palaearctic	T
Euclidium syriacum (L.) R. Br.	Mediterranean	T
Goldbachia laevigata (Bieb.) DC.	Mediterranean	T
Isatis minima Bunge	Irano-Turanian	T
Isatis violascens Bunge	Turanian	T
Lachnoloma lehmannii Bunge	Turanian	T
Lepidium latifolium L.	Palaearctic	H
Lepidium obtusum Basin.	Turanian	H
Lepidium perfoliatum L.	Palaearctic	T
Lepidium ruderale L.	Holarctic	T
Leptaleum filifolium (Willd.) DC.	Eastern Mediterranean	T
Litwinowia tenuissima (Pall.) Woronow ex Pavl.	Eastern Mediterranean	T
Matthiola stoddartii Bunge	Turanian	T
Megacarpaea megalocarpa (Fisch. ex DC.) B. Fedtsch.	Eastern Mediterranean	GR
Meniocus linifolius (Steph.) DC.	Mediterranean	T
Octoceras lehmannianum Bunge	Irano-Turanian	T
Pachypterygium multicaule (Kar. & Kir.) Bunge	Turanian	T
Sameraria armena (L.) Desv.	Northern Turanian	T
Streptoloma desertorum Bunge	Turanian	T
Strigosella africana (L.) Botsch.	Mediterranean	T
Strigosella brevipes (Bunge) Botsch.	Irano-Turanian	T
Strigosella circinata (Bunge) Botsch.	Mediterranean	T
Strigosella scorpioides (Bunge) Botsch.	Irano-Turanian	T
Syrenia montana (Pall.) Klok.	Palaearctic	T
Tauscheria lasiocarpa Fisch. ex DC.	Eastern Mediterranean	T
Tetracme quadricornis (Steph.) Bunge	Mediterranean	T
Tetracme recurvata Bunge	Eastern Mediterranean	T
Caryophyllaceae Juss.		
Acanthophyllum borsczowii Litv.	Turanian	Ch
Acanthophyllum pungens (Bunge) Boiss	Eastern Mediterranean	H
Gypsophila paniculata L.	Palaearctic	H
Gypsophila perfoliata L.	Palaearctic	H
Silene nana Kar. & Kir.	Irano-Turanian	T
Silene odoratissima Bunge	Turanian	H
Ceratophyllaceae S.F.Gray		
Ceratophyllum demersum L.	Palaearctic	Hy
Chenopodiaceae Vent. (85)		
Agriophyllum minus Fisch. & C. A. Mey.	Irano-Turanian	T
Agriophyllum squarrosum (L.) Moq.	Eastern Mediterranean	T
Anabasis aphylla L.	Eastern Mediterranean	Ch
Anabasis salsa (C. A. Mey.) Benth. ex Volkens	Turanian–Dzungarian	Ch
Anabasis truncata (Schrenk) Bunge	Turanian–Dzungarian	H

(continued)

Table 8.2 (continued)

Species	Geoelement	Life form
Arthrophytum lehmannianum Bunge	Turanian	T
Atriplex aucheri Moq.	Irano-Turanian	T
Atriplex cana C. A. Mey.	Northern Turanian–Dzungarian	T
Atriplex dimorphostegia Kar. & Kir.	Mediterranean	T
Atriplex littoralis L.	Palaearctic	T
Atriplex micrantha C. A. Mey.	Palaearctic	T
Atriplex moneta Bunge	Irano-Turanian	T
Atriplex patula L.	Holarctic	T
Atriplex pratovii Suchor.	Aral endemic	T
Atriplex pungens Trautv.	Kazakhstan endemic	T
Atriplex sagittata Borkh.	Palaearctic	T
Atriplex sphaeromorpha Iljin	Pontic–northern Turanian	T
Atriplex tatarica L.	Mediterranean	T
Bassia hyssopifolia (Pall.) O. Kuntze	Eastern Mediterranean	T
Bassia sedoides (Pall.) Aschers.	Eastern Mediterranean	T
Bienertia cycloptera Bunge	Irano-Turanian	T
Chenopodium glaucum L.	Holarctic	T
Chenopodium rubrum L.	Holarctic	T
Ceratocarpus arenarius L.	Eastern Mediterranean	T
Climacoptera affinis (C. A. Mey.) Botsch.	Northern Turanian–Dzungarian	T
Climacoptera aralensis (Iljin) Botsch.	Northern Turanian	T
Climacoptera brachiata (Pall.) Botsch.	Eastern Mediterranean	T
Climacoptera ferganica (Drob.) Botsch.	Irano-Turanian	T
Climacoptera lanata (Pall.) Botsch.	Eastern Mediterranean	T
Corispermum aralo-caspicum Iljin	Aralo-Caspian	T
Corispermum hyssopifolium L.	Pontic–northern Turanian	T
Corispermum laxiflorum Schrenk	Aral endemic	T
Corispermum lehmannianum Bunge	Mediterranean	T
Corispermum orientale Lam.	Mediterranean	T
Gamanthus gamocarpus (Moq.) Bunge	Irano-Turanian	T
Girgensohnia oppositiflora (Pall.) Fenzl	Irano-Turanian	T
Halimione pedunculata (L.) Aell.	Palaearctic	T
Halimione verrucifera (Bieb.) Aell.	Palaearctic	Ch
Halimocnemis karelinii Moq.	Turanian	T
Halimocnemis longifolia Bunge	Turanian	T
Halimocnemis sclerosperma (Pall.) C. A. Mey.	Turanian	T
Halimocnemis villosa Kar. & Kir.	Turanian	T
Halocnemum strobilaceum (Pall.) Bieb.	Mediterranean	Ch
Halogeton glomeratus C. A. Mey.	Eastern Mediterranean	T
Halostachys belangeriana (Moq.) Botsch.	Mediterranean	Ph-n
Halothamnus subaphyllus (C. A. Mey.) Botsch.	Irano-Turanian	Ch
Haloxylon aphyllum (Minkw.) Iljin	Irano-Turanian	Ph-m
Haloxylon persicum Bunge ex Boiss. & Buhse	Irano-Turanian	Ph-m
Horaninovia anomala (C. A. Mey.) Moq.	Irano-Turanian	T
Horaninovia excellens Iljin	Southern Turanian	T
Horaninovia minor Schrenk	Eastern Mediterranean	T

(continued)

Table 8.2 (continued)

Species	Geoelement	Life form
Horaninovia ulicina Fisch. & C. A. Mey.	Irano-Turanian	T
Kalidium caspicum (L.) Ung.- Sternb.	Turanian–Dzungarian	Ch
Kalidium foliatum (Pall.) Moq.	Eastern Mediterranean	Ch
Kirilowia eriantha Bunge	Turanian	T
Kochia iranica Bornm.	Eastern Mediterranean	T
Kochia odontoptera Schrenk	Eastern Mediterranean	T
Kochia prostrata (L.) Schrad.	Mediterranean	Ch
Krascheninnikovia ceratoides (L.) Gueldenst.	Mediterranean	Ch
Londesia eriantha Fisch. & C. A. Mey.	Irano-Turanian	T
Nanophytum erinaceum (Pall.) Bunge	Turanian–Dzungarian	Ch
Ofaiston monandrum (Pall.) Moq.	Irano-Turanian	T
Petrosimonia brachiata (Pall.) Bunge	Mediterranean	T
Petrosimonia glaucescens (Bunge) Iljin	Northern Turanian	T
Petrosimonia hirsutissima (Bunge) Iljin	Kazakhstan endemic	T
Petrosimonia squarrosa (Schrenk) Bunge	Northern Turanian	T
Petrosimonia triandra (Pall.) Simonk.	Pontic–northern Turanian	T
Salicornia europaea L. s. l.	Cosmopolitan	T
Salsola arbuscula Pall.	Mediterranean	Ph-n
Salsola australis (R.) Br.	Palaearctic	T
Salsola chiwensis M. Pop.	Turanian	Ch
Salsola dendroides Pall.	Irano-Turanian	Ch
Salsola foliosa (L.) Schrad.	Eastern Mediterranean	T
Salsola implicata Botsch.	Southern Turanian	T
Salsola micranthera Botsch.	Southern Turanian	T
Salsola nitraria Pall.	Mediterranean	T
Salsola orientalis S. G. Gmel.	Mediterranean	Ch
Salsola paletzkiana Litv.	Southern Turanian	Ph-n
Salsola paulsenii Litv.	Eastern Mediterranean	T
Salsola richteri (Moq.) Kar. ex Litv.	Irano-Turanian	Ph-n
Salsola tamariscina Pall.	Eastern Mediterranean	T
Suaeda acuminata (C. A. Mey.) Moq.	Eastern Mediterranean	T
Suaeda altissima (L.) Pall.	Mediterranean	T
Suaeda arcuata Bunge	Irano-Turanian	T
Suaeda crassifolia Pall.	Aralo-Caspian	T
Suaeda heterophylla (Kar. et Kir.) Bunge	Eastern Mediterranean	T
Suaeda microphylla Pall.	Irano-Turanian	Ch
Suaeda microsperma (C. A. Mey.) Fenzl.	Turanian	T
Suaeda physophora Pall.	Irano-Turanian	Ch
Suaeda salsa (L.) Pall.	Aralo-Caspian	T
Convolvulaceae Juss.		
Convolvulus arvensis L.	Cosmopolitan	H
Convolvulus erinaceus Ledeb.	Irano-Turanian	Ch
Convolvulus subsericeus Schrenk	Turanian	Ch
Cyperaceae Juss.		
Bolboschoenus maritimus (L.) Palla	Cosmopolitan	GB
Carex pachystylis J. Gay	Irano-Turanian	H

(continued)

Table 8.2 (continued)

Species	Geoelement	Life form
Carex physodes Bieb.	Irano-Turanian	H
Scirpus lacustris L.	Palaearctic	Hy
Scirpus tabernaemontani C. C. Gmel.	Palaearctic	Hy
Elaeagnaceae Juss.		
Elaeagnus oxycarpa Schlecht.	Northern Turanian	Ph-m
Ephedraceae Dumort.		
Ephedra distachya L.	Mediterranean	Ch
Ephedra intermedia Schrenk & C. A. Mey.	Mediterranean	Ch
Ephedra strobilacea Bunge	Irano-Turanian	Ch
Equisetaceae Rich. ex DC.		
Equisetum ramosissimum Desf.	Holarctic	H
Euphorbiaceae Juss.		
Euphorbia inderiensis Less. Kar. et Kir.	Turanian	H
Euphorbia seguierana Neck.	Palaearctic	H
Euphorbia turczaninowii Kar. & Kir.	Eastern Mediterranean	H
Euphorbia undulata Bieb.	Pontic–northern Turanian	H
Fabaceae Lindl.		
Alhagi pseudalhagi (Bieb.) Fisch.	Northern Turanian	H
Ammodendron bifolium (Pall.) Yakovl.	Eastern Mediterranean	Ph-n
Ammodendron conollyi Bunge	Turanian	Ph-m
Ammodendron karelinii Fisch. et Mey.	Turanian	Ph-n
Astragalus amarus Pall.	Aralo-Caspian	H
Astragalus ammodendron Bunge	Turanian	Ph-n
Astragalus brachypus Schrenk	Kazakhstan endemic	Ph-n
Astragalus campylorrhynchus Fisch. & C. A. Mey.	Irano-Turanian	T
Astragalus lehmannianus Bunge	Turanian	H
Astragalus longipetalus Chater	Northern Turanian	H
Astragalus ninae Pavl.	Kazakhstan endemic	H
Astragalus oxyglottis Stev. ex Bieb.	Mediterranean	T
Astragalus testiculatus	Eastern Mediterranean	H
Astragalus villosissimus Bunge	Southern Turanian	Ph-n
Astragalus vulpinus Willd.	Northern Turanian–western Siberian	H
Eremosparton aphyllum (Pall.) Fisch. et Mey.	Northern Turanian	Ph-n
Glycyrrhiza aspera Pall.	Palaearctic	H
Glycyrrhiza glabra L.	Mediterranean	H
Halimodendron halodendron (Pall.) Voss.	Mediterranean	Ph-n
Pseudosophora alopecuroides (L.) Sweet	Eastern Mediterranean	H
Sphaerophysa salsola (Pall.) DC.	Eastern Mediterranean	H
Trigonella arcuata C. A. Mey.	Mediterranean	T
Trigonella orthoceras Kar. et Kir.	Mediterranean	T
Frankeniaceae S. F. Gray		
Frankenia hirsuta L.	Mediterranean	H
Fumariaceae DC.		
Fumaria vaillantii Loisel.	Mediterranean	T

(continued)

Table 8.2 (continued)

Species	Geoelement	Life form
Geraniaceae Juss.		
Erodium oxyrhynchum Bieb.	Mediterranean	T
Geranium transversale (Kar. & Kir.) Vved.	Eastern Mediterranean	GR
Hypecoaceae Nakai		
Hypecoum parviflorum Kar. et Kir.	Eastern Mediterranean	T
Iridaceae Juss.		
Iris longiscapa Ledeb.	Turanian	GR
Iris songarica Schrenk	Eastern Mediterranean	GR
Iris tenuifolia Pall.	Eastern Mediterranean	GR
Juncaceae Juss.		
Juncus gerardii Loisel.	Palaearctic	H
Lamiaceae Lindl.		
Chamaesphacos ilicifolius Schrenk	Irano-Turanian	T
Eremostachys tuberosa (Pall.) Bunge	Northern Turanian–western Siberian	GR
Lallemantia royleana (Benth.) Benth.	Irano-Turanian	T
Liliaceae Juss.		
Gagea reticulata (Pall.) Schult. & Schult. fil.	Turanian	GB
Rhinopetalum karelinii Fisch. ex Alexand.	Turanian	GB
Tulipa buhseana Boiss.	Turanian	GB
Limoniaceae Lincz.		
Limonium caspium (Willd.) Gams.	Pontic–northern Turanian	H
Limonium gmelinii Willd. O. Kuntze	Mediterranean	H
Limonium otolepis (Schrenk) O. Kuntze	Eastern Mediterranean	H
Limonium suffruticosum (L.) O. Kuntze	Mediterranean	Ch
Malvaceae Juss.		
Malva neglecta Wallr.	Palaearctic	H/T
Nitrariaceae Bercht. & J. Presl.		
Nitraria schoberi L.	Mediterranean	Ph-n
Nitraria sibirica Pall.	Irano-Turanian	Ph-n
Orobanchaceae Vent.		
Cistanche salsa (G. A. Mey.) G. Beck	Northern Turanian	GP
Orobanche cernua Loefl.	Mediterranean	GP
Papaveraceae Juss.		
Roemeria hybrida (L.) DC.	Eastern Mediterranean	T
Roemeria refracta DC.	Irano-Turanian	T
Peganaceae (Engl.) Tiegh. ex Takht.		
Peganum harmala L.	Mediterranean	H
Plantaginaceae Juss.		
Plantago tenuiflora Waldst. & Kit.	Palaearctic	T
Poaceae Barnhart		
Aeluropus littoralis (Gouan) Parl.	Mediterranean	H
Agropyron desertorum (Fisch. ex Link) Schult.	Eastern Mediterranean	H
Agropyron fragile (Roth) P. Candargy	Eastern Mediterranean	H
Anisantha tectorum (L.) Nevski	Mediterranean	T
Calamagrostis dubia Bunge	Palaearctic	H
Catabrosella humilis (Bieb.) Tzvel.	Mediterranean	H
Crypsis schoenoides (L.) Lam.	Palaearctic	T

(continued)

Table 8.2 (continued)

Species	Geoelement	Life form
Eremopyrum bonaepartis (Spreng.) Nevski	Mediterranean	T
Eremopyrum distans (C.Koch) Nevski	Mediterranean	T
Eremopyrum orientale (L.) Jaub. et Spach.	Mediterranean	T
Eremopyrum triticeum (Gaertn.) Nevski	Mediterranean	T
Leymus racemosus (Lam.) Tzvel.	Palaearctic	H
Phragmites australis (Cav.) Trin. ex Steud.	Cosmopolitan	H
Poa bulbosa L.	Holarctic	H
Puccinellia distans (Jacq.) Parl.	Palaearctic	H
Puccinellia dolicholepis V. Krecz.	Aralo-Caspian	H
Puccinellia gigantea (Grossh.) Grossh.	Turanian	H
Schismus arabicus Nees	Mediterranean	T
Stipa caspia C. Koch	Eastern Mediterranean	H
Stipa sareptana Beck.	Pontic–northern Turanian	H
Stipagrostis karelinii (Trin.& Rupr.)Tzvl.	Turanian	H
S. pennata (Trin.) de Winter	Irano-Turanian	H
Polygonaceae Juss.		
Atraphaxis replicata Lam.	Mediterranean	Ph-n
Atraphaxis spinosa L.	Mediterranean	Ph-n
Calligonum acanthopterum Borszcz.	Turanian	Ph-n
Calligonum alatiforme Pavl.	Turanian	Ph-n
Calligonum alatum Litv.	Northern Turanian	Ph-n
Calligonum androssovii Litv.	Kazakhstan endemic	Ph-n
Calligonum aphyllum (Pall.) Guerke	Northern Turanian	Ph-n
Calligonum aralense Borszcz.	Turanian–Dzungarian	Ph-n
Calligonum borszczowii Litv.	Turanian	Ph-n
Calligonum cancellatum Mattei	Turanian	Ph-n
Calligonum caput-medusae Schrenk	Turanian–Dzungarian	Ph-n
Calligonum colubrinum Borszcz.	Kazakhstan endemic	Ph-n
Calligonum commune (Litv.) Mattei	Northern Turanian	Ph-n
Calligonum crispatum (Litv.) Mattei	Kazakhstan endemic	Ph-n
Calligonum cristatum Pavl.	Turanian	Ph-n
Calligonum densum Borszcz.	Turanian	Ph-n
Calligonum dubjanskyi Litv.	Turanian	Ph-n
Calligonum elatum Litv.	Turanian	Ph-n
Calligonum erinaceum Borszcz.	Kazakhstan endemic	Ph-n
Calligonum eriopodum Bunge	Turanian	Ph-n
Calligonum humile Litv.	Kazakhstan endemic	Ph-n
Calligonum junceum (Fisch.& C.A.May.) Litv.	Eastern Mediterranean	Ph-n
Calligonum lamellatum (Litv.) Mattei	Kazakhstan endemic	Ph-n
Calligonum leucocladum (Schrenk) Bunge	Northern Turanian	Ph-n
Calligonum macrocarpum Borszcz.	Turanian	Ph-n
Calligonum membranaceum (Borszcz.) Litv.	Turanian	Ph-n
Calligonum microcarpum Borszcz.	Turanian	Ph-n
Calligonum minimum Lipsky	Turanian	Ph-n
Calligonum muravljanskyi Pavl.	Turanian	Ph-n
Calligonum palibinii Mattei	Kazakhstan endemic	Ph-n
Calligonum platyacanthum Borszcz.	Turanian	Ph-n

(continued)

Table 8.2 (continued)

Species	Geoelement	Life form
Calligonum pseudohumile Drob.	Kazakhstan endemic	Ph-n
Calligonum rotula Borszcz.	Turanian	Ph-n
Calligonum rubens Mattei	Turanian	Ph-n
Calligonum spinulosum Drob.	Kazakhstan endemic	Ph-n
Calligonum squarrosum Pavl.	Turanian	Ph-n
Calligonum undulatum Litv.	Turanian	Ph-n
Polygonum arenarium Waldst. Ed Scit.	Palaearctic	T
Polygonum monspeliense Thieb. ex Pers.	Palaearctic	T
Rheum tataricum L.	Turanian	GR
Rumex marschallianus Reichenb.	Mediterranean	T
Ranunculaceae Juss.		
Adonis parviflora Fisch. ex DC.	Turanian	T
Ceratocephala falcata (L.) Pers.	Mediterranean	T
Ceratocephala testiculata (Grantz.) Bess.	Mediterranean	T
Clematis orientalis L.	Eastern Mediterranean	Ph-n
Consolida rugulosa (Boiss.) Schröding.	Mediterranean	T
Myosurus minimus L.	Cosmopolitan	T
Ranunculus platyspermus Fisch. ex DC.	Northern Turanian	GR
Thalictrum isopyroides C. A. Mey.	Eastern Mediterranean	GR
Rosaceae Juss.		
Hulthemia persica (Michx. ex Juss.) Bornm.	Irano-Turanian	Ch
Rubiaceae Juss.		
Asperula danilewskiana Basin.	Northern Turanian–western Siberean	Ch
Galium spurium L.	Palaearctic	T
Rutaceae Juss.		
Haplophyllum perforatum Kar. et Kir.	Turanian	H
Salicaceae Mirb.		
Populus euphratica Olivier/P. diversifolia	Turanian	Ph-m
Scrophulariaceae Juss.		
Linaria dolichoceras Kuprian.	Aralo-Caspian	H
Veronica campylopoda Boiss.	Mediterranean	T
Solanaceae Juss.		
Hyoscyamus pusillus L.	Palaearctic	T
Lycium ruthenicum Murr.	Mediterranean	Ph-n
Tamaricaceae Link.		
Tamarix aralensis Bunge	Mediterranean	Ph-n
Tamarix elongata Ledeb.	Eastern Mediterranean	Ph-n
Tamarix florida Bunge	Irano-Turanian	Ph-n
Tamarix hispida Willd	Mediterranean	Ph-n
Tamarix hohenackeri Bge	Mediterranean	Ph-n
Tamarix karelinii Bunge	Mediterranean	Ph-n
Tamarix laxa Willd.	Mediterranean	Ph-n
Tamarix leptostachys Bunge	Eastern Mediterranean	Ph-n
Tamarix litwinowii Gorschk.	Mediterranean	Ph-n
Tamarix ramosissima Ledeb.	Mediterranean	Ph-n
Typhaceae Juss.		
Typha angustifolia L.	Holarctic	Hy

(continued)

Table 8.2 (continued)

Species	Geoelement	Life form
Zygophyllaceae R. Br.		
Zygophyllum eichwaldii C. A. Mey.	Southern Turanian	Ch
Zygophyllum fabago L.	Mediterranean	H
Zygophyllum macropterum Boriss.	Irano-Turanian	H
Zygophyllum oxianum Boriss.	Turanian	H

Ch chamaephytes, *G* geophytes, *B* with bulbs, *P* parasitic, *R* with rhizomes, *H* hemicryptophytes, *Hy* hydrophytes, *Ph-m* microphanerophytes, *Ph-n* nanophanerophytes, *T* therophytes
New additions from august 2011 by L.Dimeyeva: *Hydrocharis morsus-ranae* L. (Hydrocharitaceae) and *Schoenoplectus kasachstanicus* (Dobrochot.) T.V.Egorova (Cyperaceae) from Syr Darya river mouth

Fig. 8.1 Life forms of the Aralkum flora (Abbreviations see above)

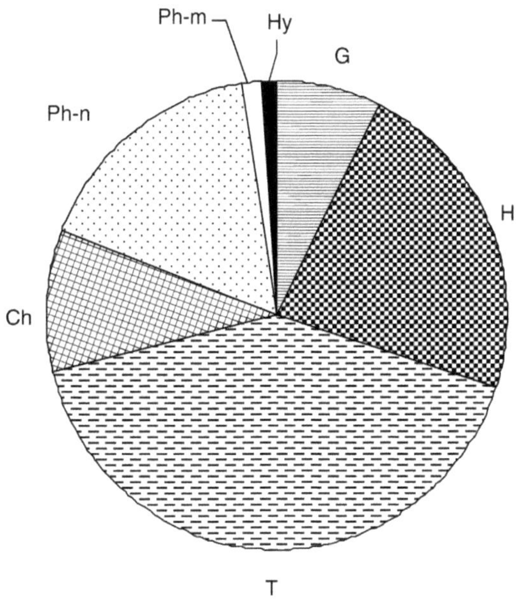

The most numerous group is therophytes (152 species; 41.3%). They are mainly represented by ephemerals and annual species from the Chenopodiaceae family. Annual chenopods can play a dominant role in plant communities (*Salicornia europaea*, *Suaeda acuminata*, *Suaeda crassifolia*, *Climacoptera aralensis*, *Climacoptera lanata*, *Petrosimonia triandra*, etc.), especially as pioneers in the first successional stages of primary succession (Chap. 10).

The group of geophytes (27 species; 7.3%) consists of ephemeroid species: geophytes with bulbs (species of genera *Tulipa*, *Allium* and *Gagea*); geophytes with rhizomes/roots or tubers (*Iris* spp., *Geranium transversale*, *Rheum tataricum*, *Megacarpaea megalocarpa*, etc.) and parasitic geophytes (*Cystanche salsa*).

Hemicryptophytes (79 species; 21.5%) are perennial herbs. Some of them can dominate in plant communities (*Agropyron fragile*, *Leymus racemosus*, *Phragmites australis*, *Stipagrostis pennata*, *Aeluropus littoralis*, *Carex physodes*, *Alhagi pseudalhagi*, *Zygophyllum oxianum*, *Limonium caspium*, *Karelinia caspia*, etc.).

Chamaephytes (35 species; 9.5%) are the most typical life form for deserts of Central Asia. They consist of dwarf subshrubs and subshrubs with annual dying-off of the generative organs. The main dominants are dwarf subshrubs of the genus *Artemisia*, species of *Kalidium*, *Suaeda microphylla*, *Suaeda physophora, Kochia prostrata*, etc.

Phanerophytes (67 species; 18.2%) are represented by small desert trees (microphanerophytes) – *Haloxylon aphyllum*, *Haloxylon persicum* and *Ammodendron conollyi* – and nanophanerophytes (species of genera *Calligonum* and *Tamarix, Lycium ruthenicum*, *Ammodendron bifolium*, etc.).

The group of hydrophytes is the smallest (*Typha angustifolia*, *Scirpus lacustris*, *Scirpus tabernaemontani*).

8.5 Geographical Analysis of the Flora

Analysis of the geographical distribution of species is based on published data (Flora of USSR 1934–1963; Flora of Kazakhstan 1956–1966; Lavrenko 1962; Kurochkina 1978; Erezhepov 1978; Rachkovskaya 1986). Schemes of botanical-geographical regionalization of the Ancient Mediterranean subdominion (Lavrenko 1962) the Eurasian steppe region (Lavrenko 1970) and desert region of Kazakhstan and Middle Asia (Rachkovskaya et al. 2003) have been used for analysis. The types of areas are not equivalent in size. Species in one area are not characterized by absolute coincidence of geographical distribution. The ratio of geoelements in the Aralkum flora is represented in Fig. 8.2.

Cosmopolitan species (five species; 1.4%) occupy humid and arid areas of the northern and southern hemispheres (*Phragmites australis*, *Bolboschoenus maritimus*, *Salicornia europaea*, *Convolvulus arvensis*).

Holarctic species (eight species; 2.2%) are distributed in the Holarctic kingdom of flora (*Typha angustifolia*, *Poa bulbosa*, *Chenopodium glaucum*, *Atriplex patula*, *Descurainia sophia*, *Equisetum ramosissimum*).

Palaearctic species (37 species; 10.1%) are spreading in temperate and subtropical regions of Europe and Asia (*Calamagrostis dubia*, *Puccinellia distans*, *Atriplex sagittata*, *Atriplex micrantha*, *Sonchus oleraceus*, *Tripolium vulgare*).

Mediterranean species (71 species; 19.3%) are distributed in the southern part of the Palaearctic (Ancient Mediterranean subdominion; Popov 1927; Lavrenko 1962). The area occupies wide desert and steppe territories of Eurasia and partly North Africa. The group includes the following species: *Aeluropus littoralis*, *Atraphaxis replicata*, *Atriplex tatarica*, *Peganum harmala*, *Zygophyllum fabago*, *Nitraria schoberi*, *Tamarix laxa*, *Tamarix litwinowii*, *Tamarix ramosissima*, and *Kochia prostrata*.

Eastern Mediterranean species (51 species; 13.9%) are distributed in steppes of Kazakhstan and Mongolia, the Turanian and Gobi deserts and mountains of Central Asia (Rachkovskaya 1986). The following species belong to this group: *Agropyron*

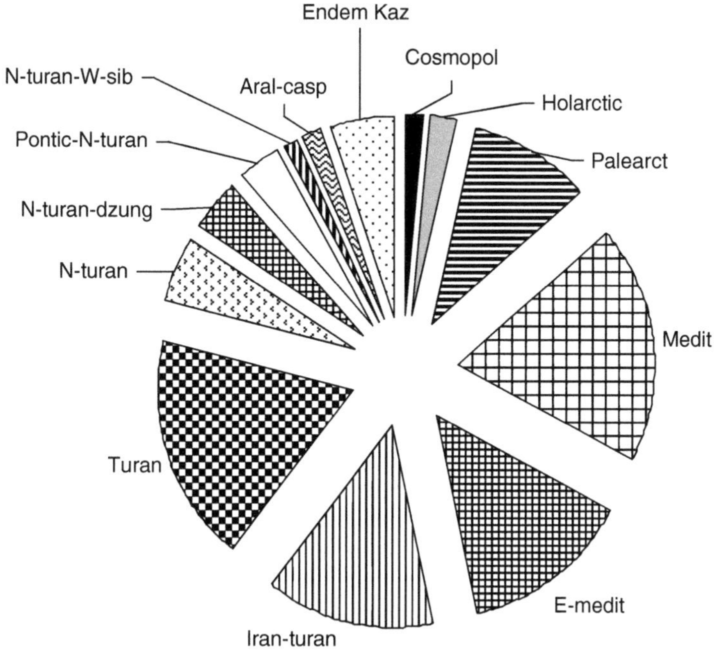

Fig. 8.2 Chorological structure of the Aralkum flora

desertorum, *A. fragile*, *Stipa caspia*, *Anabasis aphylla*, *Artemisia scoparia*, *Artemisia schrenkiana*, *Tamarix elongata*.

Irano-Turanian species (51 species; 13.9%). They are distributed in the limits of Irano-Turanian subregion of the Sahara–Gobi desert region of the Ancient Mediterranean subdominion (Lavrenko 1962; Rachkovskaya et al. 2003). The group includes the following species: *Stipagrostis pennata*, *Carex physodes*, *Carex pachystylis*, *Allium schubertii*, *Atriplex aucheri*, *Suaeda microphylla*, *Suaeda arcuata*, *Suaeda physophora*, *Haloxylon aphyllum*, *Haloxylon persicum* and *Mausolea eriocarpa*.

Turanian species (67 species; 18.2%) occupy the southern and northern deserts of the Turanian lowland (*Calligonum rotula*, *Calligonum rubens*, *Calligonum macrocarpum*, *Arthrophytum lehmannianum*, *Halimocnemis karelinii*, *Halimocnemis longifolia*, *Ammodendron conollyi*, *Ammodendron karelinii*, *Zygophyllum oxianum*).

Northern Turanian species (20 species; 5.4%) are spreading in the limits of the northern Turanian desert province (the greater part of Kazakhstan deserts and the Caspian lowland including Russian territory). The following species belong to this group: *Climacoptera aralensis*, *Elaeagnus oxycarpa*, *Artemisia pauciflora*, *Chondrilla brevirostris*, *Calligonum leucocladum*, *Calligonum commune*, *Eremosparton aphyllum* and *Alhagi pseudalhagi*.

Northern Turanian–Dzungarian species (15 species; 4.1%) are distributed mostly in the northern Turanian region with extensions to Dzungar province

(*Nanophytum erinaceum*, *Artemisia songarica*, *Artemisia terrae-albae*, *Tragopogon ruber*, *Tragopogon sabulosus*).

Pontic–northern Turanian species (13 species; 3.5%) occupy a wide territory from the Black Sea to western Siberia [Black Sea (Pontic)–Kazakhstan steppe subregion according to Lavrenko 1970] and the northern Turanian desert province (*Stipa sareptana*, *Limonium caspium*, *Atriplex sphaeromorpha*, *Petrosimonia triandra*, *Ferula caspica*, *Ferula canescens*).

Northern Turanian–western Siberian species (four species; 1.1%) are distributed in the limits of the northern Turanian desert province and the western Siberian block of provinces of the Black Sea—Kazakhstan steppe subregion (*Astragalus vulpinus*, *Eremostachys tuberosa*, *Asperula danilewskiana*).

Aralo-Caspian species (seven species; 1.9%) consist of *Corispermum aralo-caspicum*, *S. crassifolia*, *Linaria dolichoceras*, *Asparagus inderiensis*, *Astragalus amarus*, and *Puccinellia dolicholepis*.

Endemics of Kazakhstan (19 species; 5.2%) are distributed only in Kazakhstan: *Atriplex pungens*, *Petrosimonia hirsutissima*, *Astragalus brachypus*, *Astragalus ninae*, *Artemisia semiarida*, *Artemisia quinqueloba*, *Artemisia aralensis*, *Artemisia scopiformis*, *Calligonum crispatum*, *Calligonum lamellatum*, *Calligonum palibinii*, *Calligonum pseudohumile*, *Calligonum humile*, *Calligonum androssovii*, *Calligonum columbrinum*, *Calligonum erinaceum* and *Calligonum spinulosum*.

The Aral region endemics are *Corispermum laxiflorum* and *Atriplex pratovii*.

The analysis of Aralkum flora shows that peculiarities of taxonomy, of life form, and of geographical structure are determined by the relevant locations within the limits of the temperate continental deserts of the Turanian lowlands. The life form spectrum demonstrates the mechanism of adaptation of species to arid desert conditions.

According to geographical analysis, most of the present species are connected with the Ancient Mediterranean territory (the eastern part): 122 species (33.2%) belong to Mediterranean and eastern Mediterranean areas. Turanian and Irano-Turanian geoelements constitute 118 species (32.1%). The northern Turanian (Kazakhstan) geographical distribution group (including endemics of Kazakhstan) consists of 39 species (10.6%). Autochthonic Aral and Aralo-Caspian species represent only a small part of the flora (nine species, 2.5%). Thus, the Aralkum flora is typical of Turanian deserts, reflecting regional biodiversity. It is an immigration flora with a low proportion of original features. However, it consists of many interesting species, typical of northern Turanian and southern Turanian desert provinces.

8.6 Conclusions

The formation of the Aralkum is not a new phenomenon in the history of the Aral Sea. Several times the sea has been desiccated and flooded back to its own coastlines following natural and anthropogenic processes. The formation of the

Aralkum flora in ancient and modern times occurs simultaneously with the development of landscapes and soils. The shoreline is a source of plant species diversity where species of wide geographical distribution (Mediterranean, Palaearctic) as well as typical Kazakhstan or Central Asian desert plants are growing in various habitats. The territory adjacent to the Aral Sea is geologically young Rubanov et al. (1987). Nevertheless, some Aral endemics are present in the area (*Atriplex pratovii*, *Corispermum laxiflorum*). The flora around the Aral Sea has not been completely investigated. Detailed investigations and retracing of expedition routes from the 1980s are necessary, especially for the southeastern and southwestern coasts. The velocity of plant colonization of the new land surface is high. Only 154 species were recorded in the 1980s; 47 species were added to the list of species in 1994, 65 in 2001 and 102 in 2008. Joint efforts of botanists from Kazakhstan, Uzbekistan and Germany in the framework of international projects could help to complete the study of the Aralkum flora, and to provide more insight into the working processes of migrations of species.

Acknowledgements This research was funded by the German Federal Ministry for Education and Research (BMBF grant 0339714 and BMBF grant 0330389).

References

Borshchov IG (1865) Materials for botanical geography of Aral-Caspian region. St Petersburg

Breckle S-W, Agachanjanz O, Wucherer W (1998) Der Aralsee: Geoökologische Probleme. Naturwiss Rundsch 9:347–355

Breckle S-W, Wucherer W, Agachanjanz O, Geldyev B (2001) The Aral sea crisis region. In: Breckle S-W, Veste M, Wucherer W (eds) Sustainable land use in deserts. Springer, Berlin, pp 27–37

Cherepanov SK (1995) Vascular plants of Russia and adjacent states (the former USSR). Cambridge University Press, 516 p

Dimeyeva LA (1990) Flora and vegetation of the Aral Sea coast (in the limits of Kazakhstan). PhD dissertation, Instute of Botany, Academy of Science, Almaty

Dimeyeva LA (2004) Flora and vegetation of the dry sea floor of the north-western bays of the Aral Sea. In: Proceedings of international conference "development of botany in central Asia and its integration into industry", Tashkent, pp 22–23

Dimeyeva LA, Alimbetova Z (2006) Flora of nature reserve Barsakelmes. In: Proceedings of the third international botanical conference, Almaty, pp 46–48

Dimeyeva LA, Breckle S-W, Wucherer W (2008) Flora of the Aralkum desert (in the limits of Kazakhstan). News Nat Acad Sci Kazakhstan 6:25–31

Dimeyeva LA, Kurochkina LYa (2005) Botanical-geographic zoning of the Aral Sea coast. Scientific reading dedicated to 100-years anniversary of VBSochava. Irkutsk, pp 125–130

Dimeyeva LA, Kuznetsov LA (1999) Flora of the Aral seaside belt. Bot J St Petersburg 4 (84):39–52

Erezhepov SE (1978) Flora of Karakalpakia. Fan, Tashkent

Flora of Kazakhstan (1956–1966) Vol 1–9 Alma-Ata

Flora of USSR (1934–1963) Vol 1–31 Academy of Science URSS, Moskov, Leningrad

Geldyeva GV, Ogar NP, Scorintseva IB, Geldyev BV, Budnikova TI, Dimeeva LA (2001): Monitoring and modeling of desertification processes in the Syr Dar'ya and Amu Dar'ya

River deltas, for GIS. Ecological research and monitoring of the Aral sea deltas. A basis for restoration. UNESCO Aral Sea Project. 1997–2000 Final Scientific reports. Book 2, pp 119–153
Hedge I, Akhani H, Freitag H, Kothe-Heinrich G, Podlech D, Rilke S, Uotila P (1997) Chenopodiaceae. In: Rechinger KH (ed) Flora Iranica 172. Akademische Druck- und Verlagsanstalt, Graz
Kurochkina L (1978) Psammophytic desert vegetation of Kazakhstan. Nauka, Alma-Ata
Kurochkina LYa, Makulbekova GB, Wucherer W, Malaisarova AN (1983) Vegetation of the dry Seafloor of the Aral Sea. State of waters and the dry seafloor of the Aral Sea. Kazak Academy of Science, Almaty, pp 91–127
Kuznetsov LA (1995) Flora of Barsakelmes Island. Biological and nature conservation problems of the Aral Sea and the coast. In: Proceedings of Russian institute of zoology. St Petersburg, pp 106–128
Lavrenko EM (1962) Main features of botanical geography of deserts in Eurasia and North Africa. Academy of Science, Moscow-Leningrad
Lavrenko EM (1970) Regionalization of Black Sea – Kazakhstan subregion of Eurasian steppe region. Bot J 55:5
Litwinov D (1905) Plants of the Aral Sea coast. Proc Turkest Russ Geogr Sci 5:1–41
Popov MG (1927) Salient features of the history of development of the flora of Middle Asia. Byull. Sredne-Asiatk. Gosud, Univ. Tashkent 15:239–292
Rachkovskaya EI (1986) Flora Gobi desert Moskow. Nauka, Moskow, pp 80–84
Rachkovskaya EI, Safronova IN (1994) The new map of botanical-geographic regionalization of Kazakhstan and Middle Asia in the limits of the desert area. Geobotanical Mapping 1992 (St Petersburg) 33–49
Rachkovskaya EI, Volkova EA, Khramtsov VN (eds) (2003) Botanical geography of Kazakhstan and Middle Asia (desert region). Komarov Botanical Institute of Russian Academy of sciences, St Petersburg
Raunkiaer C (1910) Statistik der lebensformen als Grundlage für die biologische Pflanzengeographie. Beih Bot Zentralbl 27(II):171–206
Rubanov IV, Ishniyazov DP, Baskakova MA, Chistyakov PA (1987) Geology of the Aral Sea. Fan, Tashkent
Sherbaev B (1982) Composition of flora in the southern coast of the Aral Sea. Bot J 67:1372–1377
Sherimbetov S (2008a) Primary analysis on floristic composition of the Southern Aralkum. Uzbek Biol J 2:37–41
Sherimbetov S (2008b) Main dominant plant species of flora of the South-eastern part of the Aralkum. Proc Acad Sci Uzbekistan 3:86–88
Sherimbetov S (2008c) Some results on floristic investigations in the south-western part of the Aral Sea dry seabed. In: Fundamental and applied problems of Botany in XXI century. Petrozavodsk Vol 3, pp 192–193
Smirnov SM (1875) Botanical study in Aral-Caspian region. Proc Russ Geogr Sci 11(3):190–219
Wucherer W, Breckle S-W, Dimeyeva LA (2001) Flora of the dry seafloor of the Aral sea. In: Breckle S-W, Veste M, Wucherer W (eds) Sustainable land use in deserts. Springer, Berlin, pp 38–51

Chapter 9
Vegetation of the Aralkum

S-W. Breckle, W. Wucherer, and L.A. Dimeyeva

9.1 Introduction

The desiccated seafloor of the Aral Sea is called the Aralkum. No terrestrial plants nor their seed banks had been there. But during the last 50 years an intensive invasion by organisms took place and is still taking place. The various vegetation units are described in this chapter. The formation of plant communities, soils, a new groundwater level and aquifers, the processes of ecosystems, are very dynamic. These dynamic succession processes are described in Chap. 10. The dry seafloor is the largest area worldwide where a primary succession is taking place. Unintentionally, mankind has created a huge experiment there.

In 1977 a research programme (by the Geographical Institutes of Moscow) was started in Middle Asia to study the negative effects of the drying process of the Aral Sea and the change in the discharge of the Amu Darya and the Syr Darya. Transects were set up at different coastal areas, stretching from the former coast to the receding coastline. The distribution of the soils and the vegetation was surveyed along these transects. Two of the authors participated in this programme (Wucherer 1979, 1984, 1986, 1990; Dimeyeva 1990; Kabulov 1990; Kurochkina 1979). This programme was stopped in the mid-1980. Presently, the ecological situation on the dry seafloor is different, the diversity of landscapes and plant communities has increased enormously, barchans and mainly salt deserts have spread out on the dry seafloor and salt-dust storms have become a common event.

In 1990 a UNESCO programme started with special emphasis on the delta areas (Geldyeva et al. 2001). The problems of the Aralkum itself had not been addressed since the middle of the 1980. Therefore, a new research programme was started, the international BMBF projects "Succession processes on the desiccated sea floor of the Aral Sea and perspectives of land-use" (1998–2001) and "Combating desertification and rehabilitation of salt deserts in the Aralkum" (2002–2005). These projects aimed to study the ecosystem dynamics in the Aralkum and to test experimental plantings to accelerate the natural colonization by plants. Results on the dynamics of flora and vegetation are given here (see Chaps. 8–10, 15).

S.-W. Breckle et al. (eds.), *Aralkum - a Man-Made Desert: The Desiccated Floor of the Aral Sea (Central Asia)*, Ecological Studies 218,
DOI 10.1007/978-3-642-21117-1_9, © Springer-Verlag Berlin Heidelberg 2012

Presently, the dry seafloor of the Aral Sea is a huge salt flat. According to several estimations, it is the source of many million tons of salt and dust blown out by wind annually (Chaps. 5, 7, 16) and transported to rather distant adjacent areas with irrigated fields and settlements. The present and future development is characterised by the creation of salt desert flats.

9.2 Zonal Vegetation and Main Vegetation Types at the Former Coastline

The Aral Sea basin is part of the temperate continental desert and semidesert belt. The area belongs to zonobiomes VIIa and VII (rIII) (Walter 1974; Breckle and Agachanjanz 1994). It is characterized by a zonal mosaic of xerophytic dwarf shrubs on zonal soils (plakor sites) with *Artemisia* species, *Anabasis salsa* and some *Salsola* species. The canopy height is between 20 and 60 cm, the coverage is between 15 and 40% and the amount of phytomass is 8.5 t ha^{-1}. This vegetation type is widespread on the cretaceous and tertiary plateaus (e.g. Ustyurt, Betpak-Dala) and on peneplains adjacent to the mountains.

The parent rocks are often rich in gypsum or salt. Typical soils are burozems (burye), sero-burozems (sero-burye) und serozems. The zonal vegetation at the northern and western coasts of the Aral Sea is often rich in *Artemisia terrae-albae* and *Anabasis salsa* plant communities, rather often intermixed with steppe elements such as *Stipa sareptana* and *Stipa richterana* (at the northern coast).

The psammobiome, a typical sandy desert with rather many plant species, is forming the euxerophytic and mesoxerophytic shrubby and semishrubby vegetation on sand. The appearance of this sand desert is dominated by the rather high growing *Haloxylon* and *Salsola* species (Chenopodiaceae), by *Calligonum* species (Polygonaceae) and by some species from Fabaceae and Poaceae. The main canopy height is between 0.5 and 2.5 m, with a coverage of 20–60%. Soil horizons are not developed, except for a thin layer of a microbiotic crust on the sandy surface forming nonmobile sand, as in many other sand deserts in the world (Breckle et al. 2008). The ongoing wind erosion causes a complicated relief. At the north-eastern coast, between Aralsk and the Syr Darya delta, we have the so-called sand desert Priaralski Karakum. Here, the vegetation is dominated by *Krascheninnikovia ceratoides*, *Calligonum aphyllum*, *Agropyron fragile*, *Artemisia terrae-albae*, *Artemisia arenaria* and several ephemeral species (*Gagea*, *Iris*, *Tulipa*, *Allium*, etc.). At the southeastern coast, typical sand deserts with saxaul plant communities (*Haloxylon persicum* and *Haloxylon aphyllum*) prevail.

The halophytic plant communities are formed by euhalophytic and hemi-halophytic vegetation on saline soils in the relief depressions on the Aral terraces (54–56 m asl). The plant species are mainly shrubs, semishrubs, perennials and annuals from the Chenopodiaceae (*Halostachys*, *Halocnemum*, *Salicornia*, *Suaeda*), some Tamaricaceae, Limoniaceae and Zygophyllaceae. The coverage

Fig. 9.1 *Populus ariana* (syn: *Populus euphratica* s.l.) remnant trees in a short side valley cut in the chinks of the northeastern part of the North Aral Sea (photo: Breckle 2004)

can be rather high, 10–100%. The great variability of this vegetation is a matter of the very varying salinity and water availability. The typical soils are typical solonchaks, as well as takyr-like solonchaks.

The typical tugai biomes are represented by the shrubs and small woods of the delta regions o the Amu Darya and Syr Darya and lake bays. The characteristic species of this vegetation type are mainly *Elaeagnus oxycarpa* and shrubs from the Tamaricaceae. Communities of *Populus* species from the subgenus *Turanga* (*Populus pruinosa*, *Populus diversifolia*) have been preserved as several fragments only in parts along the middle and lower Amu Darya and Syr Darya. A small site with several trees is still in rather good condition in a small hidden side valley between the northeastern chinks of the Small Aral Sea (Fig. 9.1). *Populus* trees depend on available groundwater and spring–summer flooding; they are only slightly resistant to salinity.

The salt meadow vegetation is mainly composed of reed communities with many perennial hemicryptophytes (*Puccinellia* and *Limonium* species, *Aeluropus littoralis*, *Karelinia caspia*, etc.), which can withstand soil salinity. It is spreading in the deltas of the Amu Darya and the Syr Darya and often forms vegetation complexes with tugai vegetation.

9.3 Characteristics of the Vegetation Distribution in the Aralkum

The vegetation of the Aralkum is in various stages of development, mainly on sandy and salty substrates, or with groundwater, and thus can be regarded as pedobiomes, differing from the zonal vegetation mosaic. Figure 9.2 indicates the distribution of the transects for studying this dynamics of the vegetation and ecosystems. The transects stretch from the old coastline to the present one. The vegetation

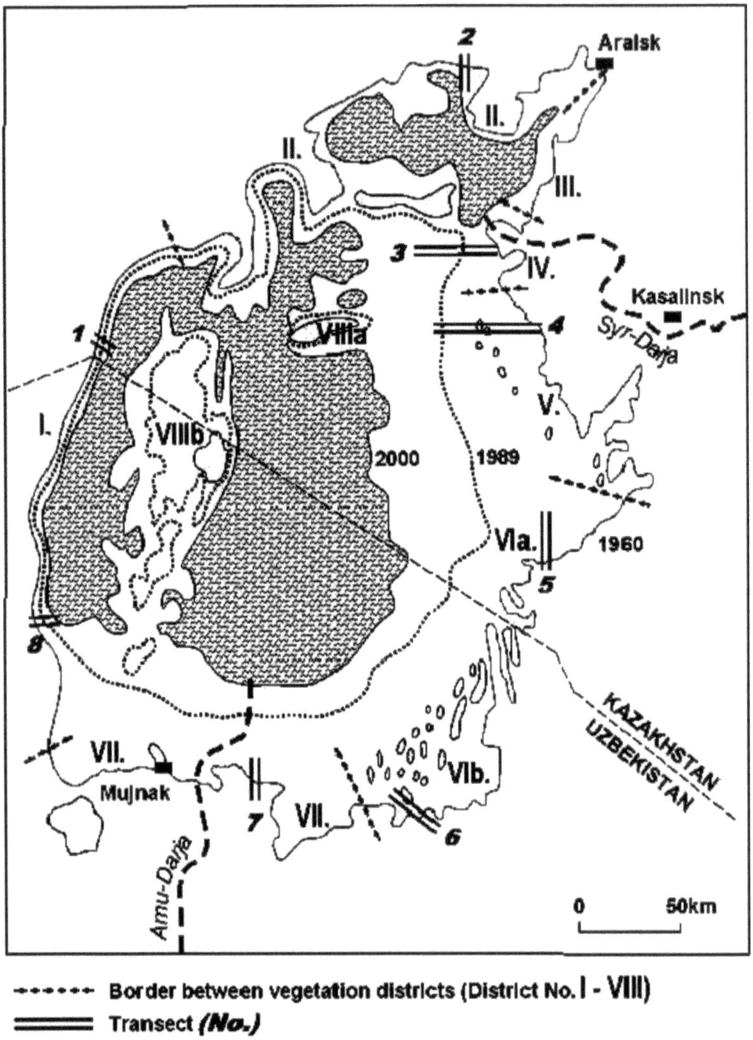

Fig. 9.2 Transects (*1–8*) investigated during recent decades and preliminary definition of vegetation districts (*I–VIII*) of the Aral Sea coastal deserts (according to Kurochkina et al. 1983; Novikova et al. 2001, modified by Wucherer):*I* Usturt district of xerophytic dwarf semishrub vegetation (*Salsola arbusculiformis*, *Artemisia terrae-albae, Anabasis salsa*). *II* Northwestern Priaralye district of haloxerophytic dwarf semishrub and psammophytic grass vegetation (*Anabasis salsa*, *Artemisia pauciflora*, *Artemisia terrae-albae*, *Agropyron fragile*). *III* Northeastern Priaralye district of psammoxerophytic dwarf semishrub and perennial vegetation (*Krascheninnikovia ceratoides, Calligonum aphyllum*, *Agropyron fragile*, *Artemisia terrae-albae*, *Artemisia arenaria*) and several ephemeral species (*Gagea, Iris*, *Tulipa*, *Allium*, etc.). *IV* Syr Darya district of meadow weed–grass, reed and tugaic vegetation (*Elaeagnus oxycarpa*, *Salix* species) in combination with xerophytic dwarf semishrub, halophytic shrub and psammophytic grasses (*Calamagrostis epigeios*, *Pseudosophora alopecuroides*, *Phragmites australis*, *Artemisia terrae-albae, Halostachys belangeriana*, *Tamarix* spp., *Agropyron fragile*). *V* Eastern district of saxaul

distribution on the transects is very different and depends on the sedimentological patterns of the dry seafloor and the geomorphological and landscape patterns of the former coast. The following three transects are typical examples of the vegetation distribution at the different coasts of the Aral Sea.

The Karabulak transect lies at the northern coast of the Aral Sea (Fig. 9.2, transect 2). The length of this transect is 6 km. The ecological conditions are very hard on the dry seafloor of the 1980 and are more favourable on the dry seafloor of the 1970 (Table 9.1). The plant coverage ranges from 20% to 90% and is especially high in the perennial plant communities with *Phragmites* and *Puccinellia* species (Fig. 9.3, numbers 1–2) and in the therophytic communities with *Salicornia* and *Suaeda* species (Fig. 9.3, numbers 18–21). The perennial plant communities occur at the former coastline. The plant communities with *Salicornia* and *Suaeda* species dominate on the marsh and coastal solonchaks close to the present coastline (as of 1994). On the transect, 56.3% of the area is salt deserts without vegetation, 37.4% is therophytic plant communities and only 6.3% is perennial plant communities (Fig. 9.4). There are 33 species on the transect. The species richness within plant communities is low and ranges from one to 14 species (Fig. 9.5). The species richness is higher on the degraded coastal solonchaks with desalinization of the soil profile. The dominant plant families are Chenopodiaceae, Brassicaceae and Asteraceae. The share of the Chenopodiaceae in the flora of the transect is 46%, that of the Brassicaceae is 24% and that of the Asteraceae is 12% (Fig. 9.6).

The Kabanbai transect lies at the southwestern coast of the Aral Sea (Fig. 9.2, transect 8) and shows the characteristics of vegetation distribution at the open chink. The length of the transect is 3.4 km. The plant coverage is high in the perennial psammophytic plant communities with *Astragalus* and *Stipagrostis* species (Fig. 9.7, numbers 1–3) at the former coastline and in the therophytic communities with *Salicornia europaea* (Fig. 9.7, numbers 9–10) at the present coastline. About 43.1% of the area of the transect is salt deserts without vegetation and 14.8% is salt deserts with a *Salicornia europaea* plant community (Fig. 9.8); 29.8% of the area is barchan deserts without vegetation, and only 12.3% of the area is perennial psammophytic vegetation with *Haloxylon aphyllum*, *Stipagrostis pennata*, and *Astragalus* species. There are 18 species on the transect. The species

---

Fig. 9.2 (continued) and xerophytic dwarf semishrub and therophyte vegetation (*Haloxylon aphyllum, Kalidium caspicum, Anabasis salsa, Artemisia terrae-albae, Halimocnemis* species). *VI* Southeastern district of saxaul and ephemeral sedge–saxaul–sagebrush vegetation (*Haloxylon aphyllum, Haloxylon persicum, Artemisia terrae-albae, Calligonum* spp., *Carex physodes*). *VIa* With perennial saltworts (*Salsola orientalis, Salsola arbuscula*). *VIb* With psammophytic shrubs (*Salsola richteri, Ammodendron conollyi*). *VII* Amu Darya district of tugaic vegetation (*Elaeagnus turcomanica, Populus ariana*), halophytic meadow vegetation (*Aeluropus littoralis, Sphaerophysa salsula*), halophytic shrub vegetation (*Halostachys belangeriana, Tamarix* spp., *Salsola dendroides*). *VIII* Island district of complex sagebrush–haloxerophytic dwarf semishrub vegetation (*Artemisia terrae-albae–Anabasis salsa*) in combination with saxaul and psammophytic shrub vegetation (*Haloxylon aphyllum, Calligonum* spp.). *VIIIa* Barsa-Kelmes. *VIIIb* Vozrozhdeniya. Transects: *1* Beigubek, *2* Karabulak, *3* Bayan, *4* Kaskakulan, *5* Bosai, *6* Akpetki, *7* Muinak, *8* Kabanbai

Table 9.1 Characteristics of the sites along the Karabulak transect (May 1998)

Plant community	Site/soil	Soil horizon (cm)	pH	EC (mS cm)	Dominant plant species
Halocnemum strobilaceum–Ofaiston monandrum community	Crusty solonchak, DSF70	0–0.3	8.49	5.05	*Ofaiston monandrum*, *Halocnemum strobilaceum*
		0.3–7	8.41	6.78	
		7–20	8.38	8.49	
		20–30	8.40	6.13	
		40–50	8.11	11.5	
Halocnemum strobilaceum–Artemisia scopaeformis–Suaeda acuminata community	Degraded coastal solonchak with takyr crust, DSF70	0–3.5	8.54	5.38	*Ofaiston monandrum*, *Climacoptera aralensis*, *Suaeda acuminata*
		3.5–7	8.40	7.00	
		7–19	8.00	2.87	
		19–42	8.27	5.23	
		42–71	8.35	4.45	
		71–100	8.36	5.52	
Suaeda acuminata–ephemeral community	Degraded coastal solonchak (upper soil grained), DSF70	0–2.5	8.53	10.1	*Suaeda acuminata*, *Climacoptera aralensis*, *Strigosella africana*, *Euclidium*
		2.5–10	8.48	1.13	
		10–19	8.05	1.10	
		19–47	8.44	0.48	
		47–80	8.35	29.3	
		80–100	8.26	13.2	
Suaeda acuminata community with *Petrosimonia hirsutissima*	Coastal solonchak (strong loamy), DSF80	0–4.5	8.19	11.3	*Suaeda acuminata*, *Petrosimonia hirsutissima*
		4.5–18	8.48	12.8	
		18–31	8.33	12.1	
		31–50	8.67	31.6	
Climacoptera aralensis (open) community with *Suaeda acuminata*	Coastal solonchak (medium loamy), DSF80	0–6	7.92	3.48	*Climacoptera aralensis*, *Suaeda acuminata*
		6–19	8.29	19.4	
		19–30	8.12	18.6	
		30–50	8.10	6.61	
Suaeda acuminata-Climacoptera aralensis (open) community	Crusty-puffy coastal solonchak (loamy), DSF90	–	–	–	*Climacoptera aralensis*, *Suaeda acuminata*
Suaeda acuminata community	Coastal solonchak (medium loamy), DSF90	0–3.5	8.15	6.82	*Suaeda acuminata*
		3.5–10	8.12	5.58	
		10–20	8.20	7.38	
Suaeda acuminata–Suaeda crassifolia community	Marshy solonchak (medium loamy) groundwater level 16 cm, DSF90	0–5	8.28	3.8	*Suaeda acuminata*, *Suaeda crassifolia*
		5–14	8.36	1.33	

EC electric conductivity of soil suspension, *DSF70* desiccated seafloor from the 1970, *DSF80* desiccated seafloor from the 1980, *DSF90* desiccated seafloor from the 1990

richness within plant communities is low – from one to nine species. The plant diversity decreases from the former to the present coastline (Fig. 9.9). The dominant plant families are Chenopodiaceae, Poaceae and Asteraceae. The share of the Chenopodiaceae in the flora is 42%, that of the Asteraceae is 16% and that of the Poaceae is 11% (Fig. 9.10).

The Kaskakulan transect (Fig. 9.2, transect 4) lies at the eastern coast of the Aral Sea between the former coastline and the former island of Kaskakulan and shows the vegetation distribution at the eastern coast of the Aral Sea on the dry seafloor of the 1960 and the 1970 in the old delta region of the Syr Darya. The length of the

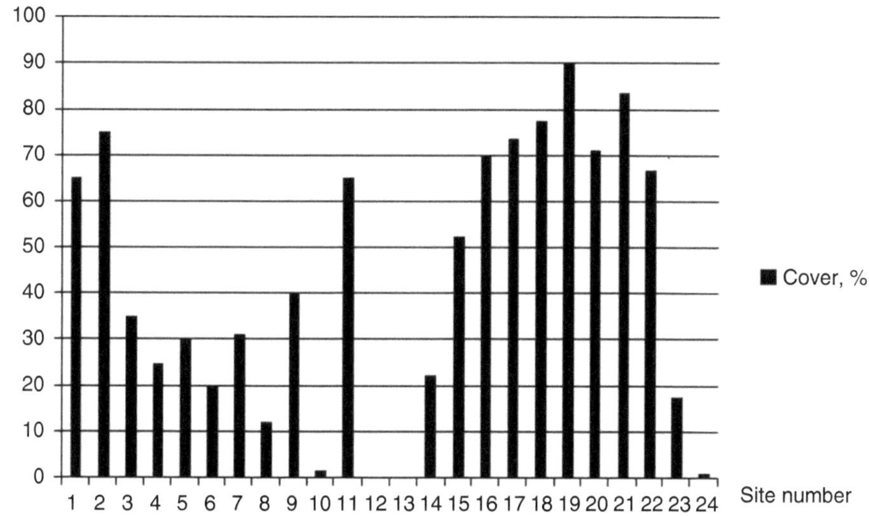

Fig. 9.3 Changes of the cover percentage in the plant communities along the Karabulak transect from the continent (site no. 1), the former coastline from 1960, to the water level of the Aral Sea (site no. *24*) (northern coast of the Aral Sea) (1994)

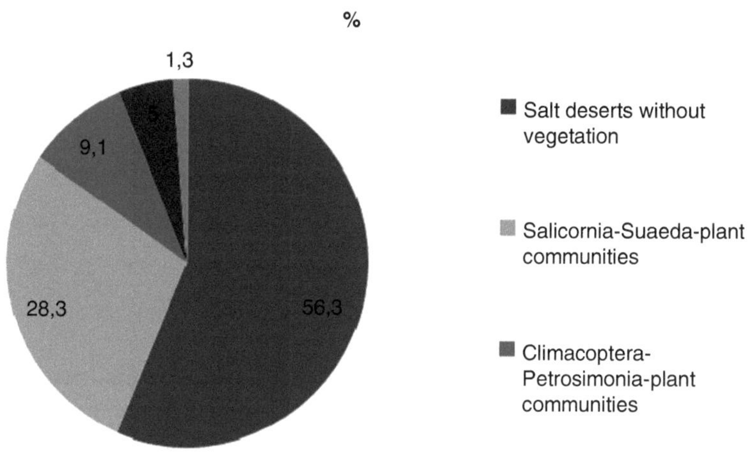

Fig. 9.4 Share of the dominant plant communities and landscapes on the Karabulak transect (northern coast of the Aral Sea) (1994)

transect is 14.9 km. Salt deserts with perennial halophyte vegetation occupy the total area on this transect. The coverage within plant communities is high and ranges from 5% to 80% (Fig. 9.11). The coverage and the species richness are low only for the habitats with high salinity (plant communities 2 and 3). *Haloxylon aphyllum* (48.3%) and *Halocnemum strobilaceum* (34.9%) plant communities dominate (Fig. 9.12). The species richness within plant communities is low and ranges from two to eight species (Fig. 9.13). The species richness is higher on the

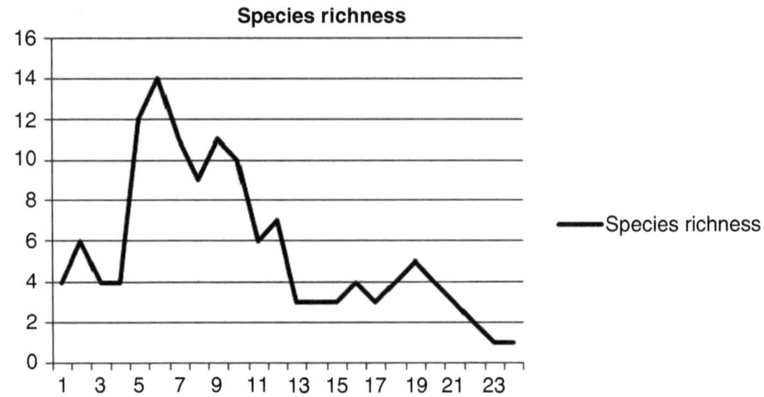

Fig. 9.5 Change of species richness (number of species) in plant communities along the Karabulak transect from the continent (site no. 1), the former coastline from 1960, to the water level of the Aral Sea (site no. 24) (northern coast of the Aral Sea) (1994)

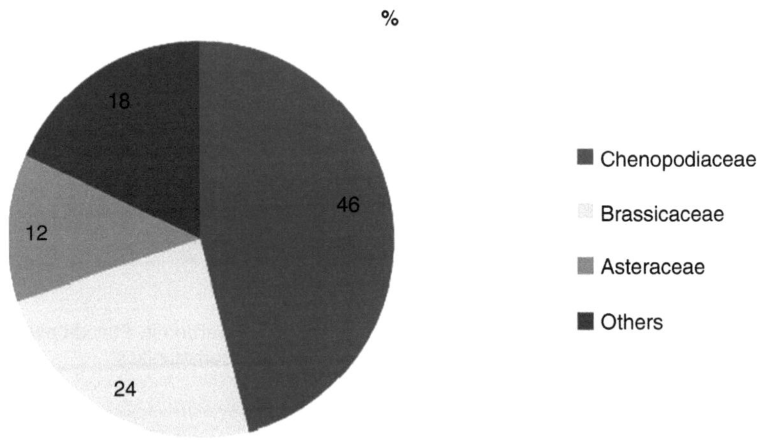

Fig. 9.6 Share of the main plant families on the Karabulak transect (northern coast of the Aral Sea) (1994)

degraded coastal solonchaks in the *Haloxylon* plant communities. The dominant plant families are Chenopodiaceae, Brassicaceae, Asteraceae and Poaceae. The share of the Chenopodiaceae in the flora of the transect is 63%, that of the Brassicaceae is 13% and that of the Poaceae is 13% (Fig. 9.14).

The distribution of the plant communities on the dry seafloor of the 1980 and the 1990 at the eastern coast of the Aral Sea of the Bayan transect is shown on the map in Fig. 9.15. The brown colour depicts the therophytic communities with *Climacoptera* and *Petrosimonia* species (Fig. 9.15). The plant coverage is often rather high, 50–100%. The green colour shows shrub vegetation with *Tamarix* and *Limonium* species and the salt meadows. The salt meadow vegetation mainly

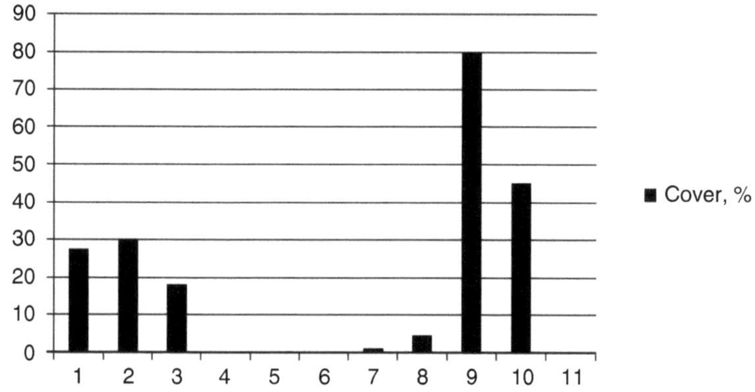

Fig. 9.7 Changes of the cover percentage in the plant communities along the Kabanbai transect from the continent (site no. 1), the former coastline from 1960, to the water level of the Aral Sea (site no. 11) (southwestern coast of the Aral Sea) (1994)

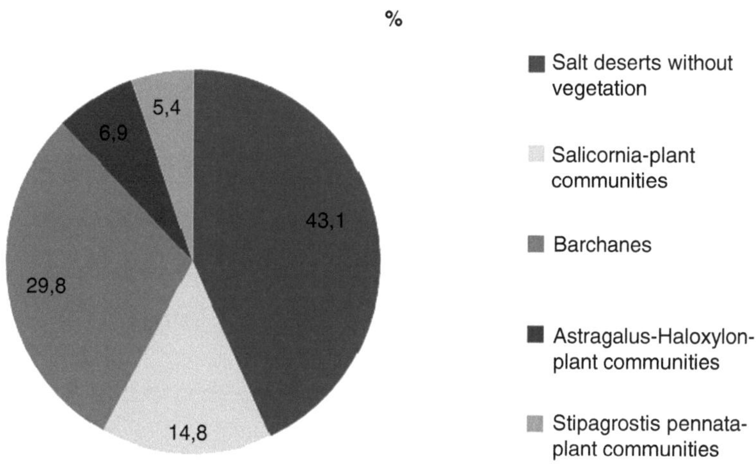

Fig. 9.8 Share of the dominant plant communities and landscapes on the Kabanbai transect (southwestern coast of the Aral Sea) (1994)

comprises reed communities with many perennial hemicryptophytes (*Puccinellia*- and *Limonium*-species, *Aeluropus littoralis*, *Karelinia caspica* etc.). The plant coverage is about 20–60%. The pink and white colours shows the salt deserts without vegetation.

The main vegetation types are as follows: halophytic, psammophytic, tugai and salt meadow communities.

The typical pattern of landscapes, vegetation and soils on the transects is striated. This banded pattern is most characteristically seen along the northern and eastern coasts of the Aral Sea (Ishankulov and Wucherer 1984).

The plant coverage on the transects in the plant communities is very high regarding the prevailing desert conditions, but the species richness is very low.

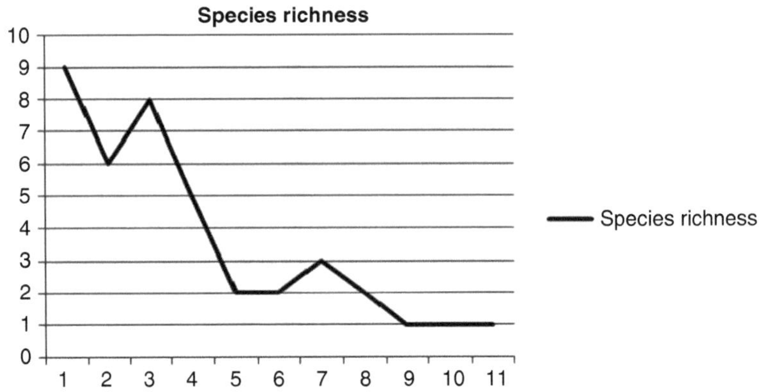

Fig. 9.9 Changes of species richness (number of species) in the plant communities along the Kabanbai transect from the continent (site no. 1), the former coastline from 1960, to the water level of the Aral Sea (site no. 11) (southwestern coast of the Aral Sea) (1994)

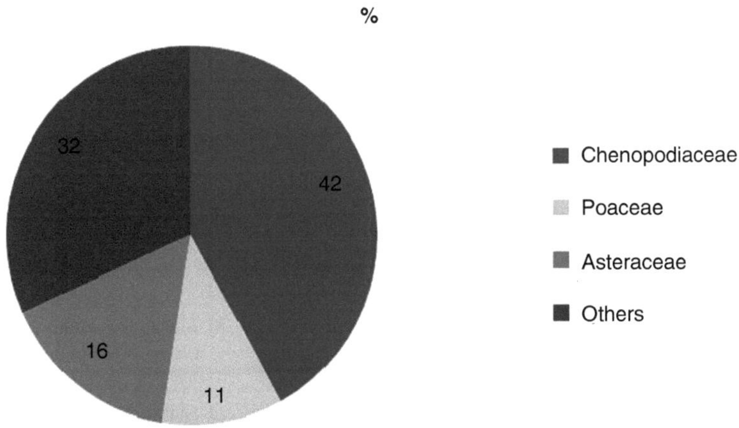

Fig. 9.10 Share of the main plant families on the Kabanbai transect (southwestern coast of the Aral Sea) (1994)

The share of Chenopodiaceae is 40–60% on the transects described. The salt and sand deserts and accordingly the halophytic and psammophytic vegetation types dominate at all types of coast. But the share of the plant communities with perennial vegetation is very low and is only 6–12% on the transects (except for the Kaskakulan transect).

The therophytes and perennials usually form uniform, monotonous stands. Another characteristic is the development of monotonous but very widespread vegetation units on the dry seafloor of the Aral Sea, which is favoured by the fact that there is a huge open flat plain. This is a perfect condition for wide-ranging plant dispersal. The therophytes may cover hundreds of square kilometres within a very short period of less than 2 or 3 years. *Petrosimonia triandra* is a very typical

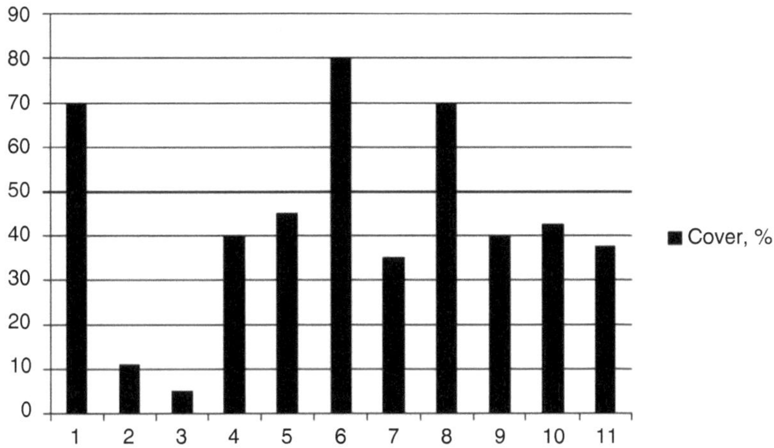

Fig. 9.11 Changes of the cover percentage in the plant communities along the Kaskakulan transect between the continent and the former island of Kaskakulan, from the continent (site no. 1), the former coastline from 1960, to the former island of Kaskakulan (site no. 11) (eastern coast of the Aral Sea) (1994)

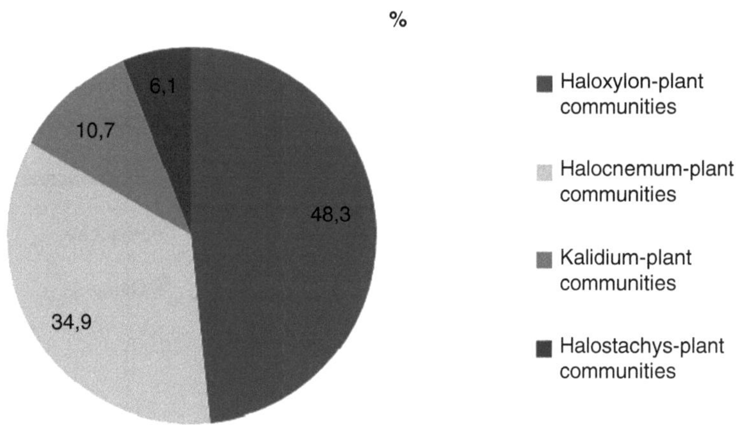

Fig. 9.12 Share of the dominant plant communities and landscapes on the Kaskakulan transect between the continent and the former island of Kaskakulan (eastern coast of the Aral Sea) (1994)

anemochorous plant; it is cut off at the base by wind and then the whole plant is dispersed like the steppe runners by wind over vast distances. Very dense stands were formed by *Petrosimonia triandra* in 1998 on the northwestern coast; *Atriplex pratovii* was extremely dense in 1994 on the southwestern coast. In 2010 *Atriplex pratovii* vegetation occupied a huge area of the dry seabed of the 1990 between the island of Barsa-Kelmes and the original coast. This territory was absolutely bare during the last 20 years. A thin sand layer on the crusty solonchak enables the hemihalophytic *Atriplex* species to overcome the toxic stress. As a rule, those units

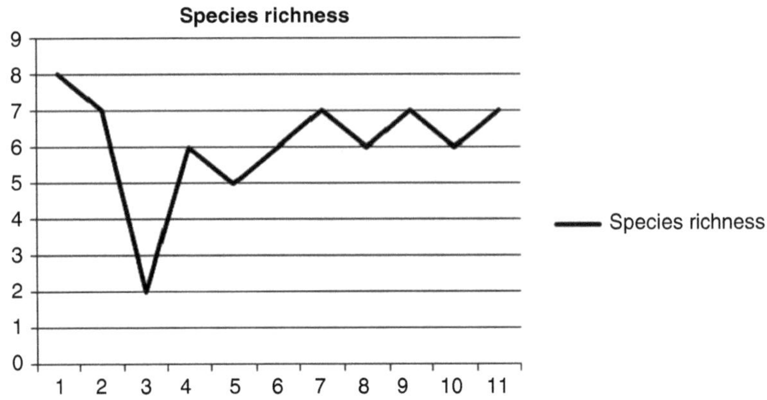

Fig. 9.13 Changes of species richness (number of species) in the plant communities along the Kaskakulan transect between the continent and the former island of Kaskakulan, from the continent (site no. 1), the former coastline from 1960, to the former island of Kaskakulan (site no. 11) (eastern coast of the Aral Sea) (1994)

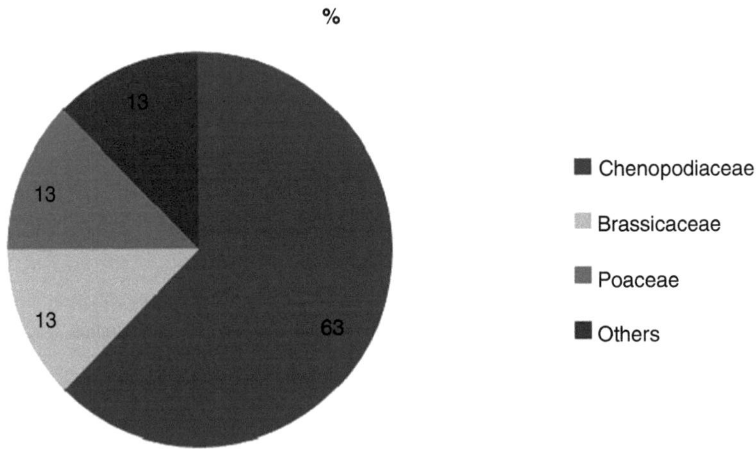

Fig. 9.14 Share of the main plant families on the Kaskakulan transect between the continent and the former island of Kaskakulan (eastern coast of the Aral Sea) (1994)

are often poor in species and over vast distances only three to seven species are found.

Halocnemum strobilaceum, as a succulent chamaephyte, forms extensive stands along the eastern coast of the Aral Sea. Such units are ideal for studies on the genetic and biological variability of these plant populations (Begon et al. 1986). On the other hand, it is almost impossible to define straight plant communities or phytosociological entities with a rather constant species pattern. This is a precondition for naming associations in the sense of phytosociological syntaxonomy. But a description of the common vegetation belts reveals some regularities (Novikova

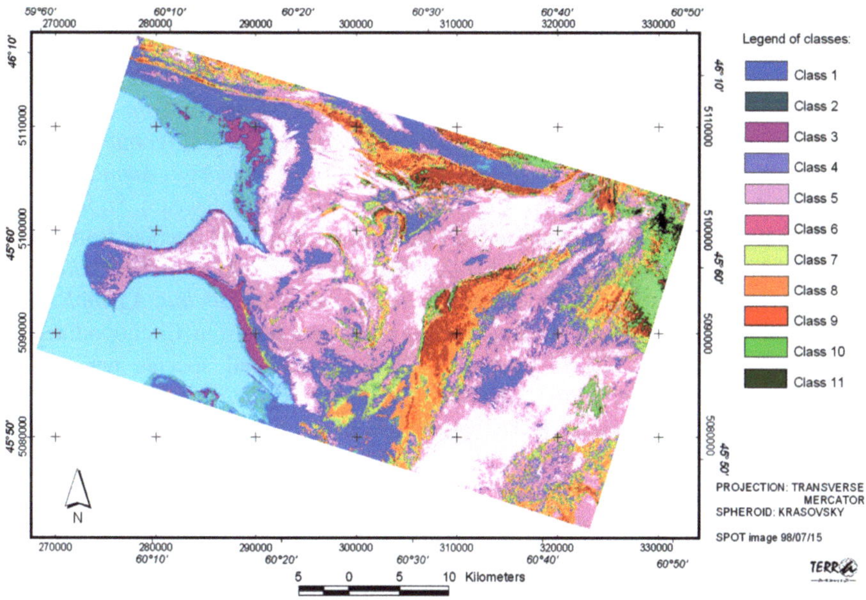

Fig. 9.15 Satellite-derived map. Class 1 sea area, classes 2 and 3 marsh solonchaks partly with *Salicornia europaea* communities, classes 4–6 salt deserts without vegetation, class 7 *Tamarix* plant communities with halophytic therophytes, classes 8 and 9 *Climacoptera aralensis* and *Petrosimonia triandra* plant communities, class 10 *Tamari*x and salt meadow plant communities, class 11 *Phragmites australis* plant community

and Kuzmina 2008), which were given by our intensive investigations (Dimeyeva 1990, Wucherer 1990, Wucherer and Breckle 2001) and are also given in this chapter.

9.4 Main Vegetation Types in the Aralkum

9.4.1 Psammophytic Vegetation

The area of sandy and sandy-loamy sediments is about 12,000 km^2 of the whole Aralkum, mainly in the oldest parts of the desiccated seafloor. This means that sandy deserts with psammophytes are the decisive vegetation around the old coastline, to a great extent at the southeastern and northeasten coasts and around the former islands of Barsa-Kelmes and Vozrozhdeniya. For the whole old coastline, sands with a particle size of about 0.1–0.5 mm are characteristic. Coarser sands are found adjacent to the Syr Darya delta. This sand is predominantly made up of quartz. However, in some parts, limestone particles from sea shells can account for 30–70%. At the eastern coast the limit of sand deposits is at about 43–46 m above sea level (asl), at the northern coast it is at 48–50 m asl and at the southern coast it goes down to about 33–36 m asl.

The appearance of this sand desert in the Aralkum is dominated by the rather high growing *Haloxylon aphyllum* and *Haloxylon persicum*, *Salsola arbuscula*, *Salsola richteri*, *Salsola paletzkiana* (Chenopodiaceae), *Calligonum* species (Polygonaceae), *Astragalus brachypus*, *Astragalus ammodendron*, *Ammodendron bifolium*, *Ammodendron conollyi*, *Eremosparton aphyllum* (Fabaceae), *Stipagrostis pennata* (Poaceae), *Artemisia arenaria* (Asteraceae) and some species from other families. The main canopy is about 1.0–2.0 m high but can reach 5 m (for *Haloxylon* species). Characteristic therophyte species are *Horaninovia ulicina*, *Agriophyllum squarrosum*, *Salsola paulsenii* and *Corispermum* species from Chenopodiaceae. The canopy is about 20–40 cm high. Typical perennial species are *Allium sabulosum*, *Artemisia santolina*, *Chondrilla brevirostris*, *Heliotropium arguzioides* and *Astragalus lehmannianus*. The plant coverage of the psammophytic plant communities is about 10–40%, rarely up to 80%. The sandy soils exhibit no or only a very slight formation of horizons. Wind erosion causes the development of a complicated relief with all kind of sand dunes. The psammophytic vegetation is most dominant on the dry surface areas from the 1960, is partly dominant on the dry surface areas from the 1970 and is rare on the dry surface areas from the 1980.

At the southeastern coast there are perennial psammophyte communities mainly with grasses such *Stipagrostis pennata* (Fig. 9.16), and also shrubs, e.g. *Eremosparton aphyllum*, *Haloxylon aphyllum*, *Astragalus brachypus* and *Calligonum* species (Wucherer and Breckle 2003).

The sandy deserts of the Aralkum receive approximately 90–120 mm of annual precipitation; the potential evaporation is about 1000 mm per year. Since sandy substrates are the most favourable substrates in arid climates in respect of water budget (Breckle et al. 2008), also these sites in the Aralkum exhibit spontaneous invasion of plants and the formation of an open plant cover. On the desiccated seafloor with sandy substrates, barchans and other sand dune types develop rather quickly by aerodynamic processes. Mainly two types can now be distinguished:

Fig. 9.16 Psammophytic vegetation on the dry sea floor of the 1960 at the southeast coast: *Stipagrostis pennata* plant community with *Haloxylon aphyllum* (photo: Wucherer)

loose open sand dunes with a barchan coverage of 10–50%, and those with a dense barchan coverage of normally over 60%. The most intensive development of dune fields is along the vast stretches of the former eastern coast, where near Kaskakulan it is obvious that they spread to the south and southeast.

9.4.2 *Halophytic Vegetation*

The halophytic vegetation is present on most of the dry seafloor of the 1970, 1980 and 1990 and especially the more recent desiccated areas. This vegetation is now the most prominent all over the Aralkum. The halobiomes (salt deserts) exhibit typical euhalophytic or hemihalophytic vegetation on more or less saline substrates. A rather considerable portion of the flora (see Chap. 8) is halophytic, mainly from Chenopodiaceae. Their ecological and ecophysiological features are characterized by the fact that they can fulfil their whole life cycle under saline conditions (see Chap. 12). Life forms are shrubs, semishrubs, dwarf shrubs and perennial and annual herbs, mainly from the Chenopodiaceae (*Halostachys*, *Halocnemum*, *Salicornia*, *Suaeda*), but also from Tamaricaceae, Limoniaceae and some other plant families. The canopy height is normally below 1 m, but for adult saxaul shrubs can exceed 3 m. The coverage also varies considerably, ranging between 10 and 100%. A rather high variability in the shape of these forms of halophytic vegetation is correlated with the very variable salinity of the soil and the various soil horizons. Typical soils for halophytic vegetation in the Aralkum are marshy and coastal solonchaks, typical and degraded solonchaks, and takyr-like solonchaks.

9.4.2.1 Annual Vegetation

Close to the retreating water level of the Aral Sea, a rather dense belt of therophytic carpets with *Salicornia europaea*, *Suaeda* species (Fig. 9.17) and *Tripolium vulgare* very often develops. The canopy is about 20–60 cm high, but for *Tripolium* can be up to 120 cm. The coverage of therophytic plant communities ranges between 10 and 100%. *Salicornia* is the main component of the annuals close to the water. But there are also mixtures of some *Suaeda* species (*Suaeda acuminata* and *Suaeda crassifolia*). *Tripolium vulgare* is only present as isolated stands. At the eastern coast near Barsa-Kelmes, aggregations of *Bassia hyssopifolia* are distributed. Adjacent to the delta regions of rivers, mixed stands of *Salicornia europaea* and *Phragmites australis* or *Bassia hyssopifolia* and *Phragmites australis* can be observed. Seeds of *Salicornia* can withstand high salinities, and germination can take place under saline conditions (Willert 1968). Close to the water level the irregular inundations of the marshy solonchaks are favourable for those annuals. The dead remnants from the preceding years can be found everywhere. The development of the therophytic vegetation belt moves according to the retreat of the water level (see Fig. 10.13).

Fig. 9.17 Annual halophytic vegetation on the dry sea floor of the 1990 at the eastern coast (Bayan transect): *Salicornia europaea* plant community with *Suaeda acuminata* (photo: Wucherer)

Fig. 9.18 Annual halophytic vegetation on the dry sea floor of the 1990 at the eastern coast (Bayan transect): *Climacoptera aralensis* plant community with *Petrosimonia triandra* (photo: Wucherer)

On the desiccated seafloor of about 3–10 years after desiccation, we can quite often find a second wave of annuals developing, mainly with *Climacoptera aralensis*, *Petrosimonia triandra*, *Bassia hyssopifolia* and *Atriplex pratovii* on the coastal solonchaks (Fig. 9.18). The canopy is about 20–60 cm high, but for *Atriplex* can be up to 100 cm. But those sites are already under the predominant influence of the zonal climatic conditions which cause rather irregular establishment and growth rates from year to year. The coverage of plant communities ranges between 10 and 100%.

9.4.2.2 Perennial Vegetation

At the eastern coast of the former Aral Sea on the older seafloor from the 1960 and 1970, salinity is moderate or low. Here plant communities with *Haloxylon aphyllum*, *Halocnemum strobilaceum*, *Kalidium caspicum*, *Limonium suffruticosum*,

Fig. 9.19 Perennial halophytic vegetation on the dry sea floor of the 1970 at the eastern coast (Kaskakulan transect): hummock-forming *Halocnemum strobilaceum* plant community (photo: Wucherer)

etc. are found (Fig. 9.19). They form stands with a coverage of 20–40%. On the oversanded coastal solonchaks, rather dense stands with coverage up to 80% can be observed. However, these are small portions in the whole mosaic and are not representative of the whole coast. *Haloxylon* (saxaul) is used by the people for fuel. In 1998 there were massive "deforestations" of saxaul.

On the desiccated seafloor of the 1980 and 1990, there are almost no perennials except for some *Tamarix* or *Halostachys belangeriana* shrubs. Invasion by perennial vegetation is definitely slower than the retreat of the sea level. The conditions for germination and establishment of the perennials is quite favourable on the marshy solonchaks a few years after desiccation, but then in the later phases it becomes much slower and often seems to be hindered.

In 1998 the perennial vegetation on the Karabulak, Bayan and Kaskakulan transects was investigated. Especially the communities with *Tamarix* were checked in the vicinity of the North Aral Sea.

9.4.2.3 Salt Deserts Without Vegetation

The whole seafloor which desiccated from 1980 can be regarded as a huge salt desert flat. It is about 70–80% of the present Aralkum. The therophytic vegetation is spatially and temporally very instable; it affects the soil surface only superficially. In the 1980, when formation of solonchaks had just started, denudation of the desiccated soil surface amounted to about 2 mm year^{-1}; in other words, within 30 years about 6 cm of loose substrate may have been blown away. The present open and often puffy salt crust surface can exhibit windblown losses of 3–7 cm per year.

This makes it important to know which annuals and which perennials are present on those rather open sites of the 1970 and 1980. Those areas are not easy accessible. Within the Bayan transect between the former islands of Barsa-Kelmes and Kokaral

Fig. 9.20 Salt desert on the dry sea floor of the 1990 at the eastern coast (Bayan transect) with crusty solonchak without any vegetation (photo: Wucherer)

it was possible to reach the actual coastline in the 1990. Here, a mosaic of pure open salt desert (Fig. 9.20) and of therophytic stands was present. The more important therophytes were *Suaeda acuminata*, *Petrosimonia triandra* and *Climacoptera aralensis*.

9.4.3 Tugai Vegetation

The tugai biome is characterized by woody vegetation on alluvial soils adjacent to river valleys or delta areas. Along the lower parts of the rivers, mainly *Salix songarica*, *Salix wilhelmsiana*, *Elaeagnus oxycarpa* and shrubs from the Tamaricaceae are dominant. The shallow groundwater and rather often periodic inundations enable good growing conditions for woody plants. Floristically those stands are rather poor. Soils can be somewhat saline, and then halophytes are mixed. The canopy height may reach 8 m, with coverage of 50–100%.

Real *Turanga* forests with adult trees of *Populus* species are rather rare nowadays. The absence of spring–summer flooding and of river-channel straightening are limiting factors for distribution of the floodplain forest on the dry seafloor. Often the original tugai forest is replaced by a dense scrub of *Elaeagnus* and *Tamarix* (*T. laxa*, *T. elongata*, *T. hispida*, *T. ramosissima*) (Fig. 9.21). *Elaeagnus oxycarpa* forms small vegetation fragments and is very local. *Tamarix laxa* and *Tamarix elongata* are more characteristic of the dry seafloor of the 1960 and 1970, *Tamarix hispida* is characteristic of the dry seafloor of the 1980 and 1990 and *Tamarix ramosissima* is characteristic of the dry seafloor of the delta areas. On the dry seafloor of the delta areas, only a *Tamarix* community is present, which also spreads to most of the dunes along the eastern and southern coasts. The canopy height may reach 3 m, with coverage of 30–80%.

Fig. 9.21 Tugai vegetation on the dry sea floor of the 1960 at the eastern coast, in the surroundings of the hot springs (Kaskakulan transect): *Tamarix hispida* plant community with *Tamarix elongata* and *Tamarix laxa* (photo: Wucherer)

9.4.4 Salt Meadows

Some intermediate salt meadows with reed vegetation and perennial hemicryptophytes (*Puccinellia* and *Limonium* species, *Phragmites australis*, *Aeluropus littoralis*, *Karelinia caspica*, etc.) can withstand rather high salinities. *Phragmites australis* forms with *Puccinellia dolicholepis* plant communities at the chink coasts of the former coastline on the dry seafloor of the 1960 with coverage of 60–80%. On the meadow solonchaks at the delta areas a very characteristic *Limonium otolepis–Aeluropus littoralis* plant community with coverage of 40–80% can be found.

9.5 Phytogeographical Zonation

9.5.1 General Remarks

Along the northern and western coasts of the former Aral Sea (Fig. 9.2, regions I and II) the terraces are almost lacking or are relatively small (10–20 m up to a few hundred metres). The main seed source is the intrazonal flora of the terraces and the steep slopes of the chinks. The zonal plant communities of the upper plateaus only marginally influence the flora and vegetation dynamics of the seafloor. In general the Brassicaceae are strongly represented. Species from Fabaceae and Polygonaceae, however, are not so frequent, in contrast to the shallow eastern coastlines (regions III and VI). Here the Aral terraces are rather broad, up to 5 km. The psammophytic vegetation of the adjacent Kyzylkum and the Priaralski Karakum are a main source of diaspores for the seafloor vegetation.

The delta regions (IV, VII) are characterized by irregular flooding from the rivers. Old seed banks are often transported and activated by floods. Tamaricaceae, Cyperaceae and Poaceae are more common, and on some terraces Chenopodiaceae form dominant halophytic associations.

The southern coast (region VII) again has narrow terraces (10–100 m), and these are not important as a seed source. The terraces and the rather low chinks are poor in species. Therophytes prevail and often form strips of vegetation or vegetation belts around lagoons. The former coastline is often a site of a rather dense vegetation belt with a coverage of often more than 20%.

According to the phytogeographical regionalization, the territory of the Aral Sea coast is situated within the Irano-Turanian subregion of the Sahara–Gobi desert region in North Turan province (Lavrenko 1962; Rachkovskaya et al. 2003). The lower units of subdivision – the plant biomes (okrugs and rayons, here called districts) are distinguished according to vegetation features depending on geomorphological features, geological features and soil conditions (Lymarev 1969; Ishankulov 1980; Ishankulov and Wucherer 1984; Kurochkina et al. 1983). The preliminary districts are shown in Fig. 9.2, and a more detailed view of the presently accepted districts is given in Fig. 9.22. A district is represented by regular combinations of vegetation units according to soil, geomorphological features and climate, and is thus somehow similar to a phytosociological unit. The floristic list of one district is typical and there are species missing in adjacent districts. In the limits of a district, all physical-geographical borders are coincident.

The characteristics of the regional subdivisions up to the level of districts were shown by Rachkovskaya and Safronova (1994) The area of the Aral Sea coast belongs to North Aral and East Aral (in the limits of Kazakhstan) and North Ustyurt, Central Ustyurt and Low Amu Darya districts (Uzbekistan). This former regionalization did not take into account the huge territory of the Aralkum. The description of the Aral coast subdivisions up to the level of the lowest units – the rayon – has already been discussed by Dimeyeva (1990) and Dimeyeva and Kurochkina (2005). A change of vegetation cover has to be distinguished especially in the dry seabed. Continued desiccation, increasing mineralization of the seawater, outcrop of marine sediments of heavy texture and salinity have led to a change of direction, speed and the mechanisms of primary successions. On the other hand, there is a process of active colonization of the eastern shore by saxaul vegetation. Future additions and changes to the scheme of zoning will depend on detailed studies especially in the Uzbekistan part of the Aral Sea coast. The shortage of published data on vegetation of the dry seabed there does not allow one to work out the zonation for the whole coastal area. The proposed present scheme of subdivisions therefore covers only the Kazakhstan territory.

The scheme of botanical-geographical zoning is based on the following unpublished maps: map of vegetation of the Aral Sea dry seafloor, scales 1:500,000 and 1:200,000 (Kurochkina et al. 1990); map of vegetation of the Aral region, scale 1:500,000 (Kurochkina et al. 1990); map of ecosystems of the Syr Darya delta (Geldyeva et al. 1998).

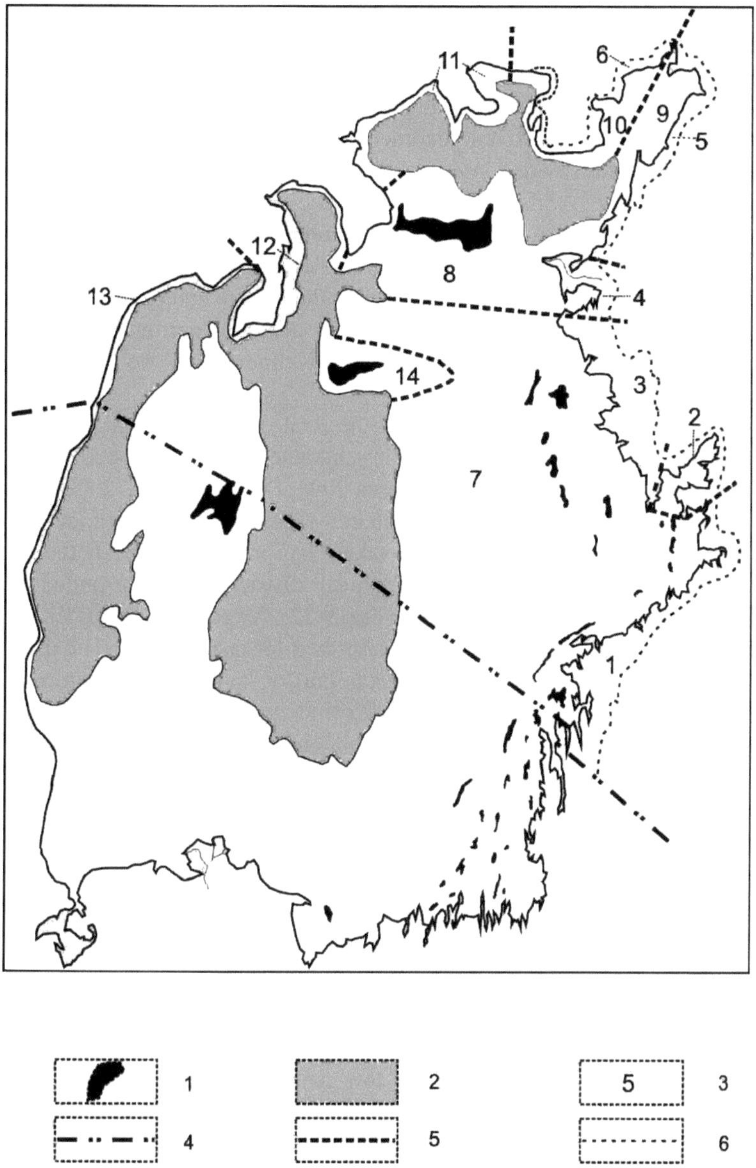

Fig. 9.22 Botanical-geographical zoning of the Aralkum in the limits of Kazakhstan, indicating the present vegetation districts by number (*1–14*). *I* Former islands, *II* the Aral Sea (in 2005), *III* number of botanical-geographical districts (1–14), *IV* boundary of Uzbekistan (Karakalpakstan), *V* boundary of districts, *VI* boundary of old Aral marine terraces

The area of the Aral Sea coast which was subjected to transgressions and regressions during the existence of the Aral Sea since the beginning of the Holocene is called the coastal region. The main features which characterise its vegetation cover are as follows: uniformity in the origin of the territory which was flooded in ancient and modern times; significant cyclical variations of the sea level; similar trends of successions; incompleteness of successional development and absence of zonal (climax) types of vegetation; depleted floristic composition; the neoendemic species *Atriplex pratovii* is not recorded outside the boundaries of the coastal region. This region is divided into two groups of districts – the ancient terraces of the Aral Sea (Ancient Aral and New Aral marine terraces, Chaps. 2, 3) and the dry seafloor of the Aral Sea (Aralkum). The division into two groups is connected with a different age of the territories and a different degree of successional development. The group of ancient terraces is determined at the eastern and partly at the northern coasts. Ancient terraces are not well defined or occupy a narrow belt on the steep western coast and around the island of Barsa-Kelmes. They were ignored in the general scheme.

Maps of vegetation, native zoning of the Aral coast (Kurochkina et al. 1983; Wucherer and Galieva 1985), floristic characteristics of modern vegetation (Dimeyeva and Kuznetsov 1999; Dimeyeva 2004; Dimeyeva et al. 2008; Wucherer et al. 2001), coastal geomorphology (Lymarev 1969) and sedimentology of marine deposits (Rubanov et al. 1987) were taken into consideration in the division of districts. Fourteen botanical-geographical districts are determined for the Kazakhstan part of the Aral Sea coast (Fig. 9.22). Vegetation characteristics and features of originality of the flora were described for each district. The difficulties accessing the dry seabed of the Kazakhstan part of Vozrozhdeniya did not allow a description of vegetation for this territory to be included. Undoubtedly, it is a separate area with original features of flora and spatial dynamics of vegetation. After investigation of this territory, it may be described as botanical-geographical district number 15.

9.5.2 Botanical-Geographical Status of the Aralkum

The territory of the Aral Sea depression is situated within the Irano-Turanian subregion of the Sahara–Gobi desert region in North Turan province (Rachkovskaya et al. 2003). Only the low Amu Darya district (Uzbekistan) as a fragment of South Turan province borders the southern coast of the Aral Sea. We can distinguish two climatic subdistricts (see Chap. 4, Fig. 4.15) in the Aralkum area – the Aralsk subdistrict in the north (North Turan) and the Muinak subdistrict in the south (South Turan). The climatic border between these two subdistricts runs from west to east through the Aralkum. From a phytogeographical viewpoint, this border is also recognisable, as already mentioned. The following species and their plant communities are the typical indicators of South Turan province:

- *Suaeda arcuata*
- *Suaeda microsperma*
- *Climacoptera turcomanica*

- *Salsola richteri*
- *Salsola paletzkiana*
- *Ephedra strobilacea*
- *Calligonum eriopodum*
- *Ammodendron conollyi.*

The last five species are psammophytes and were mentioned by Kurochkina (2003) as typical southern Turanian indicators.

Three species of halophytes were added to this list by the authors of this chapters after years of research in Central Asia. Furthermore, Wucherer visited the southern coast of the Aral Sea in 1994 and 2000. On the southwestern coast (Kabanbai) and on the southern coast (Adzhibai and Muinak) transects were defined and studied.

Suaeda arcuata is rather common in the delta area and is often found on secondary solonchaks forming small vegetation units in combination with *Suaeda paradoxa*. *Climacoptera turcomanica*, found between Nukus and Kungrad, seems to be also rather typical for South Turan.

In 1994 Wucherer described large-area plant communities of *Suaeda microsperma* in the former bay of Adzhibai and northwest of Muinak on the dry seafloor of the Aral Sea (unpublished data). As a southern Turanian indicator, this species is especially interesting, as it is closely related to *Suaeda acuminata*, which belongs to the section *Conosperma* and is found in the steppe, as well as in northern Turan. *Suaeda acuminata* exceptionally exists and forms plant communities in the Aralsk subdistrict, and *Suaeda microsperma* exists only in the Muinak subdistrict. We also found *Suaeda microsperma* in the large salt depression of Barsa-Kelmes to the west of the southern coast of the Aral Sea. According to Kabulov (1990), this species was also recorded from the Akpetki Archipelago and the adjacent seafloor (southwestern coast). *Suaeda acuminata* cannot be found there.

Salsola richteri and *Salsola paletzkiana*, which are present in plant communities on the dry seafloor at the southern and southeastern coasts, the delta area northwestern Kyzylkum (adjacent to the Akpetki Archipelago) also confirm a southern Turanian character of the Muinak subdistrict. Both species of *Salsola* are very characteristic of the Karakum and are typical southern Turanian species. They do not exist in the Aralsk subdistrict, nor in the coastal area from the former island of Ujaly to Kashkynsu. *Calligonum eriopodum* and *Ephedra strobilacea* are common in northwestern Kyzylkum. *Ammodendron conollyi* is, as already mentioned, regarded as a purely southern Turanian species, even though it also appears in areas further north. It is characteristic of the southeastern coast of the Aral Sea, namely the Akpetki Archipelago, Ujaly and northwestern Kyzylkum.

The southern and southwestern parts of the Aralkum including the dry seafloor and the former island of Ujaly belong to the southern Turanian area. The coast from the former island of Ujaly to the former Kaschkynsu Bay shows a transitional character. These facts led to the conclusion that the border between the northern and the southern Turanian areas to the east of the southeastern coast of the Aral Sea had to be modified.

9.5.3 Vegetation Districts of Ancient Marine Terraces of the Aral Sea

The southeastern (Biktau-Bozkol) district is situated on the southern coast of the Aral Sea (Fig. 19.22, District 1).

It mainly consists of saxaul vegetation with ephemerals, annual saltworts and ephedra (*Haloxylon aphyllum, Haloxylon persicum, Eremopyrum orientale, Salsola nitraria, Ephedra distachya*); sagebrush vegetation with ephemerals, teresken, perennial saltworts and black saxaul (*Artemisia terrae-albae, Anisantha tectorum, Poa bulbosa, Krascheninnikovia ceratoides, Salsola arbuscula, S.orientalis, Haloxylon aphyllum*) dominate most plains and hillocky sands. Hilly-ridge sands are vegetated by psammophytic shrub communities (*Calligonum* spp., *Astragalus brachypus, Ammodendron conollyi*). Annual and perennial saltwort and tamarisk (*Halocnemum strobilaceum, Tamarix laxa, Climacoptera aralensis, Suaeda acuminata*) vegetation is distributed in the form of a belt, a mosaic-belt according to the ecological range of communities around solonchaks. *Ephedra strobilacea* and *Ammodendron conollyi* are floristic elements of the southern deserts; the presence of representatives of *Eucalligonum* and *Pterigobasis* sections of the genus *Calligonum* (*Calligonum caput-medusae, Calligonum erinaceum, Calligonum rubens, Calligonum muravljanskyi*) is characteristic of the flora.

Bozkol district is located at the limits of the Kuan Darya delta of the Syr Darya (Fig. 19.22, District 2). It includes the territory of the former bay of Bozkol with the surrounding mainland. Periodically, the former bay is inundated by the river's water. The following plant communities are usually found there: reed with tamarisk (*Phragmites australis, Tamarix ramosissima, Tamarix laxa*,); perennial saltwort (*Halostachys belangeriana, Halocnemum strobilaceum, Kalidium caspicum*); and annual saltwort (*Suaeda acuminata, Suaeda altissima, Climacoptera aralensis, Petrosimonia brachiata*). *Climacoptera brachiata – Anabasis aphylla* and *Artemisia terrae-albae – Salsola orientalis* vegetation occupies the takyr-like soils. A typical ecological-dynamic range of communities in the Kuan Darya low stream is as follows: reed (*Phragmites australis*) → liquorice reed (*Phragmites australis, Glycyrrhiza glabra, Karelinia caspia*) → *Halimodendron halodendron* → annual saltwort (*Suaeda corniculata, Atriplex littoralis, Atriplex tatarica, Petrosimonia brachiata, Zygophyllum oxianum*). Vegetation with bog-meadow species (*Bolboschoenus maritimus, Eleocharis argyrolepis, Eleocharis acicularis, Lythrum salicaria, Butomus umbellatus, Alisma plantago-aquatica*) frequently occupies wetlands.

Agurme-Akkol district (Fig. 19.22, District 3) is situated on the eastern coast from the former Agurme peninsula to the former bay of Akkol. The vegetation cover is characterized by a complex patchiness. Solonchaks are vegetated by communities of *Halocnemum strobilaceum, Kalidium caspicum, Nitraria schoberi, Tamarix laxa, Suaeda physophora*. Communities of *Anabasis aphylla, Anabasis salsa* and *Artemisia terrae-albae* are distributed in undulating plains. Sagebrush communities with ephemerals and perennial saltwort (*Artemisia terrae-albae,*

Salsola orientalis, *Eremopyrum orientale*, *Poa bulbosa*) usually occur in a complex with annual saltworts (*Salsola nitraria*, *Climacoptera brachiata*).

Syr Darya district (Fig. 19.22, District 4) is located at the mouth of the Syr Darya. The vegetation is represented by meadow, tugaic and psammophytic plant communities. The following ecological-dynamic ranges of communities are distributed in the delta: reed (*Phragmites australis*) → rush (*Crypsis schoenoides* - *Juncus gerardii*) → halophytic grasses (*Aeluropus littoralis*); reed–annual saltwort (*Climacoptera aralensis*, *Suaeda altissima*, *Atriplex littoralis*, *Phragmites australis*) → shrubs (*Elaeagnus oxicarpa*, *Tamarix ramosissima*, *Tamarix laxa*, *Tamarix hispida*, *Halimodendron halodendron*, *Lycium ruthenicum*) → forbs (*Calamagrostis epigeios–Glycyrrhyza glabra; Karelinia caspia–Alhagi pseudalhagi*). In plains with sandy areas, communities of psammophilous shrubs, subshrubs (*Ammodendron bifolium*, *Eremosparton aphyllum*, *Convolvulus subsericeus*) and grasses (*Leymus racemosus*, *Stipagrostis pennata*) are distributed. Hillocky sands are occupied by communities of wheat grass–ephemeroid–sagebrush (*Artemisia terrae-albae*, *Poa bulbosa*, *Agropyron fragile*), sagebrush–teresken (*Artemisia terrae-albae*, *Krascheninnikovia ceratoides*) and psammophytic shrubs (*Calligonum* spp., *Ammodendron bifolium*, *Salsola arbuscula*). Original features of the flora are the local distribution of the rare species *Asparagus brachyphyllus* with endemics of Kazakhstan, bulrush (*Scirpus kasachstanicus*) and species from the *Pterococcus* section of *Calligonum* (*Calligonum crispatum*, *Calligonum pseudohumile*, *Calligonum palibinii*).

The northeastern (Bugun-Aralsk) district is situated at the northeastern coast of the Aral Sea from Bugun village to Aralsk town (Fig. 19.22, District 5). The vegetation cover consists of psammophytic communities with an insignificant portion of halophytes (*Tamarix laxa*, *Tamarix hispida*). The typical vegetation of hilly and hilly-range sands is represented by communities of psammophytic sagebrush (*Artemisia terrae-albae*, *Krascheninnikovia ceratoides*, *Carex physodes*) and psammophytic shrubs (*Calligonum aphyllum*, *Salsola arbuscula*, *Atraphaxis spinosa*) on slopes and tops and wheat grass–sagebrush (*Artemisia terrae-albae*, *Agropyron fragile*) communities in hollows. Anthropogenic modifications of the natural vegetation with invasions of *Peganum harmala*, *Heliotropium arguzioides* and *Euphorbia seguierana* are widely distributed around villages and barns in eroded sands. The original feature of the flora is the occurrence of species only from the *Pterococcus* section of *Calligonum*.

Koktyrnak district is located at the northen coast of the Aral Sea (Koktyrnak peninsula) from Aralsk town to the eastern border of Butakov Bay (Fig. 19.22, District 6). Complexes of perennial saltwort (*Anabasis salsa*, *Suaeda physophora*) and sagebrush (*Artemisia pauciflora*, *Artemisia scopiformis*, *Artemisia terrae-albae*) locally with *Atriplex cana*, *Anabasis aphylla* and *Salsola orientalis* communities are common here. Gentle slopes of ridges are occupied by ephemeral–sagebrush with *Halocnemum strobilaceum*, *Artemisia scopiformis*, *Eremopyrum orientale*, *Rheum tataricum*, and *Lepidium perfoliatum* and ephemeral–sagebrush (*Artemisia terrae-albae*, *Eremopyrum orientale*, *Lepidium perfoliatum*, *Strigosella africana*, *Meniocus linifolius*) vegetation. Sagebrush communities with teresken, wheat grass and ephemerals as subdominants (*Artemisia terrae-albae, Krascheninnikovia ceratoides*, *Agropyron fragile*, *Alyssum turkestanicum*, *Poa bulbosa*) are spread in aeolian-active sandy planes. The original feature of the flora is the local occurrence of the relict species *Calligonum bykovii* in breaks, on slopes of ridges and in stony deserts.

9.5.4 Vegetation Districts of the Former Seafloor (Aralkum)

The eastern (Uyaly-Tokpan) district is situated at the eastern coast of the Aral Sea (Fig. 19.22, District 7) from the frontier of Uzbekistan to former village of Tokpan. Characteristic features are a wide distribution of solonchaks and barchan bare lands without any vegetation through high speed of desiccation of the flat seabed. The irregularity of the coastline and multiple former bays determine the formation of solonchaks with perennial saltwort (*Halocnemum strobilaceum*, *Halostachys belangeriana*) vegetation in combination with psammophytic shrub and black saxaul (*Astragalus brachypus*, *Eremosparton aphyllum*, *Calligonum caput-medusae*, *Calligonum densum*, *Calligonum erinaceum*, *Haloxylon aphyllum*) communities. The prevailing spatial sequence of primary succession is a halosere (halophytic shrub and halophytic dwarf subshrubs in late seral stages): *Salicornia europaea* (*Suaeda acuminata*, *Suaeda crassifolia*) → *Atriplex pratovii* (*Climacoptera aralensis*, *Climacoptera ferganica*) → *Halocnemum strobilaceum* (*Nitraria schoberi*). The psammosere is distributed on sandy sediments of the southeastern coast. The spatial sequence is as follows: *Suaeda acuminata*, *Suaeda crassifolia* (*Salicornia europaea*) → *Atriplex pratovii* → *Halocnemum strobilaceum* → psammophyllous shrubs (*Calligonum* spp., *Eremosparton aphyllum*) → *Haloxylon aphyllum*. The psammophytic vegetation of sandy islands was a source of seeds for spreading of psammophilous species to the dry seabed. Sixteen species of *Calligonum* mostly from *Eucalligonum* and *Pterigobasis* sections grown in coastal areas and former islands now occupy the sandy belt of the Aralkum. The territory between the mainland and the islands does not exhibit wide barren lands. It is overgrown with black saxaul, halophilous shrubs, subshrubs (*Haloxylon aphyllum*, *Halostachys belangeriana*, *Kalidium caspicum*, *Suaeda microphylla*) and annual saltworts (*Climacoptera aralensis*, *Halogeton glomeratus*, *Salsola nitraria*, *Atriplex pratovii*). The areas to the west of the former islands are presently bare of any vegetation.

Akkol–Kokaral district is located below the broad delta of the Syr Darya area and surrounds the Kokaral peninsula (Fig. 19.22, District 8). Reed and tamarisk–reed vegetation is distributed in the former bay of Akkol, where discharge of the river water takes place at times. Dry-off surfaces are occupied by *Halostachys belangeriana*. In the avandelta (the smoothly sloping, subhorizontal former underwater accumulative surface being a continuation of a delta area), the forbs (*Aeluropus littoralis*, *Xanthium strumarium*, *Lactuca tatarica*, *Karelinia caspia*, *Puccinellia distans*) and woodreed–willow (*Salix songarica*, *Salix wilhelmsiana*,

Calamagrostis pseudophragmites, *Calamagrostis epigeios*) communities have been formed. Halophytic meadows with annual saltworts and sea aster (*Phragmites australis*, *Bolboschoenus maritimus*, *Suaeda acuminata*, *Salicornia europaea*, *Tripolium vulgare* = *Aster tripolium*) are distributed nearby the river mouth. A territory to the south of the Berg Strait (now Kokaral Dam) is occupied by tamarisk communities (*Tamarix ramosissima*, *Tamarix elongata*, *Atriplex littoralis*, *Bolboschoenus maritimus*, *Tripolium vulgare*, *Crypsis schoenoides*).

At 10 km from the dam, the vegetation is characterized by heterogeneity and diversity of plant communities. Psammophytic shrub and psammophytic grass communities (*Calligonum* spp., *Astragalus brachypus*, *Eremosparton aphyllum*, *Stipagrostis pennata*) with meadow–tugaic species (*Elaeagnus oxycarpa*, *Lycium ruthenicum*, *Glycyrrhiza glabra*) occupy a sand belt of the coast of the 1960. The spatial distribution of seral plant communities on heavy marine sediments is as follows: *Salicornia europaea* (*Suaeda acuminata*) → *Climacoptera aralensis* (*Petrosimonia triandra*) → *Tamarix laxa* (*Tamarix hispida*) → *Halostachys belangeriana* (*Halocnemum strobilaceum*). The northern coast of the Kokaral peninsula is mainly occupied by tamarisk communities (*Tamarix elongata*, *Tamarix hohenackeri*) in combination with psammophytic shrubs (*Astragalus brachypus*, *Eremosparton aphyllum*); the southen coast is characterized by other tamarisks (*Tamarix laxa*, *Tamarix elongata*), sea lavender (*Limonium suffruticosum*) and reed vegetation with some halophilous herbs (*Frankenia hirsuta*, *Karelinia caspia*, *Alhagi pseudalhagi*). Often the dry seabed joins the steep original coast and has *Stipa sareptana* plant communities and aggregations. Communities of *Halocnemum strobilaceum*, *Tamarix laxa*, *Tamarix elongata* and *Lycium ruthenicum* occupy areas to the southwest of the Karatyup peninsula. After the construction of the new Kokaral Dam and the increasing level of the North Aral Sea up to 42 m asl, the former bays of Akkol and Karashalan were flooded. Reed, bulrush and pondweed vegetation (*Phragmites australis*, *Scirpus littoralis*, *Potamogeton perfoliatus*) has developed in those new lakes.

Bugun-Sarychaganak district (Fig. 19.22, District 9) is located at the dry seafloor of the northeastern coast, to the north of Lake Karashalan including the eastern part of Sarychaganak Bay. The typical spatial distribution of seral plant communities is as follows: *Suaeda acuminata* (*Salicornia europaea*) → *Climacoptera aralensis* (*Petrosimonia triandra*) → *Tamarix* spp. (*Halostachys belangeriana*) → psammophytic shrubs and grasses (*Calligonum aphyllum*, *Astragalus brachypus*, *Eremosparton aphyllum*, *Ammodendron bifolium*, *Atraphaxis spinosa*, *Stipagrostis pennata*). The former bays of Bugun, Karatyup and Sarychaganak are vegetated by sarsazan vegetation (*Halocnemum strobilaceum*) in combination with psammophytic shrubs and *Peganum harmala* in Bugun Bay; and with *Atraphaxis spinosa*, *Calligonum aphyllum*, *Limonium suffruticosum*, and *Haloxylon aphyllum* in Karatyup Bay.

Sarychaganak-Butakov district (Fig. 19.22, District 10) is situated at the dry seafloor of the northern coast (the western part of Sarychaganak Bay, the area around the Koktyrnak peninsula and the eastern part of Butakov Bay). The prevailing spatial row of primary succession is a halosere (halophytic shrub and halophytic dwarf subshrubs in late seral stages): *Salicornia europaea* (*Suaeda acuminata*) → *Climacoptera aralensis* (*Petrosimonia triandra*) → *Halocnemum strobilaceum* (*Tamarix* spp., *Suaeda physophora*, *Limonium suffruticosum*). The psammophytic shrubs (*Calligonum aphyllum*, *Eremosparton aphyllum*, *Salsola arbuscula*) occur in combination with sarsazan and tamarisk communities. Reed, salt grasses, and sagebrush communities (*Phragmites australis*, *Puccinellia distans*, *Artemisia scopiformis*) are distributed in the dry sea belt of the 1960. Indicator species for disturbances (*Anabasis aphylla*, *Peganum harmala*, *Salsola nitraria*) occupy overgrazed areas. The original feature of local conditions is the existence of the territory named Mynbulak ("thousand springs"), where brackish groundwater is pinching out to the surface. Mynbulak is located in the northeastern part of Butakov Bay and has an area of 0.45 ha. The vegetation cover there is composed of meadow–tugaic and many halophilous species (*Tamarix* spp., *Limonium suffruticosum*, *Limonium gmelinii*, *Aeluropus littoralis*, *Atriplex pungens*, *Atriplex pedunculata*, *Atriplex littoralis*, *Phragmites australis*, *Typha angustifolia*, *Scirpus tabernaemontani*, *Juncus gerardii*, *Crypsis schoenoides*, etc.).

Akespe-Shevchenko district (Fig. 19.22, District 11) begins to the east of Akespe village at Butakov Bay, surrounds the Shubartarauz peninsula and continues to Shevchenko Bay. The prevailing spatial row of primary succession is a psammosere (with a few psammophytic shrubs): *Suaeda acuminata* (*Salicornia europaea*) → *Climacoptera aralensis* (*Petrosimonia triandra*, *Bassia hyssopifolia*) → *Tamarix* spp. → *Calligonum alatum* (*Artemisia arenaria*). The late seral stage on sandy loam and loam sediments is represented by ephemeral–saltwort (*Suaeda physophora*, *Londesia eriantha*, *Climacoptera aralensis*, *Atriplex aucheri*, *Lepidium perfoliatum*, *Senecio noeanus*) and forb–sagebrush (*Artemisia scopiformis*, *Salsola nitraria*, *Eremopyrum orientale*, *Amberboa turanica*), sometimes with black saxaul (*Haloxylon aphyllum*) or with *Kochia prostrata* (in Shevchenko Bay) plant communities. Reed vegetation occurs often around the Shubartarauz peninsula. Former small bays are vegetated by *Halocnemum strobilaceum*. Overgrazed areas surrounding all villages exhibit only poisonous weed species (*Peganum harmala*, *Anabasis aphylla*) and annual saltworts. The original feature of the flora is a local domination of some endemic species: *Atriplex pungens*, *Petrosimonia hirsutissima* and *Artemisia aralensis*. The eroded sands are occupied by *Artemisia santolina*.

Tshebas-Kulandy district (Fig. 19.22, District 12) is situated at the dry seafloor of Tshebas Bay and surrounds the Kulandy peninsula. The prevailing spatial row of primary succession is a psammosere (with psammophytic shrubs and dwarf subshrubs): *Salicornia europaea* (*Tripolium vulgare*) → *Tamarix laxa* (*Calligonum aphyllum*) → *Calligonum aphyllum* (*Eremosparton aphyllum*, *Atraphaxis spinosa*), *Artemisia terrae-albae* (*Artemisia arenaria*). The late seral stage on sandy loam sediments and coastal soils with a blown-on sand cover is represented by ephemeral–saltwort (*Suaeda physophora*, *Climacoptera aralensis*, *Eremopyrum orientale*) and ephemeral–black saxaul and with *Calligonum aphyllum* communities. There is sometimes a reed belt which often reaches the shoreline.

Aggregations and communities of *Halocnemum strobilaceum* and *Peganum harmala* surround Kulandy village.

The western (Chernyshev) district (Fig. 19.22, District 13) begins at Chernyshev Bay and continues up to the frontier of Uzbekistan. The prevailing spatial sequence of the primary succession ends with halophytic shrubs and dwarf subshrubs in late seral stages: *Suaeda acuminata* → *Tamarix elongata* (*Tamarix laxa*) → *Atriplex aucheri* (*Suaeda acuminata*, *Petrosimonia hirsutissima*) → *Tamarix laxa* (*Artemisia scopiformis*, *Limonium otolepis*). The phytocenotic characterstic is the distribution of dense tamarisk thickets alternating with annual saltworts (*Atriplex aucheri, Petrosimonia hirsutissima*) in typical belts. Psammophytic vegetation (*Eremosparton aphyllum*, *Calligonum spp.*, *Stipagrostis pennata*) in combination with tamarisk (*Tamarix laxa*) is distributed close to the frontier. The dry sea belt of the 1960 is often occupied by black saxaul and tamarisk–reed vegetation. The original feature of flora is a local domination of *Mausolea eriocarpa*.

Barsa-Kelmes district (Fig. 19.22, District 14) surrounds the former island of Barsa-Kelmes. The prevailing spatial row of primary succession is again a psammosere (up to psammophytic shrubs): *Salicornia europaea* → *Atriplex pratovii* → *Stipagrostis pennata* → *Calligonum spp.* (*Haloxylon aphyllum*). The black saxaul communities are distributed especially at the eastern coast. The southern coast adjoining the cliffs of the island plateau has reed vegetation. The northern coast is occupied by *Halocnemum strobilaceum* communities, often with black saxaul (Dimeyeva and Alimbetova 2006). The western coast is vegetated by communities of *Stipagrostis pennata* and psammophytic shrubs with black saxaul (*Eremosparton aphyllum*, *Calligonum* spp., *Haloxylon aphyllum*).

9.6 Human Impact on Vegetation

The small village of Tastjubek and the Akkuduk camel farm are located on the northeastern coast of Butakov Bay. The camel flock consists of 50 animals. Along this coastline the inhabitants are using the dry seafloor as grazing area and for hay production. The annual *Atriplex* and *Suaeda* species are ideal, as are monotonous salty meadows. The salinization of the upper soil surface, grazing and mowing are preventing the perennial species from spreading out.

As already mentioned, the overgrazed areas surrounding all villages exhibit only poisonous weed species (*Peganum harmala, Anabasis aphylla*, *Salsola nitraria*) and other annual saltworts. A few places which are less saline have been conquered by *Convolvulus arvensis*, *Glycyrrhiza* and other weeds.

Along the former coastline, the inhabitants of adjacent villages are using the dry seafloor more and more as grazing area and for hay production, but also for collecting fuelwood, especially older saxaul and *Tamarix* stems. This could be managed in a way that growth and biomass use reach a high-level equilibrium: thus,

Fig. 9.23 Saxaul (*Haloxylon aphyllum*) community, indicating distinct generations of plants, according to size (photo: Wucherer)

rehabilitation of the environment would also minimise desertification constraints (Breckle 2003). In all these attempts for restoring parts of the dry seafloor and creating a dense vegetation cover, the processes of natural succession have to be maintained and could even accelerated (Wucherer and Breckle 2001).

The wetlands, mainly the tugai forests along the two main rivers Amu Darya and Syr Darya, were greatly used also in the past, but because of sinking water levels their productivity decreased and during further desiccation most of the tugai forests disappeared (Ogar 2001; Novikova 2001). There are good chances to conserve remnants and to restore new areas within the river deltas (Novikova et al. 2001; Novikova and Kuzmina 2008).

9.7 Conclusions

The vegetation of the Aralkum is very dynamic. Rapid changes in composition within 2 years are common, but also subsequent dry years can cause an almost total dieback of pioneer species and the development of open salt desert flats with puffy salt crusts for several years.

The main vegetation types in the Aralkum are halophytic, psammophytic, tugai and salt meadow communities. In general, the canopy pattern is totally dependent on the prevailing climatic situation and the establishment of perennials during subsequent favourable years. Saxaul (*Haloxylon aphyllum*) and *Halocnemum strobilaceum* play a major role in several plant communities on degraded coastal solonschaks. *Haloxylon aphyllum* on the dry seafloor of the 1970 may exhibit about five or six successful establishment events (Fig. 9.23) under natural conditions. This is about every sixth or seventh year on average. Here germination and establishment and subsequent growth leads to a rather rapid conquering of sites.

The psammophytic vegetation (*Haloxylon*, *Calligonum*, *Astragalus* and *Salsola* species) is most dominant on the dry surface areas from the 1960, partly dominant the dry surface areas from the 1970 and rare the dry surface areas from the 1980. The whole seafloor which desiccated from 1980 can be regarded as a huge salt desert flat. It covers about 80% of the present Aralkum. The typical pattern of landscapes, vegetation and soils is striated. The southern and southeastern coasts of the former Aral Sea including the dry seafloor belong to the southern Turanian phytogeographical area, and the western, northern and northeastern coasts belong to the northern Turanian area.

The various regions around the Aral Sea are considerably different in their vegetation history and dynamics, depending also on the adjacent source of diaspores.

Acknowledgements This research was funded by the German Federal Ministry for Education and Research (BMBF grant 0339714 and BMBF grant 0330389).

References

Begon M, Harper JL, Townsend CR (1986) Ecology. Individuals, populations and communities, vols 1–2. Blackwell, Oxford

Breckle S-W (2003) Rehabilitation of the Aral Sea environment, Kazakhstan. Proceedings of the international workshop (Aleppo, May 2002). In: Combating desertification – rehabilitation of degraded drylands and biosphere reserves. UNESCO-MAB dryland series no 2, pp 47–57

Breckle S-W, Agachanjanz OE (1994) Ökologie der Erde, vol 3 (Spezielle Ökologie der Gemäßigten und Arktischen Zonen Eurasiens), 2nd edn. Fischer, Stuttgart

Breckle S-W, Veste M, Yair A (2008) General conclusions – sand dune deserts, rehabilitation and conservation (ch 30). In: Breckle S-W, Veste M, Yair A (eds) Arid dune ecosystems: the Nizzana sands in the Negev desert, vol 200, Ecol Stud 200:441–459 (Springer/Berlin)

Dimeyeva LA (1990) Flora and vegetation of the Aral Sea coast (in the limits of Kazakhstan). PhD dissertation, Univ of Almaty

Dimeyeva LA (2004) Flora and vegetation of the dry sea floor of the north-western bays of the Aral Sea. In: Proceedings of international conference in Development of botany in Central Asia and its integration into industry, Tashkent, pp 22–23

Dimeyeva LA, Alimbetova Z (2006) Flora of nature reserve Barsakelmes. In: Proceedings of the third international botanical conference, Almaty, pp 46–48

Dimeyeva LA, Breckle S-W, Wucherer W (2008) Flora of the Aralkum desert (in the limits of Kazakhstan). News Nat Acad Sci Kazakhstan 6:25–31

Dimeyeva LA, Kurochkina LYa (2005) Botanical-geographic zoning of the Aral Sea coast. Scientific reading dedicated to 100-years anniversary of VBsochava, Institute of Geography, Irkutsk, pp 125–130

Dimeyeva LA, Kuznetsov LA (1999) Flora of the Aral seaside belt. Bot J St Petersburg 4(84): 39–52

Geldyeva GV, Budnikova T, Gobernik I et al (1998) Assessment of desertification processes in natural complexes of the Syr-Dar'ya delta//Ecological research and monitoring of the Aral Sea deltas. UNESCO Aral Sea Project, pp 15–41

Geldyeva GV, Ogar NP, Scorintseva IB et al (2001) Monitoring and modeling of desertification processes in the Syr Dar'ya and Amu Dar'ya River deltas, for GIS. Ecological research and monitoring of the Aral sea deltas. A basis for restoration. UNESCO Aral Sea Project. 1997–2000 Final Scientific reports. Book 2, pp 119–153

Ishankulov MSh (1980) The classification of the landscapes of the actual coasts of the Aral Sea. Probl Osvo Pustyn 5:18–24 (in Russian)

Ishankulov MSh, Wucherer W (1984) Nature complexes of the east coast of the Aral Sea in 1982. Probl Osvo Pustyn 1:52–58 (Russian)

Kabulov SK (1990) Changes of arid plant communities by the aridisation. FAN, Tashkent (Russian)

Kurochkina LYa (1979) Botanical survey in the Aral Sea basin. Probl Osvo Pustyn 2:25–33

Kurochkina LYa (2003) Psammophyte shrub deserts. In: Rachkovskaya EI, Volkova EA, Khramtsov VN (eds) Botanical geography of Kazakhstan and Middle Asia (desert region). Komarov Botanical Institute of Russian Academy of Sciences, St Petersburg, pp 285–292

Kurochkina LYa, Makulbekova GB, Wucherer W, Malaisarova A (1983) Vegetation of the dry sea floor of the Aral Sea. In: Aralskoe more. Nauka, Alma–Ata, pp 91–144

Kurochkina LYa, Makulbekova GB, Wucherer W, Dimeyeva LA (1990) Solonchakous rangelands. In: Complex characteristic of rangelands in desert zone of Kazakhstan, Ylym, Alma-Ata, pp 139–182

Lavrenko EM (1962) Main features of botanical geography of deserts in Eurasia and North Africa. Academy of Science, Moscow-Leningrad

Lymarev VI (1969) The coasts of the Aral Sea. Nauka, Moscow (Russian)

Novikova N (2001) Ecological basis for botanical diversity conservation within the Amudarya and Syrdarya river deltas. In: Breckle S-W, Veste M, Wucherer W (eds) Sustainable land use in deserts. Springer, Berlin, Heidelberg, pp 84–94

Novikova N, Kuzmina Zh (2008) Monitoring of the vegetation in conditions of the Aral Sea ecological crisis. Moscow Russian Academy of Sciences, Water Problems Institute

Novikova NM, Kuz'mina ZhV, Dilkareva TV, Trofimova NU (2001) Preservation of the tugai bio-complex diversity within Amu-Dar'ya and Syr-Dar'ya river deltas in aridization conditions// Ecological research and monitoring of the Aral sea deltas. A basis for restoration. UNESCO Aral Sea Project. 1997–2000 Final Scientific reports. Book 2, pp 155–188

Ogar NP (2001) Vegetation dynamics on the Syrdarya delta and modern land use. In: Breckle S-W, Veste M, Wucherer W (eds) Sustainable land use in deserts. Springer, Berlin, pp 74–83

Rachkovskaya EI, Safronova IN (1994) The new map of botanical-geographic regionalization of Kazakhstan and Middle Asia in the limits of the desert area. Geobotanical mapping 1992, Komarov Botanical Institute, St Petersburg, pp 33–49

Rachkovskaya EI, Volkova EA, Khramtsov VN (eds) (2003) Botanical geography of Kazakhstan and Middle Asia (desert region). Komarov Botanical Institute of Russian Academy of Sciences, St Petersburg

Rubanov IV, Ishniyazov DP, Baskakova MA, Chistyakov PA (1987) Geology of the Aral Sea. FAN, Tashkent

Walter H (1974) Die Vegetation Osteuropas, Nord- und Zentralasiens, vol VII, Vegtationsmonographien. Fischer, Stuttgart

Willert DJv (1968) Tagesschwankungen des Ionengehalts in *Salicornia europaea* in Abhängigkeit vom Standort und von der Überflutung. Ber Dtsch Bot Ges 81:442–449

Wucherer W (1979) Primary colonisation of the dry sea floor of the Aral Sea. Probl Osvo Pustyn 3:9–18

Wucherer W (1984) Vegetation formation on the dry sea floor of the Aral Sea. In: Problemy Aralskogo morya i delty Amudar'i. FAN, Tashkent, pp 155–161 (Russian)

Wucherer W (1986) Singenez rastitel'nosti osushennogo dna morya (Vegetation dynamics on the dry sea floor of the Aral Sea). PhD dissertation, University of Tomsk (Russian)

Wucherer W (1990) Vegetation development on the new habitats in the desert. Gylym, Alma–Ata (Russian)

Wucherer W, Galieva L (1985) Vegetation mapping of the formed dynamic ecosystems. Isv AN KasSSR Ser Biol 4: 3–13 (Russian)

Wucherer W, Breckle SW (2001) Vegetation dynamics on the dry sea floor of the Aral Sea. In: Breckle SW, Veste M, Wucherer W (eds) Sustainable land-use in deserts. Springer, Heidelberg, pp 52–68

Wucherer W, Breckle SW (2003) Psammophytic succession at the south-east coast of the Aral Sea. In: Rachkovskaya EI, Volkova EA, Khramcov VN (eds) Botanical geography of Kazakhstan and Middle Asia. Komarov Botanical Institute of Russian Academy of Sciences, St Petersburg, pp 340–347

Wucherer W, Breckle SW, Dimeyeva L (2001) Flora of the dry sea floor of the Aral Sea. In: Breckle SW, Veste M, Wucherer W (eds) Sustainable land-use in deserts. Springer, Heidelberg, pp 38–51

Chapter 10
Primary Succession in the Aralkum

W. Wucherer, S-W. Breckle, and A. Buras

10.1 Introduction

On the Aralkum a surface area of about 60,000 km^2 is now exposed. The dry seafloor of the Aral Sea is a new area, where land plants (including seed banks) and animals have not existed before. It is actively invaded by organisms. The formation of plant communities, soils, a new groundwater level, aquifers, and all components and processes of ecosystems is occurring more or less simultaneously. It is a typical primary succession. All new surfaces are initially devoid of life, so primary succession has been crucial throughout Earth's history. Today, all communities of plants, animals and soils are the result of primary succession (Walker and del Moral 2003).

The rapidly growing human population is the main reason for human impacts and disturbances which vary with the scale, nature and severity of the disturbance as well as with the cultural context (Barrow 1999). The fast desiccation of the Aral Sea and the formation of the new desert, the Aralkum, is a consequence of human impact. Unintentionally, mankind has created a huge experiment, an experimental set, a laboratory of nature with thousands of local events of succession with an open end.

The processes on the dry seafloor have a natural character. The dynamics on the drying seafloor is unique. The dry seafloor is the largest area worldwide where a primary succession is taking place. For our purposes, here we will define succession as the colonization of the new surface and the replacement of one ecosystem by another. Knowledge of vegetation dynamics in the Aralkum, which is a mosaic of sand and salt desert ecosystems, is important for understanding the ecosystem dynamics in the whole Central Asian area. The dry seafloor is a huge open area (see Fig. 2.7), open to invasion by plants and animals. For such an invasion process, two different terms were proposed: autotropic succession (Begon et al. 1986) and syngenetic succession (Sukachev 1938, 1954). Sukachev emphasized syngenesis as being the main mechanism for a succession, driven by the vitality and competitive force of the plants. Following Sukachev, succession is a complex process, interrelated with the exogenesis, endoexogenesis and syngenesis of plant communities.

S.-W. Breckle et al. (eds.), *Aralkum - a Man-Made Desert: The Desiccated Floor of the Aral Sea (Central Asia)*, Ecological Studies 218,
DOI 10.1007/978-3-642-21117-1_10,

In the first few years after the drying of the seafloor, the exogenic factors (geological, geomorphological and edaphic processes) dominate the ecosystem dynamics. But the start of primary succession on a wide range of substrates and climates is rather similar, starting with cyanobacteria and annuals (Breckle 2002), thus forming a kind of microbiotic or biogenic crust (Belnap and Lange 2001; Mees and Singer 2006). Study of the mechanisms of the ecosystem dynamics and the ecological attributes of the dominant species is of great importance for clarifying the open questions. The first results of the investigation of the vegetation dynamics on the dry seafloor were published in the 1980s (Wucherer 1979, 1984, 1986, 1990; Kurochkina 1979; Kurochkina et al. 1983; Kabulov 1990). They have enabled an assessment of the state of vegetation development, aided by plot-based sampling. In later publications, some key features of the vegetation dynamics of the dry seafloor were analysed (Wucherer and Breckle 2001, 2003) and models of plant succession were developed (Groeneveld et al. 2005; see also Chap. 13).

10.2 Methods and Data Collection

The investigation of vegetation was carried out as part of an interdisciplinary research programme (1977–1985) in Middle Asia to study the effects of the drying process of the Aral Sea and the change in discharge of the Amu Darya and the Syr Darya. At the end of the 1990s an international project started, supported by the German Federal Ministry for Education and Research: "Succession processes on the desiccated sea floor of the Aral Sea and perspectives of land-use" (Department of Ecology, University of Bielefeld, Germany). The last investigation of the Bayan and Kaskakulan transects (2010) took place in the framework of the project "Investigation of the over- and underground biomass of *Haloxylon* species in Kazakhstan" (Michael Succow Foundation for Nature Protection, University of Greifswald, Germany), supported by the German Federal Ministry of Environment, Nature Protection and Nuclear Safety (BMU).

Figure 9.2 illustrates the transects used for studying the dynamics of the vegetation and ecosystems. The transects stretch from the old coastline to the present one. The long-term transects were set up at the end of the 1970s and at the beginning of the 1980s. The data set now covers 30–34 years. The main vegetation types are halophytic, psammophytic, tugai and salt meadow communities. Some of the spatial and temporal series of vegetation development in the various districts were mentioned in Chap. 9. Sample plots along transects were established at different coastal areas, stretching from the former to the present coastline. The distribution of the soils and the vegetation was surveyed along these transects. There is no better approach to studying successional change at the same site over many years.

The usual standard size of sample plots for desert annual and shrub vegetation is 100 m^2. However, the vegetation on the dry seafloor is generally not very dense, with the result that this sample area is insufficient for the purposes of recording the composition of species. In earlier investigations, an intuitive sample area of 400 m^2 was used for the research. In the course of the current project, the short transect method is frequently used. Five sample areas measuring 100 m^2 were selected adjacent to each other and were then described. In this way, it is possible to interpret the results for sample areas measuring 100 m^2, 200 m^2, 300 m^2, 400 m^2 and 500 m^2. As the species–area curves show, the test area of 400–500 m^2 is adequate for measuring the composition of species and the vegetation at sandy and salty locations on the dry seafloor (Figs. 10.1 and 10.2). As an example, three psammophytic and one halophytic plant communities were selected: a homogenous *Stipagrostis* community with a low density of species (age of locality 15–20 years),

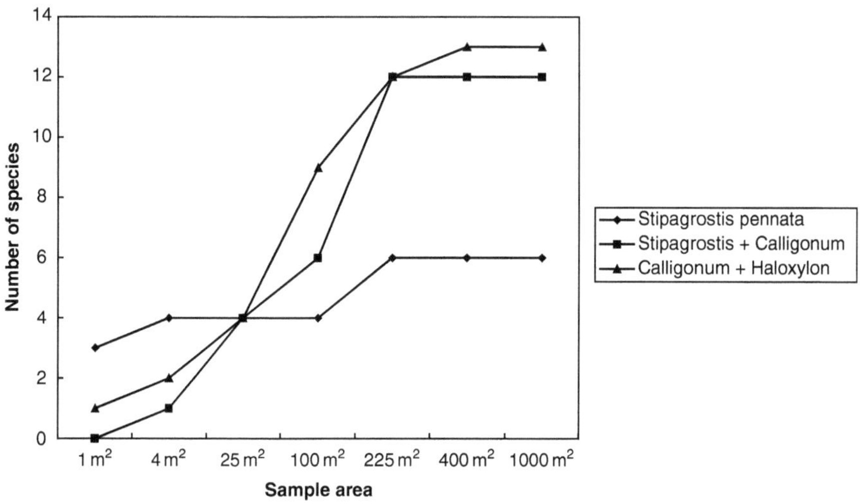

Fig. 10.1 Species–area curves for the different psammophytic plant communities

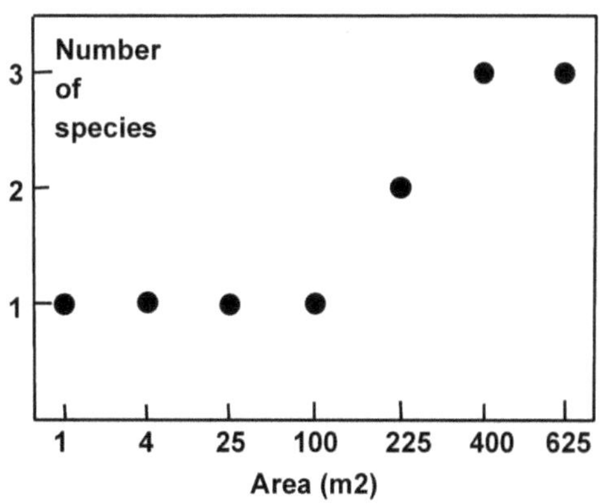

Fig. 10.2 Species–area curve for the halophytic *Halocnemum strobilaceum* plant community

a developing *Stipagrostis–Calligonum* community (age of locality 20–25 years), a relatively stable *Calligonum–Haloxylon* community (age of locality 30–35 years) and a homogenous *Halocnemum strobilaceum* plant community.

Species cover of all species was used to calculate the Shannon–Weaver Index (e.g. Magnussen and Boyle 1995) for each plot and year. In addition rank-abundance diagrams (Aoki 1995) were plotted to obtain an impression of rank-related dominance. A time series analysis using linear models was performed to determine significant trends of single species with time (see also Crawley 2007). Here, only significant trends above $r^2 = 0.4$ were considered as important. Furthermore, correlation between species development was tested using Pearson's product moment correlation. Only significant correlations above $r = 0.7$ were considered as relevant. All data were analysed using R 2.11 (R - Foundation for Statistical Computing 2010).

10.3 Results

10.3.1 Observations on the Permanent Plots on the Bayan Transect

The Bayan transect is at the eastern coast of the Aral Sea, 10 km to the south of the mouth of the Syr Darya in 1981 (Fig. 9.2, transect 3). It extends more than 30 km from the former coastline towards today's coastline. The transect's length as well as the formation of new land have accelerated greatly during the past 30 years. Research along the transect was performed in 1981–1984, 1986–1987, 1989–1990, 1992, 1994, 1998–1999, 2003 and 2010. The dry seafloor can be divided into four parts that show geographical as well as ecological differences and therefore also differ in the process of succession. The desiccated seafloor of the 1960s consists mainly of sand sediments, which are almost free of salt. Well-developed psammophytic communities of *Stipagrostis pennata*, *Eremosparton aphyllum* and *Calligonum* species formed on these sandy sites. The seafloor of the 1970s forms the second part. The soil profile in this coastal area consists of alternating layers of sand and clay. These soils are mostly encrusted puffy or degrading coastal solonchaks on which loose populations of halophytes grow consisting of *Halostachys caspica* and *Halocnemum strobilaceum*, mixed with *Tamarix* species and therophytes. Clayey oversalted substrates (so-called crusty or encrusted puffy coastal solonchaks) of the 1980s and 1990s up to 2010 form the third and fourth parts. Here, therophytic communities of *Atriplex*, *Petrosimonia* and *Climacoptera* species or open salt deserts developed. Along the transect, spatial diversity and biological diversity decrease towards today's coastline. Three plant associations were chosen for analysis:

- *Calligonum aphyllum–Eremosparton aphyllum–Stypagrostis pennata* plant community on the dry seafloor of the 1960s (plot 1)
- *Halocnemum strobilaceum–Halostachys caspica* plant community on the dry seafloor of the 1970s (plot 2)
- Therophytic plant community with *Halostachys caspica* on the dry seafloor of the 1980s: *Halostachys caspica–Petrosimonia brachiata–Climacoptera aralensis* plant community (plot 3)

Plot 1: The time of succession on the dry seafloor of the 1960s is 50 years (up to 2010). The transect grew on average in the 1960s by 15.4 m/year (Wucherer 1990). The sand beach was formed successively and only reached today's limits in the late 1960s. The low concentration of salt in the seawater in the 1960s (10–11 g/L) led to a modest salinity of the soil, and its desalinization took about 8–10 years. The presence of all three first-generation pioneer species (*Salicornia europaea*, *Suaeda acuminata*, *Atriplex pratovii*) on the Bayan transect and the absence of reed remnants allow a clear analysis of the beginning of the succession. In the first 3–4 years after the retreat of the seawater, a beach without vegetation had formed. The therophytes needed several years to realize their potential populations. The first stage in the mid-1960s was certainly a mixed *Suaeda acuminata–Salicornia europaea* community with dominating *Suaeda*. *Salicornia* disappeared with further retreat of the sea and the lowering groundwater level. A mixed plant community of *Atriplex pratovii–Suaeda acuminata* with dominant *Atriplex* (second stage of succession) was formed.

At the beginning of the 1970s, the occupation of the sand beach by *Stipagrostis pennata* began. There was a loose *Stipagrostis* community in the mid-1970s on the beach. Then, in the second half of the 1970s, the occupation by *Eremosparton aphyllum* followed. In the first year of research (1981), a mixed plant community of *Stipagrostis pennata–Eremosparton aphyllum* consisting of ten species (third stage of succession) was found on the desiccated seafloor of the 1960s. The cover rate of *Stipagrostis* was 20% and that of *Eremosparton* was 2%. During the following 30 years, the cover rate varied between 22% and 34% (Fig. 10.3) and in 2010 was 23% (comparable with the level in 1981). The diversity of species increased (Fig. 10.4), and there were 14 species in the last year of research. Saturation of species has not been reached yet. The vegetation of the dried-out floor is still developing, as confirmed of both named factors. The density of *Eremosparton aphyllum* increased in the 1980s (Fig. 10.5). A change of dominance happened in 1992, and a plant community of *Eremosparton aphyllum–Stipagrostis pennata* was formed (fourth stage of succession). The *Eremosparton* population was steady in the 1990s, but the cover rate decreased up to 2003 and 2010 (now only a cover rate of 4%). *Chondrilla brevirostris* and *Calligonum* species (Fig. 10.5) populated greater parts in the last 11 years (since 1999). The positive trend of the development of these species is also confirmed by the calculation of $r^2 - 0.56$ for *Calligonum* and 0.71 for *Chondrilla* (Table 10.1). Significant high positive correlations of 0.70 and more are exhibited by the following species (Table 10.2): *Leymus racemosus* and *Stipagrostis pennata* (early succession stage), *Eremosparton aphyllum* and

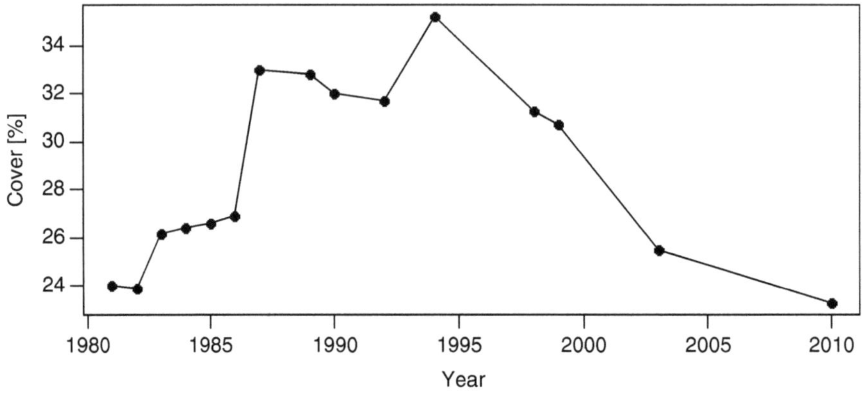

Fig. 10.3 Changes of plant cover percentage in the psammophytic community (*Calligonum aphyllum–Eremosparton aphyllum–Stipagrostis pennata*) on the dry seafloor of the 1960s, Bayan transect (plot 1)

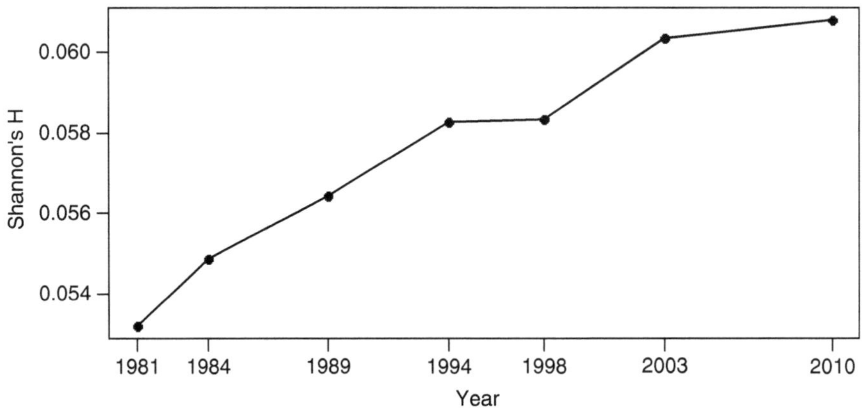

Fig. 10.4 Dynamics of species diversity in the psammophytic community (*Calligonum aphyllum–Eremosparton aphyllum–Stipagrostis pennata*) on the dry seafloor of the 1960s expressed by Shannon's diversity index *H*, Bayan transect (plot 1)

Anisantha tectorum (middle succession stage), and *Calligonum aphyllum* and *Chondrilla brevirostris* (later succession stage). All in all, there have been five different stages of succession since the beginning of the 1960s:

1. *Suaeda acuminata–Salicornia europaea* plant community (duration of 3–4 years, 1964–1967)
2. *Atriplex pratovii–Suaeda acuminata* plant community (duration of 6–7 years, 1968–1974)
3. *Stipagrostis pennata–Eremosparton aphyllum* plant community (duration of 17–18 years, 1975–1991)

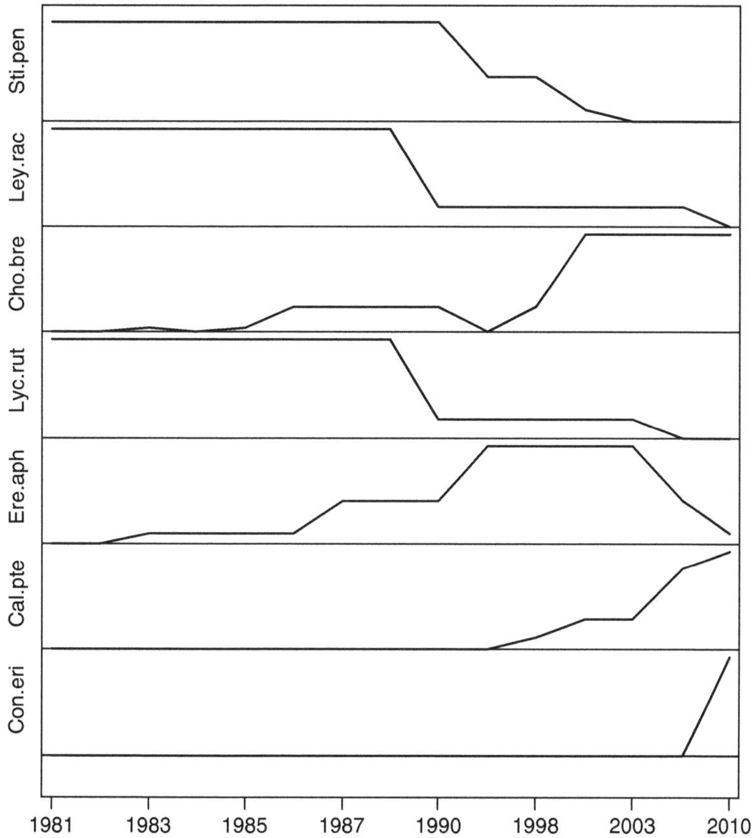

Fig. 10.5 Changes of the cover percentage of different plant species in the psammophytic community (*Calligonum aphyllum–Eremosparton aphyllum–Stipagrostis pennata*) on the dry seafloor of the 1960s, Bayan transect (plot 1). *Sti.pen*, *Stipagrostis pennata*; *Ley.rac*, *Leymus racemosus*; *Cho.bre*, *Chondrilla brevirostris*; *Lyc.rut*, *Lycium ruthenicum*; *Ere.aph*, *Eremosparton aphyllum*; *Cal.pte*, *Calligonum* sect. *Pterococcus* (*Calligonum aphyllum*); *Con.eri*, *Convolvulus erinaceus*

Table 10.1 Trend analysis of some species development (*n* negative trend – constant decrease of abundance, *p* positive trend – constant increase of abundance) in the psammophytic plant community (*Calligonum aphyllum–Eremosparton aphyllum–Stipagrostis pennata*) on the dry seafloor of the 1960s (Bayan transect, plot 1)

Species	r^2	p
Stipagrostis pennata	0.77 (n)	<0.001
Leymus racemosus	0.78 (n)	<0.001
Chondrilla brevirostris	0.71 (p)	<0.001
Lycium ruthenicum	0.80 (n)	<0.001
Eremosparton aphyllum	0.41 (p)	0.009
Calligonum aphyllum	0.56 (p)	0.001

Table 10.2 Correlations between species in the psammophytic plant community (*Calligonum aphyllum–Eremosparton aphyllum–Stipagrostis pennata*) on the dry seafloor of the 1960s (Bayan transect, plot 1)

	Stipagrostis pennata	Leymus racemosus	Chondrilla brevirostris	Lycium ruthenicum	Calligonum sect. Pterigobasis (Calligonum aphyllum)	Anisantha tectorum	Corispermum aralo-caspicum	Senecio noeanus
Leymus racemosus	0.82**							
Lycium ruthenicum	0.84**	0.99**						
Chondrilla brevirostris	−0.85**							
Calligonum sect. Pterigobasis (C. aphyllum)	−0.81**		0.83**					
Eremosparton aphyllum	−0.77*	−0.83**		−0.79*		0.70*		
Anisantha tectorum	−0.71*							
Lappula semiglabra	−0.90**		0.84**	−0.75*	0.78*		0.78*	0.83**
Atriplex pratovii		−0.73*						
Strigosella circinnata			0.75*		0.82**			0.78*
Senecio noeanus					0.72*			
Centaurea pulchella					0.88*			
Corispermum sp.							0.94**	

$*p < 0.01$, $**p < 0.001$

4. *Eremosparton aphyllum–Stipagrostis pennata* plant community with *Calligonum* (duration of 11 years, 1992–2002)
5. *Calligonum aphyllum–Eremosparton aphyllum–Chondrilla brevirostris* plant community (duration of 8 years, 2003–2010)

The changes from the first to the second phase and from the second to the third phase are mainly governed by exogenous conditions because of the decline of the groundwater level and the desalinization of the floor's profile. The changes from the third to the fourth phase and from the fourth to the fifth stages are caused by endogenous factors.

Plot 2: This seafloor site dried out in 1977. The duration of succession here is up to 34 years (until 2010). From 1981 until 1987, the cover rate of plant communities varied between 5% and 40% (fluctuations of therophyte cover). Since 1988 the cover rate has varied between 10% and 25% (Fig. 10.6), and in 2010 it was 15%. In the first phase a *Salicornia europaea* plant community was mixed with *Suaeda* species on marsh solonchaks. This stage had a duration of 5 years. The good development of *Salicornia* in the fifth year of succession was influenced by the rich rainfall in 1981. The feature of this location is a quick growth of the diversity of species in the first years after the seafloor dried out (there were 17 species registered in 1981) and an abrupt decrease at the end of the 1990s (Fig. 10.7). The soils alternated between encrusted puffy solonchaks and degrading coastal solonchaks. The second therophyte stage (between 1982 and 1987) was less distinct. It was rather a vacuum of colonization and establishment. Between 1987 and 1999 a *Halostachys caspica* plant community developed (Fig. 10.8). In the late 1990s the salinity of the soil increased and an encrusted puffy solonchak formed (as a result of a higher groundwater level and deflation). This led to a fast distribution of *Halocnemum strobilaceum* and a loss of *Halostachys caspica.* A new plant community dominated by *Halocnemum strobilaceum* developed (Fig. 10.8), and the *Tamarix* species could not establish themselves. *Halostachys caspica* shows

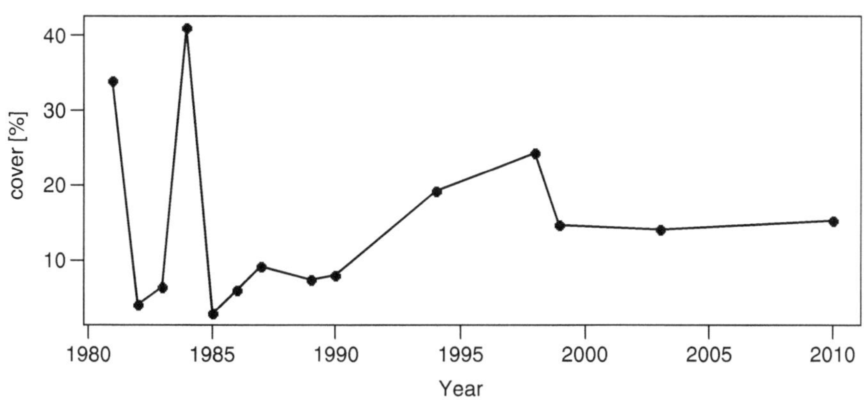

Fig. 10.6 Changes of plant cover percentage in the halophytic community (*Halocnemum strobilaceum–Halostachys caspica*) on the dry seafloor of the 1970s, Bayan transect (plot 2)

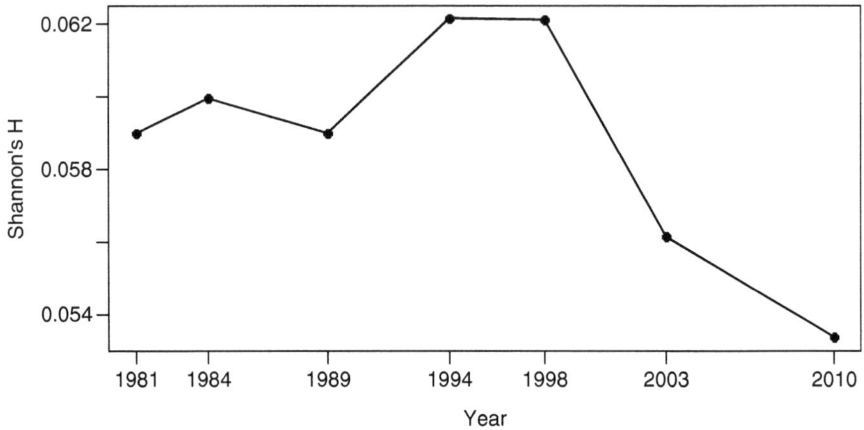

Fig. 10.7 Dynamics of species diversity in the halophytic community (*Halocnemum strobilaceum–Halostachys caspica*) on the dry seafloor of the 1970s, expressed by Shannon's diversity index *H*, Bayan transect (plot 2)

a positive development trend (0.66), as does *Halocnemum* (0.82) (Table 10.3), but the vitality of *Halostachys* has to be considered as very low.

As a result of the research, five stages of succession were distinguished in this plot:

1. *Salicornia europaea* plant community (duration of 5 years, 1977–1981)
2. Loose *Climacoptera aralensis* plant community with perennials (duration of 6 years, 1982–1987)
3. *Halostachys caspica–Halocnemum strobilaceum* plant community with therophytes (duration of 15 years, 1988–2002)
4. *Halocnemum strobilaceum–Halostachys caspica* plant community (duration of 7 years, 2003–2009)
5. *Halocnemum strobilaceum* plant community (duration of 1 year, 2010)

The fast establishment of *Halocnemum* happened rather coincidentally as a result of exogenous factors (high groundwater level and deflation).

Plot 3: This area dried out in 1981. The time of succession is 30 years (up to 2010). In the first stage, a *Salicornia europaea* community developed. It was mixed with *Suaeda crassifolia* on marshland solonchaks, and with *Climacoptera* on coastal solonchaks (Fig. 10.9). This stage lasted for 3 years. Thereafter, encrusted puffy solonchaks formed. The second therophyte phase with *Climacoptera aralensis* and *Petrosimonia triandra* lasted longer (20 years, from 1984 to 2003). Since 1994, however, a slight tendency towards a gain of *Halostachys caspica* can be observed (Fig. 10.9, Table 10.4). Only *Halostachys* shows a clear positive development trend. The cover percentage was relatively high in the 1980s and low in the 1990s (Fig. 10.10). Species diversity varied during succession; in general it increased relatively (Fig. 10.11). The correlation chart (Table 10.5) shows interesting positive relations and connections between different species, e.g. between *Halostachys*

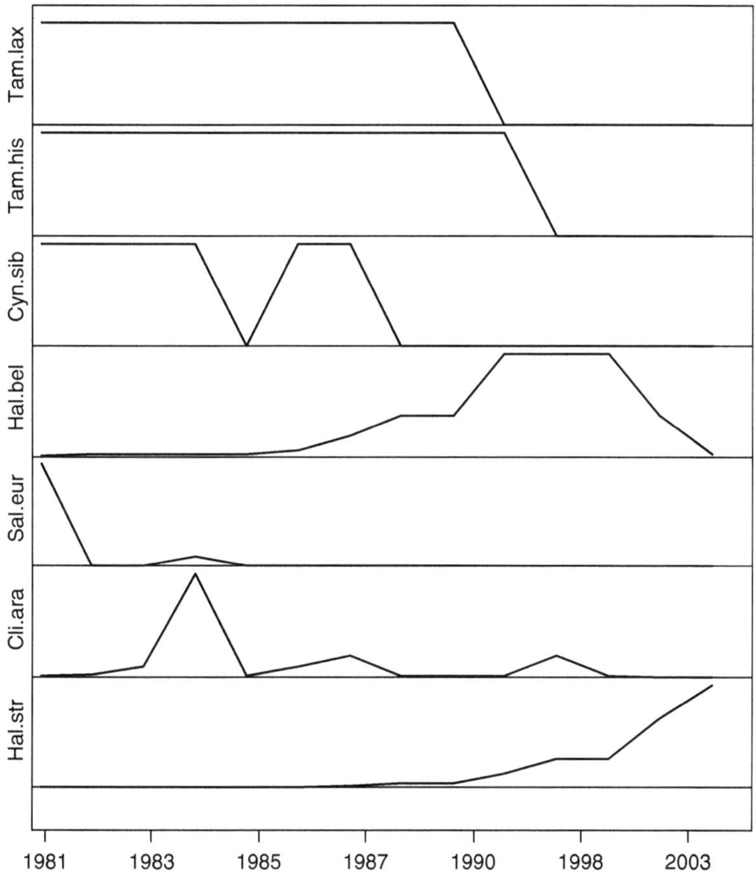

Fig. 10.8 Changes of the cover percentage of different plant species in the halophytic community (*Halocnemum strobilaceum–Halostachys caspica*) on the dry seafloor of the 1970s, Bayan transect (plot 2). *Tam.lax, Tamarix laxa*; *Tam.his, Tamarix hispida*; *Cyn.sib, Cynanchum sibiricum*; *Hal. ber, Halostachys caspica (belangeriana)*; *Sal.eur, Salicornia europaea*; *Cli.ara, Climacoptera aralensis*; *Hal.str, Halocnemum strobilaceum*

Table 10.3 Trend analysis of the some species development (*n* negative trend – constant decrease of abundance, *p* positive trend – constant increase of abundance) in the halophytic community (*Halocnemum strobilaceum–Halostachys caspica*) on the dry seafloor of the 1970s, Bayan transect (plot 2)

Species	r^2	p
Tamarix laxa	0.77 (n)	<0.001
Tamarix hispida	0.77 (n)	<0.001
Cynanchum sibiricum	0.50 (n)	<0.001
Halostachys caspica	0.66 (p)	<0.001
Halocnemum strobilaceum	0.82 (p)	<0.001
Salicornia europaea	0.23 (n)	0.08
Climacoptera aralensis	0.06 (n)	0.38

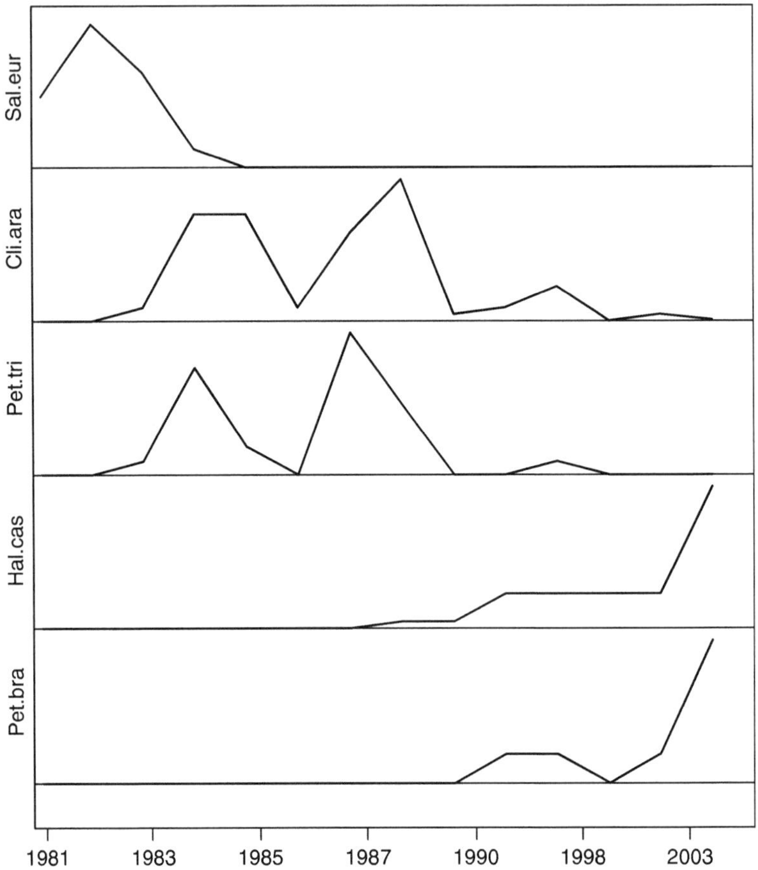

Fig. 10.9 Changes of the cover percentage of different plant species in the loose halophytic community (*Halostachys caspica–Petrosimonia brachiata–Climacoptera aralensis*) on the dry seafloor of the 1980s, Bayan transect (plot 3). *Sal.eur*, *Salicornia europaea*; *Cli.ara*, *Climacoptera aralensis*; *Pet.tri*, *Petrosimonia triandra*; *Hal.cas*, *Halostachys caspica*; *Pet.bra*, *Petrosimonia brachiata*

Table 10.4 Trend analysis of the some species development in the loose halophytic community (*Halostachys caspica–Petrosimonia brachiata–Climacoptera aralensis*) on the dry seafloor of the 1980s, Bayan transect (plot 3)

Species	r^2	p
Halostachys caspica	0.73	<0.001
Salicornia europaea	0.52	0.006
Phragmites australis	0.49	0.007
Frankenia hirsuta	0.40	0.02
Petrosimonia brachiata	0.40	0.02

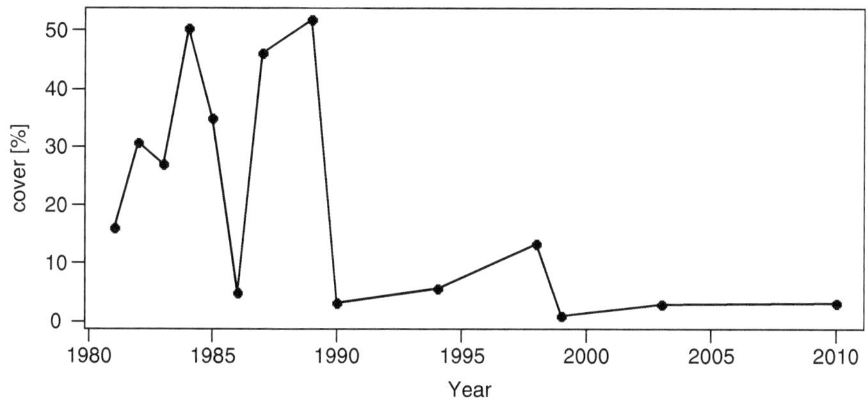

Fig. 10.10 Changes of the total cover percentage in the loose halophytic community (*Halostachys caspica–Petrosimonia brachiata–Climacoptera aralensis*) on the dry seafloor of the 1980s, Bayan transect (plot 3)

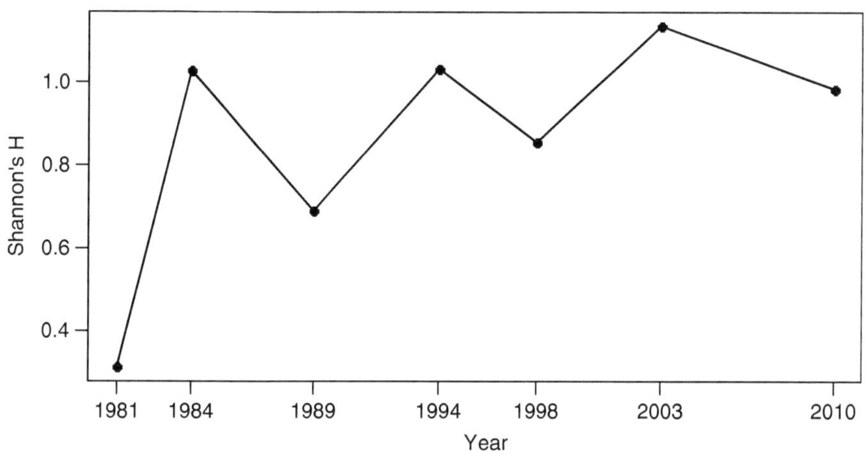

Fig. 10.11 Dynamics of species diversity in the halophytic community (*Halostachys caspica–-Petrosimonia brachiata–Climacoptera aralensis*) on the dry seafloor of the 1980s, Bayan transect (plot 3)

caspica and *Petrosimonia brachiata*, between *Climacoptera aralensis* and *Petrosimonia triandra* and between *Bassia hyssopifolia* and *Suaeda acuminata*.

In this plot, three succession stages can be distinguished:

1. *Salicornia europaea* plant community (duration of 3 years, 1981–1983)
2. Loose *Climacoptera aralensis* plant community with *Halostachys caspica* (duration of 19 years, 1984–2002)
3. Loose *Halostachys caspica* plant community with therophytes (duration of 8 years, 2003–2010)

Table 10.5 Correlations between species in the halophytic plant community (*Halostachys caspica–Petrosimonia brachiata–Climacoptera aralensis*) on the dry seafloor of the 1980ies (*$p < 0.01$, **$p < 0.001$), Bayan transect (plot 3)

	Phragmites australis	Petrosimonia triandra	Senecio noeanus	Halostachys caspica	Atriplex sphaeromorpha	Karelinia caspica	Suaeda acuminata	Descurainia sophia	Petrosimonia brachiata
Salicornia europaea	0.77*	–	–	–	–	–	–	–	–
Climacoptera aralensis	–	0.78*	0.72*	–	–	–	–	–	–
Tamarix laxa	–	–	–	–0.82**	–	–	–	–	–0.70*
Atriplex pratovii	–	–	–	–	0.96**	–	–	–	–
Petrosimonia triandra	–	xxx	–	–	–	0.72*	–	–	–
Lycium ruthenicum	–	–	0.93**	–	–	–	–	–	–
Bassia hyssopifolia	–	–	–	–	–	–	0.82**	–	–
Puccinellia distans	–	–	–	–	–	–	–	1**	–
Petrosimonia brachiata	–	–	–	0.81**	–	–	–	–	–

The long-living species (perennials) spread slowly and the conditions on the dried areas of the 1980s and 1990s are unfavourable for occupation and establishment. As can be seen, the process of succession differs greatly between the plots.

10.3.2 Chronosequences of the Psammophytic Plant Communities

Chronosequence studies are popular because they give immediate results for minimal effort and greatly extend the potential time span of a study (Walker and del Moral 2003). The characteristics of the psammophyte succession were investigated at the southeastern coast of the Aral Sea between the Syr Darya delta and the Akpetki Archipelago (Fig. 10.12). Along the banks of the mainland there is a series of sand islands or bars, such as Kenderly, Ujaly, Akpasty, Kaskakulan and Uzun-Kair. The islands and the banks are surrounded by beaches of dunes composed of sand and abraded shell debris. These sand bars or sand islands are around 7–18 km long and between several hundred metres and 2,000 m wide. They are approximately 3–7 m above the local soil level. The soil is composed of bright-

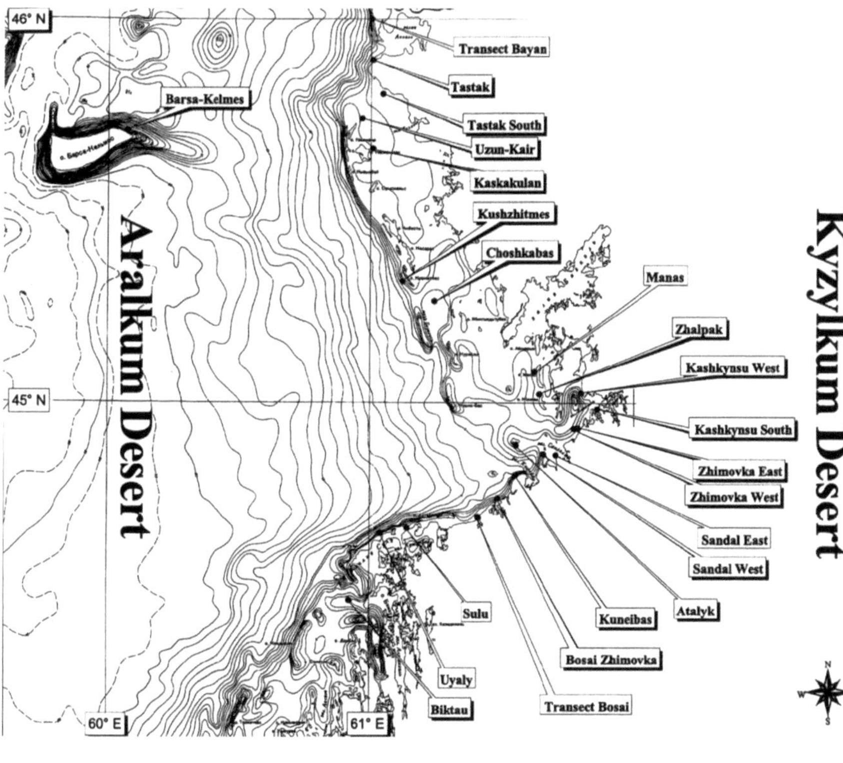

Fig. 10.12 Transects investigated at the southeastern coast of the Aral Sea

yellow quartz sand with a significant proportion of abraded shells. The sand sediments are usually distributed on the dry seafloor at altitudes of 53–46 (rarely to 43) m asl. The stretch of dry seafloor between the bar islands and the mainland shore is approximately 5–15 km wide and is composed of silty sediments. At the points where the mainland of the coast is not blocked by the islands, the waves reached the bank and formed escarpments in the wash as well as a narrow beach. As soon as the sea had withdrawn, the active development of exogenous processes began.

The geographical coordinates for the localities of our studies are based partly on GPS measurements (data from the 1990s); some are from topographic maps only (data from the 1980s). Cover percentage is given as the percentage according to the decimal scale of Londo (1975, 1984). A cover value of the Londo scale below 1 is given as: + = 0.5; *r* = 0.1. For purposes of determining the age of the habitats after desiccation, 11 age categories were defined: 1, 2, 3, 4–5, 6–7, 8–10, 11–15, 16–20, 21–25, 26–30, and 31–40 years. The plant communities and habitats from different localities with determined age after desiccation were used for the analysis of the vegetation changes. The syntaxonomical analysis of vegetation is not yet finished. Therefore, the plant communities discussed in this chapter are without syntaxonomical status. Most probably there is no stable and repeated species pattern during the successional phases; thus, syntaxonomical analysis may be useless in this case.

We distinguish the following stages of the vegetation changes on the sand habitats of the dry seafloor.

10.3.2.1 Stages of Annual Halopsammotherophytes or the *Atriplex Pratovii* Stages

Atriplex pratovii is a Turanian endemic which is found on the coast of the Aral Sea. This species occurred only sporadically. But *Atriplex pratovii* has proliferated quite rapidly on the dry seafloor and by the end of the 1970s it was found at various locations around the sea with high cover percentages. The average size of the plant is between 30–70 cm. *Atriplex pratovii* plant communities were found at three habitats (Table 10.6).

1. *Atriplex pratovii–Suaeda acuminata* plant communities on the sand marsh solonchaks.
   These plant communities were only described in 1978 directly at the water level on the coast of the island of Ujaly (Table 10.6), at the Sulu peninsula and in 1983 on the eastern coast of the island of Barsa-Kelmes. *Atriplex pratovii* has absolute dominance in those plant communities. The only pioneer species was *Suaeda acuminata*. The groundwater level during the first half of the year was less than 70 cm below the surface and the salinity of the upper soil was 1–1.5% (dry mass) and rarely rose above this. This means that the seeds from *Atriplex pratovii* survived the flooding and the effects of the moderately saline seawater on the

Table 10.6 Species composition and cover of the *Atriplex pratovii* plant communities on the habitats of the dry seafloor of the Aral Sea (with different age after the drying out)

	Sand marsh solonchaks					Sandy coastal and degraded solonchaks					Loamy coastal solonchaks with sand layer				
	Ba	U	Su	Su	Tb	Tb	Ku	Aw	Ba	Bo	Sw	K	Ma	Uz	Ka
Time after desiccation (years)	1	1	1	2	4–5	4–5	4–5	4–5	6–7	6–7	6–7	8–10	8–10	21–25	26–30
Number of species	1	2	2	2	3	6	4	4	4	6	6	5	8	6	6
Therophyte cover (%)	2	21	11	30	80	30	10	30	4	40	13	4	15	45	88
Perennial cover (%)						0.2	0.3	0.1	0.2	0.3	2	1	3	1	0.1
Atriplex pratovii	**2**	**20**	**10**	**20**	**50**	**30**	**10**	**20**	**2**	**30**	**10**	**4**	**10**	**40**	**80**
Suaeda acuminata		0.5	0.5	10	30	0.1									0.1
Salicornia europaea					0.1	0.1									
Astragalus brachypus						0.1	0.1			0.1					0.1
Stipagrostis pennata									0.1						
Calligonum (Pterigobasis, Eucalligonum)										0.1					
Eremosparton aphyllum							0.1								
Salsola paulsenii						0.1			2	10	0.5	0.1	0.5	4	4
Agriophyllum squarrosum								0.1		0.1	2	0.1			
Corispermum aralo-caspicum								10			0.1				2
Halocnemum strobilaceum								0.1		0.1	2				
Tamarix laxa						0.1	0.1		0.1				0.5		
Nitraria schoberi												0.5	2	0.5	
Limonium suffruticosum												0.5	0.5		
Salsola nitraria													4	0.1	
Haloxylon aphyllum														0.1	
Tamarix elongata											2				
Calligonum (Pterococcus)													0.1		
Eremopyrum orientale													0.1		
Climacoptera ferganica														0.5	
Senecio noeanus															2

Aw Aralsk West, *Ba* Barsa-Kelmes, *Bo* Bosai, *K* Kuneibas, *Ka* Kaskakulan, *Ku* Kushzhitmes, *Ma* Manas, Su Sulu, *Sw* Sandal West, *U* Ujaly, *Uz* Uzun-Kair

upper soil. One can assume from this that during the course of the 1960s and until the end of the 1970s, this was an important pioneer species in the colonization of the sandy marsh solonchaks along with *Suaeda acuminata* and *Salicornia europaea*. On the island coastlines which open onto the sea and where wave and surf action were particularly intensive, it is completely dominant. This stage lasts for approximately 1–2 years and also includes the first stage in the formation of coastal solonchaks.

The remnants of the sea grass (*Zostera minor*) which are present in large quantities on the dry seafloor reduce evaporation and the accumulation of salt in the soil while maintaining the moisture of the upper soil profile. *Suaeda acuminata* forms its own synusium together with *Atriplex* on the sea grass covering the coastal solonchaks. On the basis of ecological conditions, this habitat is comparable to the marsh solonchaks.

2. *Atriplex pratovii* plant communities with annual psammophytes on the sand coastal and degraded solonchaks.
   This plant community was described on the Bosai transect of the southeastern coast (Table 10.6) and on the coast of the island of Barsa-Kelmes. The coastal and degraded solonchaks indicate the absence of sea storm flooding and a drop in the groundwater level (to 1.2–1.3 m). The sand surface (10–20 cm) of the degraded solonchaks is dry. Deflation is beginning to take place. A number of small dunes or "tongue dunes" measuring 10–30 cm high are formed around the plants. This is comparable to the formation of the low primary dunes with *Agropyrum junceum* and *Elymus arenarius* at the North Sea coast. The difference is that the plants which form low dunes in our case are therophytes. This stage lasts 4–5 years. Here, we can distinguish between two different stages. In this stage, the typical psammophytic pioneer species *Salsola paulsenii* and *Agriophyllum squarrosum* (similarly therophytes) can also form their own synusia.
3. *Atriplex pratovii* plant communities on loamy coastal solonchaks with a sand layer.
   This habitat illustrates a loam soil profile which is covered by a layer of sand which is 10–50 cm thick. This covering of sand may have a natural origin or it may be caused by secondary means as a result of aeolian processes. The conditions required for germination of plants and their establishment are more favourable because of this type of sand layer. This soil type occurs at the most remote boundaries of the sand deposits on the dry seafloor and at the edge of the newly created crescentic dune fields. The composition of the pioneer species is relatively heterogeneous (Table 10.6) and is dependent on the thickness of the sand layer and the level of salinity of the loamy subsoil. It can include typical halophytes such as *Halocnemum strobilaceum* and *Climacoptera ferganica* or tugai species such as *Tamarix laxa* and *Limonium otolepis*.
   The conclusion is that *Atriplex pratovii* is an important pioneer species on the sandy marsh, coastal and degraded solonchaks and represents two succession stages with different pioneer species. The loamy coastal solonchaks with sand layers are the optimal habitat for this species.

10.3.2.2 *Haloxylon aphyllum* and *Nitraria schoberii* Stages on the Loamy Coastal Solonchaks with a Sand Layer

Haloxylon aphyllum, as an Irano-Turanian species, has a broad ecological range and occurs on salty soils, on sands and even in the Hamada deserts. The sand-covered loamy coastal solonchaks as a habitat provide the optimum conditions for germination and establishment of this species on the dry Aral Sea floor. The low amount of coverage (less than 20%) in most of the plant communities (Table 10.7) means that the plant communities are still in the development phase. On the other hand, it has to be taken into consideration that the root systems in saxaul are much more extensive and have greater biomass than the above-ground organs (Walter

Table 10.7 Species composition and cover of the *Haloxylon aphyllum* and *Nitraria schoberi* plant communities on the habitats of the dry seafloor of the Aral Sea (with different age after desiccation)

	Sand degraded solonchaks and sandy soils					Loamy coastal solonchaks with sand layer					Loamy coastal solonchaks with sand layer				
	Ks	Sw	Bi	Bi	U	T	Tb	Ka	Zu	Ka	T	Ts	Zh	Ma	Ak
Time after desiccation (years)	8–10	8–10	21–25	26–30	16–20	6–7	8–10	21–25	21–25	26–30	11–15	11–15	11–15	11–15	16–20
Number of species	5	6	5	7	9	8	7	5	6	7	4	7	6	5	8
Therophyte cover (%)	11	1	0.2	0.3	0.2	1	0.5	1	3	21	4	1	11	0	10
Perennial cover (%)	4	4	14	13	20	7	11	10	12	54	20	10	10	6	21
Haloxylon aphyllum	**4**	**4**	**10**	**10**	**10**	**4**	**10**	**10**	**10**	**50**	0.1	0.1			
Stipagrostis pennata		0.1	0.1	0.5	4										
Calligonum (*Pterigobasis*, *Eucalligonum*)		0.1	4	2	4										
Corispermum aralo-caspicum		0.1	0.1	0.1											
Agriophyllum squarrosum	0.5		0.1	0.1									0.1		
Ephedra lomatolepis				0.1	0.1										
Salsola nitraria					0.1	0.1		0.1	2	20					
Tamarix hispida					0.1	0.1	0.1								
Halocnemum strobilaceum						0.5	0.1								
Halostachys caspica						0.5			2						
Nitraria schoberi						2	0.1	0.1			**20**	**10**	**10**	**4**	**20**
Tamarix laxa					4							0.1	0.1	0.1	0.1
Phragmites australis												0.1	0.1	0.1	0.1
Limonium otolepis							0.1							0.1	
Salsola paulsenii	0.5	0.5			0.1		0.5	0.5	0.5	0.5	0.1	0.1	0.5		
Atriplex pratovii	0.1	0.1				0.5					4	0.5	10		
Horaninovia ulicina	10			0.1						0.5		0.1			10
Astragalus brachypus					2					4				2	0.1
Tamarix elongata					4										
Kalidium foliatum						0.1									
Zygophyllum oxianum							0.1								

(continued)

Table 10.7 (continued)

	Sand degraded solonchaks and sandy soils					Loamy coastal solonchaks with sand layer					Loamy coastal solonchaks with sand layer				
	Ks	Sw	Bi	Bi	U	T	Tb	Ka	Zu	Ka	T	Ts	Zh	Ma	Ak
Lepidium perfoliatum										0.1					
Strigosella circinnata										0.1					
Peganum harmala									0.1						
Girgensohnia oppositiflora									0.1						
Climacoptera ferganica								0.1							
Calligonum (*Pterococcus*)															0.1
Convolvulus subsericeus															0.1
Eremosparton aphyllum															0.1

Ak Akbasti, *Bi* Biktau, *Ka* Kaskakulan, *Ks* Kashkynsu South, *Ma* Manas, *Sw* Sandal West, *T* Tastak, *Tb* Bosai transect, *Ts* Tastak South, *U* Uyaly, *Zh* Zhalpak

1974; Miroshnichenko 1975), and thus it may be a completely rooted substrate already.

In the Kaskakulan transect (Fig. 9.2, transect 4) there are plant communities where the level of coverage is up to 50% and more. Such a dense *Haloxylon* vegetation is only found on the old flood terraces. This may be an indication of a large groundwater supply. *Nitraria schoberi*, which is also an Irano-Turanian species, has its optimum proliferation conditions on the sand-covered loamy coastal solonchaks. The coverage here is between 10% and 20%. The germination of this species is strongly influenced by the amount of precipitation.

Both species occupy the dry seafloor habitats 7–8 years after desiccation. The following spreading is controlled by local habitat conditions.

10.3.2.3 Grass Stage or the *Stipagrostis pennata* Stages

This stage can be observed in virtually every desert on the planet and indicates the colonization of the dunes and wind-eroded sand areas (natural or anthropogenic deflation). In North Africa and the Middle East, there are various *Stipagrostis* (*Aristida*) species (Danin 1996), and in China there is *Psammochloa villosa* (Petrov 1973). In Central Asia there is a classic vicariance of species. From the north to the south, the vegetation changes from *Leymus racemosus* to *Stipagrostis pennata* to *Stipagrostis karelinii* plant communities. In the area of the Aral Sea on the dry seafloor, this stage is characterized by *Stipagrostis pennata* on the southeastern coast of the Aral Sea to the south of the Syr Darya delta. We distinguish three steps of colonization by *Stipagrostis pennata*.

1. *Stipagrostis pennata* plant communities on degraded coastal solonchaks.
   The first examples of *Stipagrostis pennata* were recorded on the dry areas aged some 4–5 years. By 8–10 years after desiccation (Table 10.8), sparse *Stipagrostis* communities form with a coverage of 10–20% (rarely more). The chokolaks (small dunes around the plants) are 10–30 cm high. The topsoil is desalinized to a depth of about 1 m. The other pioneer species are generally therophytes such as *Atriplex pratovii* and *Horaninovia ulicina*, all of which form their own synusia.
2. *Stipagrostis pennata* plant communities on sandy soils.
   Once the groundwater level is below 1–2 m, the topsoil becomes almost entirely desalinized. The coverage of *Stipagrostis pennata* increases (20–30%, maximum 50%). The therophytes only occur in isolated cases or are not present. Other hardy psammophytes such as *Calligonum* species and *Haloxylon aphyllum* occur in small groups and form their own synusia (Table 10.8). The age of the habitats is 20–30 years.
3. *Stipagrostis pennata* plant communities with halophytes and tugai species.
   This type of transitory plant community occurs more in lake areas, where the groundwater level decreases slowly. First, typical halophytic or tugai vegetation develops. Later, during desalinization of the topsoil these areas are colonized by

Table 10.8 Species composition and cover of the *Stipagrostis* and *Calligonum* psammophytic plant communities on the habitats of the dry seafloor of the Aral Sea (with different age after the drying out)

	Sand degraded solonchaks					Sandy soils					Sandy soils				
	Ba	At	Ze	Zw	Tb	Bi	Sw	Ze	U	B	Se	Se	Bi	Kw	Tb
Time after desiccation (years)	8–10	8–10	11–15	11–15	11–15	16–20	16–20	16–20	21–25	26–30	11–15	16–20	16–20	26–30	31–40
Number of species	5	6	7	6	4	4	11	7	6	10	7	11	6	7	12
Therophyte cover (%)	1	31	4	12	5	0.2	0.4	0.2	0.1	0.3	2	0.2	0.4	0.2	0.2
Perennial cover (%)	10	11	11	30	10	24	36	55	31	31	4	21	12	15	49
Stipagrostis pennata	**10**	**10**	**10**	**30**	**10**	**20**	**30**	**50**	**20**	**20**		**10**	**2**	**4**	**2**
Calligonum (*Pterigobasis*, *Eucalligonum*)		**0.5**	**0.1**	**0.1**		**4**	**2**	**0.1**	**10**		**4**	**10**	**10**		**30**
Salsola paulsenii	0.5	0.5	0.1	0.1	0.5	0.1	0.1	0.1	0.1	0.1	0.5	0.1	0.1		
Agriophyllum squarrosum			4	2	4	0.1	0.1	0.1			0.5		0.1	0.1	
Astragalus brachypus	0.1	0.5	0.1				2	0.5	0.5	0.5		0.1		0.5	2
Corispermum aralo-caspicum		30		0.1			0.1			0.1					0.1
Atriplex pratovii	0.5		0.1		0.5						0.5	0.1	0.1		
Horaninovia ulicina				10			0.1						0.1	0.1	
Eremosparton aphyllum	0.1									10					
Haloxylon aphyllum		0.1	0.5					4	0.1						10
Nitraria schoberi							0.1				0.1	0.1			0.1
Calligonum (*Pterococcus*)										0.5	0.1	0.1		10	
Heliotropium arguzioides							2		0.5			0.5			0.1
Astragalus lehmannianus							0.1								0.1
Alhagi pseudalhagi												0.1			2
Ammodendron conollyi							0.1								
Convolvulus subsericeus								0.1							
Ephedra distachya															2
Ephedra lomatolepis															0.5
Leymus racemosus										0.1					
Corispermum hyssopifolium										0.1					

Chondrilla brevirostris						0.5					
Lycium ruthenicum						0.1					
Limonium otolepis							0.1	0.1			
Tamarix laxa								0.1			
Limonium suffruticosum										0.1	
Halocnemum strobilaceum										0.1	
Salsola australis											0.1

At Atalyk, *B* Bayan, *Ba* Barsa-Kelmes, *Bi* Biktau, *Kw* Kashkynsu West, *Se* Sandal East, *Sw* Sandal West, *Tb* Bosai transect, *U* Uyaly, *Ze* Zhimovka East, *Zw* Zhimovka West

psammophytes. The combinations of typical psammophytic and halophytic species is very unusual. When the groundwater level stabilizes, proper communities can establish themselves.

It is obvious that *Stipagrostis pennata* is a typical pioneer species on the sands of the dry seafloor and that in conditions where there is no competition plant communities can develop with a relatively high level of coverage.

10.3.2.4 Stage of Shrubby Psammophytes

The genus *Calligonum* is distributed from North Africa to the Middle East to Mongolia and China. However, only in the sand deserts of Central Asia does *Calligonum* have its evolutionary centre, with many partly abundant species. It is estimated that about 50–90 species of *Calligonum* occur there (Khalkusiev 1990; Kurochkina 1978). On the dry seafloor of the Aral Sea about 30 species of *Calligonum* have been recorded (Wucherer et al. 2001) see Chap. 8. *Calligonum* species are exclusively shrubs 1–2 (rarely up to 3) m in height, which exhibit typical leafless, fragile nodded branches. The *Calligonum* taxonomy is derived from the fruit structure: there are winged fruits (section *Pterococcus*), narrow winged fruits with short bristles (section *Pterigobasis*) and unwinged fruits, which are densely covered with long bristles (section *Eucalligonum*). Along the southeastern coast of the Aral Sea south of the former bight of Kashkynsu, plant communities occur which are dominated by species from the sections *Pterigobasis* and *Eucalligonum*: *Calligonum caput-medusae*, *Calligonum macrocarpum*, *Calligonum squarrosum*, *Calligonum microcarpum*, etc. North of the former bight of Kashkynsu, species from the section *Pterococcus* dominate: *Calligonum aphyllum*, *Calligonum membranaceum*, *Calligonum borszczowii*, *Calligonum palibinii*. The *Calligonum* species are often typical pioneer species or are invading already existing *Stipagrostis* and other semishrub plant communities (Table 10.8). There are also mixed variants known with *Stipagrostis*, *Astragalus* and *Haloxylon* species, and rarely with *Ammodendron* and *Eremosparton* species. The coverage is around 10–30%. There are about 7–12 species per community. The earliest invasion was found on areas with an age of about 4–5 years.

10.3.3 Spreading Out and Distribution of Pioneer Species Around the Small Aral Sea

In 1998 the dry seafloor around the Small Aral Sea (12 transects) was surveyed. The new hydrological situation caused by the construction of a dam, in contrast to that of the Large Aral Sea, has influenced the colonization of the dry seafloor in two ways. The therophytic vegetation with *Salicornia europaea* and *Suaeda* species was inundated and is almost lacking now on most parts of the present coastline because of the rapid rise of the water level of the Small Aral Sea. On some coastal parts also perennial and woody vegetation was inundated. The first observations show that the resistance of the various species to inundation varies greatly.

Pioneer species such as *Salicornia europaea*, *Suaeda crassifolia*, *Suaeda salsa* and *Suaeda acuminata*, which normally form plant associations close to the coastline, entered a new ecological situation (Wucherer and Breckle 2001). With the rise of the water level, the therophytic belt also drifted backwards to the former coastline. The rise of the water level occurred so rapidly that some pioneer species from the first colonization wave simply disappeared. The distribution of the pioneer species around the Small Aral Sea is shown in Table 10.9.

The small bights (Butakov, Dzhideli; Table 10.9, 1–4), where wave activity is low and salinization slightly enhanced, exhibit a well-developed plant cover with typical pioneer species such as *Salicornia europaea* and *Suaeda acuminata*. *Suaeda crassifolia* is sparse in this area. Around the larger, more open bights (Shevchenko; Table 10.9, 5–8) the typical dominant pioneer species on the marsh solonchaks is *Salicornia europaea*; the other species are only subdominants or members of other plant associations. The third group (Table 10.9, 9–12) is the open coastlines at the water level, where the halosucculent therophytes, such as *Salicornia* and *Suaeda* species, have disappeared and have been replaced by pioneers from the second generation (*Climacoptera aralenis*, *Atriplex pratovii* and *Petrosimonia* species). On the remaining dry seafloor of the Small Aral Sea, conditions are favourable for a second flush of therophytes (with several *Climacoptera*, *Salsola* and *Petrosimonia* species) as well as for some woody species from the tugai vegetation, mainly *Tamarix*. The overlapping of the ecological amplitudes of several species seems to have become wider. In the future, this may result in more stochastic processes during the formation of vegetation units and greater interspecific competition.

Table 10.9 Distribution of the pioneer species around the Small Aral Sea arranged according to different localities. The small bights (Butakov, Dzhideli, *1–4*); the larger, more open bights (Shevchenko, *5–8*); the open coastlines (*9–12*)

	Locality											
Species	1	2	3	4	5	6	7	8	9	10	11	12
Salicornia europaea	++	++	++	++	++	++	++	+				
Suaeda acuminata	++	++	++	+	–	+	–	+				
Suaeda crassifolia	+	+	+	++	+	+	+	+				
Suaeda salsa				+	–	–	–	–				
Petrosimonia triandra					–	+	+		+	+	+	+
Climacoptera aralensis									+	++	–	+
Atriplex pratovii									–	+	+	+
Phragmites australis									+	–	+	–

++ dominant species, + sparse, – missing

10.4 Discussion

10.4.1 Halophyte Succession on the Dry Seafloor

The development of vegetation units is different on sandy and loamy or even clay soils. Primary succession starts rather soon with the formation of microbiotic crusts, as in many other examples of primary succession (Breckle 2002). Then pioneer species occupy the area, normally for a rather short time. Pioneer species such as *Salicornia europaea*, *Suaeda crassifolia*, *Suaeda salsa* and *Suaeda acuminata* are found over vast stretches close to the retreating coastline. They form plant associations which are often almost monospecific. But during further desiccation, they disappear, and often bare plains with salt crusts develop, open salt deserts, mainly on loamy, clayey substrates.

On marsh solonchaks at the present coastline (stage I), dominant plant communities are found with the following species: *Salicornia europaea*, *Suaeda crassifolia* and *Tripolium vulgare* (*Aster tripolium*). These are the pioneer plants that directly follow the retreating seawater level. During the vegetation period, this area is often flooded, because of the fluctuating retreat of the water level, creating ecological conditions favourable for the development of annuals. Figure 10.13 illustrates the development of the *Salicornia* unit on the marsh solonchaks from year to year on the Bayan transect on the eastern coast. This process of a migrating *Salicornia* belt along the retreating water level can be called replacement succession. The density of the plant population varies tremendously from year to year. The plants are normally only 30–50 cm high (Fig. 10.14). When the water level sinks,

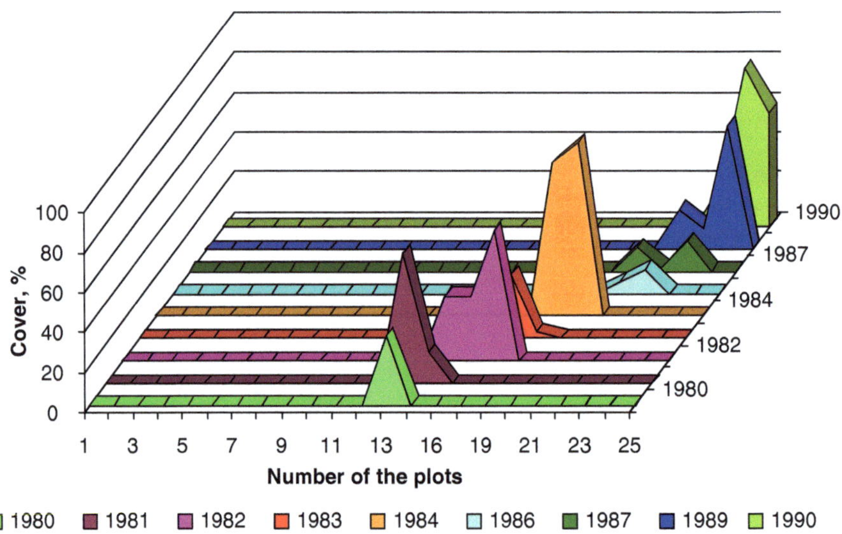

Fig. 10.13 Replacement succession of *Salicornia* belts on the marshy solonchaks from year to year on the Bayan transect, indicating the migrating *Salicornia* belt along the retreating water level

Fig. 10.14 *Salicornia europaea* plant community on the loamy marsh solonchaks (stage I), Bayan transect (photo: Wucherer)

Table 10.10 Succession on loamy and clayey substrates in the Aralkum

Phase	Time (years)	Dominant species	Main ecosystems	Substrate type
0	0	*Zostera*, cyanobacteria	Wet mud	Wet marshy solonchak
1	1–3	*Salicornia europaea*, *Suaeda crassifolia*, *Tripolium vulgare*	Often dense annual cover, very variable from year to year	Marshy solonchak
2	3–10	*Climacoptera aralensis*, *Petrosimonia triandra*	Open annual cover, mixed with bare open salt desert	Coastal solonchak
3	10–20	*Climacoptera aralensis*, *Climacoptera ferganica*, *Halocnemum strobilaceum*, *Halostachys caspica*	Large patches with open salt desert, mixed with annual vegetation and islands of perennials	Coastal solonchak, degraded solonchak
4	20–50	*Halocnemum strobilaceum*, *Halostachys caspica*, *Haloxylon aphyllum*, *Climacoptera ferganica*, *Climacoptera brachiata*, *Petrosimonia brachiata*	Shrubby perennial vegetation with therophytes mixed with bare open salt desert	Degraded solonchak, takyr soil
5	>40	*Artemisia terrae-alba*, *Anabasis salsa* (*Haloxylon aphyllum*)	Open, partly desalinized semidesert with scattered perennials (only fragments and local)	Xerosol (burozem)

the pioneer plants die and the coastal solonchaks develop (stage II). The crusty and puffy coastal solonchaks are formed when the groundwater level goes down to about 1–2 m and no further inundations occur. The strong capillary upwards movement of water leads to an extremely strong salinization of the upper moist soil surface and the formation of a salt crust, containing up to 16–20% salt. The ecological conditions for colonization and establishment of plant species are then very severe in comparison with the marshy solonchaks (Table 10.10).

Only a few species are able to colonize such stands. These are the pioneer plants of the second generation (Fig. 10.15) colonizing the dry seafloor (*Climacoptera aralensis*, *Petrosimonia triandra*, *Bassia hyssopifolia*) or successional open salt deserts develop. Those successional deserts can be defined as stands, where for lack of a seed bank, colonization by plants has not yet taken place. Thus, during further succession, mosaics with huge areas of open salt deserts are often formed, and can exist for 10–20 years or more. They are the main area and source of the salt-dust output from the dry seafloor (see Chaps. 5–7). The third stage of succession is a loose mixture of annuals and perennials such as *Halocnemum strobilaceum* and *Halostachys caspica* on degraded solonchaks (Fig. 10.16).

Degraded coastal solonchaks develop from coastal solonchaks by further lowering of the groundwater table and thus drying of the surface. The capillary threads no longer reach the surface, but end below the surface at a depth of about 10–30 cm. With time, communities with a coverage up to 20% or more are formed (stage IV). On the Kaskakulan transect under moderate salinities, permanent communities

Fig. 10.15 *Bassia hyssopifolia–Petrosimonia triandra* plant community on the loamy coastal solonchaks (stage II), Bayan transect (photo: Wucherer)

Fig. 10.16 *Halostachys caspica–Petrosimonia brachiata* plant community on the degraded coastal solonchaks (stage III), Bayan transect (photo: Wucherer)

of *Halocnemum strobilaceum* and *Haloxylon aphyllum* are widespread throughout the eastern coast (see Fig. 9.19).

10.4.2 Psammophyte Succession on the Dry Seafloor

The psammosere of vegetation is most perfectly developed around the island of Barsa-Kelmes and on the southeastern coast of the Aral Sea. The soil profile is sandy up to at least 2 m in depth. The succession is also influenced by the sinking groundwater level as well as by the desalinization by means of a good water percolation along the soil profile (Table 10.11). On the marsh solonchaks (stage I) the floristic composition is slightly different. *Salicornia europaea*, together with *Suaeda acuminata* and *Atriplex pratovii*, forms pioneer units. On the coastal solonchaks (stage II) the species compositions and the vegetation units are richer and more variable. Already during this stage deflation events start. The nanorelief with small hills (5–10 cm high) is already distinct. Plants accumulate sand around them and the small hills increase in height (15–20 cm high). These stands are optimal for the development of *Atriplex pratovii* and *Horaninovia ulicina* units (Fig. 10.17). Further decrease of the groundwater table to 1.2–1.7 m and desalinization of the topsoil (down to 50–60 cm depth) is the precondition for the following colonization with annual (*Horaninovia ulicina*, *Agriophyllum* spp.) and seldom also *Atriplex pratovii* and perennial psammophytes (stage III, Fig. 10.18). With time, psammophytic communities with *Stipagrostis pennata*, *Calligonum*, *Haloxylon* and *Astragalus* species thus develop on the sandy soils (stage IV, Fig. 10.19).

On average, the succession on loamy stands can be described by two to four stages, and on sandy soils by three to five stages. The existence of a distinct stage is a consequence of the ecological conditions and stability, and thus might range between 2 to 30 years. In the course of further desiccation, ecosystems with new characteristics arise: the granulometric composition, salt content and microrelief may be different. Particularly striking is the contrast between the dry areas of the 1960s and 1970s and those of the 1980s and 1990s and later. Accordingly, the temporal change of the vegetation and that of the soils often do not match the spatial position of plant communities.

The sequence of succession differs between the pure sandy soils and the loamy coastal solonchaks covered with a sand layer. The latter exhibit a distinctly higher proportion of halophytic and tugai species. The desalinization of the topsoil is enough for psammotherophytes, however it is not sufficient for perennial psammophytes. This explains why the sequence of succession is very much dependent on the thickness of the sandy layer with its water-storing capacity and on the annual precipitation as the main cause of salt leaching of the topsoil and is thus very variable. The final result of the succession, however, can be predicted. As a rule the final stage is formed by *Haloxylon aphyllum* and more rarely by *Nitraria schoberi*, plant communities. The low cover percentage as well as the low number of species number (five to seven) gives hints that these plant communities are still developing. *Salsola nitraria* is a species of late successional stages.

Table 10.11 Succession on sandy substrates in the Aralkum

Phase	Time (years)	Depth of water table (m)	Dominant species	Main ecosystems	Substrate type
0	0	0	*Zostera*, cyanobacteria	Wet mud	Wet marshy solonchak
1	1–3	0.3–0.7	*Salicornia europaea*, *Suaeda acuminata*, *Atriplex pratovii*	Dense annual cover, very variable from year to year	Marshy solonchak
2	3–7	0.7–1.2	*Atriplex pratovii*, *Suaeda acuminata*, *Climacoptera aralensis*	Open annual cover, mixed with bare open sand desert and small mobile dunes	Coastal solonchak, degraded solonchak
3	7–20	1.2–1.7	*Horaninovia ulicina*, *Salsola paulsenii*, *Agriophyllum* spp., *Stipagrostis pennata*, *Calligonum* spp., *Haloxylon aphyllum*	Patches with open mobile sand dunes, mixed with vegetation islands of perennials in dune valleys	Coastal solonchak, degraded solonchak
4	20–50	1.7–3	*Calligonum* spp., *Astragalus* spp., *Eremosparton aphyllum*, *Haloxylon aphyllum*, *Corispermum* spp., *Heliotropium* spp.	Open shrubby perennial vegetation between bare open sand dune desert	Sandy soils and salt hummocks
5	>40	>3	*Haloxylon aphyllum*, *Haloxylon persicum*, *Calligonum* spp., *Carex physodes*	Open, sandy desalinized semidesert with scattered, sometimes dense perennials (only fragments and local)	Arenosol

On the partly or totally desalinized sands, three stages of psammophyte succession can be distinguished:

1. Stage of annual halopsammotherophytes and psammotherophytes (*Atriplex pratovii*, *Salsola paulsenii*, *Horaninovia ulicina*) on degraded coastal solonchaks
2. Stage of grasses (*Stipagrostis pennata*) on degraded coastal solonchaks and sand soils
3. Stage of shrubby psammophytes (*Calligonum*, *Astragalus* and *Haloxylon* species) on degraded coastal solonchaks and sand soils.

Fig. 10.17 *Horaninovia ulicina–Atriplex pratovii* plant community on the sand degraded coastal solonchaks (stage II), Kaskakulan, eastern coast (photo: Wucherer)

Fig. 10.18 *Atriplex pratovii* plant community with *Haloxylon aphyllum* on the sand degraded coastal solonchaks (stage III), Kaskakulan, eastern coast (photo: Wucherer)

Fig. 10.19 *Astragalus brachypus–Stipagrostis pennata–Haloxylon aphyllum* plant community here with red fruiting *Ephedra distachya* on the sand soils (stage IV), the dry seafloor of the 1960s at the former island of Barsa-Kelmes (photo: Wucherer)

The pattern of genera and species is rather similar to that of the Kyzylkum sand desert. It is obvious that the stage with Fabaceae is not a real successional stage. The *Astragalus* species are colonizing the sand habitats simultaneously with *Stipagrostis* and *Calligonum*.

10.4.2.1 Dispersal of Perennial Species

The conditions for the colonization by perennial psammophytes are rather similar (the big open areas); however, the missing seeds, the empty seed bank, is the main bottleneck. Thus during the first 7 years after desiccation, mostly therophytic communities develop. Later, 8–10 years after desiccation, the stands are totally governed by the increasing aeolian activities. The groundwater table dropped to 1.3–1.7 m. The capillary fringe is located at about 0.5–1.0 m depth. The small dunes grow to about 30–70 cm. Typical psammophytes, e.g. *Stipagrostis pennata*, *Calligonum* species and *Astragalus* species, then form plant communities on the degraded coastal solonchaks. In this phase the coverage of *Stipagrostis pennata* amounts to 10–20%. In any case, the number of species is still low (four to seven). During the ongoing succession the cover percentage of *Stipagrostis pennata* and the ratio of perennials increases and the ratio of annuals decreases. This stage can take 8–12 years or longer. This depends primarily on the arrival of the perennial shrubs within the *Stipagrostis* community. On the dry seafloor, all around the Aral Sea there are *Stipagrostis* plant communities. This species has also exhibited an explosion-like dispersal – similar to *Atriplex pratovii*.

10.4.2.2 Mechanisms of Succession

The succession on the sands in the starting phase is caused by exogenous factors. With the retreat of the seawater level, subsequently the groundwater level decreases. This leads to desalinization of the upper soil layer down to a depth of 1 m after about 10 years. It is surprising that with a moderate precipitation of 100 mm, the desalinization of the sandy soils proceeds so quickly. The replacement of the *Salicornia* and *Suaeda* phases by annual psammophytes is caused by exogenous factors. The further process of the succession after the settlement of *Stipagrostis pennata* is endogenous and caused by biological mechanisms (speed of the dispersal of the annual and perennial species, the ability to establish itself on the open sand). According to our results, the primary settlement of the open sands by *Stipagrostis pennata* is not necessary for the dominant species of the following generation as *Calligonum* or *Astragalus*, but the dispersal of these species is faster if a settlement by *Stipagrostis* has already taken place. It can be shown that *Stipagrostis*, *Calligonum* species and *Haloxylon* form their own plant communities on equally aged areas. *Stipagrostis pennata* with its effective strategy of dispersal is much faster in colonizing open areas. At the same time, the primary settlement by *Stipagrostis* and *Calligonum* is obligatory for several perennial species with low abundance, e.g. for species such as *Chondrilla brevirostris*, *Convolvulus subsericeus*, and *Heliotropium arguzioides*. It is an autogenic succession, in which the mechanisms of facilitation, inhibition and tolerance play a crucial role (Drury and Nisbet 1973; Connell and Slatyer 1977; Glenn-Lewin et al. 1992). The ecological conditions for the whole psammophytic complex converge within 7–8 years after desiccation. The sequence of succession, however, can be different and has to be regarded as stochastic. But the dune activity by aeolian processes with the changing relief and the increasing effects of plants themselves on their stands then control the ecosystem dynamics more and more in the later successional stages.

10.4.3 Factors Influencing Primary Succession in the Aralkum

The primary succession can be considered from the viewpoint of time, habitat and the plants (Bradshaw 1993).

10.4.3.1 The Time Factor

The succession on the dry seafloor has been active for about 50 years. The time factor plays no active role, but influences indirectly the vegetation development. Species which arrived rather late are often not able to establish themselves in reasonable numbers. Either the ecological conditions have changed and became unfavourable or other species have already occupied the relevant ecological niche. The stabilization of the sea level of the Small Aral Sea has demonstrated that a whole set of species established themselves on the coastal solonchaks very rapidly, especially under conditions of a stabilized groundwater level. The change in the ecological factors leads to a change in colonization rate and thus to a change in the time sequence of the succession. We know the annual seawater level since 1961, so we are able to determine almost exactly the age of the drying process of a distinct site of the seafloor and thus the development of the relevant ecosystems. This is important to identify major mechanisms that determine the rate and direction of ecosystem changes on the dry seafloor.

10.4.3.2 Exogenic Factors

These are the sea and river floods, the geological and geomorphological, climatic, edaphic and aeolian factors. The areas adjacent to the coastline are subject to inundations. The pattern of sedimentation, as well as the former geological history of the Aral Sea, determines the particle size distribution and sedimentation layers of the new soils of the dry seafloor. The geomorphological structure of the Aral Sea basin is complicated. Plains predominate in the eastern part of the depression, with an inclination of 0.1–0.6°. Therefore, the present coastline is situated more than

100 km away from the former eastern coastline. On the western coast, between Ustyurt Plateau and the islands of Barsa-Kelmes and Vozrozhdeniya, the inclination of the plain is steeper and is to 2–5°. Correspondingly, the dry seafloor belt is only 4–10 km wide.

The new alluvial deposits of the retreating Aral Sea cover 1–6 m older layers. The salinization of the substrate varies to a great extent, causing a wide variety of saline soil types: marshy solonchaks, crusty and puffy solonchaks, solonchaks slightly covered by sand, degraded coastal solonchaks, takyr solonchaks and takyr soils, and sandy soils.

On sandy or sandy-loamy soils, deflation of soil particles takes place. Barchans or complexes of barchans and salt deserts develop. They are widespread on the eastern coast. Since the middle of the 1980s, open salt deserts have developed on a large scale. Some plant populations have decreased according to the salt desert formation. The fast increase of the salt desert areas, the changing of soil texture and increase in salt content of the soils have caused the absolute dominance of halophytes as pioneer plants, mainly species from the Chenopodiaceae, to the exclusion of most other life forms.

Their invasion, however, is often different from year to year according to climatic conditions during the preceding winter. Germination and establishment thus only occur in favourable years. The number of successful establishment event years of saxaul (*Haloxylon aphyllum*) on substrates of the 1970s can rather easily be counted by the "generations" present. Normally about five or six size groups can be distinguished, which is equal to about five to six favourable years (or sequence of years) within the last 30 years. On average, every 5th to 7th year has been favourable (Fig. 9.23).

10.4.3.3 Plant Sources

On the dry seafloor alone, to date about 370 plant species have been reported (Chap. 8). The precondition for colonization by plants is the surrounding flora and its seed production. This is mainly the flora of the old Aral terraces along the old coastline and the adjacent mainland all around the Aral Sea. The scheme in Tables 10.10 and 10.11 indicate the main directions and specialities of the vegetation development on the dry seafloor. It depends on the geomorphological and landscape pattern as well as on the pattern of the vegetation units along the coast. Districts I and II (Fig. 9.2) on the northern and western coasts of the Aral Sea lack the old Aral terraces almost completely, or these are very narrow belts (only 10–20 m up to some 100 m). The azonal flora of the Aral terraces and the steep slopes of the chinks are the main sources for the seed bank. It is remarkable that on the dry seafloor area from the 1960s there is an essential proportion of Brassicaceae and Chenopodiaceae but a sparse representation of Fabaceae and Polygonaceae. In districts III and VI, the belt of the Aral terraces is continuous and relatively broad (some 100 m to 3–5 km). The azonal vegetation of the Aral terraces and the zonal psammophytic plant communities of the Kyzylkum and Priaralski Karakum sand deserts are the sources for the plant species for the dry seafloor. The Polygonaceae and Fabaceae are significantly richer on those areas of the dry seafloor from the 1960s and 1970s than the Brassicaceae and Chenopodiaceae. In the delta districts IV and VII the irregular floods of the rivers are an important factor. These inundations lead to an extensive distribution of seeds and an activation of the seed bank already present on the dry seafloor. Mainly the Tamaricaceae, Limoniaceae, Asteraceae and Chenopodiaceae are conspicuous. Within district V only the halophytic communities and the zonal communities (*Artemisia terrae-albae*, *Anabasis salsa*, *Haloxylon aphyllum*) are present on the Aral terraces. Here, mainly members of Chenopodiaceae dominate the dry seafloor from the 1960s and 1970s. Within the island district VIII the belt of the Aral terraces is interrupted. Again, here the zonal vegetation units are of no importance for the colonization processes. The flora of the Aral terraces and of the distinctly lower chinks is poor in comparison with that of the mainland. The composition of flora of the former coast is different in each district and influences the course of the succession on the dry seafloor.

10.4.3.4 Contrasting Fluctuation Dynamics

Another phenomenon is the contrasting fluctuating dynamics. The great annual variability in cover percentages of spring annuals in deserts is well known. In most cases, in the Aralkum there are halophytic therophytes which have a life cycle of about 6–8 months. Again, their density varies tremendously from year to year. The greatest contrast was observed between the 1981 and 1982 and between 1998 and 1999. The first case was described from the northern coast of the Aral Sea (Karabulak transect) and was mapped by Wucherer and Galieva (1985).

The second case can be illustrated by an example from the Bayan transect. In 1998 along the transect therophyte units had developed with a coverage of about 80–100% with many species, mainly *Suaeda acuminata*, *Suaeda crassifolia*, *Petrosimonia triandra* and *Climacoptera aralensis*. One year later, on the very same areas an absolutely barren salt desert could be found without any plants. The soil surface, however, was covered by an incredible number of seeds, which had no chance to germinate. About 30–450 seeds per 100 cm^2 on the surface of the soil below dead *Suaeda acuminata* plants from the previous year were counted. The reasons for such a regressive dynamics are mainly too dry spring months and too warm winters, poor in snow. Each year is very different in appearance depending on the climatic conditions of the cold season.

10.4.3.5 Unexpected and Unique Plant Combinations

The Aralkum vegetation is very dynamic in composition. This leads to unexpected combinations and interrelationships. There are many examples of unique plant communities and ecosystems with unique composition formed by various species from the psammophytic, halophytic and hygrophytic units. Species which under natural conditions cannot be found together with other species despite an apparently similar ecological behaviour form new vegetation units here. As an example, *Climacoptera aralensis* occurs on crusty and puffy solonchaks as well as on secondary solonchaks in fields in the Kyzyl–Orda district. It forms isolated, monotonous units, sometimes together with a few salt meadow halophytes. *Petrosimonia triandra* is distributed very locally only and prefers moderate marsh conditions or slightly saline meadows. Both species do not occur together at the same localities. However, on the dry seafloor they meet and form extensive stands. *Petrosimonia triandra* is dispersed faster and colonizes the coastal solonchaks first. One to 2 years later, *Climacoptera aralensis* follows. Because of the winged fruits, the latter species has a high potential for wide range dispersal, but in comparison with *Petrosimonia* always remains subdominant. Both species form mixed stands with a high coverage of about 60–80%. If the salinity of the topsoil increases, then *Petrosimonia* frequency decreases, but *Climacoptera* frequency remains constant. The ecological range of both species apparently overlaps broadly, but some of their life strategies are different. *Petrosimonia triandra* has a more rapid dispersal ability, *Climacoptera aralensis* is more salt-resistant.

On the sandy substrate around the island of Barsa-Kelmes and at the southeastern coast a plant unit either with *Stipagrostis pennata* and *Halocnemum strobilaceum* or with *Stipagrostis pennata* and *Phragmites australis* can often be found. Such a combination is very peculiar and lacks an explanation. The sandy coastal solonchaks are first colonized by *Halocnemum strobilaceum*. With further decrease of the groundwater table, the upper sandy layer becomes slowly desalinized within 8–10 years down to about 1-m depth. The main active roots of *Halocnemum strobilaceum* are at a depth of 1–2 m. *Stipagrostis pennata*, however, has its main roots in the upper sandy soil. This vegetation unit is an intermediate vegetation type. With the stabilization of the seawater level, it might be possible for such a vegetation unit to become rather permanent. Thus, the differentiation in root systems at different soil depth is a precondition for this coexistence.

10.5 Conclusions

The primary succession on the dry seafloor has been active for about 50 years. The geological, geomorphological, climatic, edaphic and aeolian factors controlled the vegetation development in the Aralkum in the first few years after the desiccation. But in the late phase, mechanisms of facilitation and tolerance have played a crucial role, especially by the psammophyte succession. The sequence of succession can be very different and has to be regarded as stochastic. The precondition for the successful plant colonization of the dry seafloor is the surrounding flora and its diaspore production. This is mainly the flora of the old Aral terraces along the old coastline and the adjacent mainland all around the Aral Sea. The psammoseres and haloseres of vegetation in the Aralkum are obviously very different and can be described by two to five stages.

Acknowledgments This research was funded by the German Federal Ministry for Education and Research (BMBF grant 0339714 and BMBF grant 0330389) and the German Federal Ministry of Environment, Nature Protection and Nuclear Safety (BMU grant).

References

Aoki I (1995) Flow-indices characterizing eutrophication in lake-ecosystems. Ecol Model 82 (3):225–232

Barrow CJ (1999) How humans respond to natural or anthropogenic disturbance. In: Walker LR (ed) Ecosystem of disturbed ground, Ecosystems of the world 16. Elsevier, Amsterdam, pp 659–671

Begon M, Harper JL, Townsend CR (1986) Ecology: Individuals, populations and communities, vols 1–2. Blackwell, Oxford

Belnap J, Lange OL (eds) (2001) Biological soil crusts: structure, function and management. Ecological Studies, vol. 150, Springer, Berlin

Bradshaw AD (1993) Introduction: understanding the fundamentals of succession. In: Miles J, Walton DWH (eds) Primary succession on land. Blackwell, Oxford, pp 1–3

Breckle S-W (2002) Primärsukzessionen – Beispiele von Sanddünen, Vulkanaschen, Salzböden und anderen Substraten. Ber Reinh-Tüxen-Ges (Hannover) 14:5–16

Connell JH, Slatyer RO (1977) Mechanisms of succession in natural communities and their role in community stability and organisation. Am Nat 111(982):1119–1144

Crawley MJ (2007) The R book. Wiley, Chichester

Danin A (1996) Plants of desert dunes. Springer, Berlin

Drury WH, Nisbet ITC (1973) Succession. J Arnold Arbor Harv Univ 54(3):331–368

Glenn-Lewin DC, Peet RK, Veblen TT (eds) (1992) Plant succession. Theory and prediction. Chapman & Hall, London

Groeneveld J, Wissel C, Wucherer W, Dimeyeva L, Jeltsch F (2005) Computer simulations orientated towards plant succession on the desiccating Aral Sea floor. In: Veste M, Wucherer W, Homeier J (eds) Ökologische Forschung im globalen Kontext. Festschrift für Prof Dr Siegmar-W Breckle. Cullivier, Göttingen, pp 129–141

Kabulov SK (1990) Izmenenie fitotsenozov pustyn pri aridizatsii. FAN, Tashkent (Russian)

Khalkusiev P (1990) Evolution connections between some of the families of desert regions. Fan, Tashkent (Russian)

Kurochkina LYa (1978) Psammophyte Vegetation of Kazakhstan. Nauka, Alma-Ata (Russian)

Kurochkina LYa (1979) Botanical survey in the Aral Sea basin. Probl Osvo Pustyn 2:9–17 (Russian)

Kurochkina LYa, Makulbekova GB, Wucherer W, Malaisarova A (1983) Vegetation of the dry sea floor of the Aral Sea. In: Aralskoe more. Nauka, Alma–Ata, pp 91–144 (Russian)

Londo G (1975) Dezimalskala für die vegetationskundliche Aufnahme von Dauerquadraten. In: Schmidt W (Red) Sukzessionsforschung. Ber Int Symp IVV 1973. Cramer, Vaduz, pp 613–617

Londo G (1984) The decimal scale for relevés of permanent quadrats. In: Knapp R (ed.) Sampling methods and taxon analysis in vegetation sciences. Handb Vef Sci 4. Junk, The Hague, pp 45–49

Magnussen S, Boyle TJB (1995) Estimating sample size for inference about the Shannon-Weaver and the Simpson indices of species diversity. For Ecol Manag 78:71–84

Mees F, Singer A (2006) Surface crusts on soils/sediments of the southern Aral Sea basin, Uzbekistan. Geoderma 136:152–159

Miroshnichenko YM (1975) Root systems of trees and bushes and their ecology in the eastern Karakum. Bot Zh 60:1776–1796 (Russian)

Petrov MP (1973) The deserts of the world. Nauka, Leningrad (Russian)
R version 2.11.1 (2010) The R Foundation for Statistical Computing, Vienna
Sukachev VN (1938) The main definitions in the vegetation science. In: Rastitelnost' SSSR, vol 1. AN SSSR Moskva, pp 15–37 (Russian)
Sukachev VN (1954) Some general theoretical problems in vegetation science. In: Problems of botany, vol 1. AN SSSR Moskva, pp 291–309 (Russian)
Walker RW, del Moral R (2003) Primary succession and ecosystem rehabilitation. Cambridge University Press, Cambridge
Walter H (1974) Die Vegetation Osteuropas, Nord- und Zentralasiens. Vegetationsmonographien der einzelnen Großräume, vol VII. Fischer, Stuttgart
Wucherer W (1979) Primary colonisation of the dry sea floor of the Aral Sea. Probl Osvo Pustyn 3:9–18 (Russian)
Wucherer W (1984) Vegetation formation on the dry sea floor of the Aral Sea. In: Problemy Aralskogo morya i delty Amudar'i. FAN, Tashkent, pp 155–161 (Russian)
Wucherer W (1986): Singenez rastitel'nosti osushennogo dna morya (Vegetation dynamics on the dry sea floor of the Aral Sea). PhD dissertation, Almaty-Tomsk
Wucherer W (1990) Vegetation development on the new habitats in the desert. Gylym, Alma–Ata (Russian)
Wucherer W, Breckle SW (2001) Vegetation dynamics on the dry sea floor of the Aral Sea. In: Breckle SW, Veste M, Wucherer W (eds) Sustainable land-use in deserts. Springer, Heidelberg, pp 52–68
Wucherer W, Breckle SW (2003) Psammophytic succession at the South-East Coast of the Aral Sea. In: Rachkovskaya EI, Volkova EA, Khramcov VN (eds) Botanical geography of Kazakhstan and Middle Asia. Komarov Botanical Institute of Russian Academy of Uzbekistan Republic, St Petersburg, pp 340–347
Wucherer W, Galieva L (1985) Vegetation mapping of the formed dynamic ecosystems. Isv AN KasSSR Ser Biol 4:3–13 (Russian)
Wucherer W, Breckle SW, Dimeyeva L (2001) Flora of the dry sea floor of the Aral Sea. In: Breckle SW, Veste M, Wucherer W (eds) Sustainable land-use in deserts. Springer, Heidelberg, pp 38–51

Chapter 11
Fauna of the Aralkum

U. Joger, T. Dujsebayeva, O.V. Belyalov, Y. Chikin, D. Guicking, Y.A. Grachev, R. Kadyrbekov, and C. Miaud

11.1 Introduction

The fauna of the Aralkum has been studied only partly. Although many faunistic data have been collected in the Aral Sea basin over the past few decades, there is no systematic monitoring of the faunal settlement of the Aralkum.

During an international project on "The present state of Aral Sea Basin herpetofauna and implications for conservation of natural habitats and biodiversity", a group composed of Kazakh, Uzbek, French, German and Italian zoologists studied mainly amphibians and reptiles, but also insects on and around the dry surface of the former Aral Sea. The project lasted from 2001 to 2005 and was financed by INTAS, DFG and CNRS. The data from this project were substantially enriched by data from resident specialists who are co-authors of this chapter.

11.2 Vertebrates

With more than 300 species of birds, 100 species of fish and 200 species of mammals, the Aral basin used to have a diversity of wildlife that was comparable with that of African biomes. Many of these animal species have now disappeared from the area or are threatened.

Before 1960 the river deltas were home to over 70 species of mammals and 319 species of birds. Presently only 32 species of mammals and 160 species of birds remain. In lowstreams of the Syr Darya, more than 100,000 ha of alluvial soils became salt marsh, and more than 500,000 ha of swamp and meadow-swamp soil became dry (see also Chap. 14). A similar situation is recorded from the Amu Darya delta (Micklin 2007).

In the cultivation and inundation of land, a number of animal species have perished. The total number of animals is decreasing, but the population density is increasing on the unploughed sites on the banks of canals. Persianova et al. (1986)

S.-W. Breckle et al. (eds.), *Aralkum - a Man-Made Desert: The Desiccated Floor of the Aral Sea (Central Asia)*, Ecological Studies 218,
DOI 10.1007/978-3-642-21117-1_11, © Springer-Verlag Berlin Heidelberg 2012

found that on irrigated land in the Dzhizak area, 9 of 27 species had disappeared and four more were endangered. Of 21 reptile species, two will probably perish completely.

As a result of the Syr Darya delta drainage, flocks of waterfowl have been displaced during migration from the lower reaches of the Syr Darya to the lakes of Turgay. Accumulations of white and Dalmatian pelicans have been observed beyond the northern boundary of the former Aral Sea (Novikova and Zaletayev 1985). Accumulations of migrating waterfowl have also appeared both in water reservoirs and in filtrational lakes formed in Central Asia (Zaletayev 1976). Meanwhile, newly formed wastewater basins have created conditions for large hibernation sites for waterfowl in a number of regions. Overall, the diversity of mammals inhabiting the Aral area has decreased from 70 to 30 species and the number of bird species has decreased from 319 to 168. The disappearance of nesting sites for many bird species has led to the disappearance of 38 of the 173 bird species nesting in the lower reaches of the Syr Darya.

Tugai communities in the deltas are also endangered. These tugais were extremely rich florally since they had 576 superior plants, including 29 endemic to Central Asia. Currently, owing to desertification, 54 species are on the verge of extinction (Sagitov, personal communication). Reed thickets, meanwhile, have perished in the Karakalpak Republic in the Amu Darya delta, and the relict tugai forests are also becoming extinct.

11.2.1 Mammals

Until the late 1940s, the mammalian fauna in the Aral Sea region remained poorly studied. Extended study of mammalian faunistic composition and distribution resulted in several publications (Varshavsky 1951; Varshavsky and Shilov 1955; Rotshild et al. 1967). Some articles were devoted to selected species of mammals, such as saiga, Asian wild ass and Persian gazelle of the island of Barsa-Kelmes (Vassenko 1950; Rashek 1966; Zhevnerov 1984) and dipodids of the Kyzylkum and the North Aral Sea region (Burdelov 1985; Garbuzov 1985; Varshavsky et al. 1985; Mirzabekov et al. 1985). The results of those studies were summarized in *Mammals of Kazakhstan*, volumes I–IV (1969–1985) and *Genetic Fund of Kazakh SSR Fauna* (1989). The list of mammals in Table 11.1 is based upon the books mentioned. The species names are cited after Pavlinov et al. (1995).

11.2.1.1 Mammals of the Kazakhstan Part of the Aral Sea Region

Table 11.1 gives a list of the mammals present in the Aral Sea region. Some representative species are shown in Fig. 11.1.

Some species listed in Table 11.1 are known by single findings, only. *Suncus etruscus* is known by osteological remains found in hair balls of birds of prey in the

Table 11.1 Mammals of Kazakhstan part of the Aral Sea region

Order Insectivora
Erinaceus (Hemiechinus) auritus Gmelin, 1770
Suncus etruscus Savi, 1822
Sorex minutus Linnaeus, 1766
Diplomesodon pulchellum Lichtenstein, 1823
Order Chiroptera
Myotis mystacinus Kuhl, 1819
Nyctalus noctula Schreber, 1775
Eptesicus serotinus Schreber, 1775
Eptesicus bottae Peters, 1869
Eptesicus bobrinskii Kuzyakin, 1935
Vespertilio murinus Linnaeus, 1758
Order Carnivora
Canis aureus Linnaeus, 1758
Canis lupus Linnaeus, 1758 (Fig. 11.1a)
Vulpes corsac Linnaeus, 1768
Vulpes vulpes Linnaeus, 1758
Mustela nivalis Linnaeus, 1766
Mustela erminea Linnaeus, 1758
Mustela eversmanni Lesson, 1827
Vormela peregusna Güldenstaedt, 1770
Meles meles Linnaeus, 1758
Felis libyca Forster, 1780
Felis margarita Loche, 1858
Order Perissodactyla
Equus hemionus Pallas, 1775 (Fig. 11.2)
Order Artiodactyla
Sus scrofa Linnaeus, 1758
Gazella subgutturosa Güldenstaedt, 1780 (Fig. 11.1b)
Saiga tatarica Linnaeus, 1766 (Fig. 11.1c)
Order Rodentia
Spermophilopsis leptodactylus Lichtenstein, 1823
Spermophilus fulvus Lichtenstein, 1823 (Fig. 11.1f)
Spermophilus pygmaeus Pallas, 1779
Allactaga elater Lichtenstein, 1825 (Fig. 11.1d)
Allactaga major Kerr, 1792
Allactaga severtzovi Vinogradov, 1925
Allactaga sibirica Forster, 1778
Pygerethmus pumilio Kerr, 1792
Pygerethmus platiurus Lichtenstein, 1823
Stylodipus telum Lichtenstein, 1823
Dipus sagitta Pallas, 1773
Paradipus ctenodactylus Vinogradov, 1929
Eremodipus lichtensteini Vinogradov, 1927
Salpingotus heptneri Vorontzov et Smirnov, 1969
Salpingotus pallidus Vorontzov et Shenbrot, 1984
Cricetulus migratorius Pallas, 1773
Allocricetulus eversmanni Brandt, 1859

(continued)

Table 11.1 (continued)

Ondatra zibethicus Linnaeus, 1776
Microtus socialis Pallas, 1773
Microtus transcaspicus Satunin, 1905
Ellobius talpinus Pallas, 1770
Meriones tamariscinus Pallas, 1773
Meriones libycus Lichtenstein, 1823
Meriones meridianus Pallas, 1773
Rhombomys opimus Lichtenstein, 1823 (Fig. 11.1e)
Mus musculus Linnaeus, 1758
Rattus norvegicus Berkenhout, 1769
Nesokia indica Gray et Hardwicke, 1830
Order Lagomorpha
Lepus tolai Pallas, 1778
Lepus europaeus Pallas, 1778

Fig. 11.1 Typical mammals of the Aral Sea basin: (**a**) Wolf (*Canis lupus*); (**b**) Persian gazelle (*Gazella subgutturosa*); (**c**) saiga (*Saiga tatarica*); (**d**) small five-toed jerboa (*Allactaga elater*); (**e**) great gerbil (*Rhombomys opimus*); (**f**) large-toothed souslik (*Spermophilus fulvus*)

northern Aral Sea region (Varshavsky and Shilov 1959). *Nyctalus noctula* was once caught near Aralsk; this specimen possibly lost its way during migration (Butovsky and Shaimardanov 1985). *Nesokia indica* is also known by osteological fragments in hair balls of birds of prey from the southeastern coast of the Aral Sea, near the border with Uzbekistan (Varshavsky 1951). The following species are also known by single findings: *Mustela erminea, Allocricetulus eversmanni*, *Paradipus ctenodactylus*, *Eptesicus bobrinskii*, *Salpingotus heptneri* and *Salpingotus pallidus* (Lobachev 1971; Stogov and Zhanabayev 1985). Some species have a margin of their ranges at the Aral Sea region, others are naturally rare.

11.2.1.2 Changes in Mammal Number and Distribution

Changes of natural conditions following the shrinking of the Aral Sea affected, and sometimes significantly, the pattern of mammal distribution in the area, and sometimes the population number as well. Many former islands are now connected to the mainland, so the suitable areas were reduced for some mammals, and, contrarily, extended for others.

The island of Barsa-Kelmes, with its famous Barsa-Kelmes Nature Reserve since 1939, was connected to the mainland in 1999 (see Chap. 14). Regular studies and observations in Barsa-Kelmes Nature Reserve have provided good data to compare the present and the previous situation and to track the changes.

There were few autochthonous species on the island: *Erinaceus (Hemiechinus) auritus* (long-eared hedgehog), *Eptesicus serotinus*, *Allactaga elater*, *Pygerethmus pumilio, Cricetulus migratorius* and saiga. Some new species were introduced in 1929–1930 (before the reserve was founded) – *Spermophilus fulvus* and *Gazella subgutturosa* – and the population of saiga has been reinforced. Asian wild ass (*Equus hemionus*, Fig. 11.2) was introduced in 1953. All new species successfully survived on the island. Once, in a very cold and snowy winter, some corsac foxes and foxes invaded the island, and became permanent inhabitants of the island.

Fig. 11.2 Asian wild ass, *Equus hemionus*

Wolves also periodically appeared at the island in winters. The maximal number of saiga reached 3,000, that of Persian gazelle reached 2,000 and that of Asian wild ass reached 200.

The shrinking of the Aral Sea resulted in the appearance of huge, desert, vegetationless areas at the former coast, lowering of the groundwater levels, drying of surrounding lakes and the disappearance of reeds, lowering of the overall bioproductivity in the region and poor conditions for mammals. The increase of water salinity became hazardous and mammal death cases occurred, when animals died after drinking. For example, 87 saiga corpses were found not far from a drinking place in 1977 (Afanasyev and Smirnov 1981). Most of the Asian wild ass population was evacuated to other regions – the right bank of the valley of the Ili River (presently Altyn-Emel National Park) and Betpak-Dala Desert in the 1980s to avoid the death of animals. The remaining population was 100 specimens in 1998, and about 40 in 1999. After 2000 the remaining population, obviously, escaped from the former island (Jaschenko 1998–1999; Shaimardanov 2007).

Persian gazelles migrated to the Kyzylkum after connection of the island and the mainland; only a few of them and, obviously, visiting saiga were documented recently. The number of *Allactaga elater* and *Pygerethmus pumilio* was extremely low. *Spermophilus fulvus* is still abundant both at the former island and at the dried bottom surrounding the island. *Lepus tolai* is now a common species of the island, although it was never documented before the shrinking of the Sea (Eliseyev 2007).

The reduction of river flow in the Syr Darya delta zone has led to a decrease of tugai forest area and wetlands, which are the biotopes of wild boar, *Felis libyca*, *Canis lupus*, *Meles meles*, *Ondatra zibethicus*, *Microtus transcaspicus*, long-eared hedgehog, and *Sorex minutus*. The observation of the delta in 1965 showed high abundance of wild boar, and muskrats were an important subject of hunting. Other species were common or even abundant (Rotshild et al. 1967). Recent drainage and reduction of reeds and tugai areas led to an abrupt decrease of the number of animals which emigrated from the area.

Species such as Persian gazelle and wild boar were abundant at the eastern coast of the Aral Sea in the 1950s–1960s In 1952, 17.7 Persian gazelles were counted per 10 km of counting route and 11.9 were counted in 1956 (Rotshild et al. 1967). Wild boars and their track were very abundant at both the western (Kulandy Peninsula) and the eastern coasts of the Aral Sea (Varshavsky and Shilov 1955; Rotshild et al. 1967). However, in the 1970s only five Persian gazelles were encountered on a 360-km route (i.e. 0.14 per 10 km). The high abundance of this species was possible because of good grass coverage and numerous shrubs. Further lowering of the groundwater level and aridification of the landscapes has resulted in a change of specific composition of vegetation and poor conditions (Afanasyev and Kenessarin 1979). Also, tracks of wild boars became very rare in the region.

11.2.2 Birds of the Kazakhstan Part of the Aral Sea Basin

11.2.2.1 Historical Background

The first data on avifauna of the region were gathered by Eversmann in November 1820 and March 1821, who passed Lake Kamyshlybash when travelling to Boukhara with the Russian ambassador (Eversmann 1823). Romanov made a bird collection in the Syr Darya delta after being dispatched by Eversmann in 1848 (Eversmann 1866). Severtsov spent several seasons (1857–1858, 1865, 1874) collecting birds (Severtzov 1873). Bogdanov made some observations while moving to Khiva along with a Russian military party (Bogdanov 1882, 1884). Nikolsky (1892) visited the Aral Sea region in 1886. Vertebrate fauna, with emphasis on fish, was studied by Berg (1908) over several years (1899–1902 and 1906). Bostanjoglo studied the northern coast of the Aral Sea during April and May 1905 (Bostanjoglo 1911). Zarudny made a ship trip from Aralsk along the eastern coast of the Aral Sea in the summer of 1914 (Zarudny 1915a, b, 1916).

During 1924–1930, seven field trips were undertaken by Spangenberg and Feigin to study the avifauna of the Syr Darya lower flow and eastern coast of the Aral Sea (Spangenberg 1941; Spangenberg and Feygin 1936). Ismagilov studied birds on Barsa-Kelmes in 1942–1946 (Ismagilov 1955; Ismagilov and Vasenko 1950). The northern coast (near Akespe village) was studied by Kuzyakin in June 1946 and April to June 1947. He visited the island of Uzun-Kair in Paskevich Bay twice: 30 May 1956 and 30 May to 4 June 1947 (Kuzyakin 1959, 2005). Simultaneously with theriological research in the northern Aral Sea region in the middle of the twentieth century, Varshavsky conducted ornithological observations (Varshavsky 1957a, b, 1959a, b, 1965a, b, 1969, 1973; Varshavsky et al. 1977). The vicinity of Aralsk was observed by Grachev in the period from 30 September 1951 to 12 May 1954. He made quite long trips (up to 100 km) from Aralsk, visiting Lake Kamyshlybash (Grachev 1954, 1956a, b, 2000). Stepanyan and Galushin studied the avifauna of Barsa-Kelmes in the summer seasons of 1953 and 1954 (Stepanyan 1969; Stepanyan and Galushin 1962). Burambayev collected data on Barsa-Kelmes birds in 1963–1968 (Ismagilov and Burambayev 1973, 1978). In the period from 1971 to 1973, the avifauna of Barsa-Kelmes was studied by Gistzov (1974, 1978a, b).

Stationary studies at Barsa-Kelmes (see Chap. 14) were undertaken by Elisseyev in 1977, 1980–1984, 1988–1992 and 2005. He also visited the eastern coast of the Aral Sea (Elisseyev 1984a, b, 1985a, b, c, 1986a, b, 1987, 1989, 1990a, b, 1996a, b, 1999, 2003, 2004, 2007). The bird migrations were studied at Bayan meteorological station by Auezov, Beresovsky, Brokhovich and Khrokov in 1978 and 1979 (Beresovsky 1978, 1980, 1981, 1983a, b, 1991; Berezovsky et al. 1982; Brokhovich 1981; Khrokov et al. 1983). Collection of birds using nets was conducted by Sedunova and Yablonkevich in the spring of 1984 and 1990 (Sedunova and Yablonkevich 1991). Koblik made a trip to the vicinities of Bugun village following the former coastline in 1988 (Koblik 1991). A few researchers undertook shorter observations in the northern and northeastern coastal zones at the beginning of the twenty-first century. Kovshar visited the Syr Darya delta in August 2000 (Kovshar 2000). Khrokov conducted ornithological observations of delta lakes and the Syr Darya's mouth in the period from 27 April to 8 May 2001 (Khrokov 2004, 2006). Erokhov visited the delta lakes and northern coast of the Aral Sea in 2003 (Erokhov 2004). Ornithological observations in the lower part of the Syr Darya and the coastal zone of the Aral Sea were undertaken by Kovshar (2001) and Kovalenko (2006). Bird counting at Barsa-Kelmes and the dried sea bottom was performed by Kovshar (2007a,b). Belyalov made several visits to the Aral Sea region in 1988, 1993, 1995, 1996, 1997 and 2005.

Fragmentary data on the distribution and ecology of the birds inhabiting or visiting the Aral Sea region can be also found in Gladkov (1941, 1949), Petrovskaya (1951), Lobachyov (1959, 1961), Markov (1965), Rashek (1965) Rashek and Rashek (1963), Lankin (1976), Gubin (1998, 1999), Kovshar (2000) and Pfander (2004).

Herein the data on 322 bird species registered in the Kazakhstan part of the Aral Sea basin within 200 years are summarized. We give the species names according to Mullarney et al. (1999). Of the total list, most species registered for the Kazakhstan part of the Aral Sea basin were mainly extracted from the works of Bostanjoglo, Zarudny, Spangenberg, Elisseyev and partially Ismagilov, who conducted large-scale (the first three authors) or local but long-term (last two authors) ornithological investigations in that region. All other works, especially those conducted during the last two to three decades, could be classified rather as local observations limited in time. The total species list cannot be considered as complete because of numerous systematic issues which have not been solved. It will be changed and completed in the future. Some published data were not confirmed by the latest studies and they remain doubtful. Since these questions were outside the scope of the present review, we do not discuss them here. Generally, the number of species registered by several generations of ornithologists shows the maximal bird diversity and reflects the state at the beginning of the twenty-first century. Recently, owing to continuing drainage of the Aral Sea, the composition of Aral Sea avifauna appears to be changing.

11.2.2.2 Status of Birds in the Aral Sea Region

The most abundant group of birds in the Kazakhstan part of the Aral Sea basin is the breeding birds – as many as 165 species. There are 125 species of migrating birds (including 35 wintering ones). The vagrant birds group has 32 species. All the groups are divided into subcategories depending on their biotopic preferences. Of course, many species could be encountered in different biotopes, and one should list such species in every biotopic category. We do not make such a division, listing every species in its most typical biotope (e.g. the ruddy shelduck would nest not only in coastal outcrops, but also in ruins, mausoleums etc.). For the specific situation of the former island of Barsa-Kelmes, see Chap. 14.

The Breeding Species Group

The breeding species group has 165 species, which are categorized within three main categories: 135 species are migratory breeding species, 13 species are migratory breeding species and rare winter visitors and 17 species are resident breeding species, i.e. these species inhabit the area throughout the entire year, even though part of the population migrates south (Tables 11.2–11.4).

Seventy-four species are known to nest at the sea coast, the islands, and the lakes in the Syr Darya delta. Those species are most affected by the drying of the Aral Sea, and they could disappear from the area. However, the only place where the dynamics of species loss has been tracked is the island of Barsa-Kelmes. According to different researchers, the total list of the birds breeding in the island at whatever time consists of about 70 species (Ismagilov and Vasenko 1950; Stepanyan and Galushin 1962; Ismagilov and Burambayev 1973; Gistzov 1974; Elisseyev 2007). Since in some cases the data were controversial, it was quite difficult to draw an exact picture of the decline of bird species. Nevertheless, Ismagilov and Burambayev (1973) recorded 46 breeding birds during the period from 1942 to the late 1960s. They wrote that at the beginning of the 1960s, when sea level lowering began, the lagoon lakes of the island of Barsa-Kelmes – the nesting places of colonial birds – appeared to become shallow. From 1967, the following birds were no longer encountered there: Dalmatian pelican, great cormorant, Pallas's gull, Caspian gull, Caspian tern and black-winged stilt. Gistzov (1974) indicated disappearance of the following birds by 1971–1972: mute swan, mallard, gadwall, teal, northern shoveler, coot, oystercatcher, common gull, black-headed gull, gull-billed tern, common tern, little tern, yellow wagtail and bearded reedling. Thirty-eight breeding birds were recorded on the island at the beginning of the 1990s, but only 16 were registered in 2005: stone curlew, nightjar, swift, lesser short-toed lark, desert grey shrike, isabelline wheatear, pied wheatear, desert wheatear, rufous bush chat, Sykes's warbler, lesser whitethroat, desert warbler, desert finch, red-headed Bunting, Indian sparrow and rock sparrow (Elisseyev 2007).

Most of the islands at the eastern shore have become part of the mainland as early as 1975. Numerous bird colonies became accessible for mammalian predators, such as wolves and foxes, which feed on eggs. On the other hand, for fish-feeding bird species the extinction of most piscine species (due to abruptly increased water salinity) has become a limiting factor (Elisseyev 1990a).

When the desert occupied up to 50 km of the former sea bottom, a number of colonial birds moved to the Syr Darya delta. At the same time some species typical of the Syr Darya delta lakes also disappeared, which was connected with shallowing and salinization of the lakes located close to the shrinking eastern coast of the sea. In 1978–1979, 32 breeding species were registered in the delta lakes. Recently, birds such as grebes, pelicans, cormorants and herons have disappeared from the area. Only three species of known 13 Anseriformes remained: greylag goose, shelduck and ferruginous duck. The number of other breeding species has also been reduced: from 17 to 6 species of waders, from 6 to 2 species of gulls and from 7 to 2 species of terns (Beresovsky et al. 1982). However, it appears necessary to draw attention to

Table 11.2 Migratory breeding bird species

Wetland-dependent birds nesting at the coast, on islands and at deltaic lakes
Little grebe (*Tachybaptus ruficollis*)
Great crested grebe (*Podiceps cristatus*)
Great white pelican (*Pelecanus onocrotalus*)
Dalmatian pelican (*Pelecanus crispus*)
Great cormorant (*Phalacrocorax carbo*)
Pygmy cormorant (*Phalacrocorax pygmaeus*)
Great bittern (*Botaurus stellaris*)
Little bittern (*Ixobrychus minutus*)
Black-crowned night heron (*Nycticorax nycticorax*)
Squacco heron (*Ardeola ralloides*)
Great egret (*Egretta alba*)
Grey heron (*Ardea cinerea*)
Purple heron (*Ardea purpurea*)
Spoonbill (*Platalea leucorodia*)
Glossy ibis (*Plegadis falcinellus*)
Greylag goose (*Anser anser*)
Garganey (*Anas querquedula*)
Northern shoveler (*Anas clypeata*)
Marbled duck (*Marmaronetta angustirostris*)
Red-crested pochard (*Netta rufina*)
Pochard (*Aythya ferina*)
Ferruginous duck (*Aythya nyroca*)
White-headed duck (*Oxyura leucocephala*)
Pallid harrier (*Circus macrourus*)
Montagu's harrier (*Circus pygargus*)
Marsh harrier (*Circus aeruginosus*)
Common crane (*Grus grus*)
Baillon's crake (*Porzana pusilla*)
Moorhen (*Gallinula chloropus*)
Coot (*Fulica atra*)
Little ringed plover (*Charadrius dubius*)
Kentish plover (*Charadrius alexandrinus*)
Nothern lapwing (*Vanellus vanellus*)
White-tailed lapwing (*Vanellochettusia leucura*)
Black-winged stilt (*Himantopus himantopus*)
Avocet (*Recurvirostra avosetta*)
Oystercatcher (*Haematopus ostralegus*)
Redshank (*Tringa totanus*)
Marsh sandpiper (*Tringa stagnatilis*)
Common sandpiper (*Actitis hypoleucos*)
Ruff (*Phylomachus pugnax*)
Curlew (*Numenius arquata*)
Black-tailed godwit (*Limosa limosa*)
Collared pratincole (*Glareola pratincola*)
Black winged pratincole (*Glareola nordmanni*)
Pallas's gull (*Larus ichthyaetus*)

(continued)

Table 11.2 (continued)

Black-headed gull (*Larus ridibundus*)
Slender-billed Gull (*Larus genei*)
Common gull (*Larus canus*)
Black tern (*Chlidonias niger*)
White-winged tern (*Chlidonias leucopterus*)
Whiskered tern (*Chlidonias hybrida*)
Gull-billed tern (*Gelochelidon nilotica*)
Caspian tern (*Hydroprogne caspia*)
Common tern (*Sterna hirundo*)
Little tern (*Sterna albifrons*)
Common kingfisher (*Alcedo atthis*)
Yellow wagtail (*Motacilla flava beema*)
Black-headed wagtail (*Motacilla feldegg melanogrisea*)
Yellow-headed wagtail (*Motacilla lutea*)
Masked wagtail (*Motacilla personata*)
Cetti's warbler (*Cettia cetti*)
Savi's warbler (*Locustella luscinioides*)
Moustached warbler (*Lusciniola melanopogon*)
Sedge warbler (*Acrocephalus schoenobaenus*)
Paddyfield warbler (*Acrocephalus agricola*)
Clamorous reed warbler (*Acrocephalus stentoreus*)
Great reed warbler (*Acrocephalus arundinaceus*)
Bearded reedling (*Panurus biarmicus*)
Black-headed penduline tit (*Remiz macronyx*)
Coastal and island plain inhabitants
Steppe eagle (*Aquila nipalensis*)
Quail (*Coturnix coturnix*)
Macqueen's bustard (*Chlamydotis macqueenii*)
Stone curlew (*Burhinus oedicnemus*)
Sociable plover (*Chettusia gregaria*)
Greater sand plover (*Charadrius leschenaultii*)
Caspian plover (*Charadrius asiaticus*)
Pin-tailed sandgrouse (*Pterocles alchata*)
Nightjar (*Caprimulgus europaeus*)
Egyptian nightjar (*Caprimulgus aegyptius*)
Eurasian bee-eater (*Merops apiaster*)
Blue-cheeked bee-eater (*Merops persicus*)
Short-toed lark (*Calandrella brachydactyla*)
Asian short-toed lark (*Calandrella leucophaea*)
Bimaculated lark (*Melanocorypha bimaculata*)
Oriental skylark (*Alauda gulgula*)
Tawny pipit (*Anthus campestris*)
Northern wheatear (*Oenanthe oenanthe*)
Finsch's wheatear (*Oenanthe finschii*)
Desert wheatear (*Oenanthe deserti*)
Isabelline wheatear (*Oenanthe isabellina*)
Corn bunting (*Emberiza calandra*)
Ortolan bunting (*Emberiza hortulana*)

(continued)

Table 11.2 (continued)

Inhabitants of dense bushes at the plains
Long-legged buzzard (*Buteo rufinus*)
Imperial eagle (*Aquila heliaca*)
Turtle dove (*Streptopelia turtur*)
Turkestan shrike (*Lanius phoenicuroides*)
Red-backed shrike (*Lanius collurio*)
Desert grey shrike (*Lanius meridionalis pallidirostris*)
Brown-necked raven (*Corvus ruficollis*)
Sykes's warbler (*Hippolais rama)*
Olivaceous warbler (*Hippolais pallida*)
Upcher's warbler (*Hippolais languida*)
Lesser whitethroat (*Sylvia curruca halimodendri*)
Menetries's warbler (*Sylvia mystacea*)
Desert warbler (*Sylvia nana*)
Streaked scrub warbler (*Scotocerca inquieta*)
Rufous bush chat (*Cercotrichas galactotes*)
Saxaul sparrow (*Passer ammodendri*)
Desert finch (*Rhodospiza obsoleta*)
Red-headed bunting (*Emberiza bruniceps*)
Riparian forest inhabitants
Shikra sparrowhawk (*Accipiter badius*)
Eversmann's dove (*Columba eversmanni*)
Cuckoo (*Cuculus canorus*)
Pallid scops owl (*Otus brucei*)
Carrion crow (*Corvus corone*)
Nightingale (*Luscinia megarhynchos*)
Turkestan penduline tit (*Remiz pendulinus jaxarticus*)
White-crowned penduline tit (*Remiz coronatus*)
Coastal and island outcrop inhabitants
Saker falcon (*Falco cherrug*)
Swift (*Apus apus*)
Sand martin (*Riparia riparia*)
Pale sand martin (*Riparia diluta*)
Pied wheatear (*Oenanthe pleschanka*)
Indian sparrow (*Passer indicus*)
Grey-necked bunting (*Emberiza buchanani*)
Settlement inhabitants
Barn swallow (*Hirundo rustica*)
Long-tailed shrike (*Lanius schach*)
Lesser grey shrike (*Lanius minor*)
Spanish sparrow (*Passer hispaniolensis*)
Inhabitants of ruins including the ancient Kazakh mausoleums and necropoles
Lesser kestrel (*Falco naumanni*)
Kestrel (*Falco tinnunculus*)
European roller (*Coracias garrulus*)
Hoppoe (*Upupa epops*)
Rose-coloured starling (*Sturnus roseus*)

Table 11.3 Migratory breeding bird species and rare winter visitors

Wetland-dependent birds of the sea coast and the lakes of Syr Darya delta
Mute swan (*Cygnus olor*)
Mallard (*Anas platyrhynchos*)
Gadwall (*Anas strepera*)
Caspian gull (*Larus cachinnans*)
Coastal and island plain inhabitants
Black-bellied sandgrouse (*Pterocles orientalis*)
Pallas's sandgrouse (*Syrrhaptes paradoxus*)
Lesser short-toed lark (*Calandrella rufescens*)
Calandra lark (*Melanocorypha calandra*)
Shore lark (*Eremophila alpestris*)
Coastal and island outcrop inhabitants
Ruddy shelduck (*Tadorna ferruginea*)
Shelduck (*Tadorna tadorna*)
Jackdaw (*Corvus monedula*)
Rock sparrow (*Petronia petronia*)

Table 11.4 Resident breeding bird species

Plain inhabitants
Gray partridge (*Perdix perdix*)
Crested lark (*Galerida cristata*)
Pander's ground jay (*Podoces panderi*)
Riparian forest inhabitants
Common pheasant (*Phasianus colchicus turcestanicus*)
White-winged woodpecker (*Dendrocopos leucopterus*)
Magpie (*Pica pica*)
Turkestan tit (*Parus bokharensis*)
Coastal outcrop inhabitants
Golden eagle (*Aquila chrysaetos*)
Barbary falcon (*Falco pelegrinoides babylonicus*)
Eagle owl (*Bubo bubo*)
Little owl (*Athene noctua*)
Raven (*Corvus corax*)
Inhabitants of settlements
Rock dove (*Columba livia*)
Collared dove (*Streptopelia decaocto*)
Laughing dove (*Streptopelia senegalensis*)
House sparrow (*Passer domesticus*)
Tree sparrow (*Passer montanus*)

the insufficient degree to which that question has been studied. In the period mentioned, no special monitoring ornithological investigations were conducted in the region, but only short, often simultaneous observations were performed. So, we do not have complete data on species composition and its dynamics at our disposal.

Forty-two breeding species were known to be associated with artesian and deltaic lakes in the lower flow of the Syr Darya in the late 1980s to the early 1990s (Koblik 1991; Gubin 1999).

The abundance of birds may remain quite stable at Barsa-Kelmes for those species, nesting in the island valleys: stone curlew, Caspian plover, nightjar, lesser short-toed lark and isabelline wheatear. Also the following species remain intact: desert grey shrike, Sykes's warbler, lesser whitethroat and red-headed bunting (bush-nesting birds); swift, jackdaw and pied wheatear (outcrop-nesting birds); and barn swallow, tree sparrow and house sparrow (synanthropic species) (Gistzov 1974). Some of those species have become inhabitants of the dried bottom.

11.2.2.3 Passage Visitors

Passage visitors (Table 11.5) are also quite an abundant group, which amounts to 125 species. Waterbirds and wetland-dependent birds represent about half of the total number of species. This group nests in steppe and steppe-forest zones of Kazakhstan and western Siberia, and even in tundra near the Arctic Ocean shoreline. Their migratory way passes the Turgay Depression and runs along the Aral Sea coast further to wintering places at the Caspian Sea and farther to Africa .

This group is subdivided into three categories. The first one is the passage visitors, which comprises 64 species, migrating through the Aral Sea region in spring and autumn. The second category is passage and summer not breeding visitors, which has 26 species. This category is also typical for spring and autumn migrations; however, these birds would be partially encountered at the Aral Sea in the summer, and sometimes quite abundantly. The most typical representatives of this category are ducks and waders. These nonbreeding, roaming birds may erroneously be considered as breeding species. The third category is passage and winter visitors. They comprise 35 species, which not only migrate through the area but also sometimes winter there (Tables 11.6 and 11.7).

A decrease of migratory species numbers was first registered in the 1950s, 10 years before the shrinking started (Stepanyan and Galushin 1962). The reduction was documented later as well, and not only in the Aral Sea region, but also in regions that are not affected by serious ecological changes.

Data collected in 1978–1979 show an abundance of migrating birds. In the spring, waders, ducks and gulls predominated (up to 50,000 birds counted for the season). In the autumn of 1979, up to 230,000 water birds were counted. The researchers consider the specific changes of birds migrating in the delta region to be minor (Beresovsky et al. 1982). The same conditions exist presently.

11.2.2.4 Vagrant Birds

Only 32 species of vagrant birds (Table 11.8) were documented during 200 years of ornithological observations. Most of them appeared in the region after being led by other bird flocks migrating along the coastline of the Aral Sea. Most of the finds are single, which could be explained by rare visits to the region by ornithologists.

Table 11.5 Passage visitors (birds)

Sea coast, lakes of delta inhabitants
Black-throated loon (*Gavia arctica*)
Black-necked grebe (*Podiceps nigricollis*)
Slavonian grebe (*Podiceps auritus*)
Red-necked grebe (*Podiceps griseigena*)
Greater flamingo (*Phoenicopterus roseus*)
Red-breasted goose (*Rufibrenta ruficollis*)
White-fronted goose (*Anser albifrons*)
Bean goose (*Anser fabalis*)
Pintail (*Anas acuta*)
Tufted duck (*Aythya fuligula*)
Long-tailed duck (*Clangula hyemalis*)
Smew (*Mergus albellus*)
Hen harrier (*Circus cyaneus*)
Peregrine falcon (*Falco peregrinus*)
Hobby (*Falco subbuteo*)
Demoiselle crane (*Anthropoides virgo*)
Water rail (*Rallus aquaticus*)
Spotted crake (*Porzana porzana*)
Little crake (*Porzana parva*)
Gray plover (*Pluvialis squatarola*)
Pacific golden plover (*Pluvialis fulva*)
European golden plover (*Pluvialis apricaria*)
Ringed plover (*Charadrius hiaticula*)
Jack snipe (*Lymnocryptes minimus*)
Slender-billed curlew (*Numenius tenuirostris*)
Bar-tailed godwit (*Limosa lapponica*)
Little gull (*Larus minutus*)
Citrine wagtail (*Motacilla citreola*)
White wagtail (*Motacilla alba*)
Coastal and island plain inhabitants
Red-footed falcon (*Falco vespertinus*)
Great bustard (*Otis tarda*)
Little bustard (*Otis tetrax*)
Eurasian dotterel (*Eudromias morinellus*)
Stock dove (*Columba oenas*)
Eastern turtle dove (*Streptopelia orientalis*)
House martin (*Delichon urbica*)
Tree pipit (*Anthus trivialis*)
Meadow pipit (*Anthus pratensis*)
Red-throated pipit (*Anthus cervinus*)
Inhabitants of dense bushes
Wood cock (*Scolopax rusticola*)
Long-eared owl (*Asio otus*)
Wryneck (*Jynx torquilla*)
Golden oriole (*Oriolus oriolus*)
Grasshopper warbler (*Locustella naevia*)
Blyth's reed warbler (*Acrocephalus dumetorum*)
Booted warbler (*Hippolais caligata*)

(continued)

Table 11.5 (continued)

Barred warbler (*Sylvia nisoria*)
Whitethroat (*Sylvia communis*)
Willow warbler (*Phylloscopus thochilus*)
Chiffchaff (*Phylloscopus collybita*)
Greenish warbler (*Phylloscopus trochiloides viridanus*)
Red-breasted flycatcher (*Ficedula parva*)
Spotted flycatcher (*Muscicapa striata*)
Whinchat (*Saxicola rubetra*)
Stonechat (*Saxicola torquata*)
Common redatart (*Phoenicurus phoenicurus*)
Thrush nightingale (*Luscinia luscinia*)
Bluethroat (*Luscinia svecica*)
Black-throated thrush (*Turdus atrogularis*)
Fieldfare (*Turdus pilaris*)
Redwing (*Turdus iliacus*)
Song thrush (*Turdus philomelos*)
Scarlet rosefinch (*Carpodacus erythrinus*)
Hawfinch (*Coccothraustes coccothraustes*)

Table 11.6 Passage and summer not breeding visitors

Sea coast, lakes in delta inhabitants
Whooper swan (*Cygnus cygnus*)
Teal (*Anas crecca*)
Wigeon (*Anas penelope*)
Osprey (*Pandion haliaetus*)
Black-eared kite (*Milvus migrans lineatus*)
Pallas's sea eagle (*Haliaeetus leucoryphus*)
Turnstone (*Arenaria interpres*)
Green sandpiper (*Tringa ochropus*)
Wood sandpiper (*Tringa glareola*)
Greenshank (*Tringa nebularia*)
Spotted redshank (*Tringa erythropus*)
Terek sandpiper (*Xenus cinereus*)
Red-necked phalarope (*Phalaropus lobatus*)
Little stint (*Calidris minuta*)
Temminck's stint (*Calidris temminckii*)
Curlew sandpiper (*Calidris ferruginea*)
Dunlin (*Calidris alpina*)
Sanderling (*Calidris alba*)
Broad-billed sandpiper (*Limicola falcinellus*)
Common snipe (*Gallinago gallinago*)
Whimbrel (*Numenius phaeopus*)
Coastal and island plains
Short-toed eagle (*Circaetus gallicus*)
Black vulture (*Aegypius monachus*)
Griffon vulture (*Gyps fulvus*)
Bush (shrubs), trees
Marsh warbler (*Acrocephalus palustris*)
Reed warbler (*Acrocephalus scirpaceus*)

Table 11.7 Passage and winter visitors

Sea coast, lakes in delta inhabitants
Golden-eye (*Bucephala clangula*)
Goosander (*Mergus merganser*)
Rough-legged buzzard (*Buteo lagopus*)
White-tailed sea eagle (*Haliaeetus albicilla*)
Merlin (*Falco columbarius*)
Heuglin's gall (*Larus heuglini*)
Short-eared owl (*Asio flammeus*)
White-winged lark (*Melanocorypha leucoptera*)
Black lark (*Melanocorypha yeltoniensis*)
Great grey shrike (*Lanius excubitor*)
Starling (*Sturnus vulgaris*)
Hooded crow (*Corvus cornix*)
Coastal and island plain inhabitants
Skylark (*Alauda arvensis*)
Rook (*Corvus frugilegus*)
Lapland bunting (*Calcarius lapponicus*)
Snow bunting (*Plectrophenax nivalis*)
Inhabitants of dense bushes
Goshawk (*Accipiter gentilis*)
Sparrowhawk (*Accipiter nisus*)
Waxwing (*Bombycilla garrulus*)
Dunnock (*Prunella modularis*)
Robin (*Erithacus rubecula*)
Blackbird (*Turdus merula*)
Mistle thrush (*Turdus viscivorus*)
Chaffinch (*Fringilla coelebs*)
Brambling (*Fringilla montifringilla*)
Siskin (*Spinus spinus*)
Goldfinch (*Carduelis carduelis*)
Linnet (*Acanthis cannabina*)
Twite (*Acanthis flavirostris*)
Mealy redpoll (*Acanthis flammea*)
Crossbill (*Loxia curvirostra*)
Bullfinch (*Pyrrhula pyrrhula*)
Yellowhammer (*Emberiza citrinella*)
Pine bunting (*Emberiza leucocephala*)
Reed bunting (*Emberiza schoeniclus*)

Some finds have never been approved by reliable information and possibly could represent a result of incorrect identification.

11.2.3 Reptiles

In 2002–2004, an international team of scientists surveyed the occurrence of reptiles (Table 11.9) and amphibians in both the Kazakhstan and Uzbekistan parts of the Aral

Table 11.8 Vagrant birds

Red-throated loon (*Gavia stellata*)
Little egret (*Egretta garzetta*)
White stork (*Ciconia ciconia*)
Lesser white-fronted goose (*Anser erythropus*)
Bewick's swan (*Cygnus bewickii*)
Scaup (*Aythya marila*)
Velvet scoter (*Melanitta fusca*)
Red-breasted merganser (*Mergus serrator*)
Levant sparrowhawk (*Accipiter brevipes*)
Eurasian buzzard (*Buteo buteo*)
Spotted eagle (*Aquila clanga*)
Gyr falcon (*Falco rusticolus*)
Siberian crane (*Grus leucogeranus*)
White-naped crane (*Grus vipio*)
Little sand plover (*Charadrius mongolus*)
Gray phalarope (*Phalaropus fulicarius*)
Swinhoe's snipe (*Gallinago megala*)
Great snipe (*Gallinago media*)
Little curlew (*Numenius minutus*)
Asian dowitcher (*Limnodromus semipalmatus*)
Pomarine skua (*Stercorarius pomarinus*)
Arctic skua (*Stercorarius parasiticus*)
Sandwich tern (*Thalasseus sandvicensis*)
Saunders's tern (*Sterna saundersi*)
Wood pigeon (*Columba palumbus*)
Alpine swift (*Apus melba*)
Grey wagtail (*Motacilla cinerea*)
Isabelline shrike (*Lanius isabellinus*)
Wood warbler (*Phylloscopus sibilatrix*)
Rufous-tailed rock thrush (*Monticola saxatilis*)
Black redatart (*Phoenicurus ochruros*)
Eversmann's redstart (*Phoenicurus erythronotus*)

Sea basin (INTAS project). During these investigations a total of 28 species were found, but six species that had been recorded previously could not be found again.

11.2.3.1 Fence Lizard (*Lacerta agilis*)

At the beginning of the examinations, it was unclear whether *Lacerta agilis* occurred in the area of the Aral Sea basin at all (see Bischoff 1988). The fence lizard is admittedly spread in the forest steppes of southern Siberia and northern Kazakhstan east of eastwards to the Mongolian border; however we could not prove its presence in the first year at the Aral Sea. Only in the second year did we find several populations in the big Barsuki sand dunes north of Shalkar (Fig. 11.3). These are the populations situated next to the Aral Sea. South of Shalkar, the ecological niche of the species becomes occupied by representatives of the genus *Eremias* (see later).

Table 11.9 Reptiles collected around the Aral Sea in 2002–2004 compared with earlier records which could not be confirmed (*red*)

Species	Family	No. of records/last record
Ablepharus deserti	Scincidae	3
Agrionemys horsfieldi	Testudinidae	*39*
Alsophylax loricatus	Gekkonidae	1997
Alsophylax pipiens	Gekkonidae	1
Coluber karelini	Colubridae	12
Coluber ravergieri	Colubridae	7
Crossobamon eversmanni	Gekkonidae	4
Cyrtopodion caspius	Gekkonidae	18
Cyrtopodion russowi	Gekkonidae	14
Echis multisquamatus	Viperidae	1968
Emys orbicularis	Emydidae	1915
Elaphe dione	Colubridae	5
Elaphe (quatuorlineata) sauromates	Colubridae	1
Eremias arguta	Lacertidae	2
Eremias grammica	Lacertidae	1
Eremias intermedia	Lacertidae	14
Eremias lineolata	Lacertidae	*25*
Eremias scripta scripta	Lacertidae	4
Eremias velox	Lacertidae	*63*
Eryx miliaris	Boidae	2
Eryx tartaricus	Boidae	2
Gloydius halys	Crotalidae	9
Natrix tessellata	Colubridae	14
Phrynocephalus guttatus	Agamidae	?
Phrynocephalus helioscopus	Agamidae	13
Phrynocephalus interscapularis	Agamidae	13
Phrynocephalus moltschanowi	Agamidae	1990
Phrynocephalus mystaceus	Agamidae	7
Phrynocephalus rossikowi	Agamidae	1997
Psammophis lineolatum	Psammophiidae	*22*
Spalerosophis diadema	Colubridae	1
Teratoscincus scincus	Gekkonidae	12
Testudo (Agrionenys) horsfieldi	Testudinidae	39
Trapelus sanguinolentus	Agamidae	*48*
Varanus griseus	Varanidae	2
Vipera renardi	Viperidae	1

Morphological and genetically, on the basis of a sequence of 1,140 bp of cytochrome *b* and the transfer RNA threonine, this population could be unequivocally assigned to the subspecies *Lacerta agilis exigua*. In the total distribution area of the fence lizards, *Lacerta agilis exigua* occupies the biggest area: It is spread from the Black Sea east to Lake Baikal. Although other subspecies, specifically in the Caucasus and in the Balkans, are genetically variable, *Lacerta agilis exigua* is genetically very homogeneous (Kalyabina et al. 2001). This speaks on behalf of the hypothesis that the gigantic distribution area of this subspecies was only populated after the last glacial, presumably from the Caucasus.

Fig. 11.3 Fence lizard, *Lacerta agilis exigua*

Fig. 11.4 Desert runner, *Eremias intermedia*

11.2.3.2 Desert Runners (*Eremias* spp.)

We collected six species of desert runners in the Aral Sea basin: *Eremias arguta*, *Eremias grammica*, *Eremias intermedia*, *Eremias lineolata*, *Eremias scripta* and *Eremias velox* (Fig. 11.4). The ecological niches of the different species and genotypes are of interest. With great likelihood, they can serve as indicator species for certain habitat types. *Eremias grammica* and *Eremias lineolata* inhabit sandy areas (dunes). Individuals from some species populate the dried sea bottom as pioneers (see later).

Eremias grammica could not be studied by molecular methods. A preliminary molecular genetic tree, using the mitochondrial cytochrome *b* gene, revealed that *Eremias intermedia* and *Eremias scripta* are sister species, whereas *Eremias arguta* is a sister of *Eremias velox* (Fig. 11.24).

11.2.3.3 Steppe Agama (*Trapelus sanguinolentus*)

This biggest lizard in the area of the Aral Sea (Fig. 11.5) occurs around the former sea and is also found on the dried sea bottom. The pioneer settlement of the

Fig. 11.5 Steppe agama, *Trapelus sanguinolentus*

Fig. 11.6 Steppe snake, *Elaphe dione*

Aralkum can therefore be documented with its help. It nourishes itself not only from insects but it takes also vegetable food, which increases its ecological pointer-value.

11.2.3.4 Steppe Snake, *Elaphe dione*, and Four-Striped Snake, *Elaphe (quatuorlineata) sauromates*

As they occupy a higher trophic level than lizards, these snakes show a special quality of the biocenosis. We reckon on a strong regional differentiation in *Elaphe dione* (Fig. 11.6), as we already determined for the four-striped snake (Lenk et al. 2001). This species probably consists of two to three subspecies or even species. For the steppe snake, this is still to be explored. The four-striped snake is much rarer than the steppe snake according our findings in the Aral Sea basin and is probably very selective ecologically.

11.2.3.5 Steppe Adders (*Vipera renardi*)

According to morphological criteria (Nilson and Andrén 2001), two subspecies of the steppe adder exist in the vicinity of the Aral Sea: *Vipera renardi renardi*, in northern Kazakhstan, and *Vipera renardi tienshanica*, in southern Kazakhstan and eastern Uzbekistan, in the Syr Darya lowlands. No Steppe adders would live in the immediate surroundings of the lake according to Nilson and Andrèn. However, we found specimens not far from the northern shore of the Small Aral Sea. There remains the question of to which subspecies these animals belong. A connection would be either possible to the main area of *Vipera renardi renardi* further north, or along the Syr Darya to *Vipera renardi tienshanica* in the southeast. The question is of animal-geographical and evolutionary historical interest. A specimen from the Barsuki sand dunes belongs to *Vipera renardi renardi* (Fig. 11.7). Steppe adders are an indicator for undisturbed steppe habitats. They are selective in reference to habitat quality and food. They have already vanished from the agriculturally used areas of Uzbekistan and Kazakhstan.

11.2.3.6 Steppe Tortoise [*Testudo* (*Agrionemys*) *horsfieldi*]

The steppe tortoise is the only tortoise in Central Asia. It remains in subterranean burrows most of the year and becomes active only during 3 months of the year (mainly in spring). Perälä (2002) proposed subdividing the species into several distinct species and distinguishing a special subspecies in the Aral Sea basin, (Perälä, unpublished work). However, these morphologically characterized forms could not be verified genetically. A phylogeographical study (Fritz et al. 2009) using mitochondrial DNA identified three main haplotype groups which do not correspond well with the morphological forms. In the Aral sea basin, as well as in most of Kazakhstan and Uzbekistan, a northern haplotype (*Testudo horsfieldi*

Fig. 11.7 Steppe viper, *Vipera renardi*

Fig. 11.8 (**a**) Steppe tortoise, *Testudo horsfieldi*, aged specimen; (**b**) European pond turtle, *Emys orbicularis*, from a tributary of the Syr Darja

kazachstanica) dominates. On Ustyurt Plateau, a less common haplotype was found, corresponding to the Caspian subspecies *Testudo horsfieldi rustamovi*.

The species is under strong pressure through collecting. Furthermore, the tortoise reacts sensitively to environmental poisons, which it accumulates in its body, since it eats considerable quantities of plants and reaches a very old age. It is therefore a pointer for an intact steppe environment.

In the Uzbekistan part of the Aral Sea basin, we found the species in large numbers, but the size distribution was biased towards large specimens (Fig. 11.8). The rarity of young specimens could be an alarming indication of reduced fertility.

11.2.3.7 European Pond Turtle (*Emys orbicularis*)

This aquatic turtle reaches its eastern distribution limit in northwestern Kazakhstan. During our research, we detected a population on the banks of Turgay River and in nearby ponds, about 300 km north of the Aral Sea. Genetically these turtles were assigned to haplotype I sensu Lenk et al. (1999). This haplotype has a wide distribution from Kazakhstan to Poland. We did not find *Emys* closer to the Aral Sea. It had been recorded from the Syr Darja in the beginning of the 20th century (Nikolsky 1915, Zarudny 1915), but this was not confirmed later. However in 2010, a new record of *Emys orbicularis* from a tributary of the Syr Darja near the town of Turkestan was published (Ryssakova and Sarzhanov 2010). Therefore we can regard this turtle as a probable former inhabitant of Aral Sea waters, at least in the Syr Darja delta.

11.2.4 Amphibians

Only two amphibian species – the lake frog *Pelophylax (Rana) ridibundus* (Fig. 11.9a) and the green toad *Bufo viridis* (Fig. 11.9b) – inhabit the region.

Our research has confirmed the presence of *Pelophylax ridibundus* in the eastern Aral Sea region including the lower part and the delta of the Syr Darya, where the frog prefers the freshwater arteries and freshwater or slightly salty constant water reservoirs such as lakes, ponds and artificial pools. The most western record of *Pelophylax ridibundus* is in Lake Tishchye-Bass in the delta region of the Syr Darya (46°09′N; 61°16′E). To the west and closer to the recent eastern coast of the Aral Sea as well as at the northern and western coasts of the sea, the lake frog was absent because of the absence of available habitats.

We did not find the frog in a few earlier known (early twentieth century) frog habitats in the Syr Darya delta (e.g., in the vicinity of Bugun village). This can be explained by the extreme salinity of the territory and drying of water reservoirs. For surrounding areas, the records of *Pelophylax ridibundus* were obtained for the Lake Shalkar system and the Mugodzhar Mountains. *Pelophylax ridibundus* was also found along the Amu Darya. Owing to its dependence on freshwater, it is not an inhabitant of the Aralkum.

In general, the pattern of distribution of *Bufo viridis* in the Aral Sea region has not changed from the time of the first known records of this amphibian (early twentieth century). Unlike the lake frog, the green toad belongs to the terrestrial amphibians with close connection to water only in the breeding season.

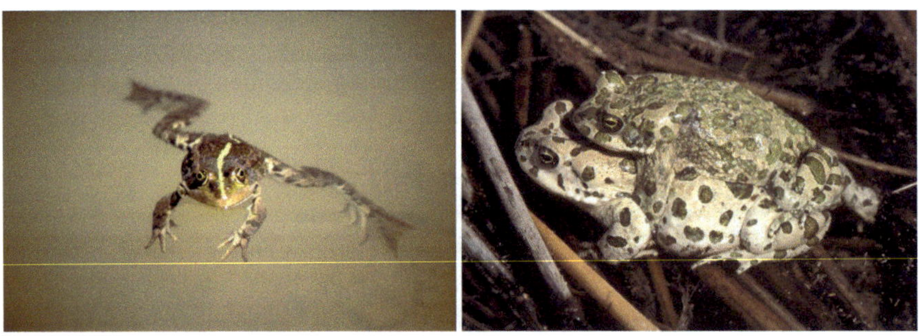

Fig. 11.9 Amphibians of the Aral Sea basin: (**a**) lake frog, *Pelophylax ridibundus*; (**b**) green toad, *Bufo viridis* complex

Being well adapted to a very short spring season suitable for breeding in the deserts and being very resistant to environmental fluctuation (Dinessman 1953), the toad successfully inhabits not only small constant water reservoirs such as artesian water holes, creeks and wells but also temporary water bodies (rain pools, floods of melted waters, etc.) even with an elevated level of salinity. We confirmed a wide distribution of the green toad at the eastern, northeastern, northwestern and southern coasts of the Aral Sea and in surrounding areas. However, we did not find the toads in the coastal lagoons with seawater from where *Bufo viridis* was known in the first half of the twentieth century (Elpatjevsky 1903; Dinessman 1953). In spite of the quite high resistance of the green toad to high water salinity (up to 10.4–11.4‰), the recent salinity of Aral Sea water (up to 15‰) seems to be too high for amphibian survival. New finds appeared for the northern Aral Sea region, including the Irgiz–Turgay interflow. For the northern part of Ustyurt Plateau, Bolshie Barsuki sands and the Mugorzhar mountains finds of *Bufo viridis* were registered for the first time. It was also found near isolated ponds between the western shore of the lake and the southern (Uzbek) part of Ustyurt Plateau.

In relation to water quality, the green toad demonstrated more plasticity than the lake frog. The pH of water in the reservoirs where the lake frog was recorded ranged from 7.35 to 8.7 and mineralization ranged from 0.7 to 4.9 g/l, whereas for the green toad these parameters were 7.7–9.8 and 0.5–5.75 g/l, respectively.

The lake frog prefers water reservoirs of the $HCO_3^- + Ca^{2+}$ group and very rarely (southwestern coasts of Lake Kamyshlybash, Lake Laikol) was recorded in the lakes with quite significant content (1.0–1.5 g/l) of SO_4^{2-} and Cl^- ions. The frogs were absent from the water reservoirs with high concentration (1.5–3.0 g/l) of SO_4^{2-} and Cl^- ions (Lake Rayim, eastern and southeastern coasts of Lake Kamyshlybash). Unlike *Pelophylax ridibundus, Bufo viridis* was recorded not only in the $HCO_3 + Ca^{2+}$ group reservoirs, but also in the reservoirs of $HCO_3^- + Cl^- + Na^+$, $HCO_3^- + Na^+$, and $SO_4^{2-} + Na^+$ groups.

11.3 Insects of the Kazakhstan Part of the Aral Sea Region

Insects play a tremendously important role in terrestrial ecosystems. Insect biomass, counted per hectare, exceeds the biomass of all other animals. Recently, insects have become the main consumer of green vegetation as the role of herbivorous wild mammals is often reduced. On the other hand, insects used to be the main food resource for most birds, reptiles, amphibians and some mammals. So, the meaning of insects in biocenoses is hard to overestimate, as this group is one of the main components forming the biocenosis.

The Aral Sea has been known for decades to be an ecological catastrophe region, and the consequences of the Aral Sea disaster are very important for Middle Asia as a whole. Enormous areas have recently been represented by the dried sea bottom, due do abrupt shrinking of the sea. Dried areas are colonized by succession communities having different stages of development. Eurybiont species, both plants and insects, appeared to be first to colonize the dried bottom.

Insects of the Aral Sea were studied in 1969–1970 and 1990 by entomologists from the Institute of Zoology (Almaty, Khazakhstan) and specialists from some other newly independent states, and by the author (R. Kadyrbekov) during his field work, supported by INTAS (2002–2003), and short trips in 1995–1996.

Previous studies of Aral entomofauna were published in Chetverikov (1906), Kozhanchikov (1937, 1950), Rodendorf (1937), Paramonov (1940), Ogloblin and Znoiko (1950), Zimin (1951), Nikolskaya (1952), Kuzin (1953), Lukyanovich and Ter-Minassyan (1957), Makhnovsky (1958), Kostin (1960, 1973), Zagulyayev (1960, 1964, 1973, 1975, 1979), Schteinberg (1962), Sinadsky (1963, 1968), Krizhanovsky (1965, 1976, 1983), Ter-Minassyan (1967, 1988), Gucevich et al. (1970), Olsufyev (1977), Guryeva (1979, 1989), Kerzhner (1982), Pessenko (1983), Leley (1985), Longvinovsky (1985), Medvedev and Nepessova (1985), Kulenova and Lopatin (1986), Nikolayev (1987), Nikolayev and Kozminikh (2002), Nikolayev and Kolov (2005), Dzhanokmen (1993), Fedotova (1993, 2000), Jashenko (1993), Kadyrbekov (1993, 2003, 2004), Kadyrbekov and Tleppaeva (2004), Kadyrbekov et al. (1997), Kazenas (1993, 2002), Kazenas et al. (1997, 1998), Konev (1993), Mitiayev (1993, 2002), Nassirova (1993), Pirjulin (1993a, b, 1995), Pirjulin and Ozersky (1995), Zhdanko (1993), Kabanova (1995), Assanova (1996), Chikin et al. (2004) and Essenbekova and Kadyrbekov (2004).

11.3.1 Taxonomic Diversity of the Insects of the Aral Sea Basin

A total of 2649 insect species belonging to 17 orders were recorded (Table 11.10): Odonata – 13 species, Blattoptera – 7 species, Ephemeroptera – 1 species, Plecoptera – 1 species, Mantoptera – 7 species, Phasmoptera – 1 species, Dermaptera

Table 11.10 Taxonomical diversity of insect orders in the Aral Sea region

Order	Number of species	Number of families	Taxonomical diversity (%)
Odonata	13	7	0.5
Blattoptera	7	2	0.26
Ephemeroptera	1	1	0.04
Plecoptera	1	1	0.04
Mantoptera	7	2	0.26
Phasmoptera	1	1	0.04
Dermaptera	2	2	0.08
Orthoptera	72	7	2.7
Homoptera	539	20	20.34
Heteroptera	133	23	5.1
Trichoptera	3	2	0.11
Thysanoptera	10	–	0.38
Coleoptera	925	28	34.9
Lepidoptera	168	14	6.3
Hymenoptera	288	15	10.9
Diptera	473	10	17.9
Neuroptera	6	3	0.23

– 2 species, Orthoptera – 72 species, Homoptera – 539 species, Trichoptera – 3 species, Heteroptera – 133 species, Thysanoptera – 10 species, Coleoptera – 925 species, Lepidoptera – 168 species, Hymenoptera – 288 species, Diptera – 473 species, Neuroptera – 6 species. Information on the other 13 orders inhabiting the Turan Depression is lacking in the available literature. Some groups remain poorly studied (some beetle families, night butterflies, most of Hymenoptera and Diptera families), but the true number of species in the region should exceed 3,000.

Dragonflies (Odonata) are a comparatively small order of insects with full (complete) metamorphosis and aquatic larval stage, which inhabit mostly tropical and subtropical areas. There are 13 species from the families Aeschnidae, Calopterygidae, Coenagrionidae, Corduliidae, Lestidae, Libellulidae and Gomphidae (Belishev and Shevtschenko 1971). These species are inhabitants of tugai forests and wetland landscapes. Imagos of some species were encountred in sandy, clay deserts. Five species are included in the *Red Data Book of Kazakhstan* (Table 11.13): *Ischnura aralensis*, *Calopteryx virgo*, *Anax imperator*, *Orthetrum sabina*, and *Celysiothemis virgo*.

Ephemeroptera is a small order of water-related insects with incomplete metamorphosis. It comprises about 1,600 species worldwide, all of them are main food resources for fishes. Only one undetermined species is known so far for the Aral Sea region.

Plecoptera is a small, mostly Holarctic order of aquatic insects with incomplete metamorphosis. It comprises 2,000 species. One species is documented in the Aral Sea region, and belongs to the genus *Isopterix* (Kussainova 1954; Kussainova and Embergenov 1971).

Trichoptera is a rather small order of water-related insects, which comprises about 6,000 species in global fauna. Three species are known from Aral Sea region (Pirjulin 1995; Kussainova 1954): *Oecetis intima, Oecetis ochracea parvula*, and *Agrypnetes crassicornis*.

Blattoptera is a small, mostly tropical order of insects with incomplete metamorphosis, comprising about 3,000 species. Seven species are known from the Aral Sea region (Pirjulin and Ozersky 1995): *Phyllodromica irinae, Phyllodromica pygmaea, Phyllodromica tartara, Ectobius semenovi, Arenivaga roseni, Polyphaga pellucida*, and *Mononychoblatta semenovi*. These species are nocturnal, and inhabit both different desert types and wet, water-related biotopes.

Dermaptera is a small, mostly tropical order of insects with incomplete metamorphosis, comprising about 1,000 species. Two species of two families are known from the Aral sea region: *Labidura riparia* (Labiduridae) and *Anechura asiatica* (Forficulidae) (Kadyrbekov 2004; Pirjulin and Ozersky 1995). Both species are associated with wet environments.

Mantoptera is a small, mostly tropical order of insects with incomplete metamorphosis, comprising about 2,000 species. Seven species from two families are known from the Aral Sea region: *Empusa pennicornis (*Empusidae*), Armene pusilla, Bolivaria brachyptera, Iris polystictica, Mantis religiosa, Oxyothespis turcomaniae*, and *Rivetina nana* (Manteidae) (Pirjulin and Ozersky 1995), all well adopted to desert conditions. Some species were encountered in tugai forest. One species – *Bolivaria brachyptera* – s included in the *Red Data Book of Kazakhstan* (2006).

Table 11.11 Taxonomical diversity of Orthoptera in the Aral Sea region

Family	Number of species	Turan endemics	Rare and relict species
Tettigoniidae	10	1	2
Gryllidae	9	2	–
Tridactylidae	2	–	–
Tetrigidae	1	–	–
Pamphagidae	3	2	–
Pyrgomorphidae	2	1	–
Acrididae	45	7	–

After Pirjulin and Ozersky 1995

Orthoptera is a large order of insects with incomplete metamorphosis, comprising about 20,000 species. Seventy-two species are known from the Aral Sea region (Pirjulin and Ozersky 1995; Nassirova 1993). These species belong to the families Tettigoniidae, Gryllidae, Tridactylidae, Tetrigidae, Pamphagidae, Pyrgomorphidae and Acrididae and inhabit different environments. The taxonomical diversity of the group is shown in Table 11.11. Two species – *Saga pedo* and *Caereocercus fuscipennis* (Tettigoniidae) – are included in the *Red Data Book of Kazakhstan* (2006).

Phasmoptera is a small, rather tropical order, which comprises about 2,500 species. One species – *Ramulus bituberculatus* (Lonchodidae) is known from the Aral Sea region. It could be found at chinks and in clayish deserts.

Homoptera is a large order of sucking phytophagous insects, comprising five suborders – Cicadinea, Aphidinea, Coccidinea, Psyllinea and Aleyrodinea – adapted to different environments.

Cicadinea is a large suborder of insects comprising about 30,000 species. Approximately 300 species are known from the Aral Sea region (Mitiayev 1993; 2002), belonging to the families Aphrophoridae, Cicadellidae, Cicadidae, Cixiidae, Delphacidae, Dictyopharidae, Flatidae, Issidae, Membracidae and Tettigometridae.

Aphidinea is a comparatively small suborder, mostly Holarctic, comprisig about 5,000 species. Eighty-five species of the Aphididae family are known from the Aral Sea (Kadyrbekov 1993, 2004). These species are divided between subfamilies as follows: Phloeomyzinae (1 species), Eriosomatinae (3 species), Saltusaphidinae (1 species), Chaitophorinae (4 species) and Aphidinae (76 species). The most diverse and richest genera are *Aphis* (7 species), *Protaphis* (7 species), *Brachyunguis* (12 species) and *Macrosiphoniella* (10 species).

Most of the species were documented in sandy deserts and tugai forests – 44 and 33, respectively. Salty marshes are inhabited by 27 species, clayish deserts by 19 species, chinks by 15 species, sagebrush-cereal semideserts and *Anabasis* plant communities by 10 species and *Salsola arbuscula* plant communities by 11 species.

19 species were documented at the dried bottom: *Aphis craccivora, Protaphis anuraphoides, Xerobion cinae, Brachyunguis atraphaxidis, Brachyunguis cynanchi, Brachyunguis harmalae, Brachyunguis zygophylli, Cryptosiphum astrachanicae, Clypeoaphis suaedae, Coloradoa heinzei, Acyrthosiphon gossypii, Pleotrichophorus persimilis, Staticobium latifoliae, Staticobium otolepidis, Staticobium suffruticosum, Macrosiphoniella kirgisica, Macrosiphoniella seriphidii, Macrosiphoniella tapuskae, and Macrosiphoniella* sp. Most of them are more abundant at the dried bottom than in zonal desert.

Coccidinea is a relatively small suborder, comprising more than 4,000 species. Seventy species from the families Coccidae, Diaspididae, Eriococcidae, Margarodidae, Ortheziidae and Pseudococcidae are known from the Aral Sea region (Jashenko 1993).

Psyllinea is a small suborder of insects, comprising 1,000 species. Nineteen species have been registered in the Aral Sea region; they belong to the Aphalaridae, Psyllidae and Triozidae families (Ivanova 1975).

The suborder Aleyrodinea has never been studied in the Aral Sea region.

Heteroptera is a large order of aquatic and terrestrial insects, comprising about 40,000 species. One hundred and thirty-three species are known from the Aral Sea region (Assanova 1996; Esenbekova and Kadyrbekov 2004; Kerzhner 1982; Pirjulin 1995), belonging to the families Anthocoridae, Cimicidae, Coreidae, Corixidae, Cydnidae, Hebridae, Hydrometridae, Gerridae, Lygaeidae, Micronectidae, Miridae, Nabidae, Naucoridae, Nepidae, Notonectidae, Pentatomidae, Pleidae, Pyrrhocoridae, Reduviidae, Rhopalidae, Saldidae, Scutelleridae and Veliidae.

Thysanoptera is a rather small and poorly known order. The exact number of species is unknown. Approximately ten undescribed species were collected in the Aral Sea region from Asteracea flowers.

Coleoptera is one of the largest orders of aquatic and terrestrial insects, with more than 250,000 species. Nine hundred and twenty-five species are known from the Aral Sea region (Guryeva 1979, 1989; Kabanova 1995; Kadyrbekov and Tleppaeva 2004; Konev 1993; Kostin 1973; Krizhanovsky 1965, 1976, 1983; Kuzin 1953; Kulenova and Lopatin 1986; Longvinovsky 1985; Lukyanovich and Ter-Minassyan 1957; Makhnovsky 1958; Nikolayev 1987; Nikolayev and Koz'minykh 2002; Nikolayev and Kolov 2005; Ogloblin and Znoiko 1950; Sinadsky 1963, 1968; Ter-Minassyan 1967, 1988), belonging to 28 families: Dytiscidae, Hydrophilidae, Carabidae, Cicindelidae, Gerrinidae, Staphylinidae, Silphidae, Histeridae, Scarabaeidae, Dermestidae, Cerambycidae, Buprestidae, Elateridae, Hydraenidae, Tenebrionidae, Meloidae, Melyridae, Coccinellidae, Chrysomelidae, Curculionidae, Anobiidae, Bruchidae, Bostrichidae, Alleculidae, Nitidulidae, Cleridae, Bruchelidae and Apionidae (Table 11.12). Tenebrionidae, Buprestidae and Meloidae are most diverse in the region, whereas Chrysomelidae, Cerambycidae and Elateridae are poorly represented, compared with mountain and forest environments. *Cicindela nox* (Cicindelidae), *Haplosoma ordinatum* (Scarabaeidae), *Hesperophanes heydeni* (Cerambycidae), *Capnodis miliaris metallica* (Buprestidae) and *Stethorus punctillum* (Coccinellidae) are included in the *Red Data Book of Kazakhstan* (2006). Some families are still poorly studied in the region, particularly of the large families Carabidae, Staphylinidae and Curculionidae.

Lepidoptera is a large insect order with complete metamorphosis, comprising as many as 100,000 species. One hundred and sixty-eight species of the families Papilionidae, Pieridae, Satyridae, Nymphalidae, Lycaenidae, Hesperiidae, Sphingidae, Zygaenidae, Noctuidae, Geometridae, Pyralidae, Tineidae, Ochsenheimeriidae and Orgyidae are known from the Aral Sea region (Chetverikov 1906; Kozhanchikov 1937, 1950; Zagulyayev 1960, 1964, 1973, 1975, 1979; Scheck 1975; Aibassov 1975; Zhdanko 1993, 2005). *Microzegris pyrothoe, Zygaena*

Table 11.12 Taxonomical diversity of Coleoptera in the Aral Sea region

Family	Number of species	Aral endemics	Rare and relict species
Carabidae	150	3	2
Histeridae	34	3	–
Cicindelidae	9	1	1
Scarabaeidae	125	13	1
Dytiscidae	28	–	–
Gerrinidae	3	–	–
Hydrophilidae	9	–	–
Cerambycidae	17	1	2
Coccinellidae	15	–	–
Buprestidae	28	4	1
Staphylinidae	132	–	–
Dermestidae	10	–	–
Meloidae	22	3	–
Chrysomelidae	130	–	–
Tenebrionidae	85	4	–
Elateridae	5	–	–
Anobiidae	7	–	–
Hydraenidae	3	–	–
Silphidae	2	–	–
Melyridae	2	–	–
Cleridae	1	–	–
Bostrichidae	1	–	–
Nitidulidae	1	–	–
Alleculidae	3	–	–
Curculionidae	91	–	–
Bruchidae	7	–	–
Bruchelidae	3	–	–
Apionidae	2	–	–

turchmena, Polyommatus elvira, Aricia chinensis myrmecias, Paragluphisia oxiana and *Catocala optima* are included in the *Red Data Book of Kazakhstan* (2006). The true number of species should be much more in the region, as most nocturnal butterfly families are poorly studied.

Hymenoptera is the largest order of insects, comprising up to 300,000 species. Two hundred and eighty-eight species of the families Crabronidae, Sphecidae, Ampulicidae, Pteromalidae, Formicidae, Mutillidae, Apoidea, Chalcididae, Callimonidae, Eucharitidae, Aphelinidae, Encyrtidae, Eupelmidae, Halictidae and Scoliidae are known from Aral Sea region (Dzhanokmen 1993; Kazenas 1993, 2002; Leley 1985; Marikovskaya 1993; Marikovsky 1974, 1979; Nikolskaya 1952; Pessenko 1983; Schteinberg 1962). *Sphex flavipennis* (Sphecidae), *Prionix macula lugens* (Crabronidae), *Scolia hirta* (Scoliidae) and *Hoplitis fulva* (Apoidea) are included in the *Red Data Book of Kazakhstan* (2006). However, the fauna of other families of this order is still poorly studied in the region.

Diptera is a large order of insects with complete metamorphosis, comprising up to 80,000 species. Four hundred and seventy-three species of the families Culicidae,

Ceratopogonidae, Chironomidae, Cecidomyiidae, Tabanidae, Bombyliidae, Muscidae, Sarcophagidae, Syrphidae and Ephydridae are known from the region (Gucevich et al. 1970; Zimin 1951; Olsufyev 1977; Paramonov 1940; Pirjulin 1995; Rodendorf 1937; Fedotova 1993, 2000). Most of the families are poorly studied, so the true number of species should be as many as three times more. Some data are known for the Chironomidae family.

Compared to earlier years (before 1971), in 1976–1977 and 1980, when the salinity rose above 15‰, only two species of Chironomids remained in the lake's bays: *Chironomus salinarius* and *Chironomus halophilus* (Andreeva 1989). Other species disappeared from the benthic communities (Aladin and Potts 1992). At the end of the 1980s, a drastic lowering of the sea level made further studies in the open sea impossible. Rare investigations took place in the littoral zones only (Filippov 1995). In 1990, the productivity of benthic assemblages in the coastal zone of the Small Aral Sea was found to be 2.4 times that in the Large Aral Sea (Filippov 2001, 2002).

Freshwater Chironomids, however, were found in the Syr Darya only and were represented by a single species, *Polypedilum nubeculosum* Meigen; it was very rare and present in very low numbers only. Currently, only *Chironomus salinarius* can be found in the western basin of the lake (Mirabdullayev et al. 2004).

Neuroptera is a small order of carnivorous insects with incomplete metamorphosis. The exact number of species in the Aral Sea region is unknown; all findings belong to the families Chrysopidae, Myrmeleonidae ad Ascalaphidae, including one species of *Nineta* (Chrysopidae), one species of *Ascalaphus macaronius* (Ascalaphidae) ad 27 species of ant lions (Myrmeleonidae). The disappearance of five ant lion species from the Aral Sea region since 1915 has been registered; a number of other species are endangered (Krivokhatskii and Piryulin 1996).

In spite of taxonomical diversity, there is another important quantitative index, namely the index of endemism, reflecting the original character of any regional fauna. One hundred and thirty-two of two thousand six hundred and forty nine Aral species are Aral Sea endemics or subendemics (5%). This index is even higher in some selected groups; for example, in well-known groups of Coccidinea (Homoptera) and Cecidomyiidae (Diptera) it reaches 35% and 22%, respectively. Thus, the Aral Sea insect fauna is rather original and possesses a number of local and relict species.

Rare species which were included in the latest *Red Data Book of Kazakhstan* (2006) are shown in Table 11.13. There are 25 *Red Data Book of Kazakhstan* species; most of them are associated with tugai forest at the lower Syr Darya. There are also several poorly known species, which, obviously, should be included in the *Red Data Book of Kazakhstan* as well. These species are *Aromia moschata vetusta* Jank. (Coleoptera, Cerambycidae) and *Eurythyrea oxyana* Sem. (Coleoptera, Buprestidae), reported from Syr Darya basin, but for a long time not collected in traps and in counting.

There are different ecological groups within the entomofauna of the Aral Sea, including psammophilous species, gallophilous species, clayish desert species, tugai mesophils and eurybiotic species.

Table 11.13 Rare insect species of the Aral Sea

Species	Tugai forest	Coastal zone	Sandy desert	Clay desert	Salty marshes	Chinks
Ischnura aralensis Harit.	+	–	–	–	–	–
Calopteryx virgo L.	+	+	–	–	–	–
Anax imperator Leach.	+	+	–	–	–	–
Orthetrum sabina Drury	+	–	–	–	–	–
Selysiothemis nigra V.d. Lind.	+	–	–	–	–	–
Bolivaria brachyptera Pall.	–	–	–	+	+	+
Saga pedo Pall.	+	+	–	–	–	–
Caereocercus fuscipennis Uv.	+	–	–	–	–	+
Cicindela nox Sem.	+	–	–	–	–	–
Haplosoma ordinatum Sem.	–	–	+	–	–	–
Hesperophanes heydeni Ball.	–	–	–	+	+	–
Capnodis miliaris melallica Ball.	+	–	–	–	–	–
Zygaena turchmena Evers.	+	–	–	–	–	–
Paragluphisia oxiana Djak.	+	–	–	–	+	–
Catocala optima Stgr.	+	–	–	–	–	–
Microzegris pyrothoe Evers.	+	+	+	–	–	+
Polyommatus elvira Evers.	+	–	–	–	–	–
Sphex flavipennis Fabr.	–	+	+	+	–	+
Aricia chinensis myrmecias Christ.	–	–	+	+	–	–
Prionix macula lugens Kohl.	+	–	+	+	+	+
Stethorus punctillum Weise	+	+	–	–	–	–
Scolia hirta Schr.	+	+	+	+	–	+
Hoplitis fulva Evers.	+	–	+	–	–	+
Ephedromyia debilopalpis Marik.	+	–	–	–	–	–
Psectrosema diversicornis B. Mam. et Beckn.	+	–	–	–	–	–

A very diverse insect fauna is characteristic of sandy deserts of the region (Bolshie Barsuki, Malye Barsuki, Barshakum, Aralkum), and the northern type of deserts is extensively penetrated with steppe insect species.

Clayish deserts have a smaller diversity of insects, with a large proportion of eurybiotic species, which could also be encountered in other environments, as salt marshes, other kinds of desert, etc. Also a small number of specific species are present.

Salsola desert and salt marshes are widely spread in the region; they possess a nondiverse, but rather specific gallophilous entomofauna.

Chinks – the former coastline of the ancient sea – possess a mixed type of entomofauna, comprising species typical for clayish and *Salsola* deserts, and a few tugai-specific elements as well.

Riverbed tugai forests are inhabited by a clearly discernible insect fauna although several desert and eurybiotic species penetrated this zone. The tugai forest fauna is highly diverse with Tertiary relicts, as tugai forests used to be refuges for Poltava and tugai floras and, consequently, fauna.

11.4 Faunistic Invasions

11.4.1 Succession of Reptiles, Amphibians and Insects

The Aral Sea is quite a young lake which appeared about 35,000 years ago within the territory of the Turan Depression (Hoeffler and Novitzkiy 2003). There is a series of obvious data (historical, geological) indicating that during the Holocene the Aral Sea underwent significant repeated changes of water level (Mayev and Mayeva 1977). The recent regression started in the 1960s. As a result of shrinking, about four million hectares of the sea bottom has become exposed (Hoeffler and Novitzkiy 2003).

The appearance of vast dried areas and the connection of former islands to the mainland is a good opportunity to study the process of invasion of terrestrial animals into these areas.

Such an investigation with insects, amphibians and reptiles (in both primary coastal and dried bottom zones) was conducted in the first half of May 2002 and in the second half of September to early October 2003. Both stations and transects were established in the northern (Butakov Bay), northeastern (Sarychaganak Bay) and southern (locality placed north of the Kazak Darya and Tiger Tail Peninsula) Aral Sea regions and in the coastal zone of Ustyurt Plateau as well (Fig. 11.10).

Drying levels from the 1980s to the 1990s were easily recognizable in these stations. The dried zones of the 1960s and 1970s were less visible, being somewhat transformed into zonal desert biotopes. We have established eight stations (or model grounds) to estimate their biotic diversity. The following methods were used in our research:

- Description of soil and vegetation was conducted after Wucherer and Breckle (2000) and Wucherer et al. (2000);
- Insect collecting using methods of mowing and soil-level traps along with soil sampling (Fasulati 1971) (for stations within Kazakhstan territory);

Fig. 11.10 Localization of the stations

- Counting of the numerical distribution of amphibians and reptiles using standard field methods (Dinessman and Kaletzkaya 1952) within transects up to 2 km long and 1–6 m wide depending upon plant coverage and relief conditions.

As is known (Odum 1986), a succession process usually passes through several stages of development. Each stage has unique biogeocenotic characters. The formation of these subsequent stages with their own characters may be explained for the Aral Sea model grounds examined in the following scheme. Of course, some deviations, depending upon the local ecological conditions, may occur.

11.4.1.1 Surf Zone

No vegetation (Fig. 11.11).

Entomofauna: Rare specimens of *Cicindela deserticola* (Cicindelidae), *Cicindela schrenki, Tachys turcestanicus, Tachys vittatus* (Carabidae), *Cataglyphis aenescens, Messor aralocaspius* (Formicidae).

Herpetofauna: Sporadically penetrated by *Eremias velox, Eremias arguta* and *Bufo viridis* seeking insects.

11.4.1.2 First 2–5 Years of Drying (from the Late 1990s to the Beginning of the Twenty-First Century)

Vegetation with *Salicornia, Suaeda*, and *Petrosimonia*.

Entomofauna: In the soil samples and traps *Pogonus luridipennis, Tachys vittatus, Daptus vittatus* (Coleoptera, Carabidae), *Anthicus* sp. (Coleoptera, Anthicidae), *Bothynoderus carinatus, Stephanophorus verrucosus, Phacephorus argyrostomus* (Coleoptera, Curculionidae), *Geocoris ater, Emblethis verbasci* (Heteroptera, Lygaeidae), *Labidura riparia* (Labiduridae), *Cataglyphis aenescens* and *Messor aralocaspius* (Formicidae). In the samples collected by plant mowing

Fig. 11.11 Surf zone of the northern coast of the Aral Sea, May 2002

Clypeoaphis suaedae (Homoptera, Aphidinea), undetermined cicads (Homoptera, Cicadinea) and *Bulaea lichatshovi* (Coccinellidae) were abundant.

Herpetofauna: Single records of *Testudo horsfieldi, Phrynocephalus helioscopus, Eremias velox, Eremias arguta* and *Bufo viridis* (Fig. 11.12).

11.4.1.3 Drying Coast of the Early 1990s (Fig. 11.13)

Vegetation: Plants – edificators (*Salicornia, Suaeda, Petrosimonia*) and first halophytic bushes *Halocnemum strobilaceum* and *Halostachys belangeriana.*

Entomofauna: in the soil samples and traps *Daptus vittatus* (Carabidae), *Bothynoderus carinatus, Bothynoderus vexatus, Phacephorus argyrostomus, Stephanophorus verrucosus (Curculionidae), Anatolica* sp., *Scleropathrum seidlitzi, Psammestus dilatatus* (Tenebrionidae), *Cataglyphis aenescens, Messor aralocaspius* (Formicidae) and species of *Oniscoidea.* In the herbosa-fruticose stratum the representatives of *Cicadinea* and *Chrysomelidae* were dominant.

Herpetofauna: Reptile associations include a higher species diversity than for younger stages. In addition to *Eremias arguta* and *Eremias velox* observed in the previous stage, the lizards *Phrynocephalus helioscopus* and *Trapelus*

Fig. 11.12 The lizards *Eremias velox* (**a**) and *Phrynocephalus helioscopus* (**b**) are among the first inhabitants of the drying zone of the northern coast of the Aral Sea

Fig. 11.13 Drying coast of the early 1990s with *Salicornia–Suaeda* plant association

Fig. 11.14 Halys pitviper, *Gloydius halys*

sanguinolentus and the snakes *Psammophis lineolatum* and *Gloydius halys* (Fig. 11.14) were registered.

It is important to note that among both insects and reptiles a few species became very abundant at this stage of succession. Among reptiles a high density was recorded for *Phrynocephalus helioscopus* (up to 10 species per hectare), *Eremias arguta* (up to 13 specimens per hectare), *Trapelus sanguinolentus* (13–22 specimens per hectare) and *Eremias velox* (up to 50 specimens per hectare). Some insect species as curculionids of the tribe Cleonini (*Stephanophorus verrucosus, Bothynoderus carinatus, Phacephorus argyrostomus*) appeared to be quite abundant as well. This succession stage could be characterized by "the principle of deviated conditions". Species diversity is still not high, but the population numbers of those species may reach quite high values. This is a feature of "extreme biotopes" when biotope conditions deviate highly from the normal ones supporting few specific species.

11.4.1.4 Drying Coast of the 1980s

A few types of succession associations can be observed:

1. Sandy dunes with psammophyto-halophytic plant association (*Tamarix ramosissima, Salsola richteri, Calligonum* sp. and several ephemerals).
   Entomofauna: Soil stratum poorly inhabited by insects; on the soil surface *Cicindela lacteola* (Cicindelidae), *Curtonotus armeniacus, Harpalodema* sp. (Carabidae), *Menecleonus lagopus* (Curculionidae), *Blaps parvicollis, Diaphanidus ferrugineus, Anatolica angustata, Microdera convexa, Psammestus dilatatus* (Coleoptera, Tenebrionidae), *Cataglyphis aenescens*, and *Messor aralocaspius (Formicidae)* were recorded. In the herbosa-fruticose stratum *Adonia variegata, Bulaea lichatshovi* (Coccinellidae), *Coniatus* sp., *Sibinia bipunctata* (Curculionidae), *Cataglyphis aenescens* (Formicidae), *Aspilaspis pallida, Artheneis alutacea* (Heteroptera, Nabidae, Lygaeidae), *Vestitohalictus* sp. (Apidae) and *Chalcidoidea* (Hymenoptera) were numerous.
   Herpetofauna: It was the first time we also registered psammophilous reptiles on a sandy substrate, such as *Phrynocephalus mystaceus* (Fig. 11.15) and

Fig. 11.15 Sandy desert lizard, *Phrynocephalus mystaceus*

Fig. 11.16 *Halocnemum strobilaceum* plant associations with ephemerals formed within salt-marshes area

Eremias grammica inhabiting the tops of dunes, and quite numerous *Testudo (Agrionemys) horsfieldi* scrawling by dune bases. Several other species were also registered: the lizards *Eremias arguta, Eremias velox* and *Trapelus sanguinolentus* and the snakes *Psammophis lineolatum* and *Elaphe dione*. In the region of Amu Darya delta we additionally registered the toad *Bufo viridis*, the gecko *Cyrtopodion russowi* and the snakes *Natrix tessellata* and *Platyceps* (*Coluber*) *karelini*. Together with increasing species diversity, the abundance of dominating species became lower than in the previous zone. For example, the number of *Phrynocephalus mystaceus* did not exceed 5–20 specimens per hectare, that of *Eremias grammica* did not exceed 4–20 specimens per hectare and that of *Trapelus sanguinolentus* did not exceed 4–8 specimens per hectare. The population of the steppe tortoise (*Testudo horsfieldi*) comprised quite large and old specimens, whereas young animals were very rare.

2. *Halocnemum strobilaceum* plant associations with several ephemerals formed within salt marsh areas (Fig. 11.16).

   Entomofauna: Of the insects found on the soil surface *Anatolica* sp., *Paranemia schroederi* (Tenebrionidae), *Daptus vittatus* (Carabidae) and *Cataglyphis aenescens* (Formicidae) were especially abundant. For the herbosa-fruticose

stratum *Bulaea lichatshovi* and *Phyllotreta ustulata* (Chrysomelidae) were dominant species.
Herpetofauna: *Eremias arguta* sporadically visited even the spots of bare salt marshes. Rarer was *Trapelus sanguinolentus*, which preferred the habitats with *H alocnemum strobilaceum*.

3. *Phragmites* succession in the salty area of lower Syr Darya (e.g. the model ground at the eastern coast of the sea).
   Entomofauna: On the soil surface *Camponotus turcestanicus* (Formicidae) and *Paranemia schroederi* (Tenebrionidae) were dominant species. *Daptus vittatus, Dyschirius salinus* (Carabidae) and *Blaps scutellaris* (Tenebrionidae) were also recorded but with few specimens. The mowing in the herbosa-fruticose stratum at the beginning of May 2003 during rainy and cold weather did not lead to successful insect collection.
   Herpetofauna: The results were poor because of the weather mentioned above. Only green toads were recorded.
4. *Limonium–Artemisia–Anabasis* plant association.
   Entomofauna: On the soil surface stratum *Cicindela lacteola* (Cicindelidae), *Pangus brachypus* (Carabidae), *Phacephorus argyrostomus, Stephanophorus verrucosus, Aralkumia deserti* (Curculionidae), *Blaps striata, Pimelia cephalotes, Pterocoma costata, Diaphanidus ferrugineus* (Tenebrionidae), *Messor aralocaspius, Camponotus turcestanus* and *Tapinoma emerianum* (Formicidae) were dominant. In the herbosa-fruticose stratum common representatives were *Bulaea lichatshovi* (Coccinellidae), a few species of Cicadinea and *Macrosiphoniella seriphidii* from Aphidinea.
   Herpetofauna: Common species were *Eremias velox, Trapelus sanguinolentus,* and *Gloydius halys;* within the takyr area *Phrynocephalus helioscopus* was common.

11.4.1.5 The Final Stage of the Succession (1960s–1970s)

This is characterized by the conversion of the former sea bottom into a typical mainland landscape, i.e. placor desert. Animal communities inhabiting zones that dried out in the 1960s and 1970s were already identical in general to those inhabiting the mainland placor deserts. As an example, the *Anabasis salsa* community of the 1970s (model ground 1) was similar in many details to that from the original coast.

Entomofauna*:* On the soil surface *Cicindela lacteola* (Cicindelidae), *Pangus brachypus, Pangus undulatus* (Carabidae), *Stephanophorus verrucosus, Aralkumia deserti, Menecleonus lagopus* (Curculionidae), *Pimelia cephalotes, Pterocoma costata, Lasiostola carinata* (Tenebrionidae), *Messor aralocaspius* and *Tapinoma emeruanum* (Formicidae) were common species. By mowing in the herbosa-fruticose stratum, a few species of Cicadinea, *Macrosiphoniella seriphidii* (Aphidinea), *Bulaea lichatshovi* (Coccinellidae) and *Phyllotreta ustulata* (Chrysomelidae) were found. It is important to note for this stage the presence of single specimens of such spring species as *Callisthenes beithseni* (Carabidae) and *Dorcadion glycerrhizae androsovi* (Cerambycidae), which are typical for stable ecosystems of the original sea coast.

As suggested (Alenitzin 1876; Bogdanov 1965), all sandy deserts of Middle Asia historically passed through the same way of progression. Here it is important to note that as far as a sea level continues to descend, environmental conditions become more extreme and the rate of biota formation processes decreases. Within the vast shallow water areas of southern and southeastern parts of the former sea bottom zones of wind soil deposits where any traces of life were absent were described. A formation of terrain isthmuses between the former islands and the mainland is a possible way for animal invasion of the islands. Thus, a quite clear relation can be discovered between the eastern islands (Kyndyrly, Seleuly, etc.) and Priaralski Kyzylkum, between the Akpetki Archipelago and the Beltau Hills region, and between the western islands (Lazareva, Vozrozhdeniya) and the Bolshie Barsuki sands. It is possible that the poor diversity of herpetofauna and the high density of single species on some islands (e.g. Barsa-Kelmes; Karpenko 1958) could be explained by disconnection of island territory from the mainland during the sea transgression in early stages of biota formation.

Our research conducted within the model grounds has shown that most of the territories of the former Aral Sea bottom were characterized by restricted variance of insect and herp species diversity typical of the original placor deserts. The dominance of the beetles of the families Tenebrionidae, Curculionidae and Carabidae and the ants (Formicidae) among the insects corresponded well to the food composition of the lizards registered within the model grounds (Paraskiv 1956; Bogdanov 1960, 1962; Darevsky 1969; Kamalova 1972; Shammakov 1981; Brushko 1995). The species complexes of lizards with dominating genera *Trapelus, Phrynocephalus* and *Eremias* together with presence of numerous rodents (*Rhombomys opimus, Meriones meridianus*) can explain the appearance of not numerous snake species (*Elaphe dione, Psammophis lineolatum, Gloydius halys*), the diet of which is based upon the above-mentioned animals (Terentyev and Chernov 1949; Karpenko 1958; Bogdanov 1965). The reptiles which invaded the former sea bottom are referred to as "inhabitants of horizontal surfaces with hard soil" (Bogdanov 1965). The pioneer species of invasion were *Eremias velox, Eremias arguta* and *Phrynocephalus helioscopus.* For most of Kazakhstan and Uzbekistan territories, in general, these species are the typical inhabitants of clay, clay and crushed stone deserts, hard sandy soils with rare vegetation and even the salty areas. All these species are able to burrow and were encountered in all the localities of the Aral Sea examined. The steppe agama (*Trapelus sanguinolentus*) could also be categorized as a pioneer species but with the only restriction that its appearance and settled inhabitation in dried areas in many respects are determined by its role as a commensal of the rodent burrows. The steppe tortoise [*Testudo* (*Agrionemys*) *horsfieldi*] could explore the dried zones in a short period of springtime. The stenotopic species *Phrynocephalus mystaceus* and *Eremias grammica* and the lizards using vertical surfaces, *Cyrtopodion russowi* (lives in bushes) and *Cyrtopodion caspius* (lives in rodent burrows), could appear within the dried zones only in certain stages of their evolution. The snakes, as high-level consumers, may follow to these areas when the food resources are already quite large.

11.4.2 On Settling by Reptiles of the New Land Bridge Between the Former Island of Barsa-Kelmes and the Eastern Coast of the Aral Sea

Of the whole territory of the eastern island and coastal zone of the Aral Sea, the herpetofauna of Barsa-Kelmes has been studied best (see also Chap. 14). The first, scanty information on amphibians and reptiles of the island and adjacent areas of the eastern Aral Sea region appeared in the nineteenth century and the first quarter of the twentieth century (Alenitzin 1876; Elpatjevsky 1903; Berg 1908; Zarudny 1915; Nikolsky 1915, 1916, 1918; Sidorov 1925). In the middle of the twentieth century Paporotny (1950) and Karpenko (1958) studied the ecology of the Halys viper (*Gloydius halys*), and simultaneously composed a list of herpetofauna of Barsa-Kelmes and collected some data on their number and biotopic distribution. From the 1970s to the 1990s, herpetological investigations in Barsa-Kelmes Nature Reserve were conducted by Russian zoologists who collected rich herpetological materials, later partially stored at the Zoological Institute of the Russian Academy of Sciences and the Zoological Museum of Moscow State University. In the early 1980s, Ismagilov undertook an investigation of the distribution and ecology of common reptile species of the island. His unpublished data were used by Brushko (1995). Certain information can be found in the herpetological review of Paraskiv (1956) and Iskakova (1959) as well as in the transactions of Barsa-Kelmes Nature Reserve (Rashek and Rashek 1963; Ismagilov et al. 1990). We carried out brief herpetological research of the drying zone adjoining to old boundaries of the reserve in 2003–2004. During the last 5 years, special herpetological observations have been conducted in the drying bottom by the reserve's zoologists.

According to data from 1936 to 1981, when Barsa-Kelmes was still separated from the mainland, the island herpetofauna composition included eight species – constant inhabitants of the island: one amphibian – *Bufo viridis* – and seven reptiles – *Testudo horsfieldi*, *Alsophylax pipiens*, *Phrynocephalus helioscopus*, *Eremias velox*, *Elaphe dione*, *Psammophis lineolatus* and *Gloydius halys* (formerly *Agkistrodon halys*). Two species – *Cyrtopodion* (*Mediodactylus*) *russowii* and *Natrix tessellata* – seemingly appeared on the island after a storm or were occasionally introduced by humans (Elisseyev 2007).

The creation of the bridge between the island and the mainland and creation of newly dried areas in the eastern Aral Sea region caused good conditions for reptiles to migrate and occupy the new land. The rapid racerunner (*Eremias velox*) and the sunwatcher (*Phrynocephalus helioscopus*) quite quickly moved from the former island of Barsa-Kelmes in an eastern direction as well as in the opposite direction. The rapid racerunner successfully settled on the dry land of the 1990s (area

25–35 km from the former island of Barsa-Kelmes), where it occupied mainly the sands blown by wind and overgrown with *Atriplex pratovii* or sandy mounds originating near *Tamarix* or *Halostachys* bushes, e.g. the psammophytic habitats with bush communities. In 2003 single lizards were also encountered in a *Suaeda–Salicornia* plant community, formed on dry land of the last decade when the creation of the land bridge between the island and the mainland had been completed (Satekeev and Chirikova 2007). The same process of rapid racerunner movement along the drying land occurred in the opposite direction (Fig. 11.11a).

Seemingly the sunwatcher (Fig. 11.11b) settled on the drying land also in both directions. However, we have reliable data only for its west-to-east movement. According to Elisseyev (2007), the number of the sunwatchers on the former island of Barsa-Kelmes has visibly fallen from the second half of the twentieth century. In the 1980s to the 1990s, there were calculations from 1–2 to 8–10 specimens of *Phrynocephalus helioscopus* per kilometre, and the density of sunwatcher settlements at those time was 82–116 specimens per hectare (Brushko 1995 according to Ismagulov's unpublished data). Now to encounter even one lizard it is necessary to cover no less than 2–4 km (Elisseyev 2007). The movement of that species along the drying sea bottom occurred quickly compared with that of *Eremias velox* because of other ecological preferences. The sunwatcher does not depend on bush vegetation for its distribution as the rapid racerunner does and it easily occupies even the areas completely lacking plants.

According to known data, the western boundary of the range of *Eremias arguta* in the eastern part of the Aral Sea region passed close to the coastline of the sea in the middle of the twentieth century (near 61–61°30′E). In 2007 *Eremias arguta* was recorded more than once 25–30 km to the east of the former island of Barsa-Kelmes within the drying zone of the 1990s, which confirms the species' dispersal in a western direction (Fig. 11.17). The habitats where the steppe runner was found were occupied by a *Suaeda–Salicornia* plant community developed on hard salty soil. We have no data on the settling on the dry bottom of the 1990s by the snakes *Psammophis lineolatus* and *Gloydius halys* except for an observation of visible

Fig. 11.17 Desert runner, *Eremias arguta*

decline of the first species on the island. Probably, this process is closely connected with migration to the mainland by the main food of that snake – the sunwatcher and the rapid racerunner (Elisseyev 2007).

So, the last data on reptile dispersal on the eastern drying bottom of the Aral Sea confirm our recent conclusion on the pioneering status of the rapid and steppe racerunners and the sunwatcher in the process of colonization of newly dried land. As we showed for northern and northeastern Aral Sea regions, these species appeared first in new biocenoses during the first 2–5 years of the succession process (Chikin et al. 2004).

11.4.3 Succession of Mammals on the Dried Sea Bottom

The eastern coast of the Aral Sea had a dried area extending up to 6–7 km from the former coastline near the Uzbekistan border (Bassay locality) and 40–50 km in the region of the former island of Kaskakulan as early as the middle of the 1970s. Observations in 1976–1977 at the dried bottom in Bassay locality showed the presence of *Rhombomys opimus, Meriones meridianus, Meriones libycus, Meriones tamariscinus, Dipus sagitta, Mus musculus, Lepus tolai, Cricetulus migratorius, Allactaga severtzovi* and *Allactaga elater* (Mazin 1979, 1981).

The first permanent rodent colonies were revealed at a distance of 1.3 km from water. The following species have predominated in the newly dried bottom (litoral zone): *Meriones meridianus* (44.3% of totally trapped animals), *Dipus sagitta* (25.7%) and *Meriones libycus* (12.6%). *Rhombomys opimus* was also quite abundant in a few areas – there were up to seven living colonies per kilometre of counting route. All these rodent species were also registered on the mainland; however, their abundance there was at least four times less than at the dried bottom. This situation can be explained by better vegetation development at the litoral conditions of the newly dried bottom compared with the dry mainland area. Abundance of rodents attracts carnivorous mammals, so the tracks of the following animals (and sometimes the animals themselves) were registered: foxes, corsac foxes, *Mustela eversmanni, Vormela peregusna* and *Felis libyca*. A *Meles meles* burrow was found. Sometimes saiga and Persian gazelles and their tracks were encountered.

The dried bottom between the former island of Kaskakulan and the mainland coast lacks vegetation, so the specific composition of mammals here is much poorer. The following were encountered: *Rhombomys opimus, Meriones meridianus, Meriones libycus, Pygerethmus pumilio, Allactaga elater, Dipus sagitta, Cricetulus migratorius, Mus musculus, Lepus tolai*, foxes, *Meles meles* and saiga. The following rodents dominated in traps: *Meriones meridianus, Mus musculus* and *Cricetulus migratorius* (62.8%, 15.3% and 14.2% of trapped animals, respectively). Living colonies were spotted about 2 km from seawater.

In 1989 the invasion of the dried bottom by *Rhombomys opimus* was documented between the former Tasboget and Bozkol bays. The colony density was 0.2–3.0 per kilometre (Mirzabekov et al. 1992).

Thus, the dried bottom is colonized by mammals from the mainland. These are desert species (sand and great gerbils, dipodids) or common widely distributed representatives of rodents (*Mus musculus, Cricetulus migratorius*). Carnivorous mammals which appeared on dried bottom belong to Mustelidae, Canidae and Felidae families. One species of Lagomorpha – *Lepus tolai* – is also a very common pioneer species (Elisseyev 2007).

Small groups of saiga were encountred at the dried bottom in 1976 (Mazin 1986). The Asian wild ass migrated from the island of Barsa-Kelmes Island and successively colonized the former island of Kaskakulan (where three freshwater artesian wells function), and adjacent areas (see also Chap. 14). In 2005 as many as 112 wild asses were counted (Shaimardanov 2007). In general, 30 species of mammals were registered on the dried bottom, 16 of them were rodents (Mazin et al. 1992). The most abundant mammalian fauna was represented in the dried bottom near the old shore, i.e. at the very early drying zone, where phytocenoses were already very similar to natural conditions.

In the Karakalpak part of the dried bottom, some additional species were found: long-eared hedgehog, *Sorex minutus*, *Pipistrellus pipistrellus*, *Paradipus ctenodactylus, Allactaga sibirica, Eremodipus lichtensteini, Nesokia indica, Ellobius talpinus, Microtus transcaspicus*, grey wolf, Asian jackal and wild boar (Zaletayev and Kosbergenov 1985; Reimov and Karabekov 1989). The authors consider *Mus musculus, Allactaga elater, Pygerethmus pumilio* and *Meles meles* to be common on the dried bottom, whereas all other species are rare.

Conclusions. As a result of the Aral Sea drying, the habitats of mammals populating wetlands, tugai and reed bushes in the Syr Darya delta as well as sea coast (*Sus scrofa, Ondatra zibethicus, Felis libyca*) were visibly reduced and aggravated, which finally led to a decrease of their numbers in the region. At the same time, the dried sea bottom has become a new habitat for many species, first the rodents (gerbils, dipodids, house mouse, grey hamsters, Aral yellow suslik), then carnivores (red fox, corsac fox, badger, steppe polecat), tolai hare, ungulates, etc. Among the colonizers, a dominant position in diversity belongs to desert species (more than 50%). There are visibly fewer species of trans-Palaearctic distribution, mesophilic and steppe ones. Of the desert fauna more than half of the species diversity is allocated to widely distributed Turanian and Kazakhstano-Mongolian species; the rest includes the representatives of other faunistic complexes.

11.4.4 Birds Occupying the Dried Bottom

The colonization of the Aral Sea bottom by birds involved several steps after the water had gone. Elisseyev (1999, 2007) provided such observations at the island of Barsa-Kelmes. For the first few years the dried bottom remained empty. Then, after the first vegetation had appeared, birds started to visit the dried bottom to feed on plants. Soon after this, some species began nesting at the dried bottom. Such a sequence was documented for Pallas's sandgrouse and the lesser short-toed lark, which firstly only fed on plant seeds, and from early 1980 began nesting. When the shrubs became more or less prominent at the dried bottom, a few other bird species appeared, such as Sykes's warbler, lesser whitethroat, desert wheatear, red-headed bunting and nightjar. Nests of Pallas's sandgrouse were found as far as 2 km from the mainland in 1983. *Tamarix* shrubs, which grew after the sea had shrunk in 1975–1976, were inhabited by Sykes's warblers as early as 1988–1989 (Elisseyev 1990).

Near Bugun village, nests of the following species were documented in June 1988: blue-cheeked bee-eater, lesser short-toed lark, Asian short-toed lark, desert grey shrike, Sykes's warbler, isabelline wheatear, rufous bush chat (Koblik 1991).

Data collected in 1989–1990 show nesting of Pallas's sandgrouse, pin-tailed sandgrouse, black-winged stilt, Kentish plover, lesser short-toed lark, Asian short-toed lark, black-headed wagtail, desert grey shrike, Sykes's warbler, desert warbler, lesser whitethroat, desert wheatear, isabelline wheatear, desert finch, and red-headed bunting in the dried area. Isabelline wheatear is associated with habitation of the dried bottom by the rodents great gerbil (*Rhombomys opimus*) and large-toothed souslik (*Spermophilus fulvus*), as it uses the holes of the rodents for nesting (Gubin 1999).

On 1–2 July we found on the dried bottom near the Kokaral Dam Sykes's warbler, lesser whitethroat, Menetries's warbler, desert finch and red-headed bunting in *Tamarix* shrubs, and oystercatcher, Kentish plover, blue-checked bee-eater, lesser short-toed lark and isabelline wheatear in the plains (Belyalov, unpublished data).

In the open areas of the dried bottom the lesser short-toed lark predominates, whereas in the shrubs Sykes's warbler is very abundant.

Table 11.14 shows the bird species found nesting on the dried former Aral Sea bed, and some of these species are depicted in Fig. 11.18.

Table 11.14 Bird species nesting on dried areas (comprises 20 species)

Kentish plover (*Charadrius alexandrinus*)
Black-winged stilt (*Himantopus himantopus*) (Fig. 11.18a)
Oystercatcher (*Haematopus ostralegus*)
Pallas's sandgrouse (*Syrrhaptes paradoxus*) (Fig. 11.18b)
Pin-tailed sandgrouse (*Pterocles alchata*)
Nightjar (*Caprimulgus europaeus*)
Blue-checked bee-eater (*Merops persicus*) (Fig. 11.18c)
Black-headed wagtail (*Motacilla feldegg melanogrisea*)
Desert grey shrike (*Lanius meridionalis pallidirostris*)
Sykes's warbler (*Hippolais rama*)
Lesser whitethroat (*Sylvia curruca halimodendri*)
Menetries's warbler (*Sylvia mystacea*)
Desert warbler (*Sylvia nana*) (Fig. 11.18d)
Desert wheatear (*Oenanthe deserti*)
Isabelline wheatear (*Oenanthe isabellina*) (Fig. 11.18e)
Rufous bush chat (*Cercotrichas galactotes*)
Desert finch (*Rhodospiza obsoleta*)
Red-headed bunting (*Emberiza bruniceps*)
Lesser short-toed lark (*Calandrella rufescens*) (Fig. 11.18f)
Asian short-toed lark (*Calandrella leucophaea*)

Fig. 11.18 Birds nesting on the dried ground of the former Aral Sea: (**a**) black-winged stilt (*Himantopus himantopus*); (**b**) Pallas' sandgrouse (*Syrrhaptes paradoxus*); (**c**) blue-cheeked bee-eater (*Merops persicus*); (**d**) desert warbler (*Sylvia nana)*; (**e**) isabelline wheatear (*Oenanthe isabellina*); (**f**) lesser short-toed lark (*Calandrella rufescens*)

11.4.5 General Conclusions

1. The temporal succession communities formed within the drying coasts of the Aral Sea are primarily penetrated by a few eurybiont insect and reptile species which very actively master the free ecological niches. These species become very abundant and they significantly outnumber the same species in the original zonal arid communities.
2. Among the insect plant feeders mastering the drying coasts, the polyphagous or widely oligophagous species are prevalent.

3. Within the drying area of the 1970s, thanks to constant progressive increase of the number of ecological niches, the taxonomical insect and vertebrate diversity is growing and interspecies regulation of individual numbers is taking place.
4. The succession communities of the 1960s and 1970s generally have contents of dominant species (plants, insects, reptiles, birds, mammals) similar to those of the original desert ecosystems.

11.5 Zoogeographical Aspects

11.5.1 Zoogeographical and Ecological Characteristics of Aral Sea Mammals

In Kazakhstan the shoreline of the Aral Sea belongs mainly to two zoogeographical zones – northern Aralo-Caspian deserts and the Kyzylkum (Afanasyev 1960).

The southern border of the first zone runs along the state border between Kyzylkum and Uzbekistan (Karakalpak Republic) and along the Syr Darya. The Kyzylkum zone comprises the Kyzylkum from the Syr Darya to the state border with Uzbekistan, and the Uzbekistan coast.

There are 55 known species of mammals in the region. Most of them are desert species (58.2%), then common Palaearctic species (18.2%), mesophilous species of southern countries (12.7%), steppe species (5.5%), forest species (2.6%) and introduced species (1.8%).

Classification of the desert theriofauna is given here according to Afanasyev (1960) and Bekenov (1988):

Widely distributed desert species. In Aral Sea regions the following species belong to this group: long-eared hedgehog, *Eptesicus bottae, Vormela peregusna, Felis libyca*, Persian gazelle and *Ellobius talpinus*. Most of them (excluding the Persian gazelle) penetrate other landscape zones (mountains, steppes); however, their origin is related to desert, and these species are more abundant in the deserts.

Kazakh desert species. This group comprises *Spermophilus fulvus, Pygerethmus platiurus, Allactaga severtzovi* and *Eptesicus bobrinskii*. The optimal conditions for their distribution are found in Kazakh deserts, from where these species originated.

Kazakh-Mongol desert and semidesert species. This group includes four species of dipodids – *Allactaga sibirica, Stylodipus telum, Pygerethmus pumilio, Dipus sagitta, Allocricetulus eversmanni* – and also Asian wild ass. The ranges of those species lie in desert and semidesert zones of both Kazakhstan and Mongolia.

The *Turan desert species* complex includes *Diplomesodon pulchellum, Spermophilopsis leptodactylus* and three species of dipodids – *Paradipus ctenodactylus, Eremodipus lichtensteini* and *Allactaga elater*. The ranges of these species fall within the area of the Middle Asian desert, only *Allactaga elater* has a wider area.

The *Turan-Kazakh desert species* complex yields recently described (Vorontzov and Shenbrot 1984) *Salpingotus pallidus* (Kazakhstan endemic) and *Salpingotus heptneri*. The Turan-Kazakh origin of the *Salpingotus* genus, formerly considered as the endemic group of Central Asia, was proved (Geptner 1984).

The *Turan-Mongol desert species* complex includes three species of gerbils: *Rhombomys opimus*, *Meriones meridianus* and *Meriones tamariscinus*. The ranges of these species stretch from the Caspian Sea to the Gobi.

The *Afghanistan-Iran mountain desert species* complex with *Meriones libycus* and *Microtus socialis*; their distribution is related to desert foothills and gravel deserts.

Felis margarita and *Lepus tolai* are *African-Western Asian desert species*. These species have wide ranges.

The biogeographical position of some species remains questionable: *Allactaga elater*, *Spermophilus* spp. and saiga (Afanasyev 1960; Kucheruk 1959).

Mustela eversmanni, *Vulpes corsac* and *Allactaga major* are representatives of steppe fauna. The optimum areas for these species lie within the steppe zone, but they penetrate the desert zone far to the south because of their ecological plasticity.

Two species of forest fauna are known from the Aral Sea region: *Nyctalus noctula*, trapped accidentally near Aralsk, and *Microtus transcaspicus*.

Except for the main faunistic complexes (desert, steppe, forests), the southern species, related to subtropical forests, oases and southern mountain forests are known for Kazakhstan. Afanasyev (1960) claimed this complex as "mesophilous species of southern countries". In the Aral Sea region, the following species are attributed to this complex: *Sorex minutus*, *Suncus etruscus*, Asian jackal, wild boar, *Cricetulus migratorius*, *Nesokia indica*, *Lepus europaeus*.

Myotis mystacinus, *Vespertilio murinus*, *Eptesicus serotinus*, grey wolf, fox, *Mustela nivalis*, *Mustela erminea*, *Meles meles*, *Mus musculus* and *Rattus norvegicus* are common Palaearctic species.

The only introduced species in the Aral Sea region is the North American muskrat.

Among the mammals of the Aral Sea region, there are 47 terrestrial species, one fossorial species, one semiaquatic species and six aerial, or flying, species. The habitat preferences are different and are regulated by feeding, breeding and other environmental parameters.

11.5.1.1 Terrestrial Mammals

Insectivora. The long-eared hedgehog is encountered in different biotopes, but usually in wet bushy and grassy regions, by river valleys, lakes and near the sea. *Sorex minutus* is also common in wet tugai and reeds at water-body shores. *Diplomesodon pulchellum* is very common in sandy environments, especially in weathered coarse sands, and in the dried riverbeds (Rotshild et al. 1967).

Rodents. *Spermophilopsis leptodactylus* is typical for sandy deserts; *Spermophilus fulvus* inhabits both sandy and clayish plains; *Allactaga elater* is quite common in open clayish–salty marsh regions with poor vegetation.

Most species of the successful group of dipodids (seven species) are known from dense clayish soil environments with mostly *Salsola* or *Artemisia* and *Anabasis* vegetation. These species are *Pygerethmus pumilio, Pygerethmus platiurus, Stylodipus telum, Allactaga sibirica, Allactaga severtzovi, Allactaga elater* and *Allactaga major*. The remaining five species (*Dipus sagitta, Paradipus ctenodactylus, Eremodipus lichtensteini, Salpingotus pallidus* and *Salpingotus heptneri*) live in sandy deserts with diverse shrubs and grasses (Burdelov 1985; Ismagilov 1985; Varshavsky et al. 1985).

One of four gerbils, *Meriones meridianus*, is distributed mainly in sandy desert in the patches of well-grown shrubs. *Meriones tamariscinus* also lives in sands, but in wetter conditions, near water bodies. *Rhombomys opimus* lives in either hilly sands or clayish or sandy-clayish environments. *Meriones libycus* inhabits clayish or gravel-clayish environments, takyrs, including salty ones. It may also be found in marginal areas of sandy massifs.

One cricetid species – *Cricetulus migratorius* – inhabits different sandy and clayish biotopes. *Allocricetulus eversmanni* prefers salty marshes and *Salsola* in gravel environments and avoids wet conditions. *Microtus socialis* would rather inhabit dry clayish-gravel desert, and *Microtus transcaspicus* prefers wet conditions related to lake coasts and riverbeds.

Muridae. *Mus musculus* inhabits both diverse natural and anthropogenic environments. The distribution of grey rat, also typically a synanthropic species, is mostly associated with anthropogenic environments. *Nesokia indica* at the southeastern coast of the Aral Sea was encountered at the gentle coast, covered with poor reeds (Varshavsky 1951).

Lagomorpha. *Lepus tolai* inhabits sandy, clayish and other kinds of deserts with patches of shrubs, *Haloxylon* and tugai forests. *Lepus europaeus* is also associated with similar conditions, usually being observed among shrubs.

Carnivora. Most of the 11 species (*Canis lupus, Vulpes vulpes, Vulpes corsac, Meles meles, Mustela erminea, Mustela nivalis, Vormela peregusna, Felis libyca*) are observable in different kinds of deserts, but breeding holes are usually placed in tugai and reeds. *Mustela eversmanni* would rather inhabit open, nonsandy deserts, but it could also be found in *Haloxylon* patches. Its distribution appears to be related to colonies of *Rhombomys opimus*. *Felis margarita* is an inhabitant of vegetated sandy dunes with shrub patches.

Ungulates. Asian wild ass and saiga both live in open plains with low vegetation, migrating to sandy massifs in the winter. Persian gazelle is typical of sandy and gravel deserts. Wild boars live in reeds and tugai along rivers and at lakes and the sea coast. In winters they may migrate to sands or saxaul forests.

11.5.1.2 Fossorial Mammals

These are represented by only one species – *Ellobius talpinus*; it inhabits mostly open places with loose clayish soils, takyrs or sometimes a sandy environment.

11.5.1.3 Semiaquatic Mammal

The muskrat inhabits freshwater or moderately salty water bodies with well-developed coastal vegetation.

11.5.1.4 Flying Mammals (Bats)

There are six species in this group. *Myotis mystacinus* inhabits diverse biotopes, usually being rather found near water bodies, but sometimes it lives in waterless regions. *Nyctalus noctula* could be found in the regions where trees are present, including human settlements, especially along riverbanks. *Eptesicus bobrinskii* prefers wet conditions, close to rivers, wells or artesian holes and lives in ruined houses, mausoleums, etc. Sometimes this species inhabits waterless landscapes. *Eptesicus serotinus* is a typical synanthropic species. *Eptesicus bottae* uses both human constructions and natural structures. *Vespertilio murinus* can be encountered at human settlements and far from them, but prefers wet places.

11.5.2 Biogeography of the Birds of the Aral Sea Basin

Darlington (1980) wrote: "...the processes that have produced the pattern [of bird distribution] – the evolution and dispersal of birds – are very difficult to trace and understand. Perhaps because the subject is so difficult, most... ornithologists... seem very little interested in the geographical relationships and sources of... birds" (p. 236). The points of view of different authors who tried to understand these processes were often in disagreement. Here, surveying the biogeographical position of the breeding birds in Aral Sea region, we are of the opinion of Shtegman (1938), who proposed an ornithological model of Palaearctic division. Of 165 breeding birds observed in the region, 67 could be classified to be widely distributed and will not be analysed below.

Twelve species belong to the European fauna: little bittern, Montagu's harrier, imperial eagle, turtle dove, swift, nightjar, European roller, lesser grey shrike, nightingale, sedge warbler, lesser whitethroat and corn bunting.

Richest in bird diversity is the Mediterranean fauna, which is represented by 37 species: pygmy cormorant, black-crowned night heron, squacco heron, purple heron, glossy ibis, marbled duck, red-crested pochard, white-headed duck, shikra sparrowhawk, Macqueen's bustard, stone curlew, white-tailed lapwing, collared oratincole, Pallas's gull, slender-billed gull, Eversmann's dove, black-bellied sandgrouse, pin-tailed sandgrouse, Egyptian nightjar, Eurasian bee-eater, blue-checked bee-eater, calandra lark, bimaculated lark, rose-coloured starling, Cetti's warbler, moustached warbler, clamorous reed warbler, Sykes's warbler, olivaceous warbler, Upcher's warbler, Menetries's warbler, desert warbler, streaked scrub warbler, rufous bush chat, Spanish sparrow, desert finch and red-headed bunting.

There are nine species of the Mongolian fauna: long-legged buzzard, saker falcon, Barbary falcon, greater sand plover, Pallas's sandgrouse, Asian short-toed lark, oriental skylark, shore lark and isabelline wheatear. There are nine species of both Mediterranean and Mongolian faunas: great white pelican, Dalmatian pelican, ruddy shelduck, shelduck, steppe eagle, lesser kestrel, black-winged stilt, avocet, little owl, pallid scops owl, crested lark, lesser short-toed lark, short-toed lark, tawny pipit, brown-necked raven, Finsch's wheatear, pied wheatear, desert wheatear, saxaul sparrow, rock sparrow, grey-necked bunting. Two species – common pheasant and long-tailed shrike – were referred by Shtegman (1938) to Chinese fauna.

We considered 11 species as elements of Middle Asian fauna that was distinguished as a separate region by Severtzov as long ago as 1873. These are the Caspian dull, white-winged woodpecker, pale sand martin, masked wagtail, black-headed wagtail Turkestan shrike, desert grey shrike, black-headed penduline tit, white-crowned penduline tit, Turkestan tit and Indian sparrow. Shtegman (1938) refused to accept autonomy of the Middle Asian region; he believed it to be rather heterogenous. The birds mentioned above were not included into his work of 1938. We think there can be another opinion on their biogeographical status since some ornithologists accept these birds no higher then subspecies and regard them within the species with wide geographic distribution. Our support of Severtzov is based on admitting these birds as full species.

Three birds are *endemics of Eurasian steppes* (mainly within limits of their Kazakhstan part) - pallid harrier, sociable plover, black-winged pratincole and one species – the Caspian plover – is an autochthonic representative of *Eurasian deserts and semideserts* and has its breeding range mainly in the territory of Kazakhstan. Dolgushin (1947) devoted a special paper to endemic elements in steppe ornithofauna. Pander's ground jay recorded on the eastern coast of the Aral Sea belongs to the genus *Podoces*, endemic for Asian deserts and including four species distributed from Iran to Mongolia.

There are no species of Siberian or Tibetan faunas breeding in the Aral Sea region.

Ornithogeographical division of Kazakhstan has been first proposed by Dolgushin (1957). He selected five provinces based on zonal landscape division and distinguished within them 16 regions. Kovshar (2006) advancing the ideas of Dolgushin classifies the territory of Kazakhstan into six provinces with their future subdivision into 24 regions. According to his classification Aral Sea Basin can be referred to Aralo-Syrdaryinsky desert region of Turanian (Desert) Province, Palearctic District. The following birds are typical for this region: little grebe, black-necked grebe, great white pelican, Dalmatian pelican, great cormorant, pygmy cormorant, great bittern, little bittern, black-crowned night heron, squacco heron, great egret, little egret, spoonbill, glossy ibis, grey heron, purple heron, marbled duck, red-crested pochard, ferruginous duck, Macqueen's bustard, stone curlew, white-tailed plover, pin-tailed sandgrouse, Eversmann's dove, laughing dove, pallid scops owl, Egyptian nightjar, oriental skylark, Pander's ground jay, Menetries' warbler, pied stonechat, saxaul sparrow (Kovshar 2006).

11.5.3 General Biogeography of Reptiles

Figure 11.19 shows the biogeographical composition of the reptile fauna of the Aral Sea basin. More than 60% of the reptile species can be classified as Turanian, thus representing an endemic fauna on the larger scale. The European-Mediterranean herpetofauna represents 21% of the species, thus indicating that the western influence is much greater than the eastern (Asian) influence. A small percentage of the species is widespread Saharo-Sindian or East Palaearctic.

11.5.4 Phylogeography of the Dice Snake Natrix tessellata

The aquatic dice snake *Natrix tessellata* (Fig. 11.20) occurs in the entire Aral Sea region. We have found the species at different water bodies in Kazakhstan and Uzbekistan and have taken blood or tissue samples from almost 30 animals from the hinterland of the Aral sea. Additionally, samples were collected in the eastern provinces of Kazakhstan, which – together with already existing samples from Europe, western and Middle Asia for comparison – were included in the research.

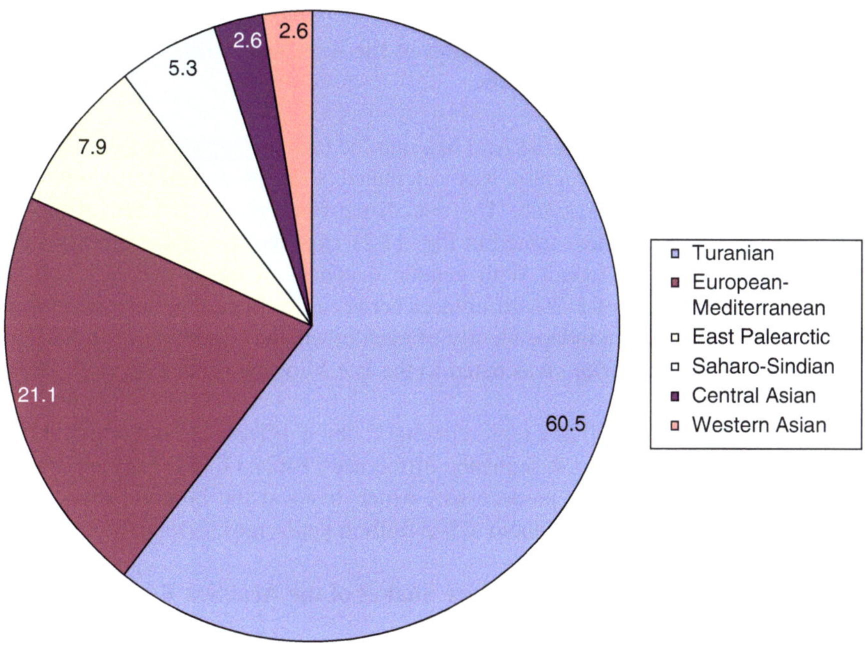

Fig. 11.19 Biogeographic composition of the reptile fauna of the Aral Sea basin

Fig. 11.20 Dice snake, *Natrix tessellata*

The material of the dice snakes was examined with two different molecular-genetic methods. In the first one, the mitochondrial cytochrome *b* gene was sequenced; the second was a genomic fingerprint analysis with help of the inter-simple sequence repeat (ISSR)–PCR fingerprint technique. Whereas the cytochrome *b* sequences as mitochondrial markers represent only the maternal heredity line, the ISSR fingerprints portray the genetic constitution of the nuclear DNA. They are also suitable for proving hybridization of different genetic lines.

11.5.4.1 Placement of the Dice Snakes of the Aral Sea in the Total Phylogeny of the Species

To classify the dice snakes of the Aral Sea into the total phylogeny of the species, a maximum likelihood phylogram was calculated with all available cytochrome *b* sequences of *Natrix tessellata*. There is strong intraspecific differentiation. At least ten big genetic clades, named in Fig. 11.21 by statement of the origin of the affiliated groups, were found. With genetic distances of up to over 8% and an assumed evolution rate of 1–2% differences between two lines, that have developed within one million years independently of each other, the beginning of the differentiation of *Natrix tessellata* was dated to the late Miocene (Guicking et al. 2006, 2009).

The animals of the Aral Sea region belong to three different clades which differ from each other by 1.7–3.1% sequence difference (Table 11.15). They obviously originated from a common predecessor, which lived at the border between the Pliocene and the Pleistocene (about 1.5–2 million years ago).

11.5.4.2 Phylogeography of the Dice Snakes of the Aral Sea Region

A more elaborate contemplation of the cytochrome *b* data of Middle Asian dice snakes makes possible statements about its descent and historic development.

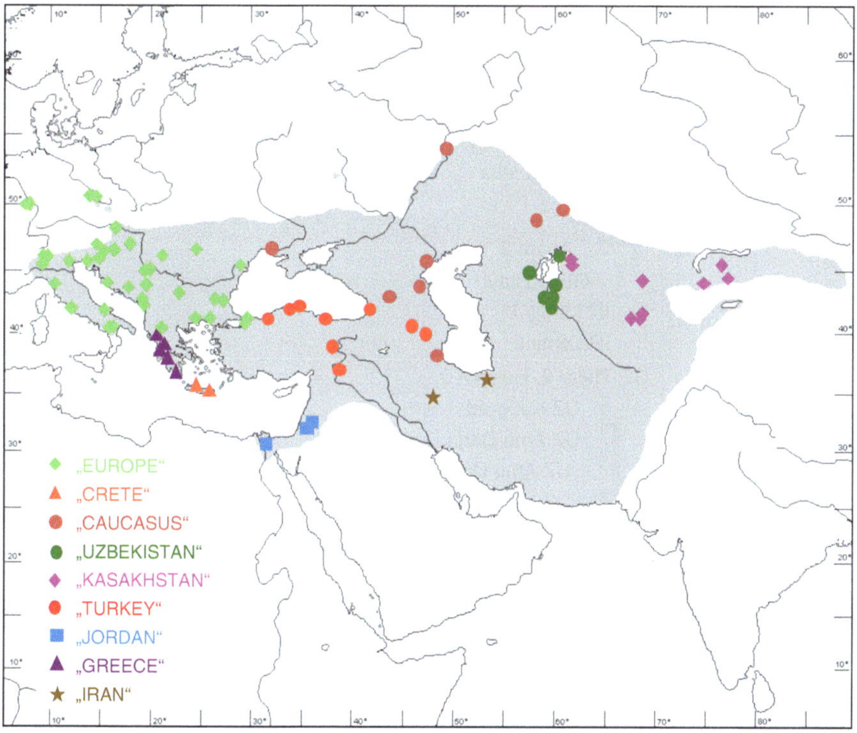

Fig. 11.21 Distribution of cytochrome *b* haplotype groups among regional populations of *Natrix tessellata*

Table 11.15 Genetic distances of the three clades of *Natrix tessellata* occurring in the Aral Sea area, based on complete sequences of the mitochondrial cytochrome *b* gene. Stated are the minimum–maximum of pairwise group comparisons (*below the diagonals*) and the maximum genetic distances within the individual groups (*in the diagonals*). As the genetic distance, the *p* distance, which represents the share of bases as a percentage, is used here

	1	2	3
1 Uzbekistan east, Kazakhstan east	1.07		
2 Uzbekistan northwest	2.33–3.13	0.72	
3 Caucasus, Ukraine, Kazakhstan northwest	1.97–2.86	1.70–2.15	0.81

The eastern clade "Uzbekistan east, Kazakhstan east" includes snakes from the regions around Almaty and Tashkent as well as from the middle Syr Darya delta of Lake Kamyshlybash and from Raim (Fig. 11.22). A subdivision of this clade essentially removes the snakes from southeastern Kazakhstan from the remaining snakes. However, snakes from Gazalkent in eastern Uzbekistan belong to both subgroups. The affiliation of snakes from the middle Syr Darya delta to the eastern clade shows the function of the Syr Darya as a means for the spreading of the dice snake.

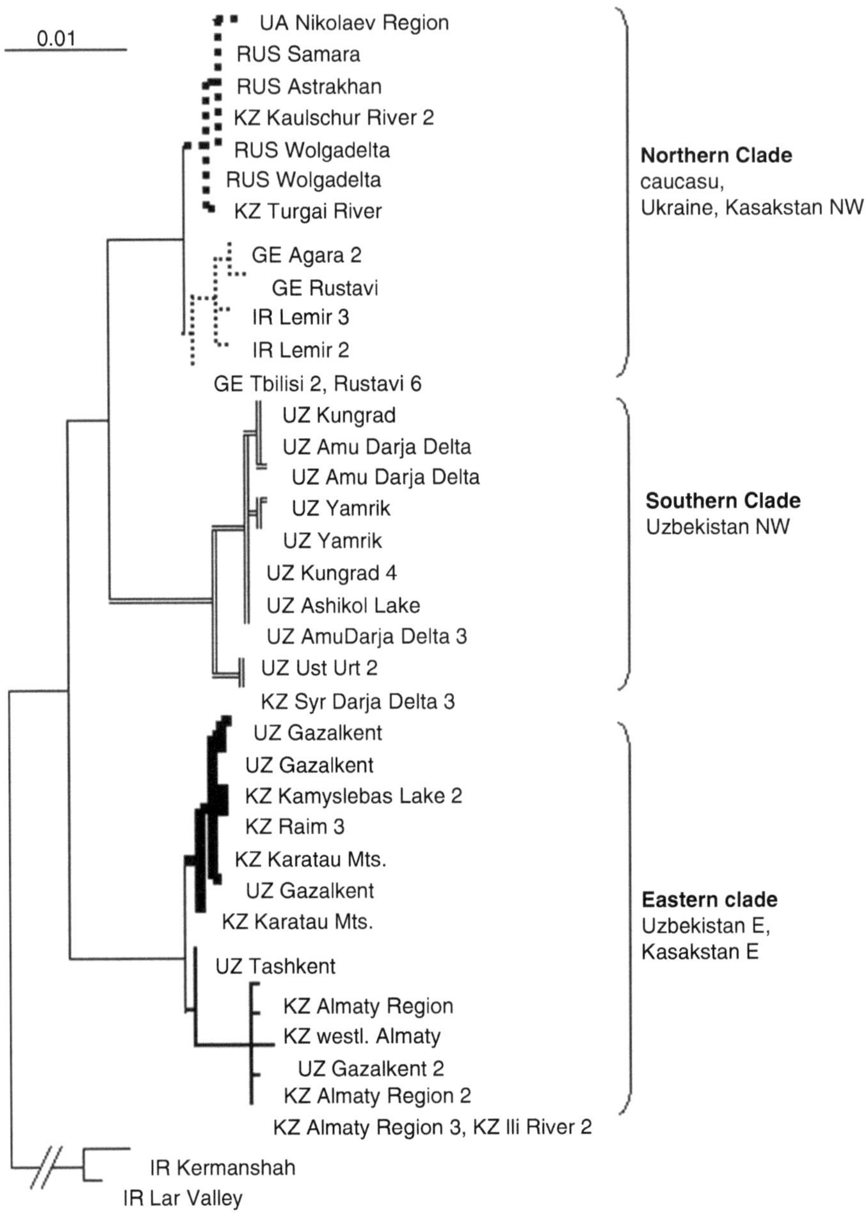

Fig. 11.22 Maximum likelihood phylogram of central Asian dice snakes, based on complete cytochrome *b* sequences. Different lines show the genetic differentiation into three clades, two of which are further subdivided. Two samples from Iran were used as outgroups. Numbers indicate that several identical individuals were found

The animals of the southern Aral Sea "Uzbekistan northwest" belong to a genetic line which is dispersed at the Amu Darya as well as in direct proximity of the Aral Sea. Interestingly, this also contains animals which originate from the estuary area of the Syr Darya at the northeastern edge of the Aral Sea. The close relationship of dice snakes of the southern and the northwestern Aral Sea clarifies that the dice snakes in the past used the Aral Sea as a habitat and obviously travelled long distances in it. An occurrence of the dice snake on the island of Vozrozhdeniya (Sidorov 1925) can be regarded as further evidence for the former existence of a dice snake population in the Aral Sea.

To the third genetic line "Caucasus, Ukraine, Kazakhstan northwest", which reaches the greater Aral Sea area, belong animals of the rivers north and northwest of the Aral Sea. They have a close relationship to animals from Russia and the Ukraine as well as to animals from the Caucasus (Fig. 11.22). Since big parts of the area inhabited by this clade were permafrost during the Pleistocene glacials (Hewitt 2000), one can assume that the ancestors of these populations survived the ice ages in the Caucasus area and advanced postglacially to the northeast and northwest. The clear differentiation into a group to the south of and one to the north of the Caucasus speaks on behalf of a scenario where the dice snakes of this genetic lines have probably outlasted the last ice ages in at least two glacial refuges on both sides of the Caucasus.

The ISSR data confirm and supplement the results obtained from cytochrome *b* data about the differentiation and historic development of the Middle Asian dice snakes.

The clear separation of the dice snakes from the northern areas from those from the direct margin of the Aral Sea is also covered by the ISSR data (Fig. 11.23). This means that no genetic exchange takes place between these lines or has taken place in the past.

In contrast, the ISSR data also show that a strong gene flow has taken place between the southern shore of the Aral Sea and the clade penetrating from the east in the recent past. Especially the animals from the middle Syr Darya delta of Lake Kamyshlybash and from Raim, which are assigned to the eastern line according to cytochrome *b* data, show a closer relationship to animals of the southern clade according to the ISSR data. This confirms that as well as over the Aral Sea, also over the Syr Darya a genetic exchange between the individual dice snake populations took place. The exchange over the Aral Sea was blocked for at least some decades, as the snake populations on the lake are completely extinct because of the strong salt content and hostility of the water to life.

11.5.4.3 Conclusions

Our data confirm the former role of the Aral Sea as a habitat of the dice snake, among other species. Through the drying of the Aral Sea for approximately 40 years, this habitat is no longer accessible for aquatic animals today. The 'Uzbekistan haplotype', apparently endemic to the region, has lost its central population, and only peripheral populations survived.

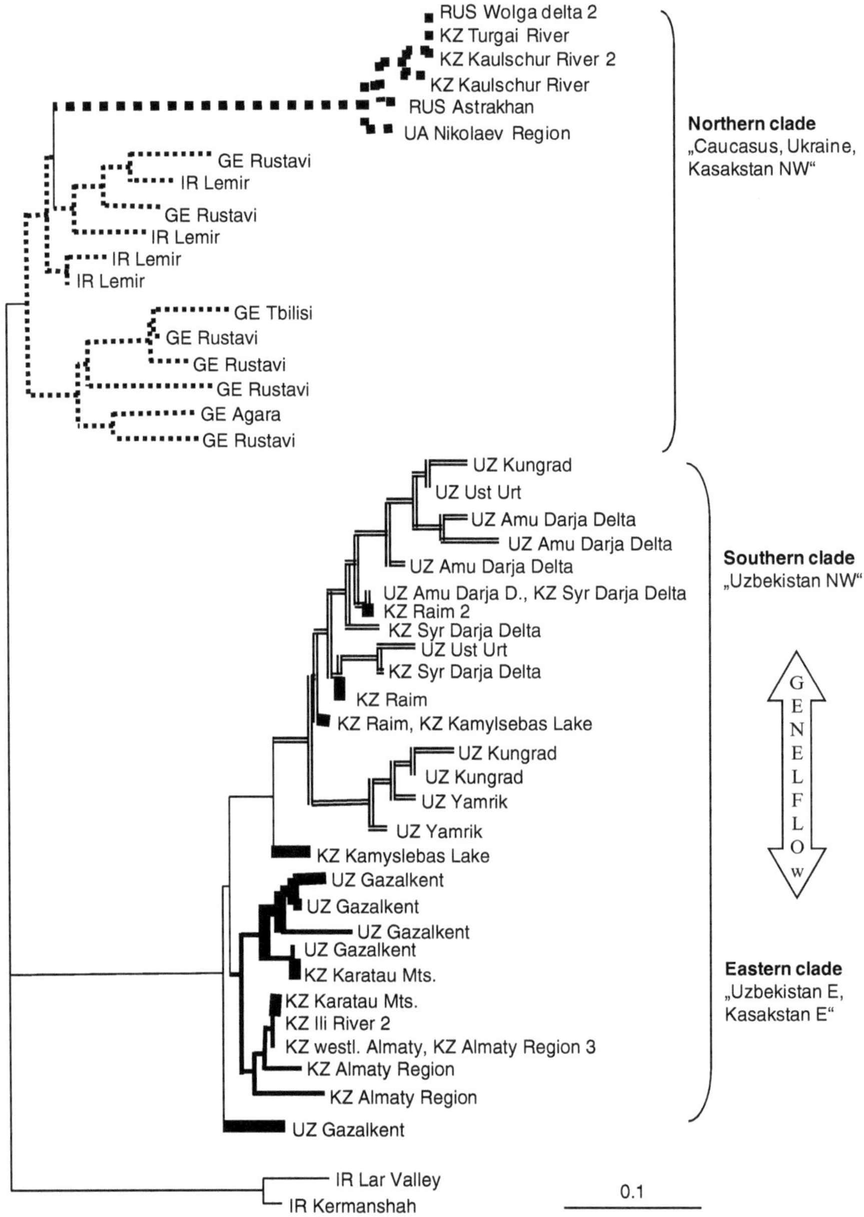

Fig. 11.23 Neighbor joining phylogram of central Asian dice snakes, based on genomic inter-simple sequence repeat PCR profiles. See Fig. 11.22 for explanations (different lines represent groups defined by mitochondrial cytochrome *b* sequences)

The example of the dice snake shows in an impressive manner that the Aral Sea catastrophe led not only to the local extinction of populations, but also had effects on the intraspecific evolution through the loss of a big region of connective water at the same time. Although an obviously quite strong exchange of individuals continued along the course of the Syr Darya, a connection between populations of the southern and northern Aral Sea region is no longer possible today. The intraspecific differentiation is accelerated by this fact.

Also, in the sense of species conservation, another consequence of the Aral Sea drying must be taken into account. On the basis of the individual exchange which obviously took place over the Aral Sea earlier, it is possible that some smaller populations were dependent particularly on the adjoining waters for immigrations from other populations in order to be able to survive. This system of a metapopulation would no longer be working today, which would have direct consequences for the populations concerned.

11.5.5 *Phylogeography of Desert Runners (*Eremias *spp.)*

The widespread species *Eremias velox* can be subdivided into six main genetic lineages, based on mitochondrial genes (Fig. 11.24, Rastegar-Pouyani et al. 2008). Three of these lineages occur in Central Asia; they are part of a monophyletic group which was probably separated from the Iranian lineages by the uplifting of the Kopet-Dag, about six million years ago. Two of these lineages occur in the Aral Sea basin; the lineage from Uzbekistan is genetically different from the lineage from the northeastern and western coasts of the Aral Sea. The latter is distributed to eastern Kazakhstan. This is consistent with morphological results and should be investigated more extensively.

Eremias intermedia and *Eremias lineolata* showed little differences between specimens from Kazakhstan and specimens from Uzbekistan.

11.5.6 *Biogeography of Aral Region Insects*

Among Homoptera, Aphidinea, 17.6% of the species have ranges which extend beyond the borders of the Tethys region (15 species). All other are Tethys species (82.4%, 70 species; Kadyrbekov 2003). Among Tethys elements there are species with wide ranges (widely Tethycal, Western Tethycal, Eastern Tethycal, Saharo-Gobian and Western Scythian) – 24.2% (17 species). The species with ranges not extending beyond the Turanian Province amount to 38.6% of species (27 species). Aral endemics form 11.7% of species (10 species).

Among xylophagous beetles inhabiting the northern Aral Sea region, the genera *Sphenoptera* (18 species), *Acmaeoderella* (3 species) and *Xylotrechus* (3 species) dominate (Kadyrbekov and Tleppaeva 2004). Two of them have a Palaearctic

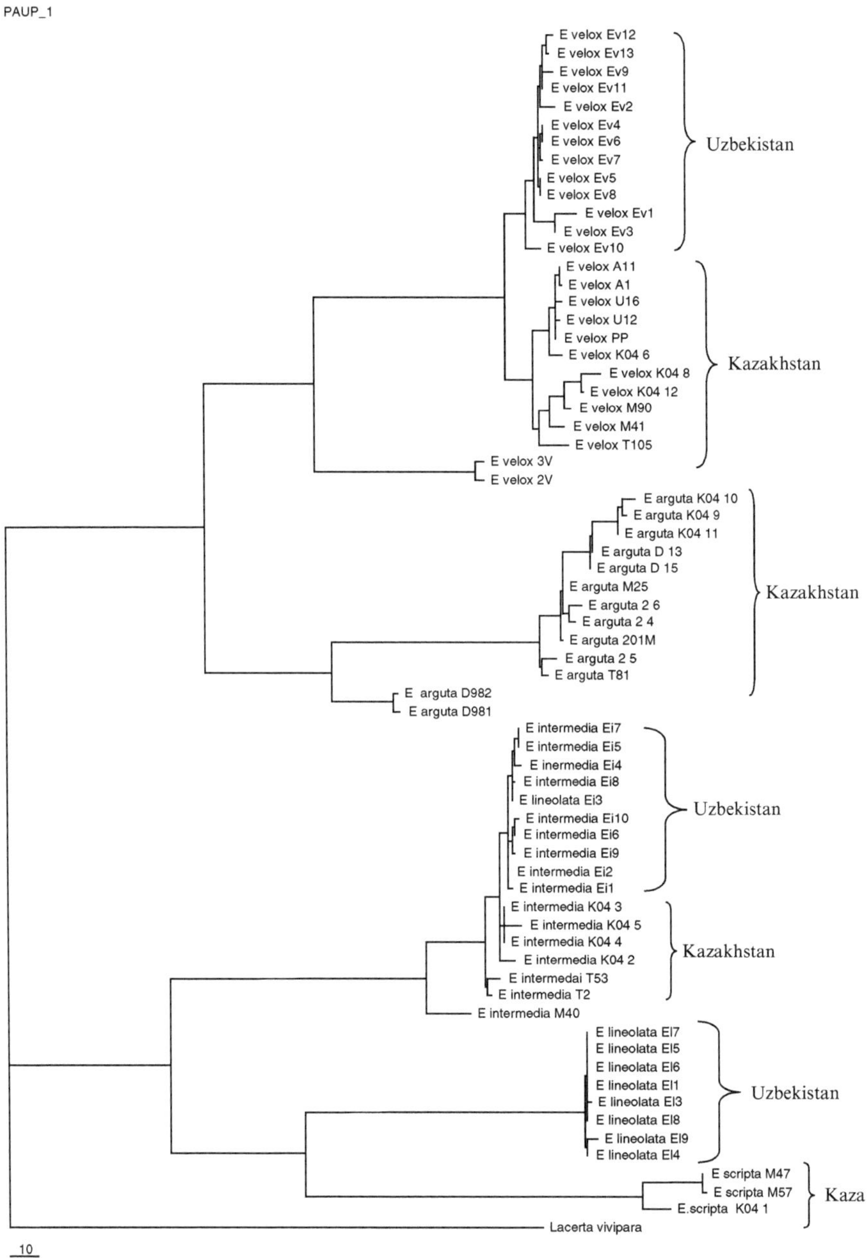

Fig. 11.24 Phylogenetic relationships (maximum parsimony) of the *Eremias* species of Central Asia, derived from cytochrome *b* sequences

range, two have a western Palaearctic range, one has a western Tethycal range, one has an eastern Tethycal range, one has an eastern Scythian range, one has a Euxino-Khorasano-Turanian rage, five have a Irano-Turanian-Gobian range, six have an Irano-Turanian range, four have a Turano-Gobian range, one has a northern Turano-Gobian range, 14 have a Turanian range, five have a northern Turanian range and two have an Aralian range. The species with wide ranges amount to only 15.5%, whereas those with ranges not extending beyond the borders of the Turanian Desert Province come to 46.6% (21 species).

These examples confirm the originality of the Aral Sea insect fauna having a core with very local and relict species.

11.6 Nature Conservation

Finally, our data clarify the special situation of the Aral Sea at the border of very different faunal areas. Both the species composition and the intraspecific differentiation of some much dispersed species, as the example of the dice snake shows, indicate a strong biogeographical influence from different directions. Here, the strong differentiation of the northwestern Aral Sea basin contrasts with the differentiation to the south of it and to the east of the Aral Sea. Whereaes the influence of the European region is conspicuous in the northwest, representatives of the Middle Asian faunal realm predominate in the south and east. A suitable concept for the protection of the biodiversity in the region has to take this into account through the establishment of several reserves in the faunistically different regions at the Aral Sea.

As a consequence of our examinations, we would like to recommend several areas for protection, in addition to the expanded Barsa-Kelmes area (see Chap. 14), and for further monitoring:

Kazakhstan:

1. Lake Kamyshlybash (western shore, 46°10′N, 61°38′E). Big, obviously intact reed communities, numerous bird species, one of the last habitats for remaining aquatic herpetofauna of the Aral Sea area, among others *Pelophylax ridibundus* and *Natrix tessellata*. Endangered through salinization (already at the eastern shore).
2. Shores of the Small Aral Sea near Akespe (46°48′N, 60°32′E). Big habitat variety with steppe, shores, sand dunes and south-exposed chink (plateau). Very diverse wildlife with typical inhabitants of the named habitats.
3. Big Barsuki sand dunes. Transition area between European fauna, in the north, and Central Asian fauna, in the south. Several parts of this spacious area should therefore be put under protection.

Uzbekistan:

1. Little lakes in hill country between Ustyurt Plateau and the western shores of the Aral Sea (44°51.4′N, 58°10.9′E) Some of the lakes still have freshwater and an

intact reed girdle. Last surviving dice snakes (*Natrix tessellata*) and green toads (*Bufo viridis*) at the western shore; of the latter species, however, only dead animals were found. Furthermore, there is very rich land reptile fauna as well as bats and a rich bird world. Endangered through fire clearance and salinization (lower lake).
2. Moist area at the southeastern shore (43°30.8′N, 60°25.1′E) through dune-barrier pent-up drainage water of the Amu Darya delta (Kok Darya). Numerous lakes of different water quality. Still viable populations of dice snakes (*Natrix tessellata*) and green toads (*Bufo viridis*). Big tortoise population [*Testudo* (*Agrionemys*) *horsfieldi*]. Rich bird world, eagle owl, pelicans, limicoles.
3. Small dune area near Raushan, between Kungrad and Ustyurt Plateau (43°08.3′N, 58°38.6′E),: Very rich tortoise population (*Testudo horsfieldi*), species-rich reptile coenose. Small ponds with aquatic fauna (dragonflies) and good avifauna. Some dice snakes (*Natrix tessellata*).

According to our results, the above-mentioned Uzbekistan populations of *Natrix tessellata* belong to an endemic genotype for the Aral Sea area.

11.7 Conclusions

The ecological disaster of Aral Sea reduced the faunistic diversity of the area in a selective manner. Aquatic and semiaquatic animal species such as fish-eating birds, waterfowl, Amphibia, water snakes and aquatic insects suffered dramatic reductions in numbers. Some freshwater species and species of riverine forest died out or left the area completely. In the hypersaline remnant water bodies, only few animals survive, mainly *Artemia*, the salt brine shrimp (Arashkevich et al. 2009), which in the future may even have some economic role as a protein source.

On the other hand, desert species and certain eurybionts were able to extend their ranges into the Aralkum, the new desert on the dried seafloor. The succession needed about three decades after drying to establish stable communities.

The situation is less simple with typical steppe animals, such as certain mammal and reptile species. Some of them (rodents, small carnivores, lizards, snakes) were able to colonize new habitats on the former seafloor, others remained in preserved steppe habitats, and a third group suffered from population reduction. The latter fact must be stated for mammals of the former island of Barsa-Kelmes, a nature reserve which used to be protected from outside influences when it was an island, but which lost its unique characteristics when it became connected to the mainland.

Further monitoring of the fauna of the Aralkum is strongly recommended.

Acknowledgements We thank numerous cooperation partners, especially those who participated in the faunistic monitoring during the INTAS project, namely Marina Chirikova, Sergio Castellano, Hilde Enting, A. Khodjaev, E. Kurshut, Alexander Levin, Seksenbay Kushekbayev, Paola Magni, Dmitry Malakhov, A. Nuridjanov, Sergey Pachin, Sergey Sokolov and E. Vashetko. U.J. is strongly indebted to Michael Wink, who supported the laboratory work, as well as Eskandar Rastegar-Pouyani, Svetlana Kalyabina-Hauf, Nikolaus Stuempel and Oleksandr Zinenko for their molecular genetic analyses. The investigations were supported by INTAS, DFG and CNRS.

References

Mammals and General References

Afanasyev AV (1960) Zoogeography of Kazakhstan. Alma-Ata: 260 pp. [in Russian]

Afanasyev YuG, Kenessarin GA (1979) Fauna changes in connection with Aral Sea shrinking. Influence of Aral Sea level decreasing on environment. Almaty, pp 206-207. [in Russian]

Afanasyev YuG, Smirnov YuA (1981) Influence of Aral Sea level decreasing to mammals of Barsakelmes Island. Shrinking of Aral Sea and desertification of Aral Sea basin. Alma-Ata, pp 196–199 [Russian]

Bekenov AB (1988) Analysis of the desert mammalian fauna of Kazakhstan. Ecology and behavior of the mammals of Kazakhstan. Alma-Ata, pp 166–183 [Russian]

Burdelov LA (1985) Distribution and number of the dipodids in northwestern Aral Sea region and northern Ustyurt Plateau. Dipodidae of USSR Fauna. Moscow, pp 29–31 [Russian]

Butovsky PM, Shaimardanov RN (1985) Common Noctule. Mammals of Kazakhstan, vol IV. Alma-Ata, pp 185–201 [Russian]

Elisseyev DO (2007) Recent statement of vertebrate fauna in Barsakelmes Island and its dynamics during last 50 years. Transactions of Barsa-Kelmes reserve. Almaty, 2, pp 107–128 [Russian]

Garbuzov VK (1985) Data on distribution, number and ecology of *Pygerethmus platiurus* and *Salpingotus crassicauda* in northern Aral Sea region. Dipodidae of USSR Fauna. Moscow, pp 143–144 [Russian]

Geptner VG (1984) Data on phylogeny and classification of dipodidae of USSR. The problems of variation and phylogeny of the mammals. Moscow, pp 37–60 [Russian]

Ismagilov MI (1985) The fauna of dipodids of Kazakhstan. Dipodidae of USSR fauna. Moscow, pp 6–9 [Russian]

Jaschenko RV (1998–1999) On recent statement of the Asian wild ass population in Aral Sea region. Selevinia 1998–1999, pp 228–229 [Russian]

Kucheruk VV (1959) The steppe faunistic complex of mammals and its place in the fauna of palearctic. Geography of terrestrial animals and methods of its study. Moscow, pp 45–87 [Russian]

Lobachev VS (1971) Record of *Salpingotus crassicauda (*Dipodidae*)* in northern Aral Sea region. Zool Zh 4(2):305–306 [Russian]

Mazin VN (1979) On fauna and number of mammals in drying bottom of Aral Sea. Influence of Aral Sea level decreasing to environment. Alma-Ata, pp 112–116 [Russian]

Mazin VN (1981) On fauna and number of mammals in drying bottom of Aral Sea. Shrinking of Aral Sea and desertification of Aral Sea basin. Alma-Ata, pp 190–193 [Russian]

Mazin VN (1986) Saiga at the eastern coast of Aral Sea. Abstr IV All-Union Meet. Theriolog Soc, Moscow (3):154–156 [Russian]

Mazin VN, Eszhanov B, Mussabekov K (1992) Mammal settling in the dried aquatorium of the Aral Sea. Medical, social and ecological problems of Aral Sea region. Alma-Ata, 123–124 [Russian]

Micklin P (2007) The Aral Sea Disaster. Ann Rev Earth Planet Sci 35:47–72

Mirzabekov ZhM, Eszhanov BE, Bekenov AB. Baytanayev OA (1985) Data on the distribution and number of the dipodids in Northern Kizilkum. Dipodidae of USSR fauna. Moscow, pp 71–73 [Russian]

Mirzabekov ZhM, Aubakirov SA, Zhubanazarov IZh, Mazin VN (1992) About *Rh. opimus* settlements within the former bottom in Northeastern Aral Sea region. Organizing of epidemical control of plaque danger and ways for its prophylactics. Alma-Ata, pp 332–334 [Russian]

Novikova NM, Zaletayev VS (1985) Water economics as a factor in the transformation of natural ecosystems. In Kibal'chich OA, Voronov AG, Vishyvkin DD (eds) Biogeographical aspects of desertification, Moscow, pp 4–12. [Russian]
Pavlinov IYa, Borissenko AV, Kruskop SV, Yakhontov EL (1995) Mammals of Eurasia, Moscow, vol 1, p 240, vol 2, p 336 [Russian]
Persianova LA, Yadgarov T, Saraeva KhM (1986) Patterns in the changes of fauna due to watering of arid territories. Prirodnye resursy pustyn' i ikh osvoenie. Ashgabad, pp 336-340. [Russian]
Rashek VA (1966) Ecology of the Asian wild ass and its acclimatization in Barsakelmes Island. Abstr Cand Biol Sci Thesis, Moscow, p 26 [Russian]
Reimov R, Karabekov M (1989) Ecological and geographical characteristics of Southern Aral Sea region. All-Union Meet. Probl. Monitoring and calculation of animals, Part 2. Ufa, pp 275–277 [Russian]
Rotshild EV, Smirnov VM, Shilov MN, Kamishev IA (1967) On the mammals of Northern Kizilkum. Fauna and ecology of the rodents 8. Moscow, pp 85–174 [Russian]
Shaimardanov RT (2007) Fate of the island population of the Asian wild ass. Transactions of Barsa-Kelmes reserve. Almaty, 2, pp 155–156 [Russian]
Stogov II, Zhanabayev MI (1985) On the distribution of *Salpingotus heptneri* in Northern Kizilkum. Dipodidae of USSR fauna. Moscow, p 69 [Russian]
Varshavsky SN (1951) Some results of the study of rodent fauna in Northern Aral Sea region (1947–1949). Transactions of Middle Asian Sci. Res. Antiplague Inst. Alma-Ata, 1, pp 33–43 [Russian]
Varshavsky SN, Shilov MN (1955) New data on distribution of some mammals, mainly rodents, in Northern Aral Sea region and surrounding areas. Bull Moscow Soc Nat Res Sec Biol 10(5): 43–58 [Russian]
Varshavsky SN, Shilov MN (1959) Some results of bird of prey belch analysis with application to study of distribution and number of micromammals in Northern Aral Sea region. Abstr II All-Union Ornithol Meet, Moscow, 3, pp 19–20 [Russian]
Varshavsky SN, Shilov MN, Krylova KN, Shilova ES, Garbuzov VK, Li AS, Gulyaevskaya NS (1985) Species composition and distribution of dipodids in Northern Aral Sea basin and surrounding regions. Dipodidae of USSR fauna. Moscow, pp 33–35 [Russian]
Vassenko EP (1950) Ecology and distribution of Saiga *(Saiga tatarica* L.). Transactions of Barsa-Kelmes state reserve 1. Alma-Ata, pp 38–115 [Russian]
Vorontzov NN, Shenbrot GI (1984) Systematic review of *Salpingotus* (rodentia, dipodidae) and description of *Salpingotus pallidus* sp.n. from Kazakhstan. Zool Zh 63:731–744 [Russian]
Zaletayev VS (1976) Life in the desert. Geographical-biogeocoenotic and ecological problems. Moscow: 271 pp. [in Russian]
Zaletayev VS, Kosbergenov MR (1985) Peculiarities of dipodid spreading within the drying bottom of the Aral Sea. Dipodidae of USSR fauna. Moscow, pp 41–42 [Russian]
Zhevnerov VV (1984) Persian gazelle in Barsakelmes Island. Alma-Ata, p 145 [Russian]

Birds

Berezovsky VG (1980) Species composition and number of waders in the eastern coast of the Aral Sea in summer 1978. Abstr. II Meet. Wader fauna and ecology. Moscow, pp 84–85 [Russian]
Berezovsky VG (1981) Number of rare and disappearing bird species during the season migration in the eastern coast of the Aral Sea. Abstr. VIII All-Union Ornithol. Meet. "Bird ecology and protection". Kishinev, pp 23–24 [Russian]
Berezovsky VG (1983a) Spring migration of the waterbirds and wetland-dependent birds at eastern coast of the Aral Sea in 1978–1979. Bird migrations in Asia 8, Alma-Ata, pp 26–34 [Russian]
Berezovsky VG (1983b) Birds of prey migration at eastern coast of the Aral Sea in 1978–1979. Birds of prey ecology. Moscow, pp 7–8 [Russian]

Berezovsky VG (1991) Short reports on the Siberian white crane in Aral Sea. Rare birds and animals of Kazakhstan. Alma-Ata, pp 137 [Russian]
Berezovsky VG, Auezov EM, Khrokov VV (1982) Recent statement of ornithofauna of the eastern coast of the Aral Sea and perspectives of its changes in connection with sea drying. Animals of Kazakhstan and problems of their protection. Alma-Ata, pp 30–31 [Russian]
Berg LS (1908) Aral Sea. An experience of physico-geographical monography. Scientific results of the Aral expedition 9. Saint Petersburg, p 580 [Russian]
Bogdanov MN (1882) The issues on nature of Khiva Oasis and Kizil-Kum desert. A description of Khivin trip in 1873 12. Tashkent [Russian]
Bogdanov MN (1884) Checklist of birds of the Russian empire. Saint-Petersburg [Russian]
Bostanjoglo NV (1911) Ornithological fauna of Aralo-Caspian steppes. Materials to the study of fauna and flora of Russian empire, zoology 11. p 410 [Russian]
Brokhovich SA (1981) Autumn migration of the waders in Aral Sea. Abstr. VIII All-Union Ornithol. Meet. "Bird ecology and protection". Kishinev: 31–32 [Russian]
Darlington PhJ Jr (1980) Zoogeography: the geographical distribution of animals. Krieger, Malabar, p 675
Dolgushin IA (194) On endemism of steppe ornithofauna of Kazakhstan. Vestn Acad Sci KazSSR 6(27):133–156 [Russian]
Dolgushin IA (1957) Ornithogeographical division of Kazakhstan. Transactions of conference on problems of land zoogeography. L'vov, pp 34–36 [Russian]
Elisseyev DO (1984a) Breeding bird fauna of vegetated sands in Barsakelmes reserve. Study and protection of the reserve objects. Alma-Ata, pp 38–39 [Russian]
Elisseyev DO (1984b) On the second reproductive cycle in the desert grey shrike. Vestn Zool 2:86–88
Elisseyev DO (1985a) Reproductive cycle ecology of common birds of ridge-and-rise sands in Barsakelmes reserve. Ecology of birds in reproductive period. Leningrad, pp 3–9 [Russian]
Elisseyev DO (1985b) Materials on breeding and feeding of the Lesser short-toed Lark. Vestn Zool 1:49–54 [Russian]
Elisseyev DO (1985c) Materials to the study of the structure of ornithofauna in vegetated sands of Barsakelmes Island during breeding period. Ecology of birds in reproductive period. Leningrad, pp 83–88 [Russian]
Elisseyev DO (1986a) Materials on ecology of *Caprimulgus europaeus unvini* (Hume 1884). Ecology and breeding of the birds. Leningrad, pp 19–28 [Russian]
Elisseyev DO (1986b) An influence of abiotic factors to elements of reproductive cycle in desert nested birds. Study of birds of USSR, protection and use. Abstr IX All-Union Ornithol Meet 1, Leningrad, pp 217–218 [Russian]
Elisseyev DO (1987) Materials on ecology of *Emberiza bruniceps* in northern Aral Sea region. Problems of formation of animal complexes in terrestrial and aquatic biocenoses. Omsk, pp 3–14 [Russian]
Elisseyev DO (1989) Structure and dynamics of ornithological complexes in vegetated sands of Barsakelmes Island. Ecology of birds in reproductive period. Leningrad, pp 109–118 [Russian]
Elisseyev DO (1990a) On the influence of Aral Sea shrinking to species composition and distribution of the birds in its surroundings. Recent statement of Aral Sea in conditions of its progressive salification. Trudy Zool Inst Acad Sci USSR Leningrad 223:147–151 [Russian]
Elisseyev DO (1990b) Recent statement of ornithofauna of Barsakelmes reserve and its dynamics during last 30 years. The reserves of USSR, their present and future. Novgorod, 3, Zoological researches, pp 225–227 [Russian]
Elisseyev DO (1996a) Ecology of Sykes's warbler in the Barsakelmes reserve. Scientific Notes of the Biology Department, Omsk, pp 175–200 [Russian]
Elisseyev DO (1996b) Materials on the biology of *Passer indicus bactrianus* Zar. et Kudasch, 1916 in northern Aral Sea region 2(1). Scientific notes of the Biology Department, Omsk, pp 79–87 [Russian]

Elisseyev DO (1999) Materials on breeding and feeding of *Oenanthe pleschanka* Lepech, 1770 in Barsakelmes Island (Aral Sea). Field and experimental researches (transactions) 4, Saint-Petersburg, Omsk, pp 122–130 [Russian]

Elisseyev DO (2003) On the biology of *Sylvia nana* in northern Aral Sea region. Functional morphology, ecology and life cycles of the animals. Transactions of Zoology Department of RSPU 3, Saint-Petersburg, pp 116–123 [Russian]

Elisseyev DO (2004) An issue on the biology of *Sylvia curruca halimodendri* Suschk, 1904 according to observations in northern Aral Sea region. Transactions of Zoology Department of RSPU 4, Saint-Petersburg, pp 179–188 [Russian]

Elisseyev DO (2007) Recent statement of vertebrate fauna in Barsakelmes Island and its dynamics during last 50 years. Transactions of Barsakelmes reserve. Almaty, 2, pp 107–128 [Russian]

Erokhov SN (2004) Ornithological observations in the northern coast of Aral Sea. Kazakhstan Ornithological Bulletin 2003, Almaty, pp 40–41 [Russian]

Eversmann E (1823) Reise von Orenburg nach Buchara, begleitet von einem naturhistorischen Anhange und einer Vorrede von H. Liechtenstein. Berlin

Eversmann E (1866) Natural history of the birds of Orenburg land. Natural history of Orenburg land. Part 3. Kazan

Gistzov AP (1974) Some changes in fauna of the birds of Barsakelmes Island during the last decades. Abstr VI All-Union Ornithol Meet, Moscow, 1, pp 189–190 [in Russian]

Gistzov AP (1978a) Migration of passeriform birds at Barsakelmes Island (Aral Sea). Biology of the birds in Kazakhstan, Alma-Ata, pp 133–136 [Russian]

Gistzov AP (1978b) Composition of wintering avifauna and its distributional patternin Barsakelmes Island. Biology of the birds in Kazakhstan, Alma-Ata, pp 147–149 [Russian]

Gladkov NA (1941) Ecological notes on bird fauna in Syr-Darya delta. Transactions Zool Mus Moscow State Univ 6:141–144 [Russian]

Gladkov NA (1949) Bird observation in Komsomolsk Island (Aral Sea). Okhrana Prir 8:29–43 [Russian]

Grachev VA (1954) The lesser sand plover in northern coast of Aral Sea. Priroda 7:117–118 [Russian]

Grachev VA (1956a) On the migration of some waders in Aralsk town vicinity. Trudy Inst Zool Acad Sci Kazakh SSR 6:197–199 [Russian]

Grachev VA (1956b) On the distribution of the Rock Sparrow in the northern coast of Aral Sea. Trudy Inst Zool Acad Sci Kazakh SSR 6:199–201 [Russian]

Grachev VA (2000) The birds of Aralsk town vicinity and Kamishlibash lake (by 1951–1954 observations). Non passeriform birds. Selevinia 1–4:95–104 [Russian]

Gubin B (1998) The present state of the North Aral Sea avifauna and its conservation problems. Sustainable use of natural resources of Central Asia, Almaty, pp 112–119

Gubin BM (1999) The birds of eastern Aral Sea region. Russky Ornithol Zhurnal, Express issue 80:3–16 [Russian]

Ismagilov MI (1955) Gills of Barsakelmes Island (Aral Sea). Trans Inst Zool Acad Sci Kazakh SSR 4:166–172 [Russian]

Ismagilov MI, Burambayev KB (1973) The conditions of existence and status of the birds in Barsakelmes reserve. Biological sciences, 5. Kazakh State Univ. named after Kirov, Alma-Ata, pp 119–125 [Russian]

Ismagilov MI, Burambayev KB (1978) On the dates of bird migration in Barsakelmes Island. Second All-Union Meet. on bird migration. Alma-Ata, 1, pp 116–118 [Russian]

Ismagilov MI, Vasenko EP (1950) The birds of Barsakelmes Island. Trans Barsakelmes Reserve 1:116–135 [Russian]

Khrokov VV (2004) On number of the birds in Syr-Darya river delta lakes and mouth in summer 2001. Bird migrations in Asia. Almaty, 12, pp 216–230 [Russian]

Khrokov VV (2006) Delta lakes of Syr-Darya river. Researches of key ornithological territories in Kazakhstan and Middle Asia. Almaty, pp 136–139 [Russian]

Khrokov VV, Auezov EM, Berezovsky VG (1983) Autumn migrations of waders in Aral Sea. Bird migrations in Asia 8. Alma-Ata, pp 157–160 [Russian]
Koblik EA (1991) The birds of pioneer plant associations of the former coastline and bottom of the Aral Sea in the region of Syr-Darya river delta. Abstr X All-Union Ornithol Meet Minsk 2(1):282–283 [Russian]
Kovalenko AV (2006) Ornithological investigations in the valley of low flow of Syr-Darya river and some other surrounding areas in 2005. Kazakhstan Ornithological Bulletin 2005, Almaty, pp 59–69 [Russian]
Kovshar AF (2000) Ornithological observations in Northern Kizilkum in spring 1984. Selevinia 2000, Almaty, pp 110–117 [Russian]
Kovshar AF (2006) Ornithogeographical division of Kazakhstan. Republic of Kazakhstan, vol 1. Nature conditions and resources. Almaty, pp 441–453 [Russian]
Kovshar AF (2007a) Ornithological observations in Barsakelmes reserve in the first decade of May 2007. Transactions of the Barsakelmes state nature reserve. Almaty, 2, pp 139–154 [Russian]
Kovshar AF (2007b) Transect car calculations of the birds in the former bottom of the Aral Sea in May 2007. Selevinia 2007, Almaty, pp 165–166 [Russian]
Kovshar V A (2001) Ornithological observations in Syr-Darya river delta and surrounding areas of the drying Aral Sea bottom at August 2000. Selevinia 2000, Almaty, pp 104–110 [Russian]
Kuzyakin AP (1959) Materials on the biology of the colonial breeding birds. 1. The birds breeding at the Zunkar Island (Aral Sea). Scientific notes of Moscow pedagog. Inst 71(4):3–23 [Russian]
Kuzyakin AP (2005) Notes on the birds of the northern coast of Aral Sea. Kazakhstan Ornithological Bulletin 2005, Almaty, pp 206–207 [Russian]
Lankin PM (1976) On the distribution of the steppe eagle in low flow of Syr-Darya river. Ornithology 12 Moscow:235–236 [Russian]
Lobachyov VS (1959) On the biology of the birds of prey from Zaaralye. Abstr II All-Union Ornithol. Meet. Part 3. Moscow, pp 31–32 [Russian]
Lobachyov VS (1961) An experience of number calculation and study of the birds of prey and their nests placement in Southern Kazakhstan. Problems of organizing and methods of calculation of resources of terrestrial vertebrate fauna. Moscow, pp 144–145 [Russian]
Markov VI (1965) Notes on the pelicans and cormorants of the Aral Sea. Fish-feeding birds and its value in fish industry. Moscow, pp 156–165 [Russian]
Mullarney K, Svensson L, Zetterstrőm D, Grant P (1999) Bird guide. The most complete field guide to the birds of Britain and Europe. HarperCollins Publ Ltd, London, p 400
Nikolsky AM (1892) On the mammal and bird fauna of the near Aral Sea steppes. Bull Moscow Soc Nat Res Sec Biol 4:477–500 [Russian]
Petrovskaya EP (1951) Feeding of the birds of prey in connection with rodent distribution in the coast of the Aral Sea. Bull Moscow Soc Nat Res Sec Biol 56(6):25–30 [Russian]
Pfander PV (2004) *Falco cherrug* in the Aral Sea. Kazakhstan Ornithological Bulletin 2003, Almaty, pp 147 [Russian]
Rashek VL (1965) Winter records of the Dunnock in Barsakelmes Island. Ornithology. Issue 7, Moscow, pp 486–487 [Russian]
Rashek VL and Rashek VA (1963) State reserve "Barsa-Kelmes Island". Alma-Ata, p 92 [Russian]
Sedunova EA, Yablonkevich ML (1991) Materials of vital investigation of passeriform birds caught during spring migration in 1984 and 1990 in Barsakelmes Island of Aral Sea. Recent statement of separating gulfs of Aral Sea. Rudy Zool Inst Acad Sci USSR. Leningrad 237:65–73 [Russian]
Severtzov VA (1873) Vertical and horizontal distribution of Turkestan animals. Proceed Soc Amateur Natur Sci, Anthropol, and Ethnography 8(2). Moscow, p 270 [Russian].
Shtegman BK (1938) Principles of ornithogeographical division of palearctic. Fauna of USSR. The Birds, 1(2). Moscow, Leningrad, p 157 [Russian]
Spangenberg EP (1941) The birds of low flow of Syr-Darya river and surrounding areas. Trans Zool Mus Moscow State Univ., Moscow, 6:77–140 [Russian]

Spangenberg EP, Feygin GA (1936) The birds of low flow of Syr-Darya river and surrounding areas. Trans Zool Mus Moscow State Univ., Moscow, 3 pp 41–184 [Russian]
Stepanyan LS (1969) On the breeding ecology of the birds of Syr-Darya river valley and Northern Kizilkum desert. Biol Sci 2:22–26 [Russian]
Stepanyan LS, Galushin VM (1962) Materials on the avifauna of Barsakelmes reserve. Ornithology 4, Moscow, pp 200–207 [Russian]
Varshavsky SN (1957a) Season moving and migration of birds mainly birds of prey in northern Aral Sea region. Proceedings III Baltic Ornithol. Meet., Vil'nus, pp 69–84 [Russian]
Varshavsky SN (1957b) Problems of methods of visual quantitative study of bird migration and some results of its approach in northern Aral Sea region. Proceedings III Baltic Ornithol. Meet., Vil'nus, pp 10–13 [Russian]
Varshavsky SN (1959a) Changes of species ranges and number of bids in northern Aral Sea basin and Aktyubinsk and Mugodzhar regions during last 50–60 years. Abstr. II All-Union Ornithol. Meet. 3, Moscow, pp 48–50 [Russian]
Varshavsky SN (1959b) An experience of ornithological characteristics of nature regions of northern Aral Sea basin and surrounding areas. Abstr. II All-Union Ornithol. Meet. 3, Moscow, pp 38–40 [Russian]
Varshavsky SN (1965a) Distribution and migration peruliarities of the anatid birds in northern Aral Sea region. Geography of waterbird sources in USSR, source species composition and rules of its reproduction and use, 2. Moscow, pp 6–9 [Russian]
Varshavsky SN (1965b) Landscapes and faunistic complexes of terrestrial vertebrates of northern Aral Sea region in connection with its value in nature centres of plaque infection. Dr. Science thesis, Saratov, p 75 [Russian]
Varshavsky SN (1969) Ecological bird complexes in different landscapes of northern desert of Aral Sea region. Ornithology in USSR, 2, Ashkhabat, pp 99–102 [Russian]
Varshavsky SN (1973) On distribution and ecology of waders in northern Aral Sea region and surrounding areas. Fauna and ecology of the waders. Moscow, 2:20–22 [Russian]
Varshavsky SN, Varshavsky BS, Gambuzov VK (1977) Some rare and endangered birds of northern Aral Sea region. Rare and endangered animals and birds of Kazakhstan. Alma-Ata, pp 146–152 [Russian]
Zarudny NA (1915a) The birds of the Kizilkum desert. Materials to the study of fauna and flora of Russian empire. Sect Zool 14:1–149 [Russian]
Zarudny NA (1915b) Trip to Aral Sea by summer 1914. Proc Turkest Dept Russ Geogr Soc 11(1) Additions to scientific results of Aral expedition 13:1–125 [Russian]
Zarudny NA (1916) The birds of the Aral Sea. Proc Turkest Depart Russ Geogr Soc 12(1):1–229 [Russian]

Reptiles and Amphibians

Alenitzin V (1876) Herpetofauna of islands and coasts of Aral Sea. – Trudy Aral-Caspian expedition, Saint-Petersburg 3, pp 1–64 [Russian]
Berg LS (1908) Aral Sea. An experience of physical and geographical monography. Saint Petersburg, p 580 [Russian]
Bischoff W (1988) Zur verbreitung und Systematik der Zauneidechse, *Lacerta agilis* Linnaeus, 1758. Mertensiella 1:11–30
Bogdanov OP (1960) The Reptiles of Turkmenistan. Ashkhabat, p 234 [Russian]
Bogdanov OP (1962) On the food of *Phrynocephalus helioscopus* in Tadjikistan. Acad Sci Tadshik SSR, Dushanbe, 22:109–117 [Russian]
Bogdanov OP (1965) Ecology of reptiles of Middle Asia. Tashkent, p 257 [Russian]
Brushko ZK (1995) The lizards of the deserts of Kazakhstan. Almaty, p 232 [Russian with english conclusion]

Chikin YuA, Dujsebayeva TN, Joger U, Kadyrbekov RKh (2004) Settlement by reptiles of the dried bottom of Aral Sea. Fauna of Kazakhstan and adjacent countries on the boundary of centuries: morphology, systematics, ecology. Almaty, pp 232–235 [Russian]

Darevsky IS (1969) Suborder lacertilia. In: The life of the animals, vol 4. Moscow, 487p [Russian]

Dinessman LG (1953) Amphibians and reptiles of south-east of Turgay Table Country and northern Aral Sea. Trudy Inst Geogr Acad Nauk USSR, Moskow, 54:384–422. [in Russian]

Dinessman LG, Kaletzkaya ML (1952) Methods of numerical counting of amphibians and reptiles. In: Methods of numerical counting and geographic distribution of terrestrial fauna. Moscow, pp 329–341 [Russian]

Elisseyev DO (2007) Recent statement of vertebrate fauna in Barsakelmes Island and its dynamics during last 50 years. Transactions of Barsakelmes reserve. Almaty, 2, pp 107–128 [Russian]

Elpatjevsky VS (1903) Herpetofauna of Aral Sea. Proceed Turkmen Dept Russ Geogr Soc "scientific results of Aral expedition" 4 [Russian]

Fritz U, Auer M, Chirikova MA, Duysebayeva TN, Eremchenko VK, Kami HGO, Kashkarov DR, Masroor R, Moodley Y, Pindrani A, Siroky P, Hundsdoerfer HK (2009) Mitochondrial diversity of the widespread Central Asian steppe tortoise (*Testudo horsfieldii* GRAY, 1844): implications for taxonomy and relocation of confiscated tortoises. Amphibia Reptilia 30 (2):245–257

Guicking D, Lawson R, Joger U, Wink M (2006) Evolution and phylogeny of the genus *Natrix* (serpentes: colubridae). Biol J Linn Soc 87:127–143

Guicking D, Joger U, Wink M (2009) Cryptic diversity in a Eurasian water snake (*Natrix tessellata*, serpentes: colubridae): evidence from mitochondrial sequence data and nuclear ISSR-PCR fingerprinting. Org Divers Evol 9:201–214

Hewitt GM (2000) The genetic legacy of the quaternary ice ages. Nature 405:907–913

Hoeffler F, Novitzkiy Z (2003) Green shield of the dried bottom of Aral Sea. Tashkent, p 76 [Russian]

Iskakova KI (1959) Amphibians of Kazakhstan. Almaty, p 92 [Russian]

Ismagilov MI, Kuznetsov LA, Rashek VL (1990) Nature sanctuary Barsa-Kelmes. In: Sokolov VE (ed) Zapovedniki of the USSR, Vol 6. Moscow, pp 42-56. [in Russian]

Kalyabina SA, Milto KD, Ananjeva NB, Legal L, Joger U, Wink M (2001) Phylogeography and systematics of *Lacerta agilis* based on mitochondrial cytochrome b gene serquences: first results. Russ J Herpetol 8:149–158

Kamalova ZYa (1972) Distribution and ecology of agamid lizards of the south of Middle Asia. Abstr PhD thesis, Tashkent, p 23 [Russian]

Karpenko VP (1958) Distribution and ecology of the Central Asian viper *Ancistrodon halys* (Pallas, 1776). Abst PhD thesis, Tashkent, p 19 [Russian]

Lenk P, Joger U, Wink M (2001) Phylogenetic relationships among European ratsnakes of the genus *Elaphe* Fitzinger based on mitochondrial DNA sequence comparisons. Amphibia-Reptilia 22:329–339

Mayev EG, Mayeva SA (1977) Paleogeographic analysis of level variation of Caspian and Aral Seas. In: Paleogeography and pleistocene deposits of the south seas of USSR. Nauka, Moscow, pp 69–74 [Russian]

Nikolsky AM (1915) Fauna of Russia and surrounding countries. Reptilia, vol 1. Chelonia and sauria, Petrograd, p 532 [Russian]

Nikolsky AM (1916) Fauna of Russia and surrounding countries. Reptilia, vol 2. Ophidia, Petrograd, p 532 [Russian]

Nikolsky AM (1918) Fauna of Russia and surrounding countries. Amphibia. Petrograd, p 310 [Russian]

Nilson G, Andrén C (2001) The meadow and steppe vipers of Europe and Asia – the *Vipera* (*Acridophaga*) *ursinii* complex. Acta Zool Acad Scient Hung 47:87–267

Odum EP (1986) Basic ecology, vol II. Mir, Moscow, p 376 [Russian]

Paporotny DI (1950) On biology of the Halys Viper in conditions of Barsa-Kelmes Island. Transactions of state reserve "Barsa-Kelmes" I. Alma-Ata, pp 136–148 [Russian]

Paraskiv KP (1956) The reptiles of Kazakhstan. Academic Science KazSSR Publishing House, Alma-Ata, p 228 [Russian]
Perälä J (2002) Biodiversity in relatively neglected taxa of *Testudo* L., 1758 s.l. Chelonii 3:40–53
Piryulin DD (1993) On the formation of enthomocomplexes of continental part of Aral Sea dried bottom. Entomologo-coenological issue. Tr Zool Inst Rus Acad Sci 250:139–153 [Russian]
Rashek VL, Rashek V A (1963) State reserve "Barsakelmes Island". Alma-Ata, p 90 [Russian]
Rastegar-Pouyani E, Kazemi Noureini S, Joger U, Wink M (2008) Molecular phylogeny and intraspecific differentiation of the *Eremias velox* complex of the Iranian Plateau and Central Asia (Reptilia Lacertidae). Systematics 2008, Göttingen, Abstracts, p 120
Satekeev GK, Chirikova MA (2007) New data on herpetofauna of Barsakelmes Island. Transactions of the Barsakelmes state nature reserve. Almaty, 2, pp 135–138 [Russian]
Sidorov SA (1925) Amphibians and reptiles of Aral Sea. Bull Moscow Nat Soc New, issue 33:188–200 [Russian]
Shammakov S (1981) Reptiles of plain areas of Turkmenistan. Ashkhabat, p 311 [Russian]
Terentyev PV, Chernov SA (1949) Guide of amphibians and reptiles. Sovetskaya Nauka, Moscow, p 340 [Russian]
Wucherer W, Breckle S-W (2000) Vegetation dynamics on the dry sea floor of the Aral Sea. In: Breckle S-W, Veste M, Wucherer W (eds) Sustainable landuse in deserts. Springer, Berlin, pp 52–68
Wucherer W, Breckle S-W, Dimeyeva L (2000) Flora of the dry sea floor of the Aral Sea. In: Breckle S-W, Veste M, Wucherer W (eds) Sustainable landuse in deserts. Springer, Berlin, pp 38–51
Zarudny NA (1915) II Herpetofauna of Aral (the amphibians and reptiles of the coasts and islands of Aral Sea, mainly of its eastern region). Trip to Aral Sea by summer 1914. Proc Turkest Dept Russ Geogr Soc 11(1):113–125 [Russian]

Invertebrates

Aibassov KA (1975) The butterflies (Insecta, Lepidoptera) of the Western Kazakhstan. Insects of Western Kazakhstan. Alma-Ata, pp 102–150 [Russian]
Aladin NV, Potts WTW (1992) Changes in the Aral Sea ecosystems during the period 1960–1990. Hydrobiologia 237:67–79. [in Russian]
Andreeva SI (1989) Macrozoobenthos of Aral Sea at the primary stage of its salinization. Proc Zool Inst, Acad Sci USSR, Leningrad 199:53–82. [in Russian]
Arashkevich EG, Sapozhnikov PV, Soloviov KA, Kudyshkin TV, Zavialov PO (2009) *Artemia parthenogenetica* (Branchiopoda: Anostraca) from the large Aral Sea: abundance, distribution, population structure and cyst production. J Mar Syst 76:359–366
Assanova RB (1996) The bugs (Heteroptera, Pentatomorpha) of Kazakhstan. Almaty, p 517 [in Russian]
Belishev BF, Shevtschenko VV (1971) Fauna of the dragonflies (Odonata, Insecta) and distribution of its components in the Kazakhstan. Biologicheskie Nauki. Kazakhsky Gosudarstvenny Universitet, Alma-Ata, 2 pp 73–77 [Russian]
Chetverikov SS (1906) List of Lepidoptera from the coast of Aral Sea. Izvestiya Turkestanskogo Otdela Russkogo Geographicheskogo Obszczestva 7:30–36 [Russian]
Dzhanokmen KA (1993) Pteromalidae (Hymenoptera) of the desert zone of Kazakhstan. Insects of the desert zone of Kazakhstan. Almaty, 3, pp 44–72 [Russian]
Essenbekova PA, Kadyrbekov RKh (2004) New data about fauna and ecology of the bugs (Insecta, Heteroptera) of Kazakhstan part of Aral Region. Vestnik Kazakhskogo National'nogo Universiteta. Ser Biol 1(22):100–102 [Russian]
Fasulati KK (1971) Field study of terrestrial invertebrates. Moscow, p 424 [Russian]

Fedotova ZA (1993) Cecidomyidae (diptera) of the desert zone of Kazakhstan. Insects of the desert zone of Kazakhstan, Almaty, 5, 131p [Russian]

Fedotova ZA (2000) Cecidomyidae – phytophagous (diptera, cecidomyidae) of the deserts and mountains of Kazakhstan: morphology, biology, distribution, phylogeny, systematics. Samara, p 803 [Russian]

Filippov AA (1995) Macrozoobenthos of the littoral zone of the Northern part of Aral Sea in modern polyhaline conditions: abundance, biomass, spatial destribution). Proc Zool Inst, Russian Acad Sci, St Peterburg 262:103–166. [in Russian]

Filippov AA (2001) Macrozoobentos of the Small Aral Sea. Biologia morya. Vladivostok 27(6): 453–455. [in Russian]

Filippov AA (2002) Productivity of the Aral Sea benthic communities. Oceanology 42(1):91–98

Gucevich AV, Monchadsky AC, Shtakelberg AA (1970) The family Culicidae. Diptera. Fauna of USSR Ser Nov, 3(4):384 [Russian]

Guryeva EL (1979) Coleoptera (elateridae). Fauna of the USSR. Leningrad 12(3):456 [Russian]

Guryeva EL (1989) Coleoptera (elateridae). Fauna of the USSR. Leningrad 12(4):296 [Russian]

Ivanova ZhA (1975) Special composition and biotopic distribution of Psyllid insects (Homoptera, Psyllidae) of Western Kazakhstan. Insects of Western Kazakhstan. Alma-Ata, pp 77–94 [Russian]

Jashenko RV (1993) Coccinea (homoptera) of the desert zone of Kazakhstan. Insects of the desert zone of Kazakhstan. Almaty, 2, pp 4–35 [Russian]

Kabanova NB (1995) Data on fauna and ecology of the beetles (Coleoptera) of Barsakelmes Island (Aral Sea). Biological and natural problems of Aral Sea and Aral Region. 2. Proceed Zool Inst Acad Sci USSR 262:145–166 [Russian]

Kadyrbekov RKh (1993) Fauna, geography and ecological particularities of Aphids (Homoptera, Aphidinea) in the desert zone of Kazakhstan. Insects of the desert zone of Kazakhstan. Almaty, 2:36–82 [Russian]

Kadyrbekov RKh (2003) Review of the aphids fauna (Homoptera, Aphididae) in the Kazakhstan part of Aral Region. Selevinia 2003:39–46 [Russian]

Kadyrbekov RKh (2004) On a problem of development of the zone of drainage in the Northern and northern-eastern coast of Aral Sea. Izvestyia Nationalnoy Akademii Nauk Respubliki Kazakhstan. Ser Biol Med 4:39–43 [Russian]

Kadyrbekov RKh, Tleppaeva AM (2004) Faunistic review of the xylophagous beetles (Coleoptera: Buprestidae, Cerambycidae) in the Kazakhstan part of Aral Region. Izvestyia Nationalnoy Akademii Nauk Respubliki Kazakhstan. Ser Biol Med 5:37–43 [Russian]

Kadyrbekov RKh, Ishkov EV, Tlepayeva AM (1997) Entomolodical research and preservation of insects in Northern Aral region. Problems of natural environment of Aral Region and adjacent territories. Almaty: 24 [Russian]

Kazenas VL (1993) The digger wasps (Hymenoptera, Sphecidae) of the desert zone of Kazakhstan. Insects of the desert zone of Kazakhstan. Almaty, 3:73–142 [Russian]

Kazenas VL (2002) The digger wasps (Hymenoptera, Sphecidae) of Kazakhstan. Tethys Entomol Res 4:173 [Russian]

Kazenas VL, Mityaev ID, Jashenko RV (1997) Ecological and zoogeographical particularities and contemporaneous conditions of the entomofauna of Aral Region. Problems of natural environment of Aral Region and Adjacent Territories. Almaty: 22–23 [Russian].

Kazenas VL, Mityaev ID, Jashenko RV, Kadyrbekov RKh, Ishkov EV, Tleppaeva AM (1998) Composition, ecology, zoogeography of insects in the North Aral Sea area. Sustainable use of natural resources of Central Asia. Almaty, pp 119–122 [Russian]

Kerzhner IM (1982) The bugs of the family Nabidae. Fauna of USSR. Ser Nov 13(2). Leningrad, p 326 [Russian]

Konev AA (1993) On the fauna of the beetles (Coleoptera) of Barasakelmes reserve. Ecological crisis in the Aral Sea. Proc Zool Inst Acad Sci USSR 250:174–187 [Russian]

Kostin IA (1960) Materials on the Ipidae fauna (coleoptera) in Kazakhstan. Proc Inst Zool Acad Sci KazSSR 11:120–136 [Russian]

Kostin IA (1973) Dendrophagous beetles of Kazakhstan. Alma-Ata, p 288 [Russian]

Kozhanchikov IV (1937) Noctuid butterflies (subfamily Agrotinae). Lepidoptera. Fauna of USSR. Ser Nov, 13(3). Moscow-Leningrad p 674 [Russian]
Kozhanchikov IV (1950) Orgyid Butterflies (Orgyidae). Lepidoptera. Fauna of USSR. Ser Nov, 12. Moscow-Leningrad, p 581 [Russian]
Krivokhatskii VA, Piryulin DD (1997) Composition, origin and current changes of Myrmeleontidae (Neuroptera) fauna in the Aral Sea Region. Entomol Rev 77:9
Krizhanovsky OL (1965) Composition and origin of the ground fauna of Middle Asia. Moscow-Leningrad, p 419 [Russian]
Krizhanovsky OL (1976) Histeridae (Coleoptera). Fauna of USSR. 5(4). Leningrad, p 436 [Russian]
Krizhanovsky OL (1983) Beetles of suborder Adephaga: Rhysodidae, Trachypachidae, Carabidae. Fauna of USSR. Leningrad, 341p [Russian]
Kulenova KZ, Lopatin IK (1986) Chrysomelid beetles (Coleoptera, Chrysomelidae) of Kazakhstan. Alma-Ata, p 200 [Russian]
Kussainova NZ (1954) Zoobenthos of the Aral Sea. Trudy laboratorii ozerovedeniya Akad. Nauk SSSR. 3. Moscow-Leningrad, pp 76–85 [Russian]
Kussainova NZ, Embergenov S (1971) Special composition of the zoobenthos in the reservoirs of Amudarya Bassin. Biologicheskie Nauki. Alma-Ata, 2:85–98 [Russian]
Kuzin BS (1953) Meloid Beetles (Coleoptera, Meloidae) of Kazakhstan. Trudy Rep Stantzii Zaszcziti Rasteny 1:72–152 [Russian]
Leley AS (1985) Mutillid Wasps (Hymenoptera, Mutillidae) of the USSR and adjacent countries. Leningrad, p 268 [Russian]
Lenk P, Joger U, Fritz U, Heidrich P, Wink M (1999): Phylogeographic patterns in the mitochondrial cytochrome b gene of the European pond turtle, Emys orbicularis (Linnaeus). Molecular Ecology 8:1911–1912
Longvinovsky VD (1985) Anobiid beetles (Anobiidae). Coleoptera. Fauna of USSR. Moscow-Leningrad, p 174 [Russian]
Lukyanovich FK, Ter-Minassyan ME (1957) Bruchid beetles (Bruchidae). Coleoptera. Fauna of USSR, Ser Nov 24(1). Moscow-Leningrad, p 209 [Russian]
Makhnovsky IK (1958) Methods of vermin control of Shrubs and Arboreous vegetation. Tashkent, 97p [Russian]
Marikovskaya TP (1993) The bees (Hymenoptera, Apoidea) of the desert zone of Kazakhstan. Insects of the desert zone of Kazakhstan. Almaty, 2, pp 82–144 [Russian]
Marikovsky PI (1974) Little hard-workers of the desert. Alma-Ata, p 159 [Russian]
Marikovsky PI (1979) The ants (Hymenoptera, Formicidae) of the deserts of Semirechye. Alma-Ata, p 264 [Russian]
Medvedev GS, Nepessova MG (1985) Key to the tenebrionid beetles (Coleoptera, Tenebrionidae) of Turkmenistan. Ashabad, p 180 [Russian]
Mirabdullayev IM, Joldasova IM, Mustafaeva ZA, Kazakhbaev S, Lyubimova SA, Tashmukhamedov BA (2004) Succession of the ecosystem of the Aral Sea during its transition from oligohaline to polyhaline water body. Journal of Marine Systems 47:101–107
Mitiayev ID (1993) The leafhoppers (Homoptera, Cicadinea) of the desert zone of Kazakhstan. Insects of the desert zone of Kazakhstan. Almaty, 1, pp 11–60 [Russian]
Mitiayev ID (2002) Fauna, ecology and zoogeography of leafhoppers (Homoptera, Cicadinea) of Kazakhstan. Tethys Entomol Res 5:168 [Russian]
Nassirova SR (1993) Orthoptera (Insecta) of the desert zone of Kazakhstan. Insects of the desert zone of Kazakhstan. Almaty, 3, pp 60–95 [Russian]
Nikolayev GV (1987) Scarabaeid beetles (Coleoptera, Scarabaeidae) of Kazakhstan and Middle Asia. Alma-Ata, p 232 [Russian]
Nikolayev G V, Kozminykh VO (2002) Carrion beetles (Coleoptera: Agyrtidae, Silphidae) of Kazakhstan, Russia and adjacent countries. Almaty, p 159 [Russian]
Nikolayev GV, Kolov SV (2005) Meloid beetles (Coleoptera, Meloidae) of Kazakhstan: biology, systematics, key. Almaty, p 165 [Russian]
Nikolskaya MN (1952) Chalcid insects of the fauna of USSR (Chalcidoidea). Keys of the fauna of USSR. Moscow-Leningrad, p 575 [Russian]

Ogloblin DA, Znoiko DV (1950) The family Alleculidae. Part 2. The subfamily Omophlinae. Coleoptera. Fauna of USSR, Ser Nov. Moscow-Leningrad 18(8):134 [Russian]
Olsufyev NG (1977) Horseflies (Tabanidae). Diptera. fauna of USSR, Ser Nov, 7(2). Leningrad, p 434 [Russian]
Paramonov SJ (1940) The family Bombyliidae (subfamily Bombyliinae). Diptera. Fauna of USSR, Ser Nov, 9(2). Moscow-Leningrad, p 414 p [Russian]
Pessenko JA (1983) Halictid bees (Halictidae) of Halictinae subfamily. Nomioidini tribe. Hymenoptera. Fauna of USSR, Ser Nov, 17(1). Leningrad, p 199 [Russian]
Pirjulin DD (1993a) Arthropods of the zone of drainage on the Barsakelmes Island. Ecological crisis in the Aral Sea. Proc Zool Inst Rus Acad Sci 250:121–138 [Russian]
Pirjulin DD (1993b) On formation of the entomological complexes in continental drainage of Aral Sea. Ecological crisis in the Aral Sea. Proc Zool Inst Rus Acad Sci 250:139–153 [Russian]
Pirjulin DD (1995) On characteristic of the temporary reservoirs and hydrofauna of insects on Barasakelmes Island and Aral Region. Biological and natural problems of Aral Sea and Aral Region 2. Proc Zool Inst Rus Acad Sci 262:3–74 [Russian]
Pirjulin DD, Ozersky PV (1995) Orthopteroidea of the Barsakelmes Island and Aral Region. Biolodical and natural problems of Aral Sea and Aral Region 2. Proc Zool Inst Rus Acad Sci 262:75–105 [Russian]
Red Data Book of Kazakhstan (2006). Rare and vanishing species of invertebrates. Almaty, p 231 [Russian]
Rodendorf BB (1937) Sarcophagidae (Part 1). Diptera. Fauna of USSR, Ser Nov, 19(1). Leningrad, p 501 [Russian]
Ryssakova NE, Sarzhanov FN (2010) First find of the European Pond Turtle, Emys orbicularis (Linnaeus, 1758) in the foothills of Karatau range. P 255 In: Dujsebayeva TN (ed) Herpetological researches in Kazakhstan and adjacent countries. Almaty, 260 pp. [Russian]
Scheck GH (1975) Noctuid butterflies (Lepidoptera, Noctuidae) infesting the plants on the fields. Alma-Ata, p 184 [Russian]
Schteinberg DM (1962) Scoliid Wasps (Scoliidae). Hymenoptera. Fauna of USSR, Ser nov, 13. Moscow-Leningrad, p 186 [Russian]
Sinadsky JV (1963) Methods of vermin control in the gallery forests in Middle Asia. Moscow-Leningrad, p 152 [Russian]
Sinadsky JV (1968) Control of dendrophagous insects of the deserts in Middle Asia. Moscow-Leningrad, p 127 [Russian]
Ter-Minassyan ME (1967) Cleoninae (Coleoptera, Curculionidae) of the USSR fauna (Lixini). Leningrad, p 114 [Russian]
Ter-Minassyan ME (1988) Cleoninae (Coleoptera, Curculionidae) of the USSR fauna (Cleonini). Leningrad, p 234 [Russian]
Zagulyayev AK (1960) Moths (Tineidae). Part 2. The subfamily Tineinae. Lepidoptera. Fauna of USSR, Ser Nov, 4(3). Moscow-Leningrad, p 267 [Russian]
Zagulyayev AK (1964) Moths (Tineidae). Part. 3. The subfamily Nenuitopogonotinae. Lepidoptera. Fauna of USSR, Ser Nov, 4(2). Moscow-Leningrad, p 424 [Russian]
Zagulyayev AK (1973) Moths (Tineidae). Part 4. The subfamily Scardiinae. Lepidoptera. Fauna of USSR, Ser Nov, 4(4). Leningrad, p 128 [Russian]
Zagulyayev AK (1975) Moths (Tineidae). Part 5. The subfamily Myrmecozelinae. Lepidoptera. Fauna of USSR, Ser Nov, 4(5). Leningrad, p 428 [Russian]
Zagulyayev AK (1979) Moths (Tineidae). Part 6. The subfamily Meessiinae. Lepidoptera. Fauna of USSR, Ser Nov, 4(6). Leningrad, p 408 [Russian]
Zhdanko AB (1993) Rhopalocera (Lepidoptera) of the desert zone of Kazakhstan. Insects of the desert zone of Kazakhstan. Almaty, 3:1–43 [Russian]
Zhdanko AB (2005) Butterflies (Lepidoptera, Papilionoidea, Hesperiidea) of Kazakhstan. Tethys Entomol Res 11:85–152 [Russian]
Zimin LS (1951) Flies. Family Muscidae, Muscini and Stomoxydini tribes. Diptera. Fauna of USSR, Ser Nov, 18(4). Moscow-Leningrad, p 285 [Russian]

Chapter 12
Halophytes and Salt Desertification in the Aralkum Area

S-W. Breckle and W. Wucherer

12.1 Introduction

The area of saline soils on the desiccated seafloor of the Aral Sea comprises about 42,000 km^2 (which is about three quarters of the dry seafloor, Chap. 2, Fig. 2.7). Within the agricultural areas with irrigated lands, a major proportion is secondarily saline; this amounts to about 22,000 km^2. In total, this means that the salt desert areas in Middle Asia have increased by more than 60,000 km^2 within the last 50 years. Salt desertification is spreading within the whole area, not only in the Aralkum. But it is a very old problem of mankind (Jacobsen and Adams 1958). All arid countries face the salinity problem (Waisel 1972, 2001; Hammer 1986; Oldeman 1994; Breckle 1982, 1989, 2002a, b; Wichelns 1999), e.g. in Australia (Dregne 1986), California (Sheridan 1981; Law and Hornsby 1982; Rhoades 1990), India (Singh 2009), China (Yang et al. 2005) and Iran (Shiati 1991). However, the Aral Sea basin is one of the most striking examples of salt desertification (Geldyeva et al. 1998; Novikova et al. 1998). The forecast that the eastern basin of the Aral Sea will have disappeared by 2010 (Breckle and Agachanjanz 1994; Agachanjanz and Breckle 1994) and huge solonchak areas will spread out was totally right, as can be seen now. A huge salt swamp has been observed already in 2009 (chap. 2).

The coast of the Aral Sea and the dry seafloor of the former Aral Sea are an excellent model where the processes of salt desertification can be seen (Glazovskii and Orlovskii 1996; Breckle and Wucherer 2007). In general, soil salinity assessments are essential for mapping land degradation in drylands as well as for agricultural surveys, and remote sensing is a helpful tool (Metternicht and Zinck 2008).

The strategies of plants for regulating salt content and for coping with salt stress are a precondition for survival, whether they are halophytes or nonhalophytes. The adaptation of plants to NaCl has to cope with the general osmotic effects of the ions, but also with the specific ionic effects of Na^+ and Cl^- on the metabolic processes (Fig. 12.1). Halophytes have evolved during long-term evolution by selection of

S.-W. Breckle et al. (eds.), *Aralkum - a Man-Made Desert: The Desiccated Floor of the Aral Sea (Central Asia)*, Ecological Studies 218,
DOI 10.1007/978-3-642-21117-1_12, © Springer-Verlag Berlin Heidelberg 2012

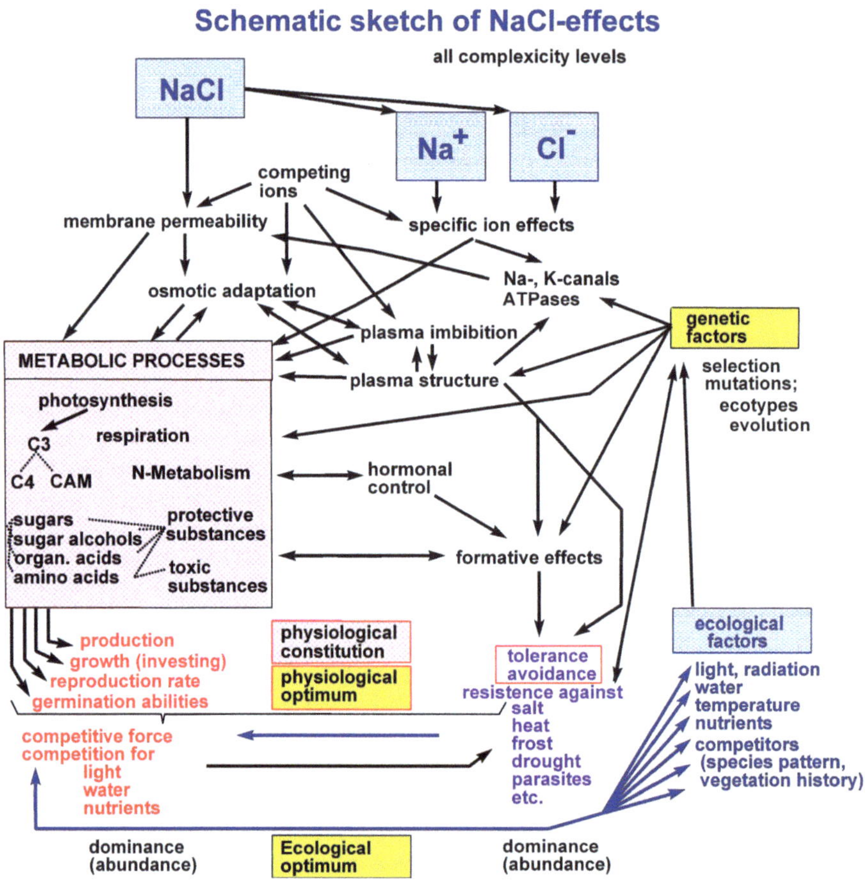

Fig. 12.1 The interrelations of NaCl effects on various complexity levels in plants (modified from Breckle 2005)

tolerant ecotypes in several plant taxa. In Central Asia, there is a biodiversity centre of halophytes (Wucherer et al. 2001). In arid sites with a continental climate, various types of salinity are known (chloride, sulphate, carbonate, magnesium and boron), more variable than along ocean coasts, depending on soil properties, climatic conditions and ecosystem processes. The presence of excessive ions in such ecosystems dominates over many other environmental factors. Only the supply of water is the other decisive factor in ecosystem development.

The invasion of the desiccated seafloor by halophytic species occurs under climatic conditions (chap. 4) which are rather variable from year to year (Breckle et al. 2001). The halophytic species, nevertheless, are on the other hand indicators of the degree of salinity at their site, and thus can be used to monitor salinity. A novel list of indicator values for salinity is presented (see below, Table 12.9). This can be used also for the necessary means of phytomelioration (Chaps. 15–17).

12.1.1 Halophyte Groups

Middle Asia and Central Asia are the evolutionary centres of many genera and species of the Chenopodiaceae. The Chenopodiaceae are characterized by their ability to accumulate inorganic ions, mainly sodium (Na^+). Only a few other angiosperm families are similarly able to withstand high soil salinities, e.g. Zygophyllaceae, Frankeniaceae, Tamaricaceae, Plumbaginaceae and a few grasses. However, there are many more genera in various angiosperm families which have evolved some degree of salt tolerance. In drylands, salinity has been such an important ecological factor that mechanisms of salt tolerance have evolved several times.

Plants have developed various mechanisms to cope with salinity. Table 12.1 gives an overview of some of the strategies which can be found in halophytes and which are sometimes even combined. Often morphological structures are typical for distinct adaptation strategies. Especially halosucculence of stems or leaves, or both, is very common in halophytes strongly adapted to salinity. Thus, succulent halophytes are either leafless and stem-succulent or have fleshy and succulent leaves. In both cases this kind of succulence has two components: the basic one is a genetically controlled succulence, whereas the second is a modifying variable

Table 12.1 Control mechanisms of halophytes to strive on saline sites (Breckle 1990, 2002a) and the main morphological strategy type

	Halophyte type
Avoidance	
Growth only during favourable seasons (time niche)	NoH, Ps, Su
Growth only on favourable sites (site niche)	Ps, NoH
Limitation of root growth and absorption activity to distinct soil horizons (site niche)	Ps, NoH
Evasion and adaptation processes	
Selectivity against Na^+ and Cl^-	NoH, Ps
Leaching of salt from shoots	NoH, Ps
Diversion of salt out of assimilating tissues	Ps
Compartmentation of salt within plant, within tissues, within cells	All plants
Accumulation of salt in xylem parenchyma in roots and shoots	All halophytes
Synthesis of organic solutes	All plants, Su
Retranslocation of salt to roots and recretion by roots	Halophytes
Disposal of older plant parts ("salt-filled organs")	Ps, all halophytes
Recretion by gland-like structures on shoots	
By salt glands	EX
By salt bladders	NX
Tolerance	
Increasing salt tolerance of tissues, cells, organelles	LSu, SSu, NX, EX, Ps
Increase in halosucculence	
increasing leaf-succulence	LSu, (Ps)
increasing stem succulence, reduction of leaves	SSu

EX exocrinohalophytes, *LSu* leaf-succulent euhalophytes, *NoH* nonhalophytes, *NX* endocrinohalophytes, *Ps* pseudohalophytes, *SSu* stem-succulent euhalophytes, *Su* xerosucculents

and can be induced by salts to a considerable degree. These types of halosucculence have to be distinguished from xerosucculence.

There are leaf-succulent euhalophytes which are annuals, e.g. some *Suaeda* species, *Halopeplis*, *Halimocnemis*, *Gamanthus*, *Girgensohnia*, etc. Other leaf succulents are herbal perennials (e.g. *Plantago*, *Aster*, *Suaeda*), and others are shrubs (e.g., some members of the genera *Salsola*, *Suaeda*, *Nitraria* and *Kochia)*. In some others, the succulence of the fruit or parts of the fruit became very pronounced (*Gamanthus)*. Regarding the adaptations of the photosynthetic pathway which have evolved, it is obvious that succulence has altered the anatomical structure dramatically, as can be seen in the various types that are exhibited by *Salsola* and *Suaeda* (Shomer-Ilan et al. 1981).

The stem-succulent euhalophyte lack leaves or have only minor scalelike leaves. The young stems are succulent, the older ones in perennial species can become rather woody. *Salicornia* and some species of *Anabasis*, for example, are annual stem-succulent species. Perennial stem-succulent halophytes are also found in *Anabasis*, *Kalidium*, *Aellenia*, *Ofaiston*, *Halostachys*, *Haloxylon*, etc. and also in the woody subshrub *Halocnemum strobilaceum* (Fig. 12.2), which is one of the most salt-tolerant species.

In contrast to halosucculents, most xerosucculents in general are very sensitive to salinity.

Many halophytes exhibit a rather rapid turnover of their leaves. The rosette leaves in *Limonium vulgare* are replaced during the vegetation period two or three times, and the leaves of *Aster tripolium* rather soon become yellow and new leaves replace them. This replacement is a mechanism of removal of large quantities of

Fig. 12.2 *Halocnemum strobilaceum*, young shoots (photo: Breckle, May 2004)

salt. Old leaves with high salt content are steadily replaced by younger leaves in many *Juncus* species. This is certainly one adaptation mechanism that enables the plant to get rid of excessive salts by shedding plant organs. A less specific adaptation is the rapid production of new leaves and dropping old leaves rich in salt. This can be observed in many pseudohalophytes. But the loss of leaves affects the supply of assimilates or hormones to the growing organs and thereby affects growth (Munns 1993; Munns et al. 1995).

But even more important in some halophytes is the existence of specific cell structures which can recrete (recretion in the sense of Frey-Wissling 1935, meaning elimination of substances not metabolically changed) inorganic ions, especially Na^+. This is done by salt glands, which have evolved several times in the angiosperms, and by bladder hairs. Salt glands eliminate salt to the outside (e.g., *Tamarix*, *Frankenia* – see Fig. 12.3 – *Glaux* and *Limonium* as well as some grasses); Bladder hairs accumulate salts in their huge vacuole (*Atriplex*, see Fig. 12.4; to a

Fig. 12.3 *Frankenia hirsuta*, in flower with many dry recreted salt crystals (photo: Breckle, May 2004)

Fig. 12.4 *Atriplex pratovii*. (**a**) Intact bladders from the lower side of leaves. (**b**) Crushed bladders from the upper side of leaves after wilting, forming a layer of salt crystals. North Aral Sea (photo: Breckle, **a** – May 2003; **b** – May 2004)

less extent *Halimione, Salsola, Chenopodium*) (Black 1954; Berger-Landefeldt 1959; Schirmer and Breckle 1982; Breckle 1992). In both cases the salts are physiologically isolated from active tissues. Here also the turnover of salt is rather rapid by recreting salt with salt glands in the exocrinohalophytes or into big bladders in the endocrinohalophytes.

Nonhalophytes exhibit almost none of these morphological adaptations. The dominant processes in the various morphological halophyte types are indicated in Table 12.1.

In general, it should be kept in mind that salt tolerance of a plant is not defined by the act of individual genes, by the individual regulation of each of them or by one specific metabolic process. Salt tolerance is a whole plant response (Hedenström and Breckle 1974; Breckle 1990, 1995; Munns 1993; Naik and Widholm 1993; Flowers and Yeo 1995; Ramani and Apte 1997), where many processes, such as efficient potassium pumping and accumulation, synthesis and transport of compatible solutes, plant signalling systems involved in tissue and in developmental regulation (Winicov and Bastola 1997), etc. are only some of many other important adaptations which are equilibrated in a harmonic way to fulfil those adaptive processes mentioned in Table 12.1.

It has to be stressed that salt tolerance has at least two quite differing aspects. One is the upper limit of salt that can be tolerated by an individual plant, which is necessary for survival. The other is the existence of a plant species that exerts successful reproduction, which is necessary for ecological success.

Salt tolerance of plants varies very much. It varies during different growth or development phases (Tobe et al. 2004, 2005), with ionic constitution of the soil solution (e.g. the presence of Ca and K as antagonists of Na), with microclimatic conditions (e.g. relative humidity), with life form and halophyte strategic type, with the plant organ affected by salinity and with the genetic variability of each species forming ecotypic varieties. Also, the effects of salinity on different growth stages and growth processes of plants have to be taken into account (Ungar 1996). Germination and seedling growth is normally more sensitive than growth of established adult plants.

For halophytes osmotic adaptation is accomplished not only by synthesis of organic compounds but also by absorbing inorganic ions, accumulated in the vacuole, counterbalanced by compatible solutes in the cytoplasm. As a rule, the osmotic potential of leaf cell sap normally differs by 0.5–1 MPa from that of the soil solution, enabling uptake of water.

12.1.2 Ion Pattern of Halophytes

For a long time, halophytes had been classified into chloride halophytes, sulphate halophytes and alkali halophytes, according to the main ions in cell sap or ash (Walter 1968). The alkali halophytes are those where a high proportion of organic acids (e.g., oxalate in *Halogeton* with up to 30% dry matter) are accumulated. It has long been known that halophytes are able to take up nutrients from the soil despite an excessive content of Na^+ and Cl^-. Most halophytes discriminate between Na^+ and K^+ and only few species are really sodiophilic (Moore et al. 1972). To demonstrate the characteristics in K^+/Na^+ discrimination, it is necessary to have the relevant soil samples from the rhizosphere of the respective plants. Then the accumulation factor for sodium in comparison with potassium can be calculated. It is easily seen that most species under a wide range of given cation ratios in the soil favour potassium uptake. The widespread Chenopodiaceae *Salicornia europaea* and *Suaeda maritima* can be termed sodiophilic, and so can *Climacoptera aralensis* and *Suaeda acuminata* (Tables 12.2, 12.3 and 12.4), whereas *Petrosimonia triandra* exhibits a rather balanced Na^+/K^+ ratio. In contrast, the grasses *Puccinellia distans* and *Stipagrostis pennata* and *Eremosparton aphyllum* very selectively accumulate potassium by a factor of 10–100 according to the soil Na^+/K^+ ratio; even in saline soils their Na^+/K^+ ratio is between 0.10 and 0.40 (Table 12.4). Slightly more sodium is accumulated in some Brassicaceae, e.g. in *Malcolmia africana*. All other Chenopodiaceae are more or less halophytic and exhibit rather high Na^+/K^+ ratios (Table 12.4), which is not really very different from the results from hot-water extracts and from acidic extracts (Tables 12.3 and 12.4). However, in the pseudohalophytes or nonhalophytes, the amount of nonvacuolar alkali ions (which are extracted additionally with the acidic extract) is considerably higher (Table 12.4). This is due to the calcium content, where by an acidic extraction up to 60 times higher amounts are analysed.

It is obvious that leaf succulents and stem succulents, such as species from the genera *Suaeda*, *Salicornia* and *Halocnemum*, accumulate considerably more Na^+ and Cl^- (3,000–5,000 mmol/kg) in comparison with other species. The ionic contents (Na^+ and Cl^-) of *Climacoptera* species and of *Ofaiston monandrum* are lower (2,000–3,500 mmol/kg) in comparison with those of species from *Salicornia* and *Suaeda*. Even lower are the values from *Petrosimonia triandra*. On the other

Table 12.2 Ion pattern of some common halophytic species of the Aralkum, analysed from hot-water extracts (*upper figure*) and from acidic extracts (*lower figure in italics*). Comparison of samples from Bayan (*Ba*) and from Karabulak (*Ka*); *n* number of samples, ion content (mmol kg^{-1} dry matter and standard deviation)

Species	*n*	Locality	Cl^-	Na^+	K^+	Ca^{2+}	Mg^{2+}	Na^+/K^+
Climacoptera aralensis	3	Ba	2,913 ± 684	4,143 ± 512	420 ± 86	2.74 ± 0.30	98 ± 38	9.9
			–	4,940 ± 480	454 ± 78	228 ± 123	259 ± 10	10.9
	5	Ka	2,700 ± 489	2,882 ± 1,126	674 ± 182	4.03 ± 1.44	168 ± 26	4.3
			–	3,850 ± 1,452	793 ± 219	246 ± 106	287 ± 36	4.9
Petrosimonia triandra	3	Ba	600 ± 109	1,016 ± 289	570 ± 54	13,0 ± 8,3	306 ± 77	1.8
			–	1,525 ± 212	685 ± 68	404 ± 54	466 ± 45	2.2
	1	Ka	603	627	521	25	233	1.2
			–	668	603	329	383	1.1
Suaeda acuminata	1	Ba	4,731	4,722	416	88	150	11.3
			–	6,500	446	232	410	14.6
	9	Ka	4,370 ± 850	4,107 ± 598	729 ± 102	6.05 ± 2.8	246 ± 66	5.6
			–	4,741 ± 1,054	842 ± 112	264 ± 47	444 ± 73	5.6

Table 12.3 Ion pattern of some common halophytic species of the Aralkum, analysed from hot-water extracts. From Bayan (*Ba*) and from Karabulak (*Ka*). *n* number of samples, ion content (mmol kg^{-1} and standard deviation). For each species the halophyte type is indicated (second column; for an explanation of the abbreviations, see Table 12.1

Species	Halophyte type	*n*	Locality	Cl^-	Na^+	K^+	Ca^{2+}	Mg^{2+}	Na^+/K^+
Climacoptera aralensis	LSu	8	3x Ba, 5x Ka	2,780 ± 531	3,356 ± 1,107	573 ± 199	3.55 ± 1.29	142 ± 46	5.9
Petrosimonia triandra	LSu	4	3x Ba, 1x Ka	600 ± 109	1,016 ± 289	570 ± 54	13.0 ± 8.3	306 ± 77	1.8
Suaeda acuminata	LSu	10	1x Ba, 9x Ka	4,406 ± 810	4,196 ± 597	697 ± 138	5.66 ± 2.9	236 ± 70	6.0
Suaeda crassifolia	LSu	2	2x Ka	4,485 ± 579	3,465 ± 298	427 ± 0.7	20.3 ± 6.3	545 ± 105	8.1
Ofaiston monandrum	SSu	2	2x Ka	2,196 ± 837	2,183 ± 1,289	423 ± 142	120 ± 149	532 ± 59	5.2
Salicornia europaea	Ssu	2	2x Ba	4,291 ± 157	3,857 ± 335	428 ± 132	5.4 ± 1.8	168 ± 23	9.0
Halocnemum strobilaceum	Ssu	1	1x Ba	3,042	3,748	506	2.02	133	7.4
Halostachys caspica	Ssu	1	1x Ba	1,095	2,088	509	2.61	60.0	4.1
Euclidium syriacum	Ps	1	1x Ka	314	127	497	142	95	0.26
Malcolmia africana	Ps	1	1x Ka	451	648	998	328	148	0.65
Eremosparton aphyllum	NoH	1	1x Ba	155	32.7	324	52.9	78.5	0.10
Stipagrostis pennata	NoH	1	1x Ba	78.0	44.7	327	141	43.4	0.32

Table 12.4 Ion pattern of some common halophytic species of the Aralkum, analysed from acidic extracts. From Bayan (*Ba*) and from Karabulak (*Ka*). *n* number of samples, ion content (mmol kg^{-1} and standard deviation), in *parentheses* factor for increased content related to hot-water extracts, see Table 12.3

Species	*n*	Locality	Na^+	K^+	Ca^{2+}	Mg^{2+}	Na^+/K^+
Climacoptera aralensis	8	3x Ba 5x Ka	4,259 ± 1,261 (1.27)	666 ± 245 (1.16)	239 ± 104 (67.3)	277 ± 31 (1.95)	6.4
Petrosimonia triandra	4	3x Ba 1x Ka	1,233 ± 533 (1.21)	664 ± 68 (1.16)	385 ± 58 (29.6)	445 ± 56 (1.45)	1.9
Suaeda acuminata	10	1x Ba 9x Ka	5,017 ± 922 (1.20)	803 ± 164 (1.15)	261 ± 46 (46.1)	441 ± 70 (1.87)	6.3
Suaeda crassifolia	2	2x Ka	4,263 ± 4.2 (1.23)	504 ± 9.9 (1.18)	321 ± 35 (15.8)	707 ± 138 (1.30)	8.5
Ofaiston monandrum	2	2x Ka	2,905 ± 1,673 (1.33)	460 ± 120 (1.09)	366 ± 190 (3.05)	659 ± 36 (1.24)	6.3
Salicornia europaea	2	2x Ba	5,310 ± 306 (1.38)	472 ± 86 (1.10)	370 ± 94 (68.5)	565 ± 116 (3.36)	11.3
Halocnemum strobilaceum	1	1x Ba	5,850 (1.56)	579 (1.14)	64 (31.9)	151 (1.14)	10.1
Halostachys caspica	1	1x Ba	2,870 (1.37)	579 (1.14)	81 (31.0)	149 (2.48)	5.0
Euclidium syriacum	1	1x Ka	130 (1.02)	557 (1.12)	333 (2.35)	140 (1.47)	0.23
Malcolmia africana	1	1x Ka	714 (1.10)	1,130 (1.13)	572 (1.74)	184 (1.24)	0.63
Eremosparton aphyllum	1	1x Ba	18 (0.55)	436 (1.35)	258 (4.9)	139 (1.77)	0.041
Stipagrostis pennata	1	1x Ba	28 (0.63)	401 (1.23)	212 (1.50)	53 (1.22)	0.070

hand, the Na^+ and Cl^- accumulation of pseudohalophytes such as *Euclidium syriacum* and *Stripagrostis pennata* is very low.

The respective data on soil from the sites studied are given in Table 9.1 along the Karabulak gradient transect. All sites are rather alkaline. Salinity is also very variable between sites and between distribution along horizons. This depends on season, as salinity changes with evaporative demands in summer along the very long capillaries in loam and clay to the upper horizons and this may form a salt crust. However, lower horizons also often have a rather high salinity level, whereas middle horizons may store less saline water from winter snow or rains. This is shown by two examples of soil profiles (Fig. 12.5). In both soil profiles it is obvious that the sulphate salinity is as high as or even higher than the chloride salinity, but differs in the horizons.

It should be briefly mentioned that the various members of the Chenopodiaceae on the Aralkum cannot be put into one group of physiotypes (Reimann and Breckle 1993). Under natural conditions the sodium levels vary very much, as do the levels of other ions. There are many articles on the chemistry of halophytes and their internal ion composition (Albert 1982), as well as on the normally taxon-specific

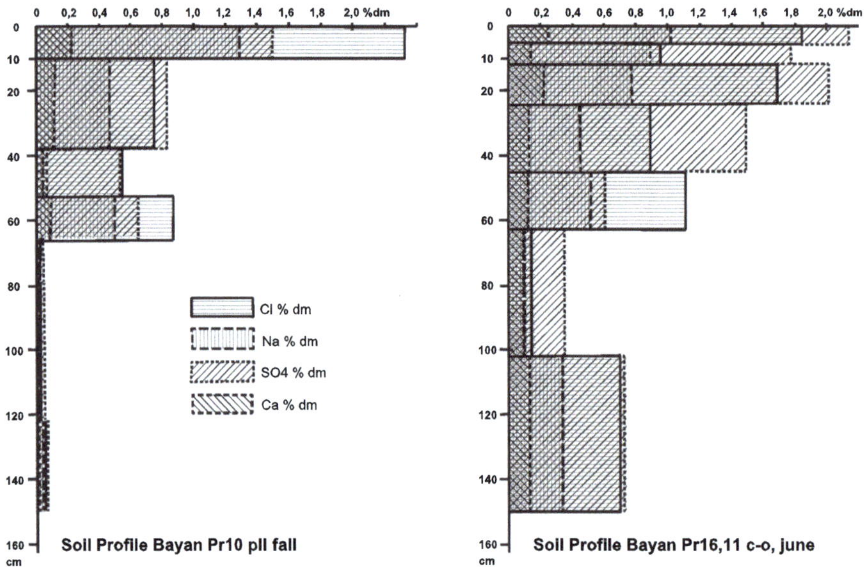

Fig. 12.5 Ion content in soil horizons of the Aralkum. *Left*: Soil profile Bayan Pr10 with main salt accumulation, mainly chloride in topsoil. *Right*: Soil profile Bayan Pr16 with salt, mainly sulphate accumulation in topsoil and in lower horizons

Fig. 12.6 *Suaeda acuminata*. Remnants from previous year, new seedlings and saplings. North Aral Sea (photo: Breckle, May 2003)

accumulation of compatible solutes (Popp 1985). The main characteristics of the physiotypes, e.g. Brassicaceae and Poaceae, are represented in the same ion pattern, as Albert (1982, 2005) extensively described.

It is always an open question to what extent the edaphic conditions influence the ionic pattern and content in plants (Mirazai and Breckle 1978). The Pontic–Irano–Turanian *Suaeda acuminata* (Fig. 12.6) is very common in Central Asia (Wucherer 1986). This species exhibits a wide ecological amplitude and thus can be found on very contrasting saline stands. Within the Karabulak transect on the northern coast of the Aral Sea, at seven localities *Suaeda acuminata* is present (Table 9.1).

It is obvious that the sodium and chloride contents of the aboveground plant organs of *Suaeda acuminata* on degraded coastal solonchaks and puffy coastal solonchaks are significantly lower. These soils contain significantly less salt in the topsoil. On these stands the sodium content is higher than the chloride content in comparison with the marshy solonchaks and crusty coastal solonchaks (Table 12.5). This example of *Suaeda* demonstrates that the ion content in halophytes growing on real solonchaks with high salinity is not distinctly influenced by the edaphic conditions.

Balnokin et al. (1991) studied the sodium, chloride and proline contents in *Salicornia europaea*, *Climacoptera aralensis* und *Petrosimonia triandra* along the Bayan transect on the eastern coast of the Aral Sea (Table 12.6). The content of proline as one of the typical compatible solutes apparently exhibits no clear correlation to the storage of salt in the plant tissues.

Table 12.5 Main ions in the stems and leaves of *Suaeda acuminata* (mol kg^{-1} dry matter) from the Karabulak transect and soil characteristics (10–20 cm) of the site. *EC* electric conductivity of soil extracts (mS cm), *DSF* dry sea floor (from the 1970s, 1980s or 1990s)

	DSF	Soil pH	Soil EC	Na^+	K^+	Mg^{2+}	Ca^{2+}	Cl^-
Degraded coastal solonchak	1970s	8.0	2.9	3.8	0.75	0.20	0.0048	3.5
Degraded coastal solonchak	1970s	8.1	1.1	5.0	0.59	0.18	0.0041	4.9
Coastal solonchak	1980s	8.5	12.8	3.5	0.76	0.19	0.0047	2.7
Crusty coastal solonchak	1980s	8.3	19.4	4.5	0.68	0.21	0.0035	5.1
Puffy coastal solonchak	1990s	–	–	4.5	0.60	0.20	0.0039	5.1
Coastal solonchak[a]	1990s	8.2	7.4	3.8	0.82	0.31	0.0118	3.9
				4.7	0.86	0.25	0.0046	5.0
Marshy solonchak[a]	1990s	8.4	1.3	3.4	0.84	0.33	0.0103	4.7
				3.6	0.66	0.36	0.0079	4.5

[a]Samples taken twice (4 and 12 May 1998)

Table 12.6 Ion content and proline content (mmol kg^{-1} fresh weight) in green tissues of halophytic plants from the Aralkum

Salicornia europaea			*Climacoptera aralensis*			*Petrosimonia triandra*		
Na^+	Cl^-	Proline	Na^+	Cl^-	Proline	Na^+	Cl^-	Proline
286	109	0.85	168	202	0.43	162	64	0.70
451	140	0.63	434	117	0.29	181	91	0.78
516	165	0.46	511	139	0.35	251	102	0.81
532	202	0.40	608	182	0.44	252	71	0.89
639	179	1.00	620	175	0.26	256	73	1.15
665	189	0.51	683	132	0.32	258	80	0.91
729	220	0.47	768	172	0.41	275	83	1.11
843	254	0.28	830	167	0.30	281	73	1.11
867	276	0.62	1,021	160	0.30	297	71	0.88
1,116	296	0.64	1,153	228	0.39	394	104	0.96

After Balnokin et al. (1991)

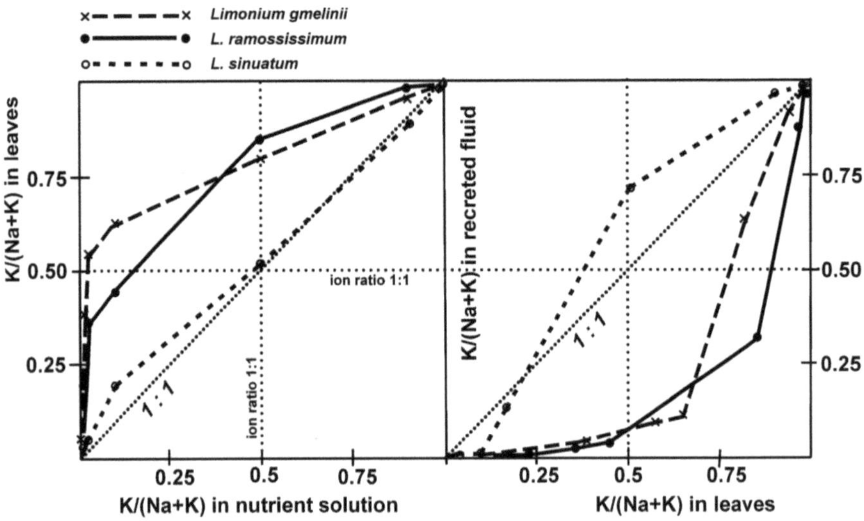

Ion conditions in recreting plant species (Limonium) in salinity cultures (acc. to Wiehe)

Fig. 12.7 Ion conditions in three *Limonium* species in salinity cultures (Wiehe and Breckle 1989) with various nutrient solutions differing in Na^+/K^+ ratio

The selectivity against ions differs in the various species according to their natural occurrence. The halophytes *Limonium gmelinii* and *Limonium ramosissimum* are very potassiophilic, as can be seen by the strong change in ion pattern (Fig. 12.7, left side) between nutrient solution and leaf cell sap. Again, there is a major change in ion composition between leaf cell sap and the recreted fluid. The ion pattern changes in such a way that the cytoplasm of the leaf cells is kept relatively low in sodium, whereas the gland fluid is rich in sodium (Fig. 12.7, right side). Such behaviour is not recognizable in *Limonium sinuatum*, a plant which inhabits slightly saline stands. In that species selectivity in both cases of transport is low (Fig. 12.7). It was also shown that the activity of the salt glands of the halophilic *Limonium* species (Wiehe and Breckle 1989) and *Aeluropus* has a threshold value and these start to recrete NaCl only after a distinct salinity level in the leaf is reached (Pollak and Waisel 1979).

There are many indications that also in the stem- and leaf-succulent halophytes, in the recretohalophytes and pseudohalophytes from the dry Aral Sea floor, different mechanisms and strategies for the adjustment and regulation of the salt concentration in the plant tissues are operating (Breckle 1995) and thus a differing salt tolerance in the various species leads to a specific pattern of species and halophyte types along salt gradients.

The sequence of species along the salt gradient in a rich halophytic area, as it is in the Central Asian deserts, reveals a typical sequence of the dominating halophyte types. Along the salt gradient (Breckle 1986), which can be derived from salinity measurements in a mosaic vegetation, it is obvious that the stem succulents and then the leaf succulents play the major role close to the saline lakes or basins, where salinity is high. The recreting halophytes (exocrinohalophytes, endocrinohalophytes) dominate in the middle part of the transect, where salinity is more variable as is water supply. This part of the transect is characterized by less water availability and often here a much higher proportion of C_4 plants occurs. This is also the case on the less saline side, where the pseudohalophytes and finally on almost salt-free substrates the nonhalophytes predominate and other ecological factors, such as water availability, water supply and nitrogen source, govern the vegetation mosaic. However, on the desiccated seafloor of the Aral Sea an equilibrium of halophyte types has not yet been reached, the dynamics of changing ecological conditions from year to year is so drastic that only by chance a mixture of more or less adapted species is found, which in part resemble the sequence of the halophyte types discussed.

12.1.3 Ecological Salinity Indicator Values for Plants of the Aralkum Region

The ecological behaviour and adaptation to distinct natural site conditions is the result of the competitive ability of a species. This depends on the floristic pattern of the region and the competitors present. Normally under natural conditions with a fluctuating climate from year to year, a dynamic equilibrium can be observed, and if the main ecological conditions vary within a rather constant range, a set of species will form a rather constant community.

Under saline conditions, the salt content of the soil plays a major decisive role for which species can compete successfully (Adam 1990). By comparing many sites with different salinities, one can evaluate the distinct ecological optimum (not ecophysiological optimum without competition, which can be rather different: many plants grow better under low salinities but are pressed to higher salinity sites by competition, where they can grow, but not optimally). This ecological optimum can be used to grade the ecological salinity tolerance by an indicator value (S value, see Table 12.7). Such indicator values are used rather widely in various regions for various ecological parameters, e.g., pH, nitrogen supply, drought tolerance and heat tolerance (Ellenberg et al. 1991). For salt tolerance a scale from 0 to 9 can be used (Table 12.7), where 9 means the highest salt tolerance. In contrast to many other indicator values, the distribution of the S values over the whole scale is oblique since most species belong to the nonhalophyte group, which has an indicator value of $S = 0$ or $S = 1$.

By long-term observations and comparing many sites, one can define S values for many species. A few species are very variable in their adaptation to saline site conditions, and those species have no definite S value ($S = X$). For others, their

Table 12.7 Definitions of the *S* value, the ecological salinity indicator value (see Breckle 1985; Ellenberg et al. 1991)

S value	
0	Not salt-tolerant, species never in brackish soils (NaCl content in soil below 0.01%), very sensitive to salt, strong nonhalophytes
1	Almost not salt-tolerant, very rare in brackish soils (NaCl content in soil below 0.05%)
2	Similar to $S = 1$, but more often in slightly brackish soils (oligohaline, about 0.05-0.3% Cl^-), slightly salt-tolerant species, which can withstand some salinity, but most frequently occuring in nonsaline soils ("pseudohalophytes", exhibiting no special morphological or anatomical features, also possible for higher *S* values)
3	Species indicating salinity; however, may also grow in soils with low salinity ("facultative halophyte", "accidental halophytes" in an ecological sense) (ß-mesohaline, about 0.3–0.5% Cl^-)
4	Similar to $S = 3$ (α/ß-mesohaline, about 0.5–0.7% Cl^-), exhibiting some salt tolerance and longer survival under salinity
5	Species normally only in saline soils (α-mesohaline, about 0.7–0.9% Cl^-), can withstand moderate salinities
6	Typical halophytes, indicating salinity, rare in nonsaline soils (α-mesohaline/polyhaline, about 0.9–1.2% Cl^-)
7	Similar to $S = 6$, but very salt-tolerant, never in nonsaline soils (often "obligatory halophytes" in ecological terms) (polyhaline, about 1.2–1.6% Cl^-], species indicating moderate to rather high salinities in soil
8	Typical halophytes, indicating high salinity, very salt-tolerant (euhaline, about 1.6–2.30% Cl^-), typical salt plants, indicating high salinities, only growing on severely saline sites (obligatory halophytes, euhalophytes)
9	Extreme halophytes, in soils with very high, during dry periods extremely high, salinity ("obligatory halophytes" in ecophysiological terms) (euhaline/hypersaline, above 2.30% Cl^-), found only on salt-crust soils and always indicating very strong salinities. Species which can fulfil their whole life cycle on highly saline sites
X	*S* value very variable, broad, indistinct, species found from nonsaline to very saline sites
–	*S* value not yet known, most probably 0 or 1

typical site conditions are not known exactly ($S = –$), and will be revealed only in the future.

It should also be kept in mind that the *S*-value list is only valid for a distinct region; it depends on the whole given flora and the respective competitive plant communities.

The Aralkum flora is very rich in halophytes; thus, the percentage of species with high *S* values (above 4) is rather high in several plant families (see Table 12.8). Other plant families are represented in the Aralkum by a quite high number of species but still mainly prefer sites with low salinity (Polygonaceae, Brassiceae, Fabaceae).

All species of the Aralkum flora are listed in Table 12.9 with their ecological salinity indicator value (*S* value), their halophytic strategy type and their life form. It is easily recognizable that the halophytic type and the *S* value are rather strongly correlated.

Table 12.8 Number of species of halophytic strategy types and related salinity indicator values for the halophytic flora of the Aralkum

Halophytic strategy type	1	2	3	4	5	6	7	8	9	*X*	Σ
Nonhalophytes	42	18	0	0	0	0	0	0	0	0	60
Pseudohalophytes	4	30	42	15	8	4	1	0	0	4	108
Xerosucculents	0	0	0	0	0	0	0	0	0	1	1
Leaf-succulent euhalophytes	0	2	2	5	14	7	22	9	1	0	62
Stem-succulent euhalophytes	0	0	0	1	1	1	1	0	2	0	6
Endocrinohalophytes	0	0	2	5	1	2	0	0	0	0	10
Exocrinohalophytes	0	0	0	0	0	9	7	4	0	0	20
Hydrohalophytes	0	0	0	1	0	0	0	0	1	0	2
Σ	46	50	46	27	24	23	31	13	4	5	269
Not determined strategy type											104
Σ Σ											373

12.1.4 Salinity

Over all the oceans, seawater is rather homogeneous in chemical composition, with a strong preponderance of NaCl. In deserts with their arid climate, salinity is caused not only by atmospheric input (cyclic salt; Teakle 1937; Breckle 1976, 1985), but also by leaching of the rocky material of the hydrotope within the endorheic basin, where the water runoff from the rare precipitation events is collected in the erosion basin, forming salt lakes, which may have accumulated also some other ions besides Na^+ and Cl^-, mainly SO_4^{2-}, HCO_3^-, Li^+, Mg^{2+}, borate, etc. (Breckle 1975a, b, 1990). Thus, in some parts of the world, salinity is caused not only by chloride, but in temperate and cold arid continental regions it may be caused by sulphate or carbonate accumulation (Curtin et al. 1993). In the Aralkum the desiccated substrate of the seafloor is rich in chloride and sulphate. This can change within deeper horizon layers (Fig. 12.5). It can also be seen indirectly by the various water sources in the region (Table 12.10) with very variable ion content and salinities. The ratio between ions is not as similar as in open-ocean water, which is rather constant within narrow limits all over the world. If the Na^+/Cl^- ratio is distinctly higher than 1, the counterbalance is normally by sulphate (sulphate salinity); if the Mg^{2+}/Na^++K^+ ratio is distinctly higher than 0.1, the water has a bitter taste. If there is a sufficient portion of potassium and alkali earth ions present, the salinity by sodium is not as severe as with pure NaCl salinity. Plants can adjust to such conditions and are able to absorb nutrients by discriminating ions. Thus, a typical halophytic community is a mixture, and is often composed of several species, where some of these species are not real halophytes. They occur only accidentally in such plant communities of oligohaline marshes, but have their optimal growth and performance in nonsaline vegetation. A typical spatial or temporal niche segregation enables nonhalophytes or pseudohalophytes to migrate and to invade halophytic stands.

Table 12.9 Species of vascular plants from the Aralkum, indicating their salinity indicator values, halophytic character and life form (*Ch* chamaephytes, *G* geophytes, *B* with bulbs, *P* parasitic, *R* with rhizomes, *H* hemicryptophytes, *Hy* hydrophytes, *Ph-m* microphanerophytes, *Ph-n* nanophanerophytes, *T* therophytes). Salinity tolerance is expressed as indicator value *S* (for definitions of *S* values, see Table 12.8; *S* = 0, nonhalophytes, are not indicated here, mainly are within the "not known" group; *S* = 9, extreme halophytes, only growing on strongly saline stands; *X* indifferent, – not known)

Species	*S* value	Halophytic strategy type	Life form
Alliaceae J. Agardh			
Allium caspium (Pall.) Bieb.	1	NoH	GB
Allium sabulosum Stev. ex Bunge	1	NoH	GB
Allium schubertii Zucc.	1	NoH	GB
Amaryllidaceae J. St.-Hil.			
Ixiolirion tataricum (Pall.) Schult. & Schult. fil.	2	Ps	GB
Apiaceae Lindl.			
Ferula canescens (Ledeb.) Ledeb.	–	–	GR
Ferula caspica Bieb.	–	–	GR
Ferula lehmannii Boiss.	2	Ps	GR
Ferula nuda Spreng.	–	–	GR
Prangos odontalgica (Pall.) Herrnst. & Heyn	1	NoH	GR
Asclepiadaceae R. Br.			
Cynanchum sibiricum Willd.	2	Ps	H
Asparagaceae Juss.			
Asparagus breslerianus Schult. & Schult. fil.	1	Ps	H
Asparagus inderiensis Blum ex Pasz.	2	Ps	H
Asparagus persicus Baker	1	Ps	H
Asteraceae Dumort.			
Acroptilon repens (L.) DC.	3	Ps	H
Amberboa turanica Iljin	2	Ps	T
Anthemis candidissima Willd. ex Spreng.	1	NoH	T
Artemisia aralensis Krasch.	2	Ps	Ch
Artemisia austriaca Jacq.	1	Ps	H
Artemisia diffusa Krasch.ex Poljak.	2	Ps	Ch
Artemisia arenaria DC.	–	–	Ch
Artemisia pauciflora Web.	–	–	Ch
Artemisia quinqueloba Trautv.	–	–	Ch
Artemisia santolina Schrenk	3	Ps	H
Artemisia schrenkiana Ledeb.	–	–	H
Artemisia scoparia Waldst. & Kit.	2	Ps	H
Artemisia scopiformis Ledeb.	–	–	H
Artemisia semiarida (Krasch. et Lavr.) Filat.	3	Ps	Ch
Artemisia songarica Schrenk	3	Ps	Ch
Artemisia terrae-albae Krasch.	X	Ps	Ch
Artemisia turanica Krasch.	–	–	Ch
Chartolepis intermedia Boiss.	–	–	H
Chondrilla ambigua Fisch. ex Kar. & Kir.	4	Ps	H
Chondrilla brevirostris Fisch. & C. A. Mey.	–	–	H
Cirsium arvense (L.) Scop.	2	Ps	H

(continued)

Table 12.9 (continued)

Species	*S* value	Halophytic strategy type	Life form
Cirsium ochrolepidium Juz.	–	–	H
Cousinia affinis Schrenk	–	–	H
Epilasia hemilasia (Bunge) Clarke	2	NoH	T
Heteracia szovitsii Fisch. & C. A. Mey.	–	–	T
Hyalea pulchella (Ledeb.) C. Koch	–	–	H
Inula caspica Blum ex Ledeb.	2	Ps	H
Inula germanica L.	1	NoH	H
Karelinia caspia (Pall.) Less.	5	Lsu	H
Koelpinia linearis Pall.	3	Ps	T
Koelpinia tenuissima Pavl. & Lipsch.	–	–	T
Koelpinia turanica Vass.	–	–	T
Lactuca serriola L.	3	Ps	H
Lactuca tatarica (L.) C. A. Mey.	3	Ps	H
Lactuca undulata Ledeb. Pojark	–	–	T
Mausolea eriocarpa (Bunge)	–	–	Ch
Saussurea salsa (Pall. ex Bieb.) Spreng.	5	Lsu	H
Scorzonera sericeolanata (Bunge) Krasch. & Lipsch.	3	Ps	GB
Senecio noeanus Rupr.	5	Lsu	T
Senecio subdentatus Ledeb.	4	Lsu	T
Sonchus oleraceus L.	3	Ps	T
Taktajaniantha pusilla (Pall.) Nazarova	–	–	GB
Tanacetum achilleifolium (Bieb.) Sch. Bip.	2	Ps	H
Taraxacum bessarabicum (Hornem.) Hand.-Mazz.	4	Ps	H
Tragopogon marginifolius Pavl.	3	Ps	GR
Tragopogon ruber S. G. Gmel.	–	–	GR
Tragopogon sabulosus Krasch. & S. Nikit.	–	–	GR
Tripolium vulgare Nees	7	Lsu	T
Berberidaceae Juss.			
Leontice incerta Pall.	1	NoH	GR
Boraginaceae Juss.			
Argusia sibirica (L.) Dandy	6	Ps	H
Arnebia decumbens (Vent.) Coss. & Kral.	1	NoH	T
Asperugo procumbens L.	1	NoH	T
Heliotropium arguzioides Kar. & Kir.	4	Lsu	H
Heliotropium dasycarpum Ledeb.	5	Ps	H
Heterocaryum rigidum A. DC.	–	–	T
Heterocaryum szovitsianum (Fisch. & C. A. Mey.) A. DC.	–	–	T
Lappula semiglabra (Ledeb.) Guerke	2	Ps	T
Lappula spinocarpos (Forssk.) Aschers.	–	–	T
Nonea caspica (Willd.) G. Don fil.	2	NoH	T
Rochelia retorta (Pall.) Lipsky	3	Ps	T
Rochelia leiocarpa Ledeb.	–	–	T
Suchtelenia calycina (C. A. Mey.) A. DC.	–	–	T
Brassicaceae Burnett			
Alyssum dasycarpum Steph.	2	NoH	T

(continued)

Table 12.9 (continued)

Species	*S* value	Halophytic strategy type	Life form
Alyssum turkestanicum Regel & Schmalh.	–	–	T
Capsella bursa-pastoris (L.) Medik.	1	NoH	T
Cardaria pubescens (C. A. Mey.) Jarm.	3	Ps	T
Chorispora tenella (Pall.) DC.	1	NoH	T
Descurainia sophia (L.) Webb ex Prantl	3	Ps	T
Diptychocarpus strictus (Fisch. ex Bieb.) Trautv.	–	–	T
Draba nemorosa L.	–	–	T
Erysimum sisymbrioides C. A. Mey.	1	NoH	T
Euclidium syriacum (L.) R. Br.	3	Ps	T
Goldbachia laevigata (Bieb.) DC.	5	Ps	T
Isatis minima Bunge	4	Ps	T
Isatis violascens Bunge	3	Ps	T
Lachnoloma lehmannii Bunge	4	Ps	T
Lepidium latifolium L.	5	LSu	H
Lepidium obtusum Basin.	3	Ps	H
Lepidium perfoliatum L.	4	Ps	T
Lepidium ruderale L.	4	Ps	T
Leptaleum filifolium (Willd.) DC.	2	NoH	T
Litwinowia tenuissima (Pall.) Woronow ex Pavl.	–	–	T
Matthiola stoddartii Bunge	2	Ps	T
Megacarpaea megalocarpa (Fisch. ex DC.) B. Fedtsch.	2	NoH	GR
Meniocus linifolius (Steph.) DC.	3	Ps	T
Octoceras lehmannianum Bunge	3	Ps	T
Pachypterygium multicaule (Kar. & Kir.) Bunge	3	Ps	T
Sameraria armena (L.) Desv.	3	Ps	T
Streptoloma desertorum Bunge	–	–	T
Strigosella africana (L.) Botsch.	–	–	T
Strigosella brevipes (Bunge) Botsch.	–	–	T
Strigosella circinata (Bunge) Botsch.	–	–	T
Strigosella scorpioides (Bunge) Botsch.	2	NoH	T
Syrenia montana (Pall.) Klok.	–	–	T
Tauscheria lasiocarpa Fisch. ex DC.	3	Ps	T
Tetracme quadricornis (Steph.) Bunge	4	Ps	T
Tetracme recurvata Bunge	–	–	T
Caryophyllaceae Juss.			
Acanthophyllum borsczowii Litv.	1	NoH	Ch
Acanthophyllum pungens (Bunge) Boiss	1	NoH	H
Gypsophila paniculata L.	3	Ps	H
Gypsophila perfoliata L.	3	Ps	H
Silene nana Kar. & Kir.	1	NoH	T
Silene odoratissima Bunge	–	–	H
Ceratophyllaceae S.F.Gray			
Ceratophyllum demersum L.	1	NoH	Hy
Chenopodiaceae Vent. (85)			
Agriophyllum minus Fisch. & C. A. Mey.	2	NoH	T

(continued)

Table 12.9 (continued)

Species	*S* value	Halophytic strategy type	Life form
Agriophyllum squarrosum (L.) Moq.	2	NoH	T
Anabasis aphylla L.	4	SSu	Ch
Anabasis salsa (C. A. Mey.) Benth. ex Volkens	6	LSu	Ch
Anabasis truncata (Schrenk) Bunge	5	LSu	H
Arthrophytum lehmannianum Bunge	5	SSu	T
Atriplex aucheri Moq.	4	NX	T
Atriplex cana C. A. Mey.	6	NX	T
Atriplex dimorphostegia Kar. & Kir.	4	NX	T
Atriplex littoralis L.	5	NX	T
Atriplex micrantha C. A. Mey.	3	NX	T
Atriplex moneta Bunge	2	Ps	T
Atriplex patula L.	2	Ps	T
Atriplex pratovii Suchor.	6	NX	T
Atriplex pungens Trautv.	4	NX	T
Atriplex sagittata Borkh.	4	NX	T
Atriplex sphaeromorpha Iljin	4	NX	T
Atriplex tatarica L.	3	Ps	T
Bassia hyssopifolia (Pall.) O. Kuntze	5	LSu	T
Bassia sedoides (Pall.) Aschers.	5	LSu	T
Bienertia cycloptera Bunge	7	LSu	T
Chenopodium glaucum L.	3	NX	T
Chenopodium rubrum L.	3	Ps	T
Ceratocarpus arenarius L.	2	Ps	T
Climacoptera affinis (C. A. Mey.) Botsch.	7	LSu	T
Climacoptera aralensis (Iljin) Botsch.	8	LSu	T
Climacoptera brachiata (Pall.) Botsch.	7	LSu	T
Climacoptera ferganica (Drob.) Botsch.	8	LSu	T
Climacoptera lanata (Pall.) Botsch.	8	LSu	T
Corispermum aralo-caspicum Iljin	3	Ps	T
Corispermum hyssopifolium L.	3	Ps	T
Corispermum laxiflorum Schrenk	2	NoH	T
Corispermum lehmannianum Bunge	2	NoH	T
Corispermum orientale Lam.	5	Ps	T
Gamanthus gamocarpus (Moq.) Bunge	8	LSu	T
Girgensohnia oppositiflora (Pall.) Fenzl	4	Ps	T
Halimione pedunculata (L.) Aell.	7	LSu	T
Halimione verrucifera (Bieb.) Aell.	7	LSu	Ch
Halimocnemis karelinii Moq.	7	LSu	T
Halimocnemis longifolia Bunge	7	LSu	T
Halimocnemis sclerosperma (Pall.) C. A. Mey.	7	LSu	T
Halimocnemis villosa Kar. & Kir.	7	LSu	T
Halocnemum strobilaceum (Pall.) Bieb.	9	SSu	Ch
Halogeton glomeratus C. A. Mey.	6	LSu	T
Halostachys belangeriana (Moq.) Botsch.	8	LSu	Ph-n
Halothamnus subaphyllus (C. A. Mey.) Botsch.	6	SSu	Ch

(continued)

Table 12.9 (continued)

Species	*S* value	Halophytic strategy type	Life form
Haloxylon aphyllum (Minkw.) Iljin	7	SSu	Ph-m
Haloxylon persicum Bunge ex Boiss. & Buhse	2 (3)	NoH (Ssu)	Ph-m
Horaninovia anomala (C. A. Mey.) Moq.	3	Ps	T
Horaninovia excellens Iljin	3	Ps	T
Horaninovia minor Schrenk	3	Ps	T
Horaninovia ulicina Fisch. & C. A. Mey.	5	Ps	T
Kalidium caspicum (L.) Ung.- Sternb.	7	LSu	Ch
Kalidium foliatum (Pall.) Moq.	7	LSu	Ch
Kirilowia eriantha Bunge	6	LSu	T
Kochia iranica Bornm.	5	Ps	T
Kochia odontoptera Schrenk	5	Ps	T
Kochia prostrata (L.) Schrad.	5	Ps	Ch
Krascheninnikovia ceratoides (L.) Gueldenst.	(2) 3 (5)	(NoH) Ps	Ch
Londesia eriantha Fisch. & C. A. Mey.	4	Ps	T
Nanophytum erinaceum (Pall.) Bunge	5	LSu	Ch
Ofaiston monandrum (Pall.) Moq.	6	LSu	T
Petrosimonia brachiata (Pall.) Bunge	7	LSu	T
Petrosimonia glaucescens (Bunge) Iljin	7	LSu	T
Petrosimonia hirsutissima (Bunge) Iljin	7	LSu	T
Petrosimonia squarrosa (Schrenk) Bunge	7	LSu	T
Petrosimonia triandra (Pall.) Simonk.	8	LSu	T
Salicornia europaea L. S. L.	9	SSu	T
Salsola arbuscula Pall.	4	LSu	Ph-n
Salsola australis (R.) Br.	–	LSu	T
Salsola chiwensis M. Pop.	–	LSu	Ch
Salsola dendroides Pall.	6	LSu	Ch
Salsola foliosa (L.) Schrad.	–	LSu	T
Salsola implicata Botsch.	–	LSu	T
Salsola micranthera Botsch.	5	LSu	T
Salsola nitraria Pall.	6	LSu	T
Salsola orientalis S. G. Gmel.	5	LSu	Ch
Salsola paletzkiana Litv.	3	LSu	Ph-n
Salsola paulsenii Litv.	3	LSu	T
Salsola richteri (Moq.) Kar. ex Litv.	4	LSu	Ph-n
Salsola tamariscina Pall.	5	LSu	T
Suaeda acuminata (C. A. Mey.) Moq.	8	LSu	T
Suaeda altissima (L.) Pall.	7	LSu	T
Suaeda arcuata Bunge	8	LSu	T
Suaeda crassifolia Pall.	9	LSu	T
Suaeda heterophylla (Kar. et Kir.) Bunge	7	LSu	T
Suaeda microphylla Pall.	8	LSu	Ch
Suaeda microsperma (C. A. Mey.) Fenzl.	8	LSu	T
Suaeda physophora Pall.	7	LSu	Ch
Suaeda salsa (L.) Pall.	8	LSu	T

(continued)

Table 12.9 (continued)

Species	*S* value	Halophytic strategy type	Life form
Convolvulaceae Juss.			
Convolvulus arvensis L.	2	Ps	H
Convolvulus erinaceus Ledeb.	1	NoH	Ch
Convolvulus subsericeus Schrenk	1	NoH	Ch
Cyperaceae Juss.			
Bolboschoenus maritimus (L.) Palla	6	Ps	GB
Carex pachystylis J. Gay	2	Ps	H
Carex physodes Bieb.	1	NoH	H
Scirpus lacustris L.	6	Ps	Hy
Scirpus tabernaemontani C. C. Gmel.	6	Ps	Hy
Elaeagnaceae Juss.			
Elaeagnus oxycarpa Schlecht.	4	Ps	Ph-m
Ephedraceae Dumort.			
Ephedra distachya L.	3	Ps	Ch
Ephedra intermedia Schrenk & C. A. Mey.	1	NoH	Ch
Ephedra strobilacea Bunge	2	NoH	Ch
Equisetaceae Rich. ex DC.			
Equisetum ramosissimum Desf.	1	NoH	H
Euphorbiaceae Juss.			
Euphorbia inderiensis Less. Kar. et Kir.	–	–	H
Euphorbia seguierana Neck.	2	Ps	H
Euphorbia turczaninowii Kar. & Kir.	–	–	H
Euphorbia undulata Bieb.	–	–	H
Fabaceae Lindl.			
Alhagi pseudalhagi (Bieb.) Fisch.	X	Ps	H
Ammodendron bifolium (Pall.) Yakovl.	–	–	Ph-n
Ammodendron conollyi Bunge	2	NoH	Ph-m
Ammodendron karelinii Fisch. et Mey.	2	NoH	Ph-n
Astragalus amarus Pall.	–	–	H
Astragalus ammodendron Bunge	–	–	Ph-n
Astragalus brachypus Schrenk	–	–	Ph-n
Astragalus campylorrhynchus Fisch. & C. A. Mey.	2	Ps	T
Astragalus lehmannianus Bunge	2	Ps	H
Astragalus longipetalus Chater	–	–	H
Astragalus ninae Pavl.	–	–	H
Astragalus oxyglottis Stev. ex Bieb.	–	–	T
Astragalus testiculatus	–	–	H
Astragalus villosissimus Bunge	–	–	Ph-n
Astragalus vulpinus Willd.	–	–	H
Eremosparton aphyllum (Pall.) Fisch. et Mey.	2	Ps	Ph-n
Glycyrrhiza aspera Pall.	4	Ps	H
Glycyrrhiza glabra L.	4	Ps	H
Halimodendron halodendron (Pall.) Voss.	5	LSu	Ph-n
Pseudosophora alopecuroides (L.) Sweet	4	Ps	H
Sphaerophysa salsola (Pall.) DC.	–	–	H

(continued)

Table 12.9 (continued)

Species	*S* value	Halophytic strategy type	Life form
Trigonella arcuata C. A. Mey.	–	–	T
Trigonella orthoceras Kar. et Kir.	–	–	T
Frankeniaceae S. F. Gray			
Frankenia hirsuta L.	8	EX	H
Fumariaceae DC.			
Fumaria vaillantii Loisel.	1	NoH	T
Geraniaceae Juss.			
Erodium oxyrhynchum Bieb.	1	NoH	T
Geranium transversale (Kar. & Kir.) Vved.	1	NoH	GR
Hypecoaceae Nakai			
Hypecoum parviflorum Kar. et Kir.	2	NoH	T
Iridaceae Juss.			
Iris longiscapa Ledeb.	1	NoH	GR
Iris songarica Schrenk	3	Ps	GR
Iris tenuifolia Pall.	1	NoH	GR
Juncaceae Juss.			
Juncus gerardii Loisel.	7	Ps	H
Lamiaceae Lindl.			
Chamaesphacos ilicifolius Schrenk	–	–	T
Eremostachys tuberosa (Pall.) Bunge	–	–	GR
Lallemantia royleana (Benth.) Benth.	1	NoH	T
Liliaceae Juss.			
Gagea reticulata (Pall.) Schult. & Schult. Fil.	1	NoH	GB
Rhinopetalum karelinii Fisch. ex Alexand.	1	NoH	GB
Tulipa buhseana Boiss.	1	NoH	GB
Limoniaceae Lincz.			
Limonium caspium (Willd.) Gams.	7	EX	H
Limonium gmelinii Willd. O. Kuntze	7	EX	H
Limonium otolepis (Schrenk) O. Kuntze	7	EX	H
Limonium suffruticosum (L.) O. Kuntze	8	EX	Ch
Malvaceae Juss.			
Malva neglecta Wallr.	3	Ps	T
Nitrariaceae Bercht. & J. Presl.			
Nitraria schoberi L.	6	LSu	Ph-n
Nitraria sibirica Pall.	6	LSu	Ph-n
Orobanchaceae Vent.			
Cistanche salsa (G. A. Mey.) G. Beck	X	Su	GP
Orobanche cernua Loefl.	1	NoH	GP
Papaveraceae Juss.			
Roemeria hybrida (L.) DC.	2	NoH	T
Roemeria refracta DC.	2	NoH	T
Peganaceae (Engl.) Tiegh. ex Takht.			
Peganum harmala L.	X	Ps	H
Plantaginaceae Juss.			
Plantago tenuiflora Waldst. & Kit.	5	Ps	T

(continued)

Table 12.9 (continued)

Species	*S* value	Halophytic strategy type	Life form
Poaceae Barnhart			
Aeluropus littoralis (Gouan) Parl.	7	EX	H
Agropyron desertorum (Fisch. ex Link) Schult.	1	Ps	H
Agropyron fragile (Roth) P. Candargy	–	–	H
Anisantha tectorum (L.) Nevski	–	–	T
Calamagrostis dubia Bunge	–	–	H
Catabrosella humilis (Bieb.) Tzvel.	–	–	H
Crypsis schoenoides (L.) Lam.	7	EX	T
Eremopyrum bonaepartis (Spreng.) Nevski	–	–	T
Eremopyrum distans (C. Koch) Nevski	3	Ps	T
Eremopyrum orientale (L.) Jaub. et Spach.	4	Ps	T
Eremopyrum triticeum (Gaertn.) Nevski	4	Ps	T
Leymus racemosus (Lam.) Tzvel.	–	–	H
Phragmites australis (Cav.) Trin. ex Steud.	X	Ps	H
Poa bulbosa L.	1	NoH	H
Puccinellia distans (Jacq.) Parl.	7	EX	H
Puccinellia dolicholepis V. Krecz.	6	EX	H
Puccinellia gigantea (Grossh.) Grossh.	6	EX	H
Schismus arabicus Nees	3	Ps	T
Stipa caspia C. Koch	1	NoH	H
Stipa sareptana Beck.	1	NoH	H
Stipagrostis karelinii (Trin. & Rupr.)Tzvl.	1	NoH	H
Stipagrostis pennata (Trin.) de Winter	1	NoH	H
Polygonaceae Juss.			
Atraphaxis replicata Lam.	1	NoH	Ph-n
Atraphaxis spinosa L.	1	NoH	Ph-n
Calligonum acanthopterum Borszcz.	–	–	Ph-n
Calligonum alatiforme Pavl.	–	–	Ph-n
Calligonum alatum Litv.	–	–	Ph-n
Calligonum androssovii Litv.	–	–	Ph-n
Calligonum aphyllum (Pall.) Guerke	–	–	Ph-n
Calligonum aralense Borszcz.	–	–	Ph-n
Calligonum borszczowii Litv.	–	–	Ph-n
Calligonum cancellatum Mattei	–	–	Ph-n
Calligonum caput-medusae Schrenk	–	–	Ph-n
Calligonum colubrinum Borszcz.	–	–	Ph-n
Calligonum commune (Litv.) Mattei	–	–	Ph-n
Calligonum crispatum (Litv.) Mattei	–	–	Ph-n
Calligonum cristatum Pavl.	–	–	Ph-n
Calligonum densum Borszcz.	–	–	Ph-n
Calligonum dubjanskyi Litv.	–	–	Ph-n
Calligonum elatum Litv.	–	–	Ph-n
Calligonum erinaceum Borszcz.	–	–	Ph-n
Calligonum eriopodum Bunge	–	–	Ph-n
Calligonum humile Litv.	–	–	Ph-n

(continued)

Table 12.9 (continued)

Species	*S* value	Halophytic strategy type	Life form
Calligonum junceum (Fisch. & C.A.May.) Litv.	–	–	Ph-n
Calligonum lamellatum (Litv.) Mattei	–	–	Ph-n
Calligonum leucocladum (Schrenk) Bunge	–	–	Ph-n
Calligonum macrocarpum Borszcz.	–	–	Ph-n
Calligonum membranaceum (Borszcz.) Litv.	–	–	Ph-n
Calligonum microcarpum Borszcz.	–	–	Ph-n
Calligonum minimum Lipsky	–	–	Ph-n
Calligonum muravljanskyi Pavl.	–	–	Ph-n
Calligonum palibinii Mattei	–	–	Ph-n
Calligonum platyacanthum Borszcz.	–	–	Ph-n
Calligonum pseudohumile Drob.	–	–	Ph-n
Calligonum rotula Borszcz.	–	–	Ph-n
Calligonum rubens Mattei	–	–	Ph-n
Calligonum spinulosum Drob.	–	–	Ph-n
Calligonum squarrosum Pavl.	–	–	Ph-n
Calligonum undulatum Litv.	–	–	Ph-n
Polygonum arenarium Waldst. Ed Scit.	2	Ps	T
Polygonum monspeliense Thieb. ex Pers.	3	Ps	T
Rheum tataricum L.	2	Ps	GR
Rumex marschallianus Reichenb.	–	–	T
Potamogetonaceae Dumort.			
Potamogeton perfoliatus L.	2	Ps	H
Ranunculaceae Juss.			
Adonis parviflora Fisch. ex DC.	2	Ps	T
Ceratocephala falcata (L.) Pers.	3	Ps	T
Ceratocephala testiculata (Grantz.) Bess.	3	Ps	T
Clematis orientalis L.	1	NoH	Ph-n
Consolida rugulosa (Boiss.) Schröding.	2	Ps	T
Myosurus minimus L.	3	Ps	T
Ranunculus platyspermus Fisch. ex DC.	2	Ps	GR
Thalictrum isopyroides C. A. Mey.	1	NoH	GR
Rosaceae Juss.			
Hulthemia persica (Michx. ex Juss.) Bornm.	3	Ps	Ch
Rubiaceae Juss.			
Asperula danilewskiana Basin.	–	–	Ch
Galium spurium L.	2	Ps	T
Rutaceae Juss.			
Haplophyllum perforatum Kar. et Kir.	2	Ps	H
Salicaceae Mirb.			
Populus euphratica Olivier/*Populus diversifolia*	(2) 3 (4)	Ps	Ph-m
Scrophulariaceae Juss.			
Linaria dolichoceras Kuprian.	–	–	H
Veronica campylopoda Boiss.	1	NoH	T
Solanaceae Juss.			
Hyoscyamus pusillus L.	1	NoH	T
Lycium ruthenicum Murr.	4	LSu	Ph-n

(continued)

Table 12.9 (continued)

Species	*S* value	Halophytic strategy type	Life form
Tamaricaceae Link.			
Tamarix aralensis Bunge	6	EX	Ph-n
Tamarix elongata Ledeb.	6	EX	Ph-n
Tamarix florida Bunge	6	EX	Ph-n
Tamarix hispida Willd	8	EX	Ph-n
Tamarix hohenackeri Bge	6	EX	Ph-n
Tamarix karelinii Bunge	6	EX	Ph-n
Tamarix laxa Willd.	6	EX	Ph-n
Tamarix leptostachys Bunge	8	EX	Ph-n
Tamarix litwinowii Gorschk.	6	EX	Ph-n
Tamarix ramosissima Ledeb.	7	EX	Ph-n
Typhaceae Juss.			
Typha angustifolia L.	2	Ps	Hy
Zannichelliaceae Dumort.			
Zannichellia palustris L.	4	HH	H
Zosteraceae Dumort.			
Zostera noltii Hornem.	9	HH	Hy
Zygophyllaceae R. Br.			
Zygophyllum eichwaldii C. A. Mey.	2	LSu	Ch
Zygophyllum fabago L.	5	LSu	H
Zygophyllum macropterum Boriss.	2	LSu	H
Zygophyllum oxianum Boriss.	5	LSu	H

12.2 Conclusions

Investigation of the adaptive mechanisms of the various halophyte types as well as succession processes is essential to obtain an adequate species composition for phytomelioration of the saline soils of the Aralkum. Ion pattern, halophytic strategy to cope with salinity and life form are very variable in halophytes; their distinction from less tolerant pseudohalophytes or nonhalophytes is only gradual. The salinization of the substrate on the dry seafloor varies to a great extent, causing a wide variety of saline soil types. Various solonchaks have developed: marshy solonchaks, crusty and puffy solonchaks, solonchaks slightly covered by sand, degraded coastal solonchaks, takyr solonchaks, etc. Studying natural halophytes is thus very important not only for all those regions where salinity has reached such a level that desalinization techniques are much too costly but also for quasi natural sites with their ecological dynamics. Understanding the adaptation of halophytes to saline sites and understanding their abilities to compete in saline communities is a good precondition for better use of halophytes. The applicability of the ecological salinity indicator value (*S* value) may be also worthwhile for adjacent agricultural areas with salinized fields and weeds for fast characterization of the sites.

Table 12.10 Ion content (g kg^{-1}) of some water samples from the Aralkum

Locality	pH	Na^+	K^+	Ca^{2+}	Mg^{2+}	Cl^-	Na^+/K^+	Na^+/Cl^-	Mg^{2+}/Na^++K^+
Water samples from August 1999									
Kambash	7.4	1.33	0.041	–	0.51	1.67	55.1	1.23	0.35
Kaskakulan warm spring	7.6	1.36	0.007	–	0.008	1.25	330	1.68	0.0057
Akbasti (Kokaral), warm spring	6.9	3.54	0.032	–	0.061	3.97	188	1.38	0.016
North Aral Sea, seawater	6.6	6.09	0.26	–	1.61	8.15	39.0	1.15	0.24
Pond for horses	6.9	0.43	0.032	–	0.074	0.31	23.2	2.17	0.16
Water samples from May 2002									
Kambash	7.3	0.26	0.011	–	0.14	0.27	41.1	1.45	0.49
North Aral Sea, Shevchenko bight, seawater	7.1	4.85	0.021	–	1.23	7.59	39.3	0.99	0.23
North Aral Sea, seawater 46°37,48′N 60°07,38′E 45 m asl	7.1	4.85	0.21	–	1.19	7.48	39.2	1.00	0.23
Akespe, warm spring, between 46°44,85′N 60°28,70′E 55 m asl	6.8	6.74	0.073	–	0.26	11.9	157	0.87	0.037
North Aral Sea, Butakov bight west side, seawater 46°46,94′N 60°34,73′E 44 m asl	7.1	7.22	0.29	–	1.80	11.0	42.2	1.01	0.23
North Aral Sea, Butakov bight east side, seawater	7.3	7.26	0.31	–	1.81	10.9	40.6	1.02	0.23
Muddy water sample 46°42.05′N 60°13,77′E 59 m asl	7.1	7.72	0.065	–	0.45	3.97	201	2.99	0.055
Bugun, new fishpond 46°09.83′N 61°11.14′E 55 m asl	7.1	0.46	0.026	–	0.28	0.94	30.0	0.75	0.56
Aralsk, irrigation water, tapwater	7.4	3.63	0.022	–	0.13	6.03	282	0.93	0.031
Aralsk, irrigation water, sandy ditch	7.1	4.53	0.023	–	0.12	5.81	340	1.20	0.026
Syr Darya, river water near Djusali 44° 51.10′N 65°22.89′E 138 m asl	7.4	0.093	0.047	–	0.052	0.041	33.8	3.52	0.51
Kyzyl-Orda, irrigation channel 44°50.41′N 65°30.18′E 135 m asl	7.2	0.11	0.051	–	0.062	0.089	37.9	1.96	0.51
Almaty, Malinki Almaty Djui, for irrigation	7.4	0.0037	0.0012	–	0.0022	<0.005	≈5.7	–	4.6
Water samples from May 2003									
Syr Darya, near Jalagash	8.1	0.11	0.0044	0.089	0.045	0.081	42.4	2.10	0.38
North of Kulandy, artesian spring 46°15.6′N 59°29.1′E	8.9	1.20	0.0067	0.014	0.0052	1.23	306	1.50	0.0040
2 km north of Kulandy, artesian spring 46°08.5′N 59°31.4′E	9.2	0.41	0.0036	0.0077	0.0018	0.57	194	1.11	0.0041
Kulandy spring, 120 m deep	9.0	0.48	0.0039	0.0093	0.0026	0.70	201	1.03	0.0051
West basin of Aral Sea 46°00.55′N 59°16.4′E	8.2	14.3	0.56	0.87	2.90	23.9	43.6	0.94	0.19
West basin of Aral Sea, muddy water sample	8.1	5.13	0.38	0.77	0.83	8.35	23.1	0.95	0.147
West basin of Aral Sea, between reeds	8.2	9.26	0.45	1.23	1.72	15.4	35.3	0.93	0.170
North Aral Sea, Shevchenko bight, seawater	8.2	4.13	0.16	0.27	0.42	5.33	147	1.20	0.092
North Aral Sea, Butakov bight, seawater	8.4	6.10	0.13	0.33	1.45	8.83	82	1.06	0.22
Seawater (for comparison)									
Open ocean water (35 g kg^{-1})	8.0	10.80	0.39	0.41	1.29	19.40	46.0	0.86	0.110

asl above sea level

Ecotypes and biogeography, germination and establishment, competition and nutrient availability under high salinity and alkalinity are subjects on the ecosystems level which have to be investigated further. Investigations of halophytic ecosystems, of the salinity process in agrarial systems and of plant strategies for salt regulation are urgently needed in the Aral Sea region, where salt desertification has become dominant.

References

Adam P (1990) Saltmarsh ecology. Cambridge University Press, Cambridge

Agachanjanz OE, Breckle S-W (1994) Umweltsituation in der ehemaligen Sowjetunion. Naturwiss Rdschau 47:99–106

Albert R (1982) Halophyten. In: Kinzel H (ed) Pflanzenökologie und Mineralstoffwechsel. Ulmer, Stuttgart, pp 33–215

Albert R (2005) Das Physiotypen-Konzept – ein Modell zur Erklärung ökophysiologischer Anpassungen? In Veste M, Wucherer W, Homeier J (eds) Ökologische Forschung im globalen Kontext. Cuvillier, Göttingen, pp 1–23

Balnokin YuV, Myasoedov NA, Baburina OK, Wucherer W (1991) Ion content of Na^+, Cl^-, S and proline in tissues from halophytes from soils with differing salinity on the dry sea floor of the Aral Sea. Probl Osv Pustyn', Ashkhabad 2:70–78 (in Russian)

Berger-Landefeldt U (1959) Beiträge zur Ökologie der Pflanzen nordafrikanischer Salzpfannen. Vegetatio 9:1–48

Black RF (1954) The leaf anatomy of Australian members of the genus *Atriplex I. Atriplex vesicaria* Heward and *A. nummularia* Lindl. Aust J Bot 2:269–286

Breckle S-W (1975a) Ionengehalte halophiler Pflanzen Spaniens. Decheniana (Bonn) 127:221–228

Breckle S-W (1975b) Wasser- und Salzverhältnisse bei Halophyten der Salzsteppe in Utah/USA. Ber Dtsch Bot Ges 87:589–600

Breckle S-W (1976) Zur Ökologie und zu den Mineralstoffverhältnissen absalzender und nichtabsalzender Xerohalophyten, Diss. Bot., Cramer, Vaduz 35, 1–176

Breckle S-W (1982) The significance of salinity. In: Spooner B, Mann HS (eds) Desertification and development: dryland ecology in social perspective. Academic, London, pp 277–292

Breckle S-W (1985) Die siebenbürgische Halophytenflora – Ökologie und ihre pflanzengeographische Einordnung. Siebenbürgisches Archiv, 3 Folge 20:53–105

Breckle S-W (1986) Studies on halophytes from Iran and Afganistan. II. Ecology of halophytes along salt gradients. Proc R Soc Edinb 89B:203–215

Breckle SW (1989) Role of salinity and alkalinity in the pollution of developed and developing countries. In: Öztürk MA (ed) International symposium on the effect of pollutants to plants in developed and developing countries in Izmir/Turkey, 22-28 August 1988, pp 389-409

Breckle S-W (1990) Salinity tolerance of different halophyte types. Plant Soil 148:167–175

Breckle S-W (1992) Salinity-stress and salt-recretion in plants. Bielefelder Ökologische Beiträge (BÖB) 6:39–52

Breckle S-W (1995) How do halophytes overcome salinity? In: Khan MA, Ungar IA (eds) Biology of salt tolerant plants. Book Graffers, Chelsea, pp 199–213

Breckle S-W (2002a) Salinity, halophytes and salt affected natural ecosystems. In: Läuchli A, Lüttge U (eds) Salinity: environment – plants – molecules. Kluwer, Dordrecht, pp 53–77

Breckle S-W (2002b) Salt deserts in Iran and Afghanistan. In: Böer B, Barth H-J (eds) Sabkha ecosystems. Kluwer, Dordrecht, pp 109–122

Breckle SW (2005) Naturwissenschaftliche Neugierde, 40 Jahre ökologisch-geobotanische Forschung – Rückblick und Ausblick. In: Veste M, Wucherer W, HomeierJ (eds) Ökologische Forschung im globalen Kontext. Festschrift zum 65. Geburtstag von Prof. Dr. S.-W. Breckle, pp 273–310
Breckle S-W, Agachanjanz OE (1994) Ökologie der Erde, vol 3, 2nd edn, Spezielle Ökologie der gemäßigten und arktischen Zonen Euro-Nordasiens. Fischer, Stuttgart
Breckle S-W, Wucherer W (2007) What will be the future of the Aral Sea? In: Lozan JL (ed) GLOBAL CHANGE: enough water for all? Wissenschaftliche Auswertungen GEO, Hamburg, pp 142–146
Breckle S-W, Wucherer W, Scheffer A (2001) Halophytes on the dry sea floor of the Aral Sea. In: Breckle S-W, Veste M, Wucherer W (eds) Sustainable land-use in deserts. Springer, Heidelberg, pp 139–146
Curtin D, Steppuhn H, Selles F (1993) Plant response to sulfate and chloride salkinity: growth and ionic relations. Soil Sci Soc Am J 57:1304–1310
Dregne HE (1986) Desertification of arid lands. In: El-Baz F, Hassan MHA (eds) Physics of desertification. Nijhoff, Dordrecht
Ellenberg H, Weber HE, Düll R et al (1991) Zeigerwerte von Pflanzen in Mitteleuropa. Scripta Geobotanica 18:248
Flowers TJ, Yeo AR (1995) Breeding for salinkty resutsnace in crop plants: where next? Aust J Plant Physiol 22:875–884
Frey-Wissling A (1935) Die Stoffausscheidung der Höheren Pflanzen. Springer, Berlin
Geldyeva GV, Budnikova TI, Gobernik IA et al (1998) Assessment of desertification processes in natural complexes of the Syrdarya delta. UNESCO Aral Sea Project, 15–41
Glazovskii NF, Orlovskii NS (1996) The problems of the desertification and droughts in GUS and the ways of their solution. Izv Akad Nauk Ser Geogr 4:7–23 (in Russian)
Hammer UT (1986) Saline lake ecosystems of the world. Junk, Dordrecht
Hedenström Hv, Breckle S-W (1974) Obligate halophytes? A test with tissue culture methods. Z Pflanzenphys 74:183–185
Jacobsen T, Adams RM (1958) Salt and silt in Ancient Mesopotamian agriculture. Science 128:1251–1258
Law JP, Hornsby AG (1982) The Colorado river salinity problem. Water Supply Manage 6:87–103
Metternicht G, Zinck A (2008) Remote sensing of soil salinization: Impact on land management. Taylor & Francis, CRC Press, Boca Raton
Mirazai NA, Breckle S-W (1978) Untersuchungen an afghanischen Halophyten I. Salzverhältnisse in Chenopodiaceen Nord-Afghanistans. Bot Jahrb Syst 99:565–578
Moore RT, Breckle S-W, Caldwell MM (1972) Mineral ion composition and osmotic relations of *Atriplex confertifolia* and *Eurotia lanata*. Oecologia 11:67–78
Munns R (1993) Physiological processes limiting plant growth in saline soils: some dogmas and hypotheses. Plant Cell Environ 16:15–24
Munns R, Schachtmann DP, Condon AG (1995) The significance of a two-phase growth response to salinity in wheat and barley. Aust J Plant Physiol 22:561–569
Naik PS, Widholm JM (1993) Comparison of tissue culture and whole plant response to salinity in potato. Plant Cell Tissue Organ Cult 33:273–280
Novikova NM, Kust GS, Kuzmina JV et al (1998) Contemporary plant and soil cover changes in the Amudarya and Syrdarya river deltas. UNESCO Aral Sea Project, 81–91
Oldeman LR (1994) The global extent of soil degradation. In: Greenland DJ, Szabolcs I (eds) Soil resilience and sustainable land use. CAB International, Wallingford, pp 99–118
Pollak G, Waisel Y (1979) Ecophysiology of salt excretion in *Aeluropus litoralis* (Gramineae). Physiol Plant 47:177–184
Popp M (1985) Osmotic adaptation in Australian mangroves. Vegetatio 61:247–253
Ramani S, Apte SK (1997) Transient expression of multiple genes in salinity-stressed young seedlings of rice (*Oryza sativa* L.) cv. Bura Rata. Biochem Biophys Res Comm 233:663–667

Reimann C, Breckle S-W (1993) Sodium relations in Chenopodiaceae, a comparative approach. Plant Cell Environ 16:323–328

Rhoades JD (1990) Soil salinity – causes and controls. In: Andrew Goudie A (ed) Techniques for desert reclamation. Wiley, New York, pp 109–134

Schirmer U, Breckle S-W (1982) The role of bladders for salt removal in some Chenopodiaceae (mainly *Atriplex* species). In: Sen DN, Rajpurohit KS (eds) Tasks for vegetation science, vol 2. Junk, The Hague, pp 215–231

Sheridan D (1981) Desertification of the United States. Council on Environmental Quality, Washington, DC

Shiati K (1991) Salinity management in river basins; modelling and management of the salt-affected Jarreh reservoir (Iran). Proefschrift, Wageningen

Shomer-Ilan A, Nissenbaum A, Waisel Y (1981) Photosynthetic pathways and the ecological distribution of the Chenopodiaceae in Israel. Oecologia 48:244–248

Singh G (2009) Salinity-related desertification and management strategies: Indian experience. Land Degradation Dev 20:367–385

Teakle LJH (1937) The salt (sodium chloride) content of rain water. J Agric Dept W Aust 14:115–133

Tobe K, Li X, Omasa K (2004) Effects of five different salts on seed germination and seedling growth of *Haloxylon ammodendron* (Chenopodiaceae). Seed Sci Res 14:345–353

Tobe K, Li X, Omasa K (2005) Effects of irrigation on seedlings emergence and seedlings survival of a desert shrub *Haloxylon ammodendron* (Chenopodiaceae). Aust J Bot 53:529–534

Ungar IA (1996) Effect of salinity on seed germination, growth, and ion accumulation of *Atriplex patula* (Chenopodiaceae). Am J Bot 83:604–607

Waisel Y (1972) Biology of halophytes. Academic, New York

Waisel Y (2001) Salinity: a major enemy of sustainable agriculture. In: Breckle SW, Veste M, Wucherer W (eds) Sustainable land use in deserts. Springer, Berlin, pp 166–173

Walter H (1968) Die Vegetation der Erde in ökophysiologischer Betrachtung. II. Die gemäßigten und arktischen Zonen. Fischer, Stuttgart

Wichelns D (1999) An economic model of waterlogging and salinization in arid regions. Ecol Econ 30:475–491

Wiehe W, Breckle S-W (1989) Die Ontogenese der Salzdrüsen von *Limonium* (Plumbaginaceae). Bot Acta 103:107–110

Winicov I, Bastola DR (1997) Salt tolerance in crop plants; new approaches through tissue culture and gene regulation. Acta Physiol Plantar 19:435–449

Wucherer W (1986) The spreading of *Suaeda acuminata* (C.A. Mey.) Moq. in Kazakhstan. In: Baitenov MS, Vasyagina MP (eds) Botanical materials of the herbarium of the Institute of Botany. Nauka, Alma-Ata, pp 38–39

Wucherer W, Breckle S-W, Dimeyeva L (2001) Flora of the dry sea floor of the Aral Sea. In: Breckle S-W, Veste M, Wucherer W (eds) Sustainable land-use in deserts. Springer, Heidelberg, pp 38–51

Yang X, Zhang K, Jia B, Ci L (2005) Desertification assessment in China: an overview. J Arid Environ 62:517–531

Chapter 13
Spatial Distribution of Plant Functional Types Along Stress Gradients – A Simulation Study Orientated Towards the Plant Succession on the Desiccating Aral Sea Floor

J. Groeneveld, W. Wucherer, L.A. Dimeyeva, F. Jeltsch, C. Wissel, and S-W. Breckle

13.1 Introduction

Since 1960 the Aral Sea has shrunk considerably, because the water of its two feeding rivers is intensively used for irrigation (Micklin and Aladin 2008). From 1960 to 2000 the irrigated area increased by approximately three million hectares and caused a negative water balance for the endorheic Aral Sea (Micklin 2007). During the desiccation a huge area of new land (approximately 54,000 km^2, Micklin and Aladin 2008) has been exposed, which is approximately 80% of the former sea surface (Micklin 2007). Plans to construct gigantic channel systems to transfer water from the major Siberian rivers into the Aral Sea have existed for decades but have not been applied (Badescu and Schuiling 2010). New satellite images (NASA) indicate a drastic retreat of the southern eastern basin in summer 2009; thus now almost 60,000 km^2 is desiccated seafloor (see Chap. 2). The new barren ground provides a complex mosaic of abiotic conditions, where soil salinity depends on the soil morphology, distance to the groundwater table and position along the gradient of the receding seawater. Sand and dust, salt dust in particular, which is blown off by strong winds (see Chaps. 5 and 7) and redistributed over the agricultural areas and villages is a serious economic (reduced crop yields) and health (respiratory illness, cancer, infant mortality) problem (O'Hara et al. 2000; Micklin 2007).

A fast and possibly dense colonization by plants could possibly help to attenuate those negative impacts of the desiccation. Therefore, a better understanding of the succession processes and the vegetation dynamics in this saline and arid environment is needed.

However, succession along environmental gradients is a complex and stochastic process, where multiple biotic and abiotic processes interact. Since the seawater has become more and more saline during the desiccation (Micklin and Aladin 2008), there is a tendency for the soil salinity to increase with distance from the former shoreline (Wucherer 1990). Continuous changes in the abiotic conditions can lead to abrupt changes in the species composition of plants and the spatial segregation of different species (Kenkel et al. 1991). Such zonation patterns have been found in

S.-W. Breckle et al. (eds.), *Aralkum - a Man-Made Desert: The Desiccated Floor of the Aral Sea (Central Asia)*, Ecological Studies 218,
DOI 10.1007/978-3-642-21117-1_13, © Springer-Verlag Berlin Heidelberg 2012

the newly established communities on the former seabed of the Aral Sea (Wucherer and Breckle 2001; see also Chaps. 9 and 10). Although zonation patterns along gradients are a well-known pattern in different ecological systems (Siccama 1974; Emery et al. 2001) it is not understood how the complex interplay of interspecific competition, seed dispersal and spatial heterogeneous abiotic conditions causes zonation. It is hypothesized that along stress gradients the lower end of a species occurrence (stressful end) is determined by its stress tolerance, whereas its upper end of occurrence (less stressful end) is determined by interspecific competition (Ungar 1998). To address these questions we have developed a spatially explicit, rule-based simulation model. In particular, we have investigated the effect of the magnitude and the scale of the spatial variability of the environmental conditions on the structure of the plant community.

13.2 Methods of Investigation – Model Description

13.2.1 General Structure

The following model description is based on previous more detailed model descriptions (Groeneveld 2003; Groeneveld et al. 2005). The main purpose of the model is to simulate the spatiotemporal vegetation dynamics along a stress gradient following the receding sea water. The spatial extension of the simulation arena (40 m × 6,000 m) was chosen to facilitate comparison with the field site Bayan transect at the northeastern coast of the Large Aral Sea. The simulation arena is a grid, whereas the size of a single grid cell is related to the space requirements of an adult individual of the tallest species (2 m × 2 m). Each grid cell is described by its position along the stress gradient and its state (empty, flooded or occupied by a dominant species). There are no shared grid cells, i.e., only one species can reach maturity in a given grid cell. Because individuals of different life forms differ in their space requirements, we assume that either a cohort of annual individuals or a single individual plant is occupying a grid cell. The grid is updated every time step (year) and typically one simulation runs for 640 time steps (years). Each time step biotic processes (germination, thinning, establishment, seed dispersal and mortality) and abiotic conditions (precipitation and stress) are modelled (see Fig. 13.1). To reduce the complexity of the model and the number of parameters, we model the dynamics of plant functional types (PFTs) instead of species. Each PFT is related to one of the dominant plant species of the study area (Bayan transect) and is defined by its longevity, its typical maximum seed dispersal distance and its stress tolerance (see Table 13.1), which is more or less the overall salinity tolerance. The names of the PFTs are a combination of the life form (*p* for "perennial" or *a* for "annual") and its stress tolerance (from 0 for no stress tolerance – nonhalophytes to pseudohalophytes – to 5 for extreme stress tolerance – euhalophytes; see Chap. 12). We have considered four annual and three perennial PFTs which are related to dominant

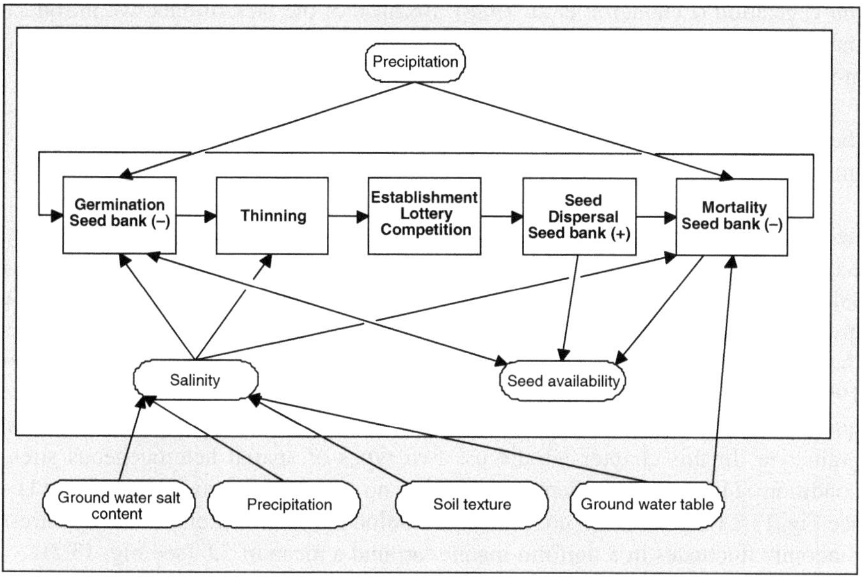

Fig. 13.1 Overview of the processes considered (*white background*). Germination, thinning, establishment, seed dispersal and mortality are simulated each year depending on precipitation, salinity and seed availability. Salinity depends on additional abiotic factors (e.g., groundwater salt content and grain size distribution in the soil – *grey background*) which are not considered explicitly in the model (Modified after Groeneveld 2003)

Table 13.1 The seven plant functional types. The functional types are named after their life form (*a* annual, *p* perennial) and their stress tolerance (5 is highest and 0 means no stress tolerance). Stress tolerance is here mainly salinity tolerance. The maximum seed dispersal distance represents the distance within which 99.9% of all seeds are dispersed

	p1	p3	p4	a2	a3	a4	a5
Stress tolerance σ	1	3	4	2	3	4	5
Maximum seed dispersal distance (m)	500	500	500	1,000	1,000	1,000	1,000
Longevity (years)	30	10	10	1	1	1	1

species (Wucherer, unpublished data): *Petrosimonia triandra* a2, *Climacoptera aralensis* a3, *Suaeda acuminata* a4, *Salicornia europaea* a5, *Tamarix laxa* p1, *Halostachys caspica* p3, and *Halocnemum strobilaceum* p4. The stress tolerance level, which is a PFT-specific attribute, ranges from 0 to 5 (stress level where the plant species starts to suffer from stress), whereas the environmental stress conditions along the stress gradient range from 0 (no stress) to 11 (maximum stress).

13.2.2 Abiotic Conditions

We consider two abiotic factors which have an impact on the spatiotemporal vegetation dynamics. In arid areas, precipitation usually has a strong impact on the vegetation (Lehouerou et al. 1988). Because of the lack of data, we introduce three abstract rainfall classes (good, average and bad), which occur randomly and are uncorrelated in time, with probabilities of 25%, 50% and 25%.

Furthermore, we consider stress as an abiotic component. Stress is presented in the model as an abstract stress index s, which combines the effects of salinity and inundation (0 for no stress, 11 for maximum stress).

Both stress factors increase along the transect at our study site. Inundation will be more frequent at the end of the gradient, where the elevation is at its minimum. Salt stress will also increase along the transect of receding seawater, because the salinity of the retreating seawater increases with time and the salinity of the groundwater increases with the distance from the former seashore. Furthermore, the average grain size of the soil particles decreases along the transect (Wucherer 1990). Therefore, the coarse-grained sand of the former beach will desalinize more quickly than the fine loamy soil of the new areas, because capillarity depends on grain size. In this chapter we discuss two types of spatial heterogeneous stress conditions: (1) a linear gradient ranging from no stress (0) to maximum stress (11) – see Fig. 13.2 for details; (2) no spatial correlation in the stress conditions, i.e., stress randomly fluctuates in a uniform manner around a mean of 5.5 (see Fig. 13.2).

13.2.3 Biotic Processes

The biological processes implemented are germination, seed bank dynamics, thinning, establishment, seed dispersal and mortality. For a complete model description, see Groeneveld (2003). A precipitation-dependent fraction of seeds germinate

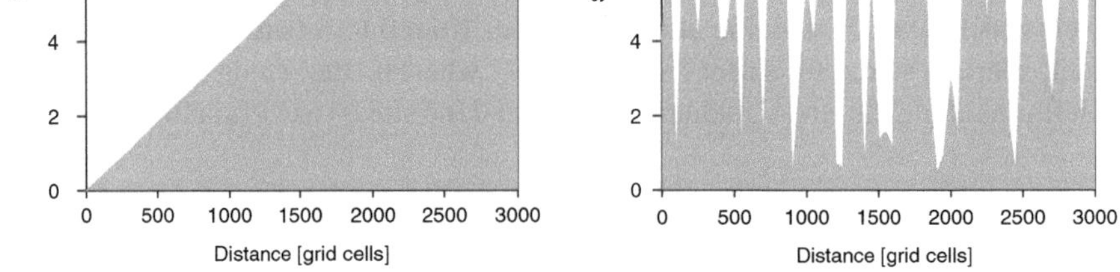

Fig. 13.2 Shapes of the stress gradients. (**a**) Linear stress gradient (0 for no stress, 11 for maximum stress). (**b**) No spatial correlation of stress along the transect

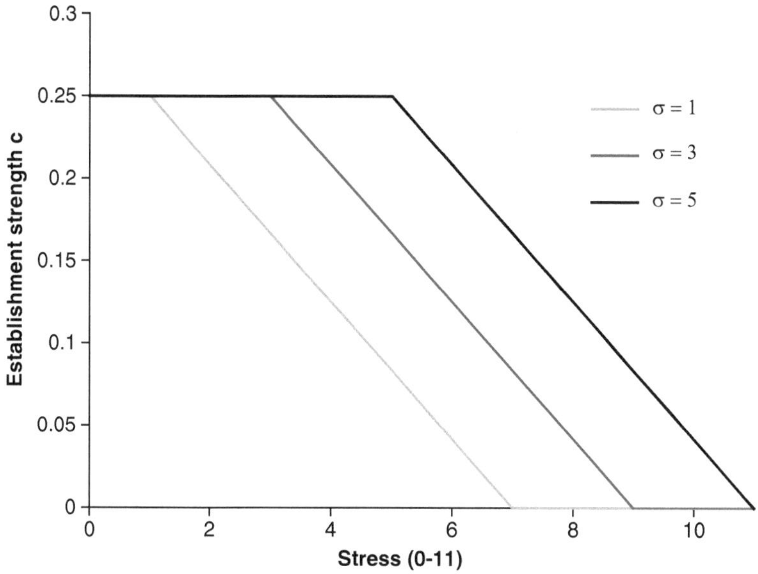

Fig. 13.3 Plant functional type (PFT)-specific reduction of the establishment strength c_i due to stress. Starting from the same seedling density, all PFTs have the same establishment strength under nonsaline conditions. As soon as stress exceeds the PFT-specific tolerance level σ, the establishment strength c_i declines linearly to zero (Modified after Groeneveld 2003)

in empty grid cells at the start of each time step (see Fig. 13.1). Intraspecific density regulation caps the local seedling densities and stress reduces the local seedling densities even further if the stress index s exceeds the PFT-specific stress tolerance level. Above the stress tolerance level, seedling densities decrease linearly with increasing stress until all seedlings die (Fig. 13.3). This reflects the assumption that stress-adapted species perform best in low-stress environments.

Finally, in all mixed grid cells, i.e., where more than one PFT is present, the winning PFT has to be chosen. This central establishment process is modelled by a lottery competition (*sensu* Chesson and Warner 1981), where one winner is chosen from several juvenile individuals competing for dominance in a grid cell. The local PFT-specific establishment probabilities ρ_i are given by Eq. 13.1.

$$\rho_i = \frac{c_i}{\sum_{j=1}^{S} c_j} \tag{13.1}$$

where ρ_i is the establishment probability of PFT i, c_i is the establishment strength of PFT i and S is the total number of PFTs.

The establishment strength of PFT i, c_i, characterizes its weight in the lottery competition and can be identified as the density of competing juveniles. Seeds, seedlings and juveniles are not modelled individually in the simulation and therefore the competition strength c_i describes not a discrete number of individuals, but a density. The absolute values of these densities and the establishment strength c_i are less important because the establishment probability is determined by the weight of the establishment strength of PFT i relative to the establishment strength of all PFTs.

After the establishment phase, all mature perennial individuals and annual cohorts disperse seeds. There are two seed dispersal scenarios:

1. Global seed bank scenario: Each fertile individual or cohort contributes seeds to the nonspatial global seed bank. The establishment probability of a PFT is proportional to its relative abundance in this global seed bank; therefore, establishment probabilities are the same for all sites.
2. Local dispersal scenario: In this scenario all individuals or cohorts disperse their seeds following a spatial dispersal kernel. Perennial PFTs and annual PFTs have dispersal kernels described by an exponential decay distribution that is parameterized such that 99.9% of all seeds are distributed within a radius of 500 m for seeds from perennial PFTs and 1,000 m for annual PFTs. Seed banks are finally determined by the superposition of all normalized dispersal kernels.

All annual cohorts and all perennial individuals which have reached their maximum age die at the end of each time step. Juvenile perennial individuals survive with an age-, stress- and precipitation-dependent probability. Adult perennial individuals always survive until they reach their longevity (see Table 13.1). Finally, only a fraction of the seeds that have not germinated survive to the next year. An overview of all processes considered is shown in Fig. 13.1 together with their dependencies on abiotic factors.

13.3 Simulation Experiments and Results

In the first scenario, environmental stress is described by a linear gradient (see Fig. 13.2a). Seed dispersal is simulated nonspatially, i.e., each fertile plant or cohort increases the density of its PFT in the global seed bank (see Sect. 13.2 for details).

In the second scenario, a linear stress gradient is also used (see Fig. 13.2a), but seed dispersal is local (see Fig. 13.2 for details).

In the third scenario, we also used localized dispersal. However, the stress conditions are not spatially correlated. Stress is allocated at random (uniform 0–11) along the transect. Environmental conditions do not change in the direction which is rectangular to the main direction of the transect (following the receding seawater). All simulations were run for 640 time steps. Abundances of PFTs were counted along 24 subplots for the first year of average rainfall conditions after 600 years of simulation time.

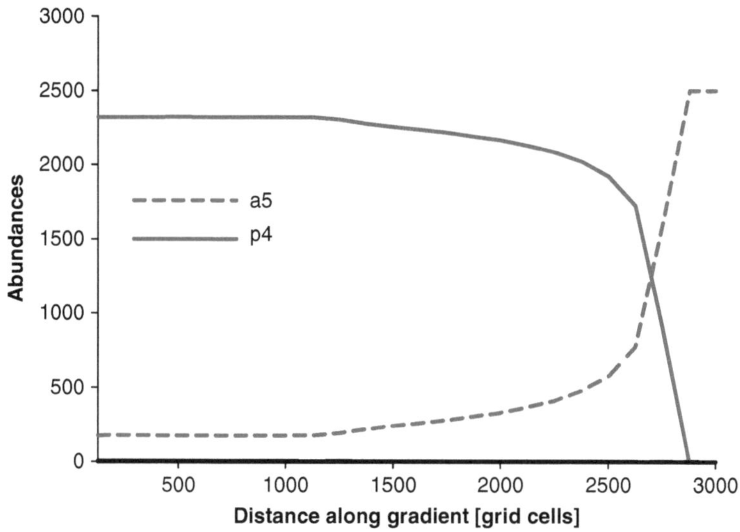

Fig. 13.4 Spatial distribution of different PFTs (*a* annual or *p* perennial and different stress tolerances: 5 is highest and 0 means no stress tolerance) after 600 years for a representative single run. For these simulations the spatially explicit dispersal process was replaced by a nonspatial global seed bank. Only two types persist for 600 years in the system and the perennial species could not outcompete the annual type at any locality (Modified after Groeneveld 2003)

13.3.1 Results for the First Scenario: Global Nonspatial Seed Bank

The spatial distribution of two dominating PFTs is shown in Fig. 13.4 for simulations after 600 years where seed dispersal was implemented as a nonspatial global seed bank. Only the best adapted PFTs of each life form, a5 and p4, persist under these conditions. Environmental stress conditions increase linearly with the distance along the gradient. The annual type a5 is present along the whole transect and cannot be outcompeted by the perennial type p4. At the stressful end of the transect, p4 cannot survive and the annual type a5 dominates.

13.3.2 Results for the Second Scenario: Local Dispersal

If seed dispersal is a spatially explicit process (Fig. 13.5a) the number of PFTs which can coexist for 600 years is doubled (four PFTs) compared with the nonspatial situation (two PFTs). Three PFTs segregate along the stress gradient (p1, p4, a5), whereas the intermediate stress-tolerant perennial type p3 can persist in the overlap zone, where p1 and p4 dominate. Compared with the nonspatial global seed

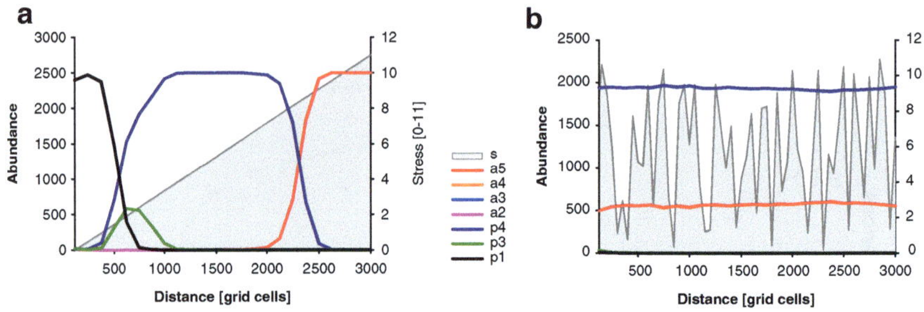

Fig. 13.5 Spatial distribution of different PFTs (*a* annual or *p* perennial and different stress tolerances: 5 is highest and 0 means no stress tolerance) after 600 years for a representative single run. (**a**) Spatially explicit seed dispersal and a linear stress gradient allow four PFTs to coexist. (**b**) No spatial correlation of stress conditions and spatially explicit seed dispersal allow only two PFTs to coexist

bank scenario (first scenario), the extremely stress tolerant annual type a5 is now restricted to the most stressful end of the gradient.

13.3.3 Results for the Third Scenario: High Spatial Variability of Stress Conditions

If the spatial stress conditions change on a smaller spatial scale (Fig. 13.2b), only two PFTs can persist (p4 and a5), whereas the perennial PFT dominates the community (Fig. 13.5b). These results are similar to those for the first scenario (nonspatial global seed dispersal and linear stress gradient), where the most-stress-tolerant perennial PFTs dominate most of the transect.

13.4 Discussion

A multi-PFT simulation model was developed to study the spatiotemporal vegetation dynamics along stress gradients. Parameterization and PFT definition were orientated towards the situation at the Aral Sea (see Chaps. 9, 10 and 12 for further details). One of the aims of the model was to study the effect of spatial processes as seed dispersal and the spatial distribution of abiotic conditions on the spatial structure and richness of the simulated plant community. Zonation as one potential spatial outcome is a common pattern along environmental gradients that has been reported for many systems (Siccama 1974; Kenkel et al. 1991), especially in marshland communities (Bertness and Hacker 1994; Emery et al. 2001; Greiner et al. 2001; see Chaps. 9 and 10 for the specific situation at the Aral Sea).

Our results show that abrupt changes in the vegetation structure (zonation) can evolve along continuous environmental gradients if the competitive abilities of PFTs differ. The zonation is facilitated by a trade-off between longevity and stress tolerance. Under low-stress conditions, the long-lived PFT p1 could outcompete all other PFTs, simply because individuals of this PFT occupy a grid cell much longer than that of the competing species and therefore had locally a higher per capita growth rate. This positive feedback results in the monodominance of this particular PFT. DeAngelis and Post (1991) also showed with a coupled partial differential equation (PDE) system that positive feedbacks can lead to the spatial segregation of species along subtle gradients. Although PDE models benefit from a compact description and sophisticated mathematical theory, PDE models of this type cannot consider stochastic processes and are hardly applicable to multispecies systems (but see Kohyama 1992). In general, equation-based models such as PDE models cannot incorporate neighbourhood interactions. In recent years substantial progress has been made to approximate neighbourhood interactions by analytical methods such as pair approximation, moment equations and coalescence methods (Dieckmann and Law 2000; Rosindell et al. 2008). Nevertheless, for some of these methods spatially explicit simulations are essential to test whether these approximations are valid and often it is not possible to include stochastic processes in these models. Equation-based models usually assume that the system will end in some sort of equilibrium, but most often ecological systems will not be in or will never reach an equilibrium. Thus, a grid-based approach seems to be more appropriate for most of the complex multispecies problems in ecology.

Spatial processes such as local competition and seed dispersal are key drivers for plant community structure and dynamics (Tilman and Kareiva 1997). We have shown here that along environmental gradients local seed dispersal resulted in zonation and therefore spatial aggregation of species, which enhanced the overall species richness. The same is true for spatially homogeneous environments, where local dispersal causes a clumped and aggregated spatial distribution of plants (Bolker and Pacala 1999). Aggregation increases intraspecific competition and reduces interspecific competition, which is a fundamental requirement for stable coexistence (Wissel 1989). Furthermore, dispersal limitation has been found to be important to explain fundamental community measures as rank abundance distributions in neutral models of biodiversity (Hubbell 1997). The spatial nature of seed dispersal is not only important on the local scale. Rare long-distance seed dispersal events are crucial for invasions, metapopulation and metacommunity dynamics (Cain et al. 2003). This highlights the importance of the consideration of spatial processes in ecological theory (Jeltsch and Moloney 2002) and casts further doubt on the applicability of so-called mean field approaches (Tilman and Kareiva 1997; Dieckmann and Law 2000).

Our study also highlights the importance of the spatial resolution of the spatial variability in abiotic conditions for species coexistence (Palmer 1992). In general, spatial variability promotes persistence and coexistence in a wide range of ecological systems (Braak and Prentice 1988). However, the spatial scale of this variability matters. If suitable habitats are becoming too small, if spatial variability is too high, the most-stress-tolerant PFTs will dominate.

The complex nature of ecological systems requires flexible modelling tools such as grid-based, stochastic and rule-based simulation models (Wissel 2000). Advances in ecological theory, computer science and statistics enable us to analyse, calibrate and perform complex simulation models efficiently to improve our understanding of the fundamental mechanisms that structure ecological systems (Grimm et al. 2005; Elith et al. 2008).

Acknowledgements We thank the German Federal Ministry of Education, Science, Research and Technology (BMBF) and the project coordinator (Research Center Jülich GmbH, PTJ Berlin) for financial support (project no. 0330389) as well as the Academy of Science, Botanical Institute (Almaty) for their overall cooperation. J.G. was also supported by the Department of Ecological Modelling, UFZ, the European Community (MOIF-CT-2006-40571) and the BMBF.

References

Badescu V, Schuiling RD (2010) Aral Sea; irretrievable loss or Irtysh imports? Water Resour Manag 24:597–616

Bertness MD, Hacker SD (1994) Physical stress and positive associations among marsh plants. Am Nat 144:363–372

Bolker BM, Pacala SW (1999) Spatial momentum equations for plant competition: understanding spatial strategies and the advantages of short dispersal. Am Nat 153:575–602

Cain ML, Nathan R, Levin SA (2003) Long-distance dispersal. Ecology 84:1943–1944

Chesson PL, Warner RR (1981) Environmental variability promotes coexistence in lottery competitive systems. Am Nat 117:923–943

DeAngelis DL, Post WM (1991) Positive feedback and ecosystem organization. In: Higashi M, Burns TP (eds) Theoretical studies of ecosystems the network perspective. Cambridge University Press, Cambridge, pp 155–178

Dieckmann U, Law R (2000) Relaxation projections and the method of moments. In: Dieckmann U, Law R, Metz JA (eds) The geometry of ecological interactions: simplifying spatial complexity. Cambridge University Press, Cambridge, pp 412–455

Elith J, Leathwick JR, Hastie T (2008) A working guide to boosted regression trees. J Anim Ecol 77:802–813

Emery NC, Ewanchuk PJ, Bertness MD (2001) Competition and salt-marsh plant zonation: stress tolerators may be dominant competitors. Ecology 82:2471–2485

Greiner La Peyre MK, Grace JB, Hahn E, Mendelssohn IA (2001) The importance of competition in regulating plant species abundance along a salinity gradient. Ecology 82:62–69

Grimm V, Revilla E, Berger U, Jeltsch F, Mooij WM, Railsback SF, Thulke HH, Weiner J, Wiegand T, DeAngelis DL (2005) Pattern-oriented modelling of agent-based complex systems: lessons from ecology. Science 310:987–991

Groeneveld J (2003) Raum-zeitliche Artenmuster entlang von Stressgradienten: Ein Simulationsmodell in Anlehnung an die Pflanzensukzession auf den trockengefallenen Aralseeböden. Ph.D. thesis, University of Marburg, Marburg

Groeneveld J, Wissel C, Wucherer W, Dimeyeva L, Jeltsch F (2005) Computer simulations orientated towards plant succession on the desiccating Aral Sea floor. In: Veste M, Wucherer W, Homeier J (eds) Ökologische Forschung im globalen Kontext. Festschrift für Prof Dr Siegmar-W Breckle. Cullivier, Göttingen, pp 129–141

Hubbell SP (1997) A unified theory of biogeography and relative species abundance and its application to tropical rain forests and coral reefs. Coral Reefs 16:S9–S21
Jeltsch F, Moloney KA (2002) Spatially-explicit vegetation models: what have we learned? Prog Bot 63:326–343
Kenkel NC, McIlraith AL, Burchill CA, Jones G (1991) Competition and the response of 3 plant-species to a salinity gradient. Can J Bot 69:2497–2502
Kohyama T (1992) Density-size dynamics of trees simulated by a one-sided competition multi-species model of rain forest stands. Ann Bot 70:451–460
Lehouerou HN, Bingham RL, Skerbek W (1988) Relationship between the variability of primary production and the variability of annual precipitation in world arid land. J Arid Environ 15:1–18
Micklin P (2007) The Aral Sea disaster. Annu Rev Earth Planetary Sc 35:47–72
Micklin P, Aladin NV (2008) Reclaiming the Aral Sea. Sci Am 298:64–71
O'Hara SL, Wiggs GFS, Mamedoc B, DavidsonG HRB (2000) Exposure to airborne dust contaminated with pesticide in the Aral Sea region. Lancet 355:627–628
Palmer MW (1992) The coexistence of species in fractal landscapes. Am Nat 139:375–397
Rosindell J, Wong Y, Etienne RS (2008) A coalescence approach to spatial neutral ecology. Ecol Inform 3:259–271
Siccama TG (1974) Vegetation, soil and climate in the Green Mountains of Vermont. Ecol Monogr 44:325–349
ter Braak CJF, Prentice IC (1988) A theory of gradient analysis. Adv Ecol Res 18:271–313
Tilman DA, Kareiva PM (1997) Spatial ecology: the role of space in population dynamics and interspecific interactions. Monographs in population biology. Princeton University Press, Princeton
Ungar IA (1998) Are biotic factors significant in influencing the distribution of halophytes in saline habitat? Bot Rev 64:176–199
Wissel C (1989) Theoretische Ökologie: Eine Einführung. Springer, Berlin
Wissel C (2000) Grid-based models as tools for ecological research. In: Dieckmann U, Law R, Metz JA (eds) The geometry of ecological interactions: simplifying spatial complexity. Cambridge University Press, Cambridge, pp 94–115
Wucherer W (1990) Vegetation development on the new habitats in the desert. Gylym, Alma–Ata (Russian) 214 pp
Wucherer W, Breckle SW (2001) Vegetation dynamics on the Dry Sea floor of the Aral Sea. In: Breckle SW, Veste M, Wucherer W (eds) Sustainable land use in deserts. Springer, Berlin, pp 52–67

Part C
Means of Present and Future Development

Chapter 14
Nature Conservation in the Aral Sea Region: Barsa-Kelmes as an Example

L.A. Dimeyeva, N.P. Ogar, Z. Alimbetova, and S.W. Breckle

14.1 General Aspects

Strict nature reserves play a most important role among all categories of protected areas. The main function of them is the conservation of genetic potentials and ecosystems, all-year-round monitoring of natural processes, the ability for reproduction and maintaining living conditions for wildlife.

There are more than 20 nature reserves, with a total area of 600,000 ha, in Uzbekistan, Turkmenistan, Tajikistan and Afghanistan, all belonging to the Aral Sea basin. Barsa-Kelmes Nature Reserve has been subjected to the greatest changes among all protected areas. The desiccation of the Aral Sea led to contact of the island with the eastern coast, and wild animals from the nature reserve could migrate to the coast. The existence Barsa-Kelmes Nature Reserve since its establishment in 1939 protected ecosystems of the northern Turan Kazakhstan deserts in natural conditions as models of landscape structure and biodiversity. Ecosystems of the island of Barsa-Kelmes had guaranteed natural successions practically without human impact. Unification of the island and the eastern coast of the Aral Sea made it very necessary to extend the protected area.

The territory of the nature reserve had expanded in 2006 almost by ten times (160,826 ha), including the territory of the dry seafloor (Wucherer and Breckle 2005). The area consists of two cluster areas: (1) the former island of Barsa-Kelmes (see Fig. 2.7) with the surrounding dry seafloor; (2) the former islands of Kaskakulan and Uzun-Kair with the surrounding dry seafloor, which are the main habitats of onager (*kulan*; *Equus hemionus onager*) and Persian gazelle (*jairan*; *Gazella subgutturosa*), to where they migrated after unification of the island with the original coast.

It is important to include in the protected area the wetlands in the mouth of the Syr Darya (Novikova 2001). The modern delta is now flooded after the construction of a dam between the North Aral Sea [Small Aral Sea, 42–43 m above sea level (asl)]; shallow waters cover the reeds – the main habitat of waterfowl. In the river mouth, a new delta has been formed with hygromesophytic meadow–tugaic

S.-W. Breckle et al. (eds.), *Aralkum - a Man-Made Desert: The Desiccated Floor of the Aral Sea (Central Asia)*, Ecological Studies 218,
DOI 10.1007/978-3-642-21117-1_14, © Springer-Verlag Berlin Heidelberg 2012

vegetation with a complex of wild animals and water and forest (tugaic) birds. Deltaic lakes and shallow water are spawning places for fish. It is the most active area for economic activity (hunting, fishing, haying, and firewood) and poaching. Protected regulations will guarantee the conservation, maintenance and reproduction of populations of commercial species of fauna for future sustainable development of the Small Aral Sea lake system.

Barsa-Kelmes Nature Reserve is situated in a region of the Aral ecological catastrophe. There are many representatives from the *Red Data Book of Kazakhstan* among animals that live in the nature reserve: onager (*Equus hemionus onager*), Persian gazelle (*Gazella subgutturosa*), saiga antelope[1] (*Saiga tatarica*), 23 species of birds and 4 species of invertebrates. Flora and vegetation represent typical combinations of plants and vegetation types of the region as well as unique trends of succession development in the Aral desiccated seafloor. The flora of vascular plants (76%) reflects the plant diversity of the Kazakhstan part of the Aral Sea coast. There are 14 endemics and 3 species from the *Red Data Book of Kazakhstan* among them.

14.2 Barsa-Kelmes Nature Reserve

14.2.1 General Information

14.2.1.1 History

A hunting farm on the island was set up in 1929 for breeding and trade of gopher (*Citellus maximus*). Several species were brought to the island, e.g., gopher (*Citellus maximus*), Persian gazelle (*Gazella subgutturosa*), saiga (*Saiga tatarica*), brown hare (*Lepus europaeus*), partridge (*Perdix perdix*) and pheasant (*Phasianus colchicus*). Barsa-Kelmes Nature Reserve was established in 1939. Fifty to sixty individuals of saiga were counted on the official opening of the nature reserve (Vasenko 1950). Onagers (*Equus hemionus onager*) were brought in 1953 from Turkmenistan. The scientific staff conducted nature observations and annual calculations of ungulate animals. The number of animals changed greatly from year to year. It was found that climate, winter conditions, availability of forage, etc. played a major role. The number of *Gazella subgutturosa* fluctuated very much. Nine individuals were brought in 1929 from Karakalpakstan. They amounted to 2,000 in 1948. After a severe winter in 1948–1949, only 60 individuals survived (Zhevnerov 1984). The mammal most adapted to island conditions was the saiga antelope. Their number varied from 900 to 3,000. In 1983 about 230 examples of *Saiga tatarica*, 160 of *Gazella subgutturosa* and of 242 *Equus hemionus onager*

[1] This species is still not included in the *Red Data Book of Kazakhstan*, but in the last decade became rare.

lived on the island (Afanasiev and Smirnov 1985). Problems with freshwater and the increasing salinity of the seawater resulted in the inevitable decision to remove the onagers gradually from the island. Some remnant animals moved to the mainland after unification of the island with the eastern coast. Inspection in 2008 showed that at the boundaries of Barsa-Kelmes Nature Reserve, 273 onagers, 170 saigas and 67 Persian gazelles were living (see also Chap. 11).

14.2.1.2 Topography

The nature reserve is situated in the plain and includes the former island of Barsa-Kelmes, the dry seafloor and the area adjacent to the former islands of Kaskakulan and Uzun-Kair. The highest point (108 m asl) is located in Barsa-Kelmes. The relief of the former island is divided into two parts: the southern – high plateau – and the northern – undulating plain crossing from south to north by valleys of temporary streams (Kuznetsov 1979). The northwestern, northern and eastern coasts are bordered by sand dune belts. The southwestern and southern shores have abrasion character. There are shallow drainless depressions at the surface where takyrs and solonchaks are formed. The original coast is separated from the gently dipping marine plain by a well-marked terrace. The primary marine plain is formed on the dry seafloor with a slightly inclined surface partly with drying cracks. The relief of the eastern coast is formed by hillock–low hummocky sand dunes oriented almost in a meridional direction. Alluvial plains differ by the slight incline with formation from erosion and accumulative processes (Veselova et al. 1987).

14.2.1.3 Geology

The basal massif of the island of Barsa-Kelmes is composed of Oligocene gypsiferous argil, aleurite and sandstone (Yanshin 1953). The northern part is composed of fine-grained sand, aleurite and loam with black silt interlayers. The marine and lacustrine sediments of the former coasts are represented mostly by medium-grained and coarse-grained sand. Quaternary and recent deposits have spread in the plain. The primary marine plains are composed of differently grained sand, aleurite and agglomerations of shells, restricted by clay and rubble – pebble material.

14.2.1.4 Soils

Grey-brown soils of different salinity and texture are represented at the original coast of the former island (Kuznetsov 1979). Most widespread are grey-brown alkaline soils; takyrs, solonchaks are represented as well. Sandy soils occur in the northern, western and eastern coasts of the former island. Marsh and coastal solonchaks, coastal sand and coastal soils with blown sand cover have formed on the dry seabed. Flood-plain meadows, meadow swamps and swampy soils are represented in the Syr Darya delta.

14.2.1.5 Climate

The climate is temperate with long hot summers, relatively cold winters, insignificant cloudiness and low precipitation that is typical of the northern deserts (see Chap. 4). The annual precipitation is low (126–128 mm), most of it falling in wintertime (see Fig. 4.1). The average air temperature in July is 25–26°C; the absolute maximum reaches 42–44°C. The average air temperature in January is −10 to −13°C; the absolute minimum is −34 to −36°C with strong winds. Northeasterly winds prevail, with average velocities of 3.5–6 m/s (the maximum reaches 20–24 m/s). The snow cover is unstable, and often strong winds blow it off. The duration of snow cover is about 80–90 days. The frost zone in soil is 45 cm, and complete defrosting is marked at the end of March.

14.2.2 Flora

According to various publications (Kuznetsov and Burambayev 1976; Kuznetsov 1979, 1995; Dimeyeva and Kuznetsov 1999; Dimeyeva and Alimbetova 2006, 2007a, b) and the last investigations on the dry seabed of the Aral Sea, the angiosperm flora of the Barsa-Kelmes peninsula and the Kaskakulan area consists of 298 species, belonging to 50 families and 176 genera. Species from Chenopodiaceae, Asteraceae, Brassicaceae, Poaceae and Polygonaceae families prevail, as on the desiccated seafloor (Chap. 8). The most important genera (Table 14.1) are

Table 14.1 Spectrum of leading vascular plant families of the flora of Barsa-Kelmes Nature Reserve

Plant families (50)	Genera (176)	Species (298)	Species (%)
Chenopodiaceae	24	60	20.1
Asteraceae	23	36	12.1
Brassicaceae	21	30	10.1
Poaceae	19	29	9.7
Polygonaceae	4	22	7.4
Fabaceae	10	18	6.0
Boraginaceae	9	12	4.0
Ranunculaceae	7	8	2.7
Liliaceae	3	7	2.3
Apiaceae	2	5	1.7
Caryophyllaceae	4	5	1.7
Cyperaceae	4	5	1.7
Additional families (38)	138	61	25.6

Calligonum (16 species), *Artemisia* (11 species), *Atriplex* (11 species), *Astragalus* (8 species), *Strigosella* (7 species), *Salsola* (6 species) and *Climacoptera* (5 species). There are 14 Kazakh endemics (*Artemisia aralensis*, *Artemisia scopiformis*, *Artemisia quinqueloba*, *Artemisia camelorum*, *Atriplex pratovii*, *Atriplex pungens*, *Petrosimonia hirsutissima*, *Astragalus brachypus*, *Tulipa borszczovii*, *Calligonum crispatum*, *Calligonum palibinii*, *Calligonum humile*, *Calligonum spinulosum*, *Corispermum laxiflorum*) and three species from the *Red Data Book of Kazakhstan* (*Tulipa biflora*, *Tulipa borszczowii*, *Atriplex pratovii*).

14.2.3 Vegetation Cover of Barsa-Kelmes

14.2.3.1 Vegetation of the Plateau and the Low Plains

The important feature of the vegetation is complexity. Zonal vegetation is composed of three basic species: *Artemisia terrae-albae*, *Anabasis salsa* and *Agropyron desertorum*. Saxaul (*Haloxylon aphyllum*) often occurs in sagebrush communities. Most widespread is the *Agropyron desertorum–Anabasis salsa–Artemisia terrae-albae* community. *Artemisia terrae-albae* and *Anabasis salsa* communities dominate in the low plains; regular alternation of them is caused by texture, degree of salinity and alkalinity of grey-brown soils (Kuznetsov 1979, 2007).

Communities of sagebrush (*Artemisia terrae-albae*) prevail on the plateau and on the low plains. They correspond to zonal grey-brown soils, sometimes alkaline or gypsiferous. The floristic composition of them comprises 60 species of vascular plants, at least 5 species of lichens and 2 species of fungi. Plant communities consist of six to seven species with 35–70% coverage. Subdominants of sagebrush communities are usually represented by *Eremopyrum orientale*, *Lepidium perfoliatum*, *Stipa lessingiana*, *Agropyron desertorum*, *Anabasis salsa*, *Anabasis aphylla* and *Haloxylon aphyllum*.

Communities of *Anabasis salsa* occur on grey-brown alkaline and solonchak soils. There are three to thirteen species in plant communities; the vegetation coverage is 10–25%. The floristic composition of them comprises 53 species of vascular plants, at least six species of lichens and two species of fungi. Subdominants of communities are usually *Artemisia terrae-albae* and *Eremopyrum orientale*. Dwarf subshrubs, ephemerals and ephemeroids are very important among life forms of these plant communities, as was shown for the other areas of the Aralkum too (see Chap. 9).

The intrazonal vegetation is distributed on solonchaks. Solonchaks of the northwestern part of Barsa-Kelmes are vegetated by *Halocnemum strobilaceum* and *Limonium suffruticosum* communities. Saltwort vegetation covers takyr-like solonchaks in the northeastern part (*Halocnemum strobilaceum*, *Climacoptera aralensis*, *Cimacoptera brachiata*, *Eremopyrum triticeum*, *Senecio noeanus*) with a coverage of 25–30%. Badlands with sulphate solonchaks are distributed in the areas of the wedging out of Tertiary argil (southeastern and western parts of the peninsula). They are almost without vegetation; only rare aggregations of annual saltworts can be found there (*Climacoptera aralensis*, *Halimocnemis sclerosperma*, *Bienertia cycloptera*).

Communities of *Caragana grandiflora* and *Stipa lessingiana* located in ravines of the plateau play a less important role in the vegetation cover of the original coast. Communities of *Krascheninnikovia ceratoides* and *Salsola arbuscula* are situated on the sand dunes bordering the western coast. Meadows with *Aeluropus littoralis* and *Phragmites australis* are connected with depressions flooded by springwater.

14.2.3.2 Vegetation of the Aral Marine Terraces

The origin of the terraces is connected with the Aral Sea transgressions and aeolian processes. The marine terraces are composed of medium-grained and fine-grained sand which developed on Tertiary gypsiferous argil occurring at a depth of 1–1.5 m in a number of cases. The sands are low in salinity or even nonsaline. The relief is mostly flat with gentle slopes and small hummocks of phytogenic origin; hillock – low hummocky sand dunes (nepkha semidesert) are situated at the northwestern and northeastern coasts of the peninsula. The terraces surround the former island as a belt from 100–200 to 2,000 m width. They are not separated at the steep southern coast.

Saxaul (*Haloxylon aphyllum*, *Haloxylon persicum*) dominates in most vegetation covers. Plant communities of *Atraphaxis spinosa*, *Ephedra distachya*, *E.intermedia*, *Calligonum aphyllum*, *Calligonum caput-medusae*, *Calligonum macrocarpum*, *Convolvulus erinaceus* and *Artemisia arenaria* are less significant. Saxaul communities have a coverage from 10% to 70%; their floristic composition includes more than 30 plant species: *Ephedra distachya*, *Atraphaxis spinosa*, *Calligonum aphyllum*, *Astragalus brachypus*, *Alhagi pseudalhagi*, *Salsola paulsenii*, *Anisantha tectorum*, *Meniocus linifolius*, *Alyssum turkestanicum*, *Eremopyrum orientale*, *Senecio noeanus*, *Kochia odonthoptera*, *Eremopyrum orientale*, and *Aeluropus littoralis*. Saxaul communities grow mostly at the eastern and western coasts. *Haloxylon aphyllum–Ephedra distachya* and *Calligonum aphyllum–Ephedra distachya* communities are typical of the eastern coast (Pankratova 2007). Communities of *Atraphaxis spinosa* are mostly found at the northern and western coasts; they have a coverage of 10–25% (sometimes up to 50%), and their floristic composition often includes fewer than 15 plant species.

Ephedra distachya communities occur at the northern coast; their microcoenoses are marked at all coasts. Their vegetation coverage is 15–25%, and their floristic composition comprises 14 species. Communities with a dominance of *Calligonum* spp. are distributed rarely at the northern and eastern coasts, but they occur everywhere as components of psammophytic plant communities.

Artemisia arenaria communities are found only in the area of the former wells (at the northern coast); their coverage is 15–20%, and their floristic composition includes nine species. *Tamarix* spp. communities occur at the northern coast. Twenty years ago, they were characterized by their high productivity and a coverage of 70–80%.

The decreasing groundwater table has caused their degradation. Communities of *Nitraria schoberi* (Demchenko 1950) had already disappeared.

Dunes of 4–6-m height and 5–20-m width adjoin the terraces from the sea direction. The vegetation of dunes is formed by *Tamarix laxa*, *Tamarix ramosissima*, *Tamarix hispida*, *Haloxylon aphyllum*, and *Calligonum* spp. The coverage in plant communities ranges from 15% to 70%, and the floristic composition comprises 15 species (*Stipagrostis pennata*, *Eremosparton aphyllum*, *Convolvulus erinaceus*, *Salsola paulsenii*, etc.). Psammophytic shrubs actively occupy sand dunes, whereas around *Tamarix* new dunes are formed on the dry seabed (nepkhas).

14.2.3.3 Vegetation of the Dry Seafloor

The dry seafloor is represented by a series of sand beaches corresponding to levels of the Aral Sea desiccation with differing ecological conditions. The zone of marsh and coastal solonchaks is characterized by a shallow groundwater table (40–45 cm). The crust horizon is highly saline (12.95%); salinity is caused by sodium chloride and sodium sulphate. Underlying horizons have low and medium salinity. Sands adjoining dunes are nonsaline; here the groundwater is at a depth of 140 cm.

The Northern and Northwestern Coasts

Aggregations of *Salicornia europaea* are distributed on marsh and coastal solonchaks in the first years of desiccation (Chap. 10). Three to four species form the floristic composition (*Salicornia europaea*, *Tamarix hispida*, *Suaeda crassifolia*, *Halocnemum strobilaceum*). Tamarisks occur as single plants or belts that are fixed by dry *Zostera noltii* coinciding with the decreasing levels of the sea. The relief is phytogenic of 30–50-cm height. The vegetation coverage is 25%. The floristic composition consists of six species: *Tamarix laxa*, *Corispermum hyssopifolium*, *Phragmites australis*, *Suaeda crassifolia*, *Salicornia europaea*, and *Atriplex pratovii*.

The desalinization of the sand and the deflation of surface horizons lead to the formation of typical psammophytic communities starting from rare aggregations of *Stipagrostis pennata* on hummocks and *Eremosparton aphyllum* in blown-out depressions. Communities of psammophytic shrubs with a coverage of 14–17% have formed on the beaches of the 1960s. Five plant species (*Eremosparton aphyllum*, *Astragalus brachypus*, *Convolvulus erinaceus*, *Calligonum* sp., *Salsola paulsenii*) represent the floristic composition. *Halocnemum strobilaceum* and *Haloxylon aphyllum–Halocnemum strobilaceum* communities are widespread along the northern coast. The vegetation coverage is 25%. Communities of *Alhagi pseudalhagi* with psammophytic shrubs of 50–55% coverage and a floristic composition of seven species have been formed on the beaches of the 1960s.

The Western Coast

The territory was widely vegetated by *Stipagrostis pennata* communities and aggregations in the last century. Nowadays, the typical spatial order of primary plant communities is as follows: *Atriplex pratovii* → rare *Halocnemum strobilaceum* (*Tamarix hispida*) → rare aggregations *of Stipagrostis pennata, Salsola paulsenii, Astragalus brachypus, Tamarix hispida* → community of psammophytic shrubs (*Astragalus brachypus, Eremosparton aphyllum, Calligonum aphyllum*) with *Haloxylon aphyllum* → tamarisk belt (*Tamarix laxa*) → community of psammophytic shrubs (*Astragalus brachypus, Eremosparton aphyllum, Calligonum aphyllum, Atraphaxis spinosa*) with *Haloxylon aphyllum.*

The Eastern Coast

The former island of Barsa-Kelmes is now joined to the original eastern coast. The soils of the dry seabed are represented mostly by coastal solonchaks and coastal soils with a blown-out sand and salt-dust cover. Vegetation cover between the island and the original coast consists of an aggregation of annual saltworts (*Salicornia europaea, Suaeda acuminata, Bassia hyssopifolia, Petrosimonia triandra, Atriplex pratovii*), sometimes with *Tamarix laxa.* Barren lands with a rare plant cover (below 1%) or without plants (open salt desert) occur frequently. Occupation by *Salicornia europaea* and *Suaeda acuminata* in some places is connected with spring–summer rainfall and respective seed banks in the soil. Saxaul communities are particularly widespread near the peninsula, where they often form a phytogenic relief; hummocks can reach 50–150 cm in height. The coverage of vegetation is up to 40%. There are five species in the floristic composition: *Haloxylon aphyllum, Salsola paulsenii, Atriplex pratovii, Halostachys belangeriana* and *Stipagrostis pennata*. Thus, it is a mixture of species from various communities, which are still in a dynamic state (see also Chap. 9).

The Southern Coast

The territory adjoins the cliffs of the plateau (chinks). The dry seabed is composed of marine deposits of different size: coarse-grained sand and rubble – pebble material. The vegetation cover is formed by communities of *Astragalus brachypus, Phragmites australis, Climacoptera aralenis,* and *Haloxylon aphyllum* with a high coverage (30–80%) and a floristic composition of three to eight species (*Stipagrostis pennata, Salsola paulsenii, Salsola foliosa, Alhagi pseudalhagi, Senecio noeanus, Atriplex pratovii, Atriplex aucheri, Descurainia sophia*).

14.2.3.4 Primary Successions in Conditions of the Island of Barsa-Kelmes

Three types of primary succession have been distinguished in the Aral Sea coast, differing by soil texture and salinity, temporal dynamics and final stages (Dimeyeva 2007a, b): the psammosere – primary succession on sand sediments towards psammophytic shrub vegetation (see also Chap. 10, Table 10.11); the halosere – on saline ground of heavy texture towards the formation of halophytic dwarf semishrub communities (Table 10.10); the potamosere – on saline soils towards the formation of a tugaic shrub vegetation with additional water supply. Each of them are controlled by different mechanisms (according to Connel and Slatyer 1977).

Two types of succession have been distinguished for the dry seabed of the island (see also Chap. 10). All insular coasts are surrounded by sands. The gradual desiccation of the sea level caused the formation of parallel dune belts with tamarisk vegetation (psammosere type). The vegetation in interdune spaces is formed according to a potamosere. As expected, the heavy marine sediments of high salinity are now the most common ones. The vegetation under such conditions will develop as a halosere.

Data on vegetation dynamics of the dry seabed at the northern coast of the island of Barsa-Kelmes are presented in Table 14.2.

A study in 2007 was conducted jointly with the staff of the nature reserve. Observations have shown that first stage of succession involved *Atriplex pratovii* aggregations and communities. The leading role of *Atriplex pratovii* remained until the beginning of the 1990s. During the first decade of this century, the salinity of marine sediments increased greatly. This caused a change of pioneer species (*Salicornia europaea*, *Suaeda crassifolia*). The first specimens of the halophytic dwarf subshrub *Halocnemum strobilaceum* occupy the surface in the fourth year of succession. The period of *Halocnemum strobilaceum* depends on the groundwater depth. If it is more than 3 m deep, the plant begins to reduce its growth, and perishes when the groundwater is at 4–5 m (Nikitin 1966). Such a phenomenon has already been registered in the plots of the profiles (N13, 14, Table 14.2). Aeolian activity leads to covering of the primary surface by blown sand, where seeds of *Haloxylon aphyllum* and *Calligonum aphyllum* can germinate during the fourth or fifth year of succession. *Halocnemum strobilaceum* vegetation can remain for more than 20 years. Psammophytic shrubs and saxaul begin to play a leading role during the next 15 years of succession (Chap. 10). The appearance of the psammophytic grass *Stipagrostis pennata* was noted during the ninth to tenth year of succession, and the appearance of shrub legumes (*Eremosparton aphyllum*, *Astragalus brachypus*) was noted in the tenth to eleventh year of desiccation.

Formation of parallel dunes is a characteristic of the island of Barsa-Kelmes caused by wave activity and aeolian processes in conditions of sandy sediments. The former dune coast (53–53.4 m asl) is represented by a sand bank more than 4 m high and vegetated mostly by tamarisks (*Tamarix laxa*, *Tamarix elongata*, *Tamarix ramosissima*, *Tamarix hispida*). Since the 1980s old dunes have been occupied by

Table 14.2 Vegetation of a detailed transect at the northern coast of the island of Barsa-Kelmes demonstrating temporal and spatial dynamics

Plot no.	Year of desiccation	Vegetation 1989	2005	2007
	1960s	*Stipagrostis* psammophytic shrub with saxaul community (*Calligonum aphyllum*, *Eremosparton aphyllum* *Astragalus brachypus*, *Stipagrostis pennata*, *Haloxylon aphyllum*)	Psammophytic shrub community with saxaul (*Calligonum aphyllum*, *Astragalus brachypus*, *Convolvulus erinaceus*, *Haloxylon aphyllum*)	Psammophytic shrub with saxaul com. (*Calligonum aphyllum*, *Atraphaxis spinosa*, *Astragalus brachypus*, *Eremosparton aphyllum* *Haloxylon aphyllum*)
0	1965–1973	Sarsazan community (*Halocnemum strobilaceum*)	Sarsazan with psammophytic shrubs and saxaul community (*Halocnemum strobilaceum*, *Calligonum aphyllum*, *Atraphaxis spinosa*, *Haloxylon aphyllum*)	Sagebrush – sarsazan with *Atraphaxis* and microcenoses of sea lavender (*Halocnemum strobilaceum*, *Artemisia scopiformis*, *Atraphaxis spinosa*, *Limonium suffruticosum*)
1		Orach community (*Atriplex pratovii*)	Sarsazan with *Calligonum* and microcenoses of sea lavender community (*Halocnemum strobilaceum*, *Calligonum aphyllum*, *Limonium suffruticosum*)	Psammophytic shrub with depressed sarsazan com. (*Astragalus brachypus*, *Eremosparton aphyllum*, *Calligonum aphyllum*, *Halocnemum strobilaceum*) Sea lavender with nitrebush com. (*Limonum suffruticosum*, *Nitraria schoberi*)
2		Orach with saxaul and *Calligonum* community (*Atriplex pratovii*, *Calligonum aphyllum*, *Haloxylon aphyllum*)	Psammophytic shrub with saxaul community (*Eremosparton aphyllum*, *Astragalus brachypus*, *Calligonum aphyllum*, *Atraphaxis spinosa*, *Haloxylon aphyllum*)	Sea lavender – sagebrush com. (*Artemisia scopiformis*, *Limonum suffruticosum*) Psammophytic shrub with saxaul com. (*Eremosparton aphyllum*, *Astragalus brachypus*, *Calligonum aphyllum*, *Atraphaxis spinosa*, *Haloxylon aphyylum*)
3	1974	*Stipagrostis* with tamarisk and *Eremosparton* community (*Stipagrostis pennata*, *Tamarix laxa*, *Eremosparton aphyllum*)	Stipagrostis – psammophytic shrub community (*Calligonum aphyllum*, *Eremosparton aphyllum*, *Stipagrostis pennata*)	Psammophytic shrub with saxaul com. (*Eremosparton aphyllum*, *Astragalus brachypus*, *Calligonum aphyllum*, *Atraphaxis spinosa*)
4–5	1975	Orach with *Stipagrostis* community (*Artriplex pratovii*, *Stipagrostis pennata*)	Aggregations of *Calligonum*, *Stipagrostis*, saxaul (*Calligonum aphyllum*, *Stipagrostis pennata*, *Haloxylon aphyllum*)	Psammophytic shrub with saxaul com. (*Eremosparton aphyllum*, *Astragalus brachypus*, *Calligonum aphyllum*, *Atraphaxis spinosa*)

6	1976	Sarsazan community (*Halocnemum strobilaceum*)	Sarsazan community (*Halocnemum strobilaceum*)	Aggregations of *Astragalus*, *Eremosparton*, sea lavender with depressed sarsazan com. (*Astragalus brachypus*, *Limonium suffruticosum*, *Halocnemum strobilaceum*)
7	1977	Aggregations of *Stipagrostis*, orach, *Eremosparton*, *Astragalus*, saxaul (*Stipagrostis pennata*, *Atriplex pratovii*, *Eremosparton ahhyllum*, *Astragalus brachypus*, *Haloxylon aphyllum*)	*Astragalus–Calligonum* with saxaul community (*Calligonum aphyllum*, *Astragalus brachypus*, *Haloxylon aphyllum*)	Psammophytic shrub with camel's thorn and depressed saxaul com. (*Calligonum aphyllum*, *Eremosparton aphyllum*, *Astragalus brachypus*, *Convolvulus erinaceus*,*Alhagi pseudalhagi*, *Haloxylon aphyllum*)
8	1978	Aggregations of orach, Stipagrostis, *Astragalus*, *Eremosparton*, saxaul and saltwort (*Atriplex pratovii*, *Stipagrostis pennata*, *Astragalus brachypus*, *Eremosparton aphyllum*, *Salsola paulsenii*)	*Stipagrostis*–psammophytic shrub with saxaul community. (*Calligonum aphyllum*, *Astragalus brachypus*, *Convolvulus erinaceus*, *Haloxylon aphyllum*)	Psammophytic shrub com. with saxaul (*Astragalus brachypus*, *Eremosparton aphyllum*, *Calligonum aphyllum*, *Haloxylon aphyllum*)
9	1979	Tamarisk dune (*Tamarix laxa*)	Tamarisk dune (*Tamarix laxa*)	Tamarisk dune (*Tamarix laxa*) with aggregations of psammophytic shrubs in inter-dune space
		Aggregations of *Stipagrostis*, tamarisk, saxaul and saltwort (*Stipagrostis pennata*, *Tamarix laxa*, *Haloxylon aphyllum*, *Salsola paulsenii*)	*Stipagrostis*–psammophytic shrub (*Astragalus brachypus*, *Eremosparton aphyllum*, *Stipagrostis pennata*)	Psammophytic shrub com. (*Calligonum aphyllum*, *Eremosparton aphyllum*, *Convolvulus erinaceus*, *Stipagrostis pennata*)
10	1980	Low dune with tamarisk and *Calligonum* (*Tamarix laxa*, *Calligonum aphyllum*) Aggregations of *Stipagrostis*, orach and saxaul (*Stipagrostis pennata*, *Atriplex pratovii*, *Haloxylon aphyllum*)	*Stipagrostis*–psammophytic shrub with tamarisk and microcoenoses of camel's thorn community (*Astragalus brachypus*, *Eremosparton aphyllum*, *Stipagrostis pennata*, *Tamarix laxa*, *Alhagi pseudalhagi*)	Psammophytic shrub com. with tamarisk (*Calligonum aphyllum*, *Eremosparton aphyllum*, *Convolvulus erinaceus*, *Stipagrosris pennata*, *Tamarix laxa*)

(continued)

Table 14.2 (continued)

Plot no.	Year of desiccation	Vegetation 1989	2005	2007
11	1981	Tamarisk dune (*Tamarix laxa*)	Tamarisk dune (*Tamarix laxa*) with aggregations of *Stipagrostis* and psammophytic shrubs in interdune space (*Stipagrostis pennata*, *Calligonum aphyllum*, *Astragalus brachypus*)	Tamarisk dune (*Tamarix laxa*) with psammophytic shrubs and saxaul in inter-dune space (*Astragalus brachypus*, *Convolvulus erinaceus*, *Eremosparton aphyllum*, *Haloxylon aphyllum*)
		Rare aggregations of saxaul, tamarisk and sarsazan (*Haloxylon aphyllum*, *Tamarix laxa*, *Halocnemum strobilaceum*)	Sarsazan–*Stipagrostis* with *Calligonum* and saxaul community (*Stipagrostis pennata*, *Halocnemum strobilaceum*, *Calligonum aphyllum*, *Haloxylon aphyllum*)	Eremosparton with psammophytic shrubs, saxaul and depressed sarsazan com. (*Eremosparton aphyllum*, *Astragalus brachypus*, *Calligonum aphyllum*, *Convolvulus erinaceus*)
12	1982	Tamarisk belt (*Tamarix laxa*)	Tamarisk dune (*Tamarix laxa*)	Tamarisk dune (*Tamarix laxa*) with saxaul and *Stipagrostis* (*Haloxylon aphyllum*, *Stipagrostis pennata*)
		Bare surface with single orach and saxaul (*Atriplex pratovii*, *Haloxylon aphyllum*)	Aggregations of *Stipagrostis* with single saxaul and saltwort (*Stipagrostis pennata*, *Haloxylon aphyllum*, *Salsola paulsenii*)	Stipagrostis – psammophytic shrub with depressed sarsazan com. (*Eremosparton aphyllum*, *Astragalus brachypus*, *Calligonum aphyllum*, *Stipagrostis pennata*, *Halocnemum strobilaceum*)
13	1982	Sarsazan with single orach community (*Halocnemum strobilaceum*, *Atriplex pratovii*)	Rare aggregations of *Stipagrostis*, nitrebush, saxaul and depressed sarsazan (*Stipagrostis pennata*, *Nitraria schoberi*, *Halocnemum strobilaceum*, *Haloxylon aphyllum*)	
13a	1983	Saxaul community (*Haloxylon aphyllum*)	Saxaul with nitrebush and depressed sarsazan community (*Haloxylon aphyllum*, *Nitraria schoberi*, *Halocnemum strobilaceum*)	Stipagrostis – saxaul with depressed sarsazan com. (*Haloxylon aphyllum*, *Stipagrostis pennata*, *Halocnemum strobilaceum*)

14	1984	Sarszan with single orach, saxaul and *Calligonum* community (*Halocnemum strobilaceum*, *Atriplex pratovii*, *Haloxylon aphyllum*, *Calligonum aphyllum*)	Rare sarsazan with saxaul community (*Halocnemum strobilaceum*, *Haloxylon aphyllum*)	
15	1985	Dune with a single tamarisk and orach (*Tamarix hispida*) Orach with single sarsazan community (*Atriplex pratovii*, *Halocnemum strobilaceum*)	Aggregations of *Stipagrostis*, saxaul, sarsazan, tamarisk (*Stipagrostis pennata*, *Haloxylon aphyllum*, *Halocnemum strobilaceum*, *Tamarix hispida*)	Tamarisk dune (low) with sarsazan and saxaul (*Tamarix hispida*, *T.laxa*, *Halocnemum strobilaceum*)
16	1986	Dune with single tamarisk (*Tamarix hispida*) Sea blite–orach community (*Atriplex pratovii*, *Suaeda crassifolia*)	Sarsazan–tamarisk with saxaul community (*Tamarix hispida*, *Halocnemum strobilaceum*, *Haloxylon aphyllum*)	Tamarisk dune with sarsazan and saxaul (*Tamarix hispida*, *T.laxa*, *Halocnemum strobilaceum*) Aggregations of nitrebush, *Kalidium*, saxaul with depressed sarsazan (*Nitraria schoberi*, *Kalidium foliatum*, *Haloxylon aphyllum*, *Halocnemum strobilaceum*)
17	1987	Bare surface with single orach (*Atriplex pratovii*)	Sarsazan with saxaul community (*Halocnemum strobilaceum*, *Haloxylon aphyllum*)	Saxaul-sarsazan com. (*Halocnemum strobilaceum*, *Haloxylon aphyllum*)
18	1988	Single orach and sea blite (*Atriplex pratovii*, *Suaeda crassifolia*)	Sarsazan with saxaul community (*Halocnemum strobilaceum*, *Haloxylon aphyllum*)	Sarsazan with nitrebush, saxaul and microcenoses of sea lavender com. (*Halocnemum strobilaceum*, *Nitraria schoberi*, *Haloxylon aphyllum*, *Limonium otolepis*)
19	1989	Single orach and sea blite (*Atriplex pratovii*, *Suaeda crassifolia*)	Sarsazan with saxaul and *Stipagrostis* community (*Halocnemum strobilaceum*, *Haloxylon aphyllum*, *Stipagrostis pennata*)	Sarsazan with saxaul and annuals com. (*Halocnemum strobilaceum*, *Haloxylon aphyllum*, *Strigosella circinata*, *Atriplex pratovii*)
20	1990		Aggregations of sarsazan (*Halocnemum strobilaceum*)	Mosaic aggregations of sarsazan, nitrebush, sea blite, *Strigosella* with saxaul (*Halocnemum strobilaceum*, *Nitraria schoberi*, *Suaeda acuminata*, *Strigosella circinata*, *Haloxylon aphyllum*)

(continued)

Table 14.2 (continued)

Plot no.	Year of desiccation	Vegetation 1989	2005	2007
				Rare aggregations of saltwort, sarsazan, saxaul, tamarisk (*Salsola nitraria*, *Halocnemum strobilaceum*, *Tamarix elongata*)
20a	1991		Sarsazan–saxaul community (*Haloxylon aphyllum*, *Halocnemunm strobilaceum*)	Saxaul with sarsazan and saltworts com. (*Haloxylon aphyllum*, *Halocnemum strobilaceum*, *Atriplex pratovii*, *Salsola paulsenii*)
			Sarsazan community (*Halocnemum strobilaceum*)	Sea lavender – sarsazan with saltwort com. (*Limonium suffruticosum*, *Halocnemum strobilaceum*, *Salsola foliosa*)
			Sarsazan community (*Halocnemum strobilaceum*)	Sarsazan and sea lavender – sarsazan with tamarisk and *Strigosella* com. (*Halocnemum strobilaceum*, *Limonium suffruticosum*, *Tamarix laxa*, *T. hispida*, *Strigosella circinata*)
			Tamarisk dune (*Tamarix laxa*)	Tamarisk dune (*Tamarix laxa*)
				Stipagrostis – camel's thorn com. in inter-dune space (*Alhagi pseudalhagi*, *Stipagrostis pennata*)
21	1992		Tamarisk dune (*Tamarix laxa*)	Tamarisk dune (*Tamarix laxa*)
	1993–1999			Aggregations of nitrebush, Astragalus, Eremosparton, sarsazan, sea lavender, orach (*Nitraria schoberi*, *Astragalus brachypus*, *Halocnemum strobilaceum*, *Limonium suffruticosum*, *Atriplex pratovii*)

psammophytes and saxaul, with some additional plants (*Haloxylon aphyllum, Calligonum aphyllum, Stipagrostis pennata, Eremosparton aphyllum, Convolvulus erinaceus, Lycium ruthenicum, Alhagi pseudalhagi, Salsola paulsenii*). The first parallel foredune started to form in 1979 (N 9), caused by single tamarisk specimens of 5% coverage. It grew to 1.5–1.8 m in the next 10 years. Nowadays, it is a tamarisk dune of 2.5-m height with *Stipagrostis* psammophytic shrub communities in the interdune spaces (*Astragalus brachypus, Eremosparton aphyllum, Calligonum aphyllum, Stipagrostis pennata*). Six dune belts had been formed by 1992. The development of vegetation is dominated by the same scheme: coverage of tamarisk and the height of phytogenic hummocks is increasing; psammophytic shrubs and saxaul are gradually occupying the interdune space.

The mechanisms of plant colonization in the dry seabed around the island have some special features. The vegetation is characterized by a poor floristic composition and less diverse spatial series in comparison with the mainland coasts. Three types of spatial dynamics were identified in the 1990s by Pankratova (2002): *Atriplex pratovii–Halocnemum strobilaceum–Haloxylon; Atriplex pratovii–Stipagrostis pennata–Haloxylon aphyllum* (eastern and western coasts); *Atriplex pratovii–Stipagrostis pennata* (southern coast). Nowadays, the eastern and northern coasts are connected to the original coast. The initial stage of succession is not marked. Within the dry seabed of the first decade of this century, on the solonchaks a single *Atriplex pratovii* cover is widely distributed. Occasionally, aggregations of *Suaeda acuminata, Salicornia europaea* and *Bassia hyssopifolia* have appeared. Saxaul vegetation moves ahead between the island and the mainland. Tamarisk dunes have formed single phytogenic hummocks up to 4 m high against a background of bare space.

Thus, insular location and uniformity of marine sediments caused low diversity of spatial dynamics. Aggregations and communities of annual chenopods (*Atriplex pratovii, Salicornia europaea, Suaeda crassifolia*) colonized a primary marine surface in the first stage of succession. Perennial species (*Halocnemum strobilaceum*) can replace annuals after 4 years of desiccation. The formation of tamarisk dunes is marked at the same time. Occupation of the territory by saxaul, psammophytic grasses and shrubs follows desalinization processes and these are covered by blown sand (fifth to tenth year of succession). Psammophytic shrub and saxaul communities begin to form after 15 years of continental development. The most diverse vegetation by floristic composition is now the plant communities of the coast of the 1960s. Saxaul and psammophytic shrub communities with microcoenoses of *Phragmites australis, Alhagi pseudalhagi, Ephedra distachya, Limonium suffruticosum, Artemisia scopiformis* and *Nitraria schoberi* have formed there.

14.2.3.5 Temporal Dynamics of Vegetation

Zonal vegetation of the plateau and the low plains (*Agropyron desertorum–Anabasis salsa–Artemisia terrae-albae* complex) has not been changed for centuries. Desiccation of the Aral Sea, aridization of the climate and the absence of wild ungulate animals during the last decade did not influence vegetation cover. It will be stable for the foreseeable future (Dimeyeva and Alimbetova 2007a, b; Kuznetsov 2007).

Vegetation cover connected with Aral Sea level, however, underwent great changes. Salt lagoons of the western and northern coasts dried up on the transformation of the ecosystems in the 1960s. Shores of lakes once vegetated by meadow–halophytic plants (*Phragmites australis*, *Typha laxmannii*, *Karelinia caspia*, *Calamagrostis epigeios*, *Aeluropus littoralis*) (Demchenko 1950) became vegetated solonchaks with *Halocnemum strobilaceum* and *Limonium suffruticosum*.

Great changes have affected the saxaul vegetation especially in marine terraces. The saxaul woodlands of the protected area are unique natural ecosystems of the Aral Sea region. Formation of saxaul communities is connected with shoreline sands. In the area of the plateau, saxaul does not form communities; sometimes it plays a subdominant role in *Artemisa terrae-albae* communities. Such a landscape looks like a savannah, but should not be called a "savannah" since this term is defined for subtropical and tropical tree and grass communities. Aggregations of saxaul can be seen in ravines of the southeastern coast and along the shoreline of the southern coast. Saxaul open woodlands were repeatedly deforested (Rashek and Rashek 1963). Wood and charcoal of saxaul were collected from the island at the end of the nineteenth century and at the beginning of the twentieth century. The protection regime since 1939 opened up the possibility for the restoration of the woodlands. However, there are presently new problems. After the unification of the island of Barsa-Kelmes with the eastern coast, wild animals migrated to the original coast. Grazing has stopped. Inspection of saxaul communities has shown disturbance of self-regeneration in some parts as a result of insufficient grazing and probably connected with the spreading of the desert moss *Tortula desertorum*.

Several communities of *Haloxylon aphyllum* have been described. Description of them is very important for understanding natural successions and development of woodlands without human impact. They are partially distinct from the formerly mentioned communities (Chaps. 9 and 10) by a more detailed mosaic and small-scale pattern.

An *Ephedra distachya–Haloxylon aphyllum* community has been described in the Aral marine terrace of the northern coast. The relief is hummocky, with a cryptogam cover of desert moss up to 90%. The vegetation coverage is 40–43%. The floristic composition comprises 12 species: *Haloxylon aphyllum* (15%), *Ephedra distachya* (20%), *Atraphaxis spinosa* (5–7%), *Calligonum aphyllum* (1%), *Astragalus brachypus* (less than 1%), *Alhagi pseudalhagi* (less than 1%), *Salsola paulsenii* (less than 1%), *Anisantha tectorum* (less than 1%), *Meniocus linifolius* (less than 1%), *Alyssum turkestanicum* (less than 1%), *Eremopyrum orientale* (less than 1%) and *Senecio noeanus* (less than 1%).

A *Haloxylon aphyllum* community has been described in the Aral marine terrace of the northeastern coast. The fine-grained sand surface is covered by litter, shells and a cryptogam cover 3–5%. The vegetation coverage is 50–60%. The floristic composition comprises nine species: *Haloxylon aphyllum* (40%), *Ephedra distachya* (3–5%), *Kochia odontoptera* (3–5%), *Senecio noeanus* (1%), *Eremopyrum orientale* (less than 1%), *Anisantha tectorum* (less than 1%), *Aeluropus littoralis* (3%), *Lepidium perfoliatum* (less than 1%) and *Atraphaxis spinosa* (1–3%).

An *Atraphaxis spinosa–Haloxylon aphyllum* community has been described in the Aral marine terrace of the northwestern coast. This is a gently dipping marine plain. Coarse-grained sand and rubble are on the surface. Moss covers from 50% to 90% of the surface. The vegetation coverage is 60%. The floristic composition comprises seven species: *Haloxylon aphyllum* (40%), *Atraphaxis spinosa* (20%), *Ephedra distachya* (less than 1%), *Anisantha tectorum* (less than 1%), *Meniocus linifolius* (less than 1%), *Strigosella circinnata* (less than 1%) and *Senecio noeanus* (less than 1%). Vitality of saxaul is satisfactory and bad. The height of saxaul ranges from 100 to 200 cm. A plot of 100 m^2 consists of 15 living depressed saxaul plants, 8 dry saxaul plants and 11 *Atraphaxis* plants. All plants are old.

A *Haloxylon aphyllum* community has been described at the northwestern coast between the Aral marine terrace and the undulating plain. It is a dry bed of a former salt lake. There are polygonal cracks on the soil surface. The crust and underlying horizons have loam soil texture. The lower horizons are coarse-grained sand. The vegetation coverage is 25–30%. The floristic composition comprises two species: *Haloxylon aphyllum* (25–30%) and *Ephedra distachya* (less than 1%). The age of saxaul is 10–20 years. Most of the plants are dry; the rest are in a depressed condition with rare assimilation shoots. Degradation of the community is natural, connected with decrease of the groundwater table.

As a result of inspection, it was discovered that the part of saxaul woodlands of the original coast is degrading. High vegetation coverage leads to insufficient water supply. Despite a high seed productivity, there are no seedlings or young plants in the communities. The undisturbed cover of desert moss is the main competitor of seedlings and ephemerals for using spring precipitation. The decreasing groundwater table connected with desiccation of the Aral Sea is the other reason for degradation of saxaul woodlands. They were formed at the western and northeastern coasts at sites of dried-up lagoons in the 1960s. The decrease of the groundwater table was gradual. The lack of moisture is expressed by an earlier appearance of an autumn yellow colour of assimilative shoots than in other habitats. Seventy percent of these woodlands have dried up in recent decades. On the other hand, active formation of saxaul communities has been continuing in the dry seabed. They were found only at the northern coast in 1980s, but nowadays saxaul spreads everywhere.

Thus, the vegetation cover of Barsa-Kelmes reflects all botanical diversity of the Aral Sea deserts (in the limits of Kazakhstan). The zonal vegetation of the plateau – *Anabasis salsa–Artemisia terrae-albae* complex – is similar to the floristic composition and structure of plant communities of the northern Aral region (Scherbinina 1971; Kuznetsov 1979). The basis for this is the common geological history of the island and the northern coast (Yanshin 1953). The vegetation of the steep southern coast is analogous to that of the northwestern bays of the Aral Sea with cliffs (Dimeyeva 2004). Similar species (*Amberboa turanica*, *Artemisia aralensis*, *Fumaria vaillantii*, *Ferula canescens*) occupy steep slopes and ravines. The psammophytic vegetation of Aral marine terraces has similarity with the eastern Aral region. Penetration of psammophytic species (*Calligonum* spp., *Haloxylon persicum*, *Ammodendron conollyi*, *Ephedra strobilacea*, *Astragalus brachypus*) into the island took place during regressive stages of the Aral Sea through land bridges between the eastern coast of Barsa-Kelmes and the mainland (as at the present time since 2000). Stems of saxaul found in marine sediments lower than the former sea level by 16 m illustrate the existence of regressions in ancient times (Eliseyev 1991).

14.3 Fauna of Barsa-Kelmes

14.3.1 Invertebrates

In addition to the general outline of the Aralkum fauna (see Chap. 11), we give here specific details on the fauna of the island of Barsa-Kelmes.

Arachnida. There are 120 species presently known from the area. Four species are rare: *Leptodrassus vina*, *Latrodectus tredecimguttata*, *Eresus niger* and *Argiope lobata*. There are two species from Solifugae – *Galeodes caspius* (Pall.) and *Desia rossica* Bir – and three species from Scorpiones (*Mesobuthus eupeus*, etc.). In Barsa-Kelmes, 107 species of spiders have been registered (Pavlenko 1985).

Insecta. The insect fauna of the island of Barsa-Kelmes is a depleted variant of entomofauna of the northern Aral shore (Piryulin 2003), with some additional species from southern and eastern parts of the Aral Sea coast and several widely distributed species (see also Table 11.3). The prevailing species are as follows: *Cicindela schrenki* Gebl. – inhabits *Anabasis salsa* plant communities; *Trigonoscelis gemmulata* Men. – prefers shrub habitats of sandy areas; *Asiotmethis muricatus* (Pall.); *Sphingonotus maculatus extimus* B.-Bien.; etc.

Species from the *Red Data Book of Kazakhstan* are as follows: *Ceraeocercus fuscipennis* Uv., *Microzegris pyrothoe* (Pieridae), *Anax imperator* Leach. and *Ischnura aralensis* Haritonov (Odonata). Among rare species are *Utetheisa pulhella* L. (Arctiidae) and *Satanas gigas* Ev. (Diptera). Some rare species inhabiting coastal areas can be found in the nature reserve: *Bolivaria brachiptera* Sauss., *Saga pedo* (Pall.), *Papilio machaon* L. and *Stethorus punctillum* Weise.

14.3.2 Vertebrates

14.3.2.1 Amphibia and Reptilia

There are two species of Amphibia – green toad (*Bufo viridis*, Fig. 11.9b) and lake frog (*Rana ridibunda* = *Pelophylax ridibundus*, Fig. 11.9a). Herpetofauna is represented by 22 species (46.9% from herpetofauna of Kazakhstan). The most diverse habitat is the sand desert (seven species): plate-tailed gecko (*Teratoscincus scincus*) and steppe agama (*Trapelus sanguinolentus*, Fig. 11.5) (Duysebayeva 2007).

Fewer species inhabit the clayey desert: sunwatcher (*Phrynocephalus helioscopus*, Fig. 11.12b), race-runner lizard (*Eremias arguta*, Fig. 11.17) and halys viper (*Gloidius halys*, Fig. 11.14). Even fewer species can live in saline biotopes: steppe agama (*Trapelus sanguionolentus*), race-runner lizards (*Eremias velox*, Fig. 11.12a, *Eremias arguta*) and sunwatcher (*Phrynocephalus helioscopus*) (Satekeev and Chirikova 2007). Some species can live in almost all types of deserts: steppe tortoise (*Agrionomys horsfieldi*, Fig. 11.8), steppe agama (*Trapelus sanguinolentus*), race-runner lizard (*Eremias velox*), sand boa (*Eryx tataricus*) and steppe ribbon snake (*Psammophis lineolatus*). Endemics of Middle Asia and Iran (Turan herpetological complex) represent the core of the herpetofauna (70%) (Duysebayeva 2007).

14.3.2.2 Aves

There were 319 species and subspecies of birds recorded from the coastal area and islands of the Aral Sea and the Syr Darya delta in the twentieth century. Among them were 173 nesting birds, 123 species were encountered during migration, and 23 species were straying birds. The habitats of birds have changed greatly in recent decades. The total number of birds has reduced to 170 species; only 68 nesting birds are still among them (see also Chap. 11). The changes affected, of course, mostly waterfowl. Among nesting birds, three species of Podicipediformes, nine species of Ciconiiformes, ten species of Aniseriformes and 20 species of Charadriiformes have disappeared. Colonies of gulls and terns in islands have disappeared or been sharply cut down. Recently, as a result of the construction of the dam and the filling of the North Aral Sea, there is a tendency of increasing bird populations. Birds of desert landscapes have been affected less by anthropogenic factors. However, there was a change in the number of nesting birds on the island of Barsa-Kelmes from 50 species in 1941–1974 to 38 species in the 1990s and to 16 species in 2005 (Eliseyev 1998, 2007).

About 175 species of birds are registered in Barsa Kelmes and adjacent areas; 23 species belong to the rare and endangered group. Among them are Dalmatian pelican (*Pelecanus crispus*), little egret (*Egretta garzetta*), squacco heron (*Ardeola ralloides*), great black-headed gull (*Larus ichthyaetus*), wooper swan (*Cygnus cygnus*), short-toed eagle (*Circaetus gallicus*), steppe eagle (*Aguila nipalensis*), imperial eagle (*Aguila heliaca*), great bustard (*Otis tarda*), Macqueen's bustard (*Chlamydotis undulata*), sociable lapwing (*Chettusia gregaria*), black-bellied sand-grouse (*Pterocles orientalis*), eagle owl (*Bubo bubo*), flamingo (*Phoenicopterus roseus*), glossy ibis (*Plegadis falcinellus*), spoonbill (*Platalea leucorodia*) and saker falcon (*Falco cherrug*) (Red Data Book of Kazakhstan 1996).

14.3.2.3 Mammalia

There are 27 species of mammals; two are threatened and thus belong to the *Red Data Book of Kazakhstan*: onager (*Equus hemionus onager*, Fig. 11.2) and Persian gazelle (*Gazella subgutturosa*, Fig. 11.1b). The saiga (*Saiga tatarica*, Fig. 11.1c) became a rare species in the last decade. The basis of mammalian species consists of desert species. Most widespread are common wolf (*Canis lupus*, Fig. 11.1a), red fox (*Vulpes vulpes*), corsac fox (*Vulpes corsak*), tolai hare (*Lepus tolai*), migratory hamster (*Cricetulus migratorius*), small five-toed jerboa (*Allactaga elater*, Fig. 11.1d), hedgehog (*Ernaceus auritus*) and gopher (*Citellus maximus*).

14.4 Vegetation of the Kaskakulan Area

The phytocoenotic biodiversity is characterized by a unique combination of plant populations of different life forms, strategies and a diversity of communities and aggregations being formed in different stages of primary succession.

The vegetation of the island of Kaskakulan is represented by saxaul–ephemeroid–sagebrush (*Artemisia terrae-albae*, *Taktajaniantha pusilla*, *Astragalus turczaninowii*, *Astragalusoxiglottis*, *Eremopyrum orientale*, *Haloxylon aphyllum*), ephemeral–*Anabasis salsa* (*Strigosella circinata*, *Trigonella arcuata*, *Leptaleum filifolium*, *Tulipa buhseana*, *Anabasis salsa*), *Halocnemum strobilaceum* communities and ephemeral–saxaul woodlands (*Haloxylon aphyllum*, *H.persicum*, *Eremopyrum orientale*, *Eremopyrum triticeum*, *Hyalea pulchela*, *Lepidium perfoliatum*, *Hypecoum parviflorm*). Adjacent to the famous warm springs there is meadow vegetation (*Phragmites australis*, *Saussurea salsa*, *Karelinia caspia*, *Atriplex littoralis*) and with ephemeral–tamarisk communities (*Tamarix elongata*, *Tamarix ramosissima*, *Tamarix laxa*, *Senecio noeanus*, *Lepidium perfoliatum*, *Anisantha tectorum*).

The vegetation of the dry seafloor is formed under the influence of an arid climate on marine sediments of light texture (sand, loam). The vegetation cover of the dry seabed bordering the eastern original coast mostly consists of communities of perennial and annual chenopods (*Anabasis aphylla*, *Kalidium caspicum*, *Nitraria schoberi*, *Haloxylon aphyllum*, *Climacoptera aralensis*, *Climacoptera ferganica*) of 15–50% coverage on takyr-like soils with a groundwater table at 4-m depth. Takyr-like solonchaks with a groundwater table lower than 3 m are occupied by *Climacoptera aralensis*–*Kalidium caspicum*, *Halocnemum strobilaceum*–*Kalidium caspicum* and *Kalidium caspicum*–*Halocnemum strobilaceum* communities. The dry seafloor between the island and the eastern coast is vegetated by *Halocnemum strobilaceum*, *Kalidium caspicum* and *Suaeda microphylla* communities on coastal soils with blown sand cover and crust-puffed solonchaks of loam texture. There are five to seven species in plant communities with a coverage of 30–70%.

Saxaul communities (*Haloxylon aphyllum*) are widespread in the areas close to the island on coastal soils with blown sand cover. *Halocnemum strobilaceum*, *Climacoptera aralensis*, *Salsola nitraria*, *Kalidium caspicum* and *Halostachys belangeriana* are subdominants in them. The plant communities consist of five to eight species with high coverage (40–80%). The dry seabed between the islands of Kaskakulan and Uzun-Kair is composed by sandy sediments; thus aeolian forms of relief are widespread there. *Atriplex pratovii*, *Corispermum aralo-caspicum*, *Eremosparton aphyllum* and *Stipagrostis pennata* species occupy hummocky sands. Deflation basins and sand plains are vegetated by *Astragalus brachypus*, *Halocnemum strobilaceum* and *Alhagi pseudalhagi* (the floristic composition is five to nine species, coverage is 30–40%). *Tamarix* spp. and *Nitraria schoberi* form phytogenic hummocks. Communities of *Haloxylon aphyllum*, *Krascheninnikovia ceratoides*, *Halocnemum strobilaceum*, *Salsola nitraria*, *Zygophyllum oxianum* and microcoenoses of *Limonium otolepis* occupy the area close to the island of Uzun-Kair. Psammophytic shrub vegetation (*Astragalus brachypus*, *Calligonum* spp., *Convolvulus erinaceus*) is distributed on the island in combination with *Tamarix hispida* phytogenic hummocks. The vegetation cover on the dry seafloor between Uzun-Kair and Barsa-Kelmes is very poor, and is not more than 5–7%. Seldom plants and aggregations of *Tamarix laxa* and *Atriplex pratovii* and sometimes *Haloxylon aphyllum* and huge surface flats without any plants occupy this area. Coastal solonchaks of heavy texture are widespread; they are gradually covered by thin blown sand layers which then are colonized by *Atriplex pratovii*.

Some plant communities described 10 years ago were inspected in 2007. Coverage of *Halocnemum strobilaceum* (sarsazan) and annual saltworts in *Halocnemum strobilaceum* with *Haloxylon aphyllum* communities has decreased; more ephemerals (*Senecio noeanus*, *Strigosella circinata*, *Strigosella africana*, *Descurainia sophia*) appeared. Bare solonchak patches have been vegetated by *Halocnemum strobilaceum*, *Haloxylon aphyllum* and *Halostachys belangeriana* (Dimeyeva 2007a, b).

14.5 Perspectives of Protected Area Development

14.5.1 Importance of Protecting Ecosystems

Nature management of the territories adjacent to Barsa-Kelmes Nature Reserve is regulated in the framework of operating environmental regulations without taking into account the ecological crisis. There is a necessity for creation of special environmental structures for solutions of the ecological problems in the region. Any intervention into the dynamic coastal ecosystems should be assessed from the point of view of possible consequences. The complexity of existing processes is caused by instability of forming ecosystems that do not allow one to calculate the long-term prognosis on the basis of traditional approaches. At the same time, the distress of the local population should be taken into account as a principal factor demanding use of the natural resources. The combination of sustainable use of natural resources with nature conservation is possible only under conditions of a special management structure and participating principles. Not the lost Aral Sea but the new developing arid and semiarid landscapes merit protection (Glantz and Figueroa 1997) as part of a world heritage site in the sense of the World Heritage Convention developed by the United Nations in November 1972.

14.5.2 Necessities for Conservation

14.5.2.1 Ecological Diversity

Landscapes are steadily changed by human activities. Studies of those changes in the Aralkum have global importance for understanding desertification processes, development of new ecosystems and biota under continental conditions, revelation of mechanisms of formation and development of aeolian processes, salinization and desalinization processes, and lastly for the formation of desert landscapes structures. An important fact for conservation should be that the links between components of landscapes are not disturbed.

14.5.2.2 Biological Diversity

This is conserved automatically by conserving the existing and enlarged areas from Barsa-Kelmes to the east towards Kaskakulan and with parts of the Syr Darya delta.

14.5.2.3 Vegetation

1. Saxaul woodlands should be conserved in natural conditions.
2. The protected area is typical (standard) of Turanian desert plant communities with three species from the *Red Data Book of Kazakhstan* and 14 endemic plants. Barsa-Kelmes ecosystems guaranteed natural successions practically without human impact. The primary vegetation of the dry seafloor has great importance. Observations of the vegetation dynamics and primary successions are an important tool for understanding the processes involved in the formation of ecosystems.
3. Meadows and shrublands of the new Syr Darya delta are an example of tugaic type of floodplain vegetation of Central Asia.
4. There should be forest plantations on the dry seafloor as positive experiments with indigenous species to speed up primary succession and to accelerate vegetation cover in order to minimize dust storms and sand deflation

14.5.2.4 Wildlife

1. Presence of rare and endangered species, which are included in the *Red Data Book of Kazakhstan*.
2. Animals brought to and acclimatized on the island of Barsa-Kelmes have expanded their own habitats. After unification of the island and the mainland, they moved to the eastern coast and even the Kyzylkum.
3. Wetlands represent great importance as a breeding site for waterfowl, as resting places of migratory birds and as spawning places for fish

14.5.3 Means of Protection and Political Actions

Two international workshops took place in Aralsk town to solve the problem of the complex development of the region and the protected areas. The first workshop, "Perspectives of sustainable development of the northeastern Aral region and nature reserve Barsa Kelmes", was held in September 2004 in the framework of the BMBF-GTZ/CCD project. Participants from local authorities and nongovernmental organizations (NGOs), leading scientists and international organizations discussed the perspectives of nature conservation and sustainable development of the region. The decision of the workshop was that the most optimal form for conservation and sustainable development of the region would be a *biosphere reserve* with a core area of Barsa-Kelmes Nature Reserve.

There are possibilities for the establishment of a biosphere reserve on the basis of Barsa-Kelmes Nature Reserve. New forms of protected areas will be an important link in the process of strengthening the protected region, the interaction with local people for a sustainable development and the future investment in the area. Realization of the "North Aral Sea" hydrological project, the Kokaral Dam and an improved canal system for irrigation, supported by the World Bank, has provided a first stage of this process – creation of optimal abiotic conditions for ecosystem existence including wetlands and coastal and marine biotopes. The next step for the formation of stable ecosystems and restoration of biodiversity is the establishment of special protected regimes that can be realized in the framework of a biosphere reserve.

The second workshop, "Possibilities of establishment of biosphere reserve in the north-eastern Aral region on a basis of Barsa Kelmes nature reserve", was held in September 2006 with the support of the International Aral Rehabilitation Fund. Outstanding scientists and researchers from Kazakhstan, Russia, Turkey and Germany, representatives from the Aral local administration, journalists, NGOs and staff from Barsa-Kelmes Nature Reserve took part. Participants of the workshop mentioned the global importance and originality of Barsa-Kelmes Nature Reserve and approved the planned nature status. Changes in the socioecological situation in the region take place in conditions of transformation of natural ecosystems in the Aral Sea dry seafloor and anthropogenic processes of rehabilitation of the North Aral Sea. A biosphere reserve is the most efficient form of regulation of nature conservation and socioeconomic activities supporting a sustainable development of the territory. It will best meet the requirements of the "Program to combat desertification in Kazakhstan, 2005–2015" and the Ramsar Convention.

This workshop considered the following to be urgent:

1. To approve the activity of the administration of Barsa-Kelmes Nature Reserve to prepare for reorganization of the nature reserve (*zapovednik*) into an enlarged biosphere reserve.
2. To recommend for the nature reserve an efficient administration to develop:
   - A programme of planning and management measures for reorganization of the nature reserve into a biosphere reserve.
   - Short-term and long-term programmes of research, nature chronicling and ecological monitoring.
   - A project on zonations of the territory, taking into account habitats of rare and endangered animals (onager, Persian gazelle, saiga) and migratory birds of wetlands to realize mapping.
   - A project on socioeconomic measures for improvement of living standards of local people: ecotourism programmes, establishment of training and education, a centre for teaching new methods of husbandry and traditional handicrafts, etc.
3. To request the Forest and Hunting Committee of the Ministry of Agriculture of the Republic of Kazakhstan support the initiative of the administration of Barsa-Kelmes Nature Reserve in reorganizing it into a biosphere territory. The territory represents great scientific importance as a unique worldwide nature laboratory for study of the dynamics of global desertification processes.
4. To entrust the administration of Barsa-Kelmes Nature Reserve and the participants of the workshop with finding funds for financing a second (terminal) stage for well-sinking of a buffer zone of the nature reserve. This will provide possibilities for ecological monitoring and the return of ungulate animals to habitats of the former islands.
5. To take into consideration the great prestige and financial support of the International Aral Rehabilitation Fund in widening the Barsa-Kelmes Nature Reserve protected area and to request it to continue to support the programme of development of Barsa-Kelmes Nature Reserve in reorganizing it into a biosphere territory as an example of a pilot project providing nature conservation and sustainable development of the area.
6. To improve activities on better social and ecological conditions of people's lives, and protection and development of Barsa-Kelmes Nature Reserve with perspectives for expanding its impact in the Aral region by the administration of the Aral rayon and the Kyzyl-Orda oblast.

There is no category of a biosphere reserve in the law on protected areas of Kazakhstan. However, according to the last version of the law (7 July 2006, N 175) "Legislative acts of Republic of Kazakhstan can provide other types of nature protected areas" (Chap. 4 Article 14). This is a possible way for future establishment of a biosphere reserve in the Aral region.

The boundaries of the biosphere reserve have already been discussed with experts. The core zone corresponds to the borders of Barsa-Kelmes Nature Reserve (Barsa-Kelmes and Kaskakulan cluster areas); the buffer zone includes the buffer zone of Barsa-Kelmes Nature Reserve and some additional places – the historical-cultural memorials Begim-Ana, Kerdery-1 and Kerdery-2; four villages with surrounding areas (Karateren, Karashalan, Zhanakurlys, Bogen) are enclosed in the zone of development. Zoning of the biosphere reserve has been discussed with local people.

There are a lot of locations on the chinks with marl banks rich in fossils, mainly on some of the northern and northwestern coastlines of the Small Aral Sea. It is necessary to protect them from elimination by illegal collectors to preserve them for scientific studies. The legal position regarding palaeontological monuments and relations to the department of conservation (regional or republic importance) should be discussed with stakeholders, and especially with professional museum staff for palaeontological items. We think they can give important palaeontological information on the early Quaternary history of the Aral Sea.

14.6 Conclusions

The island of Barsa-Kelmes was the fourth protected area in Kazakhstan in 1939. Since that time, typical ecosystems of northern Turanian deserts have been kept in natural conditions. Starting from the 1960s, the area of the island has been involved in an Aral ecological crisis.

Barsa-Kelmes Nature Reserve is the only nature reserve in the world located in a zone of ecological catastrophe on a global scale. Monitoring studies are a basis for long-term observation of natural processes. It is important for Barsa-Kelmes Nature Reserve to monitor the territory of the Aral Sea dry seabed. It represents great scientific importance as a unique worldwide nature laboratory for study of global desertification processes. The ecosystem approach is a modern method for research, monitoring and preservation of the environment. Ecosystems of the new land should be distinguished as an area of extraordinary scientific interest. It has been determined that ecosystems have been forming towards the northern Turanian zonal desert type. The rate of desiccation, the change of the hydrological regime, the salinity of sediments, the presence of a seed bank and a gene fund of biota are limiting conditions for formation of new land ecosystems. The stability of ecosystems (degree of equilibrium) is a task for discussion in a framework of biological self-regulation. It is dependent on the stage of formation of biological and soil components of ecosystems during the change of the hydromorphic regime into an automorphic one. The time of formation of stable ecosystems and their dynamics are associated with creeping environmental problems in climatic regimes of deserts (Glantz 1999).

Future study should continue on the basis of maps of different scale. Mapping on a large scale will give the spatial pattern of distribution of the ecosystems. The most important aspect of nature chronicling should be selection of key species and communities. Each protected area existing in isolation from human impact is automatically compared with similar contiguous ecosystems, first of all with vegetation. Key objects should be chosen according to their significance, value of resources – rare, vulnerable, zonal – and representative species according to global, national and local importance.

Advanced ecological monitoring is related to reorganization of the nature reserve into a biosphere reserve as the most efficient form of regulation of nature conservation and socioeconomic activities in the territory. The direction towards assessment and realization of sustainable functioning of wetlands will best meet the requirements of the Ramsar Convention after the area in the mouth of the Syr Darya has joined the nature reserve as a cluster core area.

Acknowledgements This research was funded partly by the German Federal Ministry for Education and Research (BMBF grant 0330389, 2002–2005). The information on vegetation of transects in Barsa-Kelmes for 1989 and 2005 was kindly provided by Dmitry Eliseyev (St. Petersburg, Russia).

References

Afanasiev YuG, Smirnov YuA (1985) Perspectives of State reserve Barsa Kelmes in connection with desiccation of the Aral Sea. In: Problems and perspectives of complex development of productive forces in Kyzylorda oblast. Nauka, Alma-Ata, pp 109–111

Connel J, Slatyer R (1977) Mechanisms of succession in natural communities and their role in community stability and organization. Am Nat 111(982):1119–1144

Demchenko LA (1950) Vegetation of Barsa-Kelmes island as forage basis for ungulate animals. In: Proceedings of State reserve Barsa-Kelmes, Alma-Ata, pp 6–37

Dimeyeva LA (2004) Ecological peculiarities of vegetation forming of the north-western coast of the Aral Sea. Probl Desert Dev 4:23–28

Dimeyeva L (2007a) Primary successions on the new Aral Seashore. Basic Appl Dryland Res 1:1–16

Dimeyeva LA (2007b) Primary successions of vegetation in Nature Reserve Barsa Kelmes. In: Proceedings of Nature Reserve Barsa Kelmes Tethys, Almaty 2:81–94

Dimeyeva LA, Alimbetova Z (2006) Flora of Nature Reserve Barsa Kelmes. In: Proceedings of III international botanical conference, Almaty, pp 46–48

Dimeyeva LA, Alimbetova Z (2007a) Natural vegetation dynamics of Barsa Kelmes peninsula (the Aral Sea). In: Proceedings of the international conference "Plant world and its conservation", Almaty/Kazakhstan, 12–14 Sept 2007, pp 124–128

Dimeyeva LA, Alimbetova Z (2007b) Analysis of flora of Nature Reserve Barsa Kelmes. In: Proceedings of Nature Reserve Barsa Kelmes Tethys, Almaty 2:10–34

Dimeyeva LA, Kuznetsov LA (1999) Flora of the Aral seaside belt. Botanical J 4(84):39–52, St Petersburg

Duysebayeva TN (2007) About changes of list of amphibians and reptilians fauna of Nature Reserve Barsa Kelmes. In: Proceedings of Nature Reserve Barsa Kelmes Tethys, Almaty 2:129–134

Eliseyev D (1991) What is the matter with the Aral Sea: is it perishing or relating to its former bounds? Salinet 5:81–82

Eliseyev D (1998) Composition and distribution of nesting birds in Barsa Kelmes island (the Aral Sea). In: Proceedings of Omsk University, Omsk, pp 99–111

Eliseyev DO (2007) Current state of vertebrates of Barsa Kelmes island and their dynamics during last 50 years. In: Proceedings of Nature Reserve Barsa Kelmes Tethys, Almaty 2:107–128

Glantz MH, Figueroa RM (1997) Does the Aral Sea merit heritage status? Global Environmental Change 7(4):357–380

Glantz MH (ed.) (1999) Creeping Environmental Problems and Sustainable Development in the Aral Sea Basin. Cambridge, UK, Cambridge University Press, 291 p

Kuznetsov LA (ed) (1979) Station study of ecosystems of North Aral region. LGPI, Leningrad

Kuznetsov LA (1995) Flora of Barsa Kelmes island. In: Biological and nature conservation problems of the Aral Sea and the coast. Proceedings of Russian Institute of Zoology, St Peterburg, pp 106–128

Kuznetsov LA (2007) Once more about zonal vegetation of Barsa Kelmes. In: Proceedings of Nature Reserve Barsa Kelmes Tethys, Almaty 2:35–44

Kuznetsov LA, Burambayev K (1976) Investigations on seasonal development of desert vegetation Barsa Kelmes island. In: Systematic, anatomy and ecology of plants in Asian part of USSR. LGPI, Leningrad, pp 171–188

Nikitin SA (1966) Tree and shrub vegetation of deserts in USSR. Nauka, Moscow

Novikova NM (2001) Ecological basis for botanical diversity conservation within the Amudarya and Syrdarya river deltas. In: Breckle S-W, Veste M, Wucherer W (eds) Sustainable land use in deserts. Springer, Berlin, pp 84–94

Pankratova IV (2002) Ecological-physiological characteristics of vegetation cover of sandy coast of the Aral Sea (Barsa Kelmes island). Dissertation thesis, BIN, St Peterburg

Pankratova IV (2007) Characteristics of sand dune ecosystems in the north-eastern coast of Barsa Kelmes island. In: Proceedings of Nature Reserve Barsa Kelmes Tethys, Almaty 2:45–80

Pavlenko TV (1985) Distribution of spiders in nature complexes of Barsa Kelmes island (the Aral Sea). In: Fauna and ecology of spiders in USSR, Leningrad, pp 147–155

Piryulin DD (2003) Arthropoda of temporary ponds and the dry seabed of the Aral Sea. Theses of dissertation, ZIN, St Petersburg

Rashek VL, Rashek VA (1963) State reserve “Barsa Kelmes island”. Nauka, Alma-Ata

Red Data Book of Kazakhstan (1996) Animals. Part 1. Vertebrates. Booksellers, Almaty, Istanbul

Satekeev G, Chirikova MA (2007) New information on herpetofauna of Nature Reserve Barsa Kelmes. In: Proceedings of Nature Reserve Barsa Kelmes Tethys, Almaty 2:135–138

Scherbinina II (1971) Main types of rangelands of North Aral region. In: Productivity of rangelands of North Aral region. Gydometizdat, Moscow, pp 96–128

Vasenko EP (1950) Ecology and distribution of saiga. In: Proceedings of State reserve Barsa Kelmes, Alma-Ata, 1st edn, pp 38–115

Veselova LK, Geldieva GV, Budnikova TI (1987) Factors of landscape heterogeneity of primary marine plains of the Aral Sea region. In: Development of desert and mountain territories of Kazakhstan. Alma-Ata, Nauka, pp 21–27

Wucherer W, Breckle S-W (2005) Desertifikationsbekämpfung und Sanierung der Salzwüsten am Aralsee. Sukzession und Phytomelioration, Naturschutz und nachhaltige Entwicklung. Bielefelder Ökologische Beiträge 19, 93 pp

Yanshin AL (1953) Geology of North Aral region. Nedra, Moscow

Zhevnerov VV (1984) Persian gazelle of Barsa Kelmes island. Nauka, Alma-Ata

Chapter 15
Phytomelioration in the Northern Aralkum

W. Wucherer, S-W. Breckle, V.S. Kaverin, L.A. Dimeyeva, and K. Zhamantikov

15.1 Introduction

The open sandy and salty soils of the dry seafloor are sources of widespread salt and dust storms (see Chaps. 5–7). This endangers the agriculture in the surrounding regions (especially the districts of Kzyl-Orda in Kazakhstan and Karakalpakstan in Uzbekistan) and leads to salt desertification (Agakhanjanz and Breckle 1994). An effective procedure to decrease the spreading of salt and dust of the desiccated seafloor and open salinized fields is the cultivation of soils and permanent plantings (Breckle 2003; Breckle and Wucherer 2006, 2007; Dimeyeva et al. 2000; Dimeyeva 2001, 2007, Dimeyeva and Ogar 2006; Hüfler and Novitskiy 2001; Kaverin and Salimov 2000; Kaverin et al. 2005; Novitskiy 1997; Meesa and Singer 2006; Wucherer 2001; Wucherer and Breckle 2005; Wucherer et al. 2005). The goal of phytomelioration must be the preferential treatment of the natural development of a vegetation cover and the creation of a natural reproduction with seeds (see also Chaps. 16 and 17). The plantings should benefit not only the cover of vegetation, but also the land use for the people.

Phytomelioration on the desiccated seafloor of the Aral Sea has been an important task for academic scientists and forest managers since the sea began to retreat. The first recommendations were made in the 1980s (Kurochkina et al. 1983; Kurochkina and Makulbekova 1984; Makulbekova and Wucherer 1990). They had theoretical character.

The first planting experiments of the institutes of forest economy of Kokchetav and Tashkent in the 1990s considered that sand and sandy-loamy plains were suitable for phytomelioration. But saline coastal soils with a heavy texture need land-reclamation procedures.

The continuing shrinkage of the Aral Sea led to the desiccation of vast areas of the seabed with highly saline heavy sediments. New ecological conditions required new approaches for phytoreclamation of all types of soils, including solonchaks of heavy texture.

The Kazakh Institute of Forest Economy conducted experiments at the eastern coast (Kaskakulan area) in 1989–1992, and obtained many basic results concerning

S.-W. Breckle et al. (eds.), *Aralkum - a Man-Made Desert: The Desiccated Floor of the Aral Sea (Central Asia)*, Ecological Studies 218,
DOI 10.1007/978-3-642-21117-1_15, © Springer-Verlag Berlin Heidelberg 2012

conditions for afforestation (Kaverin et al. 1994; Kaverin and Salimov 2000; see Sect. 15.2). Some experiments on phytomelioration were conducted through a UNESCO project (Dimeyeva et al. 2000; Geldyeva et al. 2001; Meirman et al. 2001a). Two experimental plots were established for identification of suitable plants for phytoreclamation of saline soils. Seeds of 17 perennial and annual halophytes were sown in autumn. Only four species of annual saltworts (*Climacoptera aralensis, Halogeton glomeratus, Atriplex pratovii, Suaeda acuminata)* germinated, survived and reached the generative stage in a plot of heavy soil texture with salt content in surface horizons up to 24.5% and 1.91–3.11% in underlying horizons. The perennial species did not survive. On a plot of light soil texture with salt content in surface horizons of 2.63% and 0.56–1.46% in underlying horizons, of six species (*Haloxylon aphyllum, Salsola nitraria, Salsolaaustralis, Petrosimonia brachiata, Climacoptera aralensis, Cimacoptera lanata*) five survived and fruited, but saxaul did not. The experiments again showed that sandy soils have more favourable conditions for implementation of phytomeliorative measures than clay soils; the aridity of the first vegetation period plays a major role in the establishment of seedlings; species from local flora are more useful for phytoreclamation than introduced species.

In the framework of two German BMBF projects, "Succession processes on the desiccated sea floor of the Aral Sea and perspectives of land-use" (1998–2001) and "Combating desertification and rehabilitation of salt deserts in the Aralkum" (2002–2005), different methodical planting approaches on the dry seafloor were tested systematically (Breckle 2003; Breckle and Wucherer 2006, 2007; Meirman et al. 2001b; Wucherer 2001; Wucherer and Breckle 2005; Yair 2001).

In the first project, minicachment methods were tested. Experimental sites were selected on the Bayan transect at the eastern coast (10 km to the southwest of the mouth of the Syr Darya). Saxaul and tamarisk saplings were planted in experimental plots. The method of planting into pits was successful for tamarisks. For saxaul, it was better to plant into deep furrows (see Sect. 15.5.1).

In the second project, large-area plantings were carried out at the eastern coast in the surroundings of the former island of Kushzhitmes. New adapted planting strategies were developed. They include changes of the spatial priorities and selection of additional species for plantings according to environmental conditions (see Sects. 15.3, 15.5, and 15.6). The development of shelterbelts for villages is particularly important (see Sect. 15.4). This work is a result of research cooperation between the University of Bielefeld (Germany), the Kazakh Institute of Forest Economy, the Institute of Botany in Almaty and the Kzyl-Orda Institute of Agroecology.

15.2 Experiments of the Institute of Forest Economy of Kokchetav on the Kaskakulan Transect

The Institute of Forest Economy of Kokchetav (Kazakhstan) joined the BMBF project (2002–2005), so as to continue the institute's work at the eastern coast of the Aral Sea and to warrant information exchange concerning the plantings.

Fig. 15.1 Plantings at the eastern coast of the Aral Sea (Satellite picture – source NASA). *1* in the environment of the former island of Kushzhitmes, *2* in the environment of the former island of Kaskakulan

V.S. Kaverin (the representative researcher and manager of the Institute of Forestry) organized the plantings and observances in the area of the Kaskakulan transect at the eastern coast (Fig. 15.1, area 2) on the dry surface of the seafloor of the 1960s and 1970s and partly that of the 1980s. The local conditions for reforestation in the area of the former island of Kaskakulan at the eastern coast were inspected in the years 1987–1989 and plantings were conducted in 1989–1992. The experimental area comprises a transect of 51-km length, which passes from the former coastline to the west in the direction of the water table of the Aral Sea. Twenty-two plots following this transect with a surface area of 400 ha on salinized soils were established. The following species were tested: *Populus diversifolia, Nitraria schoberi, Haloxylon aphyllum, Tamarix ramosissima, Tamarix hispida, Calligonum caput-medusae, Calligonum rotula, Astragalus brachypus, Eremosparton aphyllum, Halimodendron halodendron, Eremosparton aphyllum, Krascheninnikovia ceratoides (=Ceratoides papposa), Alhagi pseudalhagi, Glycyrrhiza glabra, Haloxylon aphyllum* and *Halocnemum strobilaceum.*

The afforestations were performed with seedlings, saplings and rhizomes as well as seed sowings. The plantings took place in autumn and in spring. The seeds were sown in autumn and winter. The highest establishment rate and the best phytomeliorative properties were seen with *Haloxylon*. The other species either did not establish themselves or had a very low establishment and growth rate.

All plantings were carried out manually.

Four types of soil conditions (habitats) for afforestation were identified (during plantings):

1. Good conditions: Coastal plains with sandy and light loamy-sandy sediments; the salt content in the soil layer from 5 to 30 cm is 0.16–1.04%; Cl^- 0.04–0.33%, groundwater is at a depth of 1–2 m, highly saline with a salt content up to 45 g L^{-1}.
2. Satisfactory conditions: Coastal plains with sandy loam and loam sediments with blown sand cover; salt content in the soil layer from 5 to 30 cm is

0.67–1.62%, Cl^- 0.10–0.88%; groundwater is at a depth of 1.8–3.2 m, highly saline with a salt content up to 65 g L^{-1}.
3. Relatively satisfactory conditions: Coastal plains with loam, clay-loam and clay sediments; salt content in the soil layer from 5 to 30 cm is 1.62–3.96%; Cl^- 0.39–1.40%, groundwater is at a depth of 1.8–2.3 m, highly saline with a salt content of 50–65 g L^{-1}.
4. Moving barchans: Salt content is up to 0.9 g L^{-1}; Cl^- up to 0.32%.

15.2.1 *Experimental Plots and Results of Plantings*

15.2.1.1 Sandy Soils (Habitat 1: Plots 3, 12, 13, 14)

Experimental Plot 3

This experimental plot is 13.6 km west of the former coastline on the dried seafloor from the 1960s. The upper horizons are moderately saline; the middle horizons rather strongly saline (Table 15.1). The soil profile consists of sand from 0 to 40 cm, and from 40 to 50 cm silty loam starts with increasing portions of clay below. The maximum salinity is at 50–60 cm. The groundwater level is at 2.1 m. On the plot very few individuals of *Halocnemum strobilaceum* and some remnants of dry reed could be observed.

In April 1989 seedlings of *Haloxylon, Tamarix, Halocnemum* and *Populus* and saplings of *Calligonum* were planted. The plantings of *Haloxylon a*nd *Tamarix* had the following technical variants:

- Removal of salt crust and upper soil up to 5 cm
- Sand heaps around the stem base of seedlings and saplings
- Plantings in furrows

Table 15.1 Soil salinity (percentage of dry matter) on plots of the Kaskakulan transect (four habitats)

Plot (habitat)	Soil horizon (cm)	Total salinity (%)	Cl^- (%)	SO_4^{2-} (%)
1 (H2)	5–30	1.62	0.41	0.82
3 (H1)	5–30	0.87	0.19	0.49
5 (H3)	5–30	2.3	1.0	0.6
6 (H3)	5–30	1.58	0.39	0.59
7 (H4)	5–30	0.2	0.06	0.06
12 (H1)	5–30	1.04	0.33	0.73
13 (H1)	5–30	0.16	0.04	0.05
14 (H1)	5–30	0.77	0.08	0.47
15 (H2)	5–30	0.67	0.11	0.35
16 (H2)	5–30	0.7	0.1	0.4
17 (H3)	5–30	1.54	0.67	–
19 (H4)	5–30	<0.9	<0.32	–

After 3 years the rate of establishment of *Haloxylon* was 17.0–37.2% for autumn plantings and it was 22.5% higher for spring plantings, with variation from 10.5% to 66.2%. The average height after 12 years reached 2.65 m and the diameter was 3.0 m × 3.44 m. Removal of salt crust had increased the rate of establishment by about 9.1–12.8%, whereas sand accumulations around the seedlings had increased the rate by only 0.9–6.7%. The rate of establishment in furrow plantings was about 28.1%.

The seedings were also performed in three technical variants:

1. Seeding below a mechanical cover of reed mats
2. Seeding under a mechanical cover with plastic mats
3. Seeding of granulated seeds in furrows with one-sided or two-sided accumulation of soil along the furrows

The seedings with granulated seeds on a snow cover were more effective than seeding on a soil surface without snow cover.

The plantings on experimental plot 3 were very successful insofar that a high natural germination rate was observed in the following years. After 6 years the new plants reached a height of 50–150 cm. The density of plants was very high, up to 20–95 individuals per square metre, which caused new establishments to be restricted to open gaps. Vitality was moderate. Seeds were observed up to 70 m from fruiting mother plants.

Experimental Plot 12

This plot is 7.5 km west of plot 5. The soils are moderately to strongly saline (Table 15.1). On the substrate surface remnants of reed were present. The whole soil profile consists of sand, some enriched with some silt at 30–60 cm. The groundwater level was at 1.65-m depth.

In 1989 this plot was devoid of any vegetation, except for very few *Haloxylon aphyllum* and *Tamarix* individuals. Between 1990 and 1992, in autumn and spring seedlings of *Haloxylon aphyllum, Tamarix, Halocnemum strobilaceum* and *Ceratoides papposa* were directly planted in the unprepared soil.

After 3 years the rate of establishment of *Haloxylon* was 1.3–64.8% for spring plantings and 1.5–23.6% for autumn plantings, that of *Tamarix* was 0–16% for autumn and spring plantings and that of *Halocnemum strobilaceum* was 0–69% for spring plantings and 0–2% for autumn plantings. *Ceratoides papposa* died off.

Experimental Plot 13

This plot is 3 km west of plot 12. The soils are degraded coastal solonchaks (Table 15.1). The soil profile exhibits sand between 0- and 70-cm depth. Below that a light loam follows. The groundwater is at a depth of 1.6 m.

In 1989 this plot was devoid of any vegetation, except for very few *Haloxylon aphyllum, Nitraria schoberi* and *Tamarix* individuals. Between 1990 and 1992, in autumn and spring seedlings of *Haloxylon aphyllum and Tamarix* were directly planted in the unprepared soil.

After 3 years the rate of establishment of *Haloxylon* was 25.5–64.1% for spring plantings and 1.4–24.1% for autumn plantings, and that of *Tamarix* sp. was 12–38% for spring plantings and 9–36% for autumn plantings.

Experimental Plot 14

This plot is 3.5 km west of plot 13. The soils are moderately saline (Table 15.1). The soil profile is entirely composed of sand. The groundwater level was at 1.55 m.

In 1989 this plot was devoid of any vegetation, except for very few *Halocnemum strobilaceum* and *Tamarix* individuals. Between 1990 and 1993, in autumn and spring seedlings of *Haloxylon aphyllum, Tamarix, Halocnemum strobilaceum a*nd *Ceratoides papposa* were directly planted in the unprepared soil.

After 3 years the rate of establishment of *Haloxylon* was 12.0–40.0% for spring plantings and 0–15.0% for autumn plantings, and that of *Tamarix* sp. was 8–32% for spring plantings and 3–34% for autumn plantings. *Halocnemum strobilaceum* and *Ceratoides papposa* died off.

15.2.1.2 Loamy-Clayey Soils, with Sand Cover (Habitat 2: Plots 1, 15, 16)

Experimental Plot 1

This plot is 3.2 km west of the former coastline, on the desiccated seafloor from the 1960s and the beginning of the 1970s. The soils are rather strongly salinized (Table 15.1). The upper soil horizon down to 30 cm is sandy, below it is loamy. The groundwater level is at about 1.8 m. A vegetation cover is totally lacking, except for very sparse individuals of *Halocnemum strobilaceum* and a few dry *Atriplex pratovii.*

In April 1989 seedlings of *Haloxylon, Tamarix* and *Ceratoides papposa* and saplings of *Tamarix, Calligonum* and *Populus* were planted. *Haloxylon* was subsequently planted until 1992, in autumn and in spring. All saplings and seedlings died in the same year as they had been planted except for *Haloxylon* saplings from spring 1989. They died after the dry summer of 1990.

Seeds from *Haloxylon* were also sown in a mixture with salt-free sand (1:5, 1:10, 1:15, 1:20). The seedings from autumn 1989 resulted in an intensive germination; most seedlings died in the second year. Seedings in 1990 and 1991 resulted in no germination. Granulated seeds (granulation hydrogel and clay) were also sown in autumn 1989. Germination was successful, but again most seedlings died in 1990 and the few remnants died in 1991. As a conclusion, the conditions for *Haloxylon aphyllum* on this experimental plot are not suitable. However, in the subsequent

Fig. 15.2 *Halocnemum strobilaceum* plant community with *Kalidium capsicum* (natural development) (Photo: Wucherer, September 2010)

years this area exhibited some invasion by *Halocnemum strobilaceum*. Intensive germination of *Halocnemum* was observed in 1990 and 1991. Five years later, most of the area was covered by *Halocnemum* up to 70% (Fig. 15.2). The bushes reached about 30–50 cm; and even a few individuals of *Halostachys caspica* were observed. Later, this habitat was also colonized by *Kalidium caspicum*.

After 3 years the rate of establishment of *Haloxylon* was 0–12.5% for spring plantings and 0% for autumn plantings.

Experimental Plot 15

This plot is 3.5 km west of plot 14. The soils are moderately saline (Table 15.1). This plot is adjacent to mobile barchan dunes. The soil profile is sandy from 0 to 30 cm, below it is loam and deeper it has increased clay content. The groundwater level was at 1.85 m.

In 1989 this plot was devoid of any vegetation, except for very few remnants of dead *Atriplex pratovii*. Between 1990 and 1993, in autumn and spring seedlings of *Haloxylon aphyllum, Tamarix* and *Halocnemum strobilaceum* were directly planted in the unprepared soil.

After 3 years the rate of establishment of *Haloxylon* was 0–16.5% for spring plantings and 0–11.5% for autumn plantings, that of *Tamarix* was 0–27.3% for spring plantings and 0% for autumn plantings, and that of *Halocnemum strobilaceum* was 0–37.3% for spring plantings and 0% for autumn plantings.

Experimental Plot 16

This plot is 43 km west of the former coastline on the desiccated seafloor of the 1970s, and west of the former island of Kaskakulan. The soil profile is sandy from

0 to 30 cm, below it is clay, both rather saline (Table 15.1). The groundwater level was at 2.0 m.

In 1989 this plot was devoid of any vegetation, except for very few remnants of dead *Atriplex pratovii*. Between 1990 and 1992, seedlings of *Haloxylon aphyllum, Tamarix, Halocnemum strobilaceum* and *Ceratoides papposa* were planted. The rate of establishment of *Haloxylon* was up to 61%, with the rate for plantings in spring being more than double that for plantings in autumn. After 10 years the plants had reached a height of about 1.72 m and a diameter of 1.86 m × 2.70 m. Vitality was very good.

After 3 years the rate of establishment of *Haloxylon* was 1.3–64.8% for spring plantings and 1.0–23.6% for autumn plantings, that of *Tamarix* was 0–15.6% for spring plantings and 0–15.7% for autumn plantings, and that of *Halocnemum strobilaceum* was 0–69.3% for spring plantings and 1.8% for autumn plantings. *Ceratoides papposa* died off.

15.2.1.3 Loamy-Clayey Soils (Habitat 3: Plots 2, 5, 6, 17)

Experimental Plot 2

This experimental plot is 11.6 km west of the former coastline on a site which became dry in the late 1960s and early 1970s. The soil profile exhibits sand in the upper 10 cm and then down to 120 cm it is loamy with an increasing content of clay. The groundwater level is at 2.3 m. Vegetation on the plot was very sparse with a few individuals of *Halocnemum strobilaceum* and *Atriplex pratovii*.

Seedlings of *Haloxylon, Tamarix and Krascheninnikovia (Ceratoides)* and saplings of *Tamarix, Calligonum,* and *Populus* were planted in April 1989. *Haloxylon* was subsequently planted until 1992, in autumn and in spring. All saplings and seedlings of *Tamarix* and *Haloxylon* died in the same year as they had been planted. The rate of establishment of *Haloxylon* was 0–38.8% for spring plantings and 0% for autumn plantings. For *Haloxylon* saplings planted in spring 1991 the establishment rate was 38.8% after 10 years and for those planted in 1992 the establishment rate was 12.5% after 10 years. Then the average height was 157 cm. In those plots there was regular seeding and germination during the last 5 years. Seedlings reached 10-cm growth height after the first year. The older saplings and young plants had a height of up to 1.1 m. Coverage reached 50–80%. Seeds from fruiting plants were detected at a distance of at least 10 m from mother plants. This successful establishment is correlated with very good weather conditions during the years 1991 and 1992 and the right time of the plantings. Other species died off.

The seedings had been done with some specific preparations of the substrate, especially a mechanical pressing of the surface. Then reed bundles were used at 3-m distance, and along those 20-cm-high plastic walls had been installed. On both sides of these protective plastic walls the seeds were sown in the following four technical variants:

1. Removal of salt crust and seedings with loosening of soil
2. Removal of salt crust and seedings with a salt-free sand mixture (1:5)
3. Seeding on salt crust with loosening of soil
4. Seeding on salt crust without other technical substrate preparations (control)

Almost all seeds germinated perfectly, but almost all seedlings (more than 99%) died the following summer. Only some single individuals could establish themselves, especially with the first two treatments. Those individuals were the germinal centre for a small but considerable stock and spreading community after 10 years.

Experimental Plot 5

This experimental plot is about 16.7 west of the former coastline. It is situated on the desiccated seafloor of the 1970s. The soils are strongly saline (Table 15.1). The upper horizon of soil substrate is composed of loam (0–10 cm), from 10 to 30 cm it is sand and below 30 cm it is mainly loam with increasing parts of clay. The groundwater level is at 1.8 m. Only very few individuals of *Haloxylon* with a height up to 1.5 m were present.

In April 1989 seedlings of *Populus* and *Elaeagnus* were planted. In autumn 1989 and the subsequent 2 years, *Haloxylon, Halocnemum* and *Tamarix* were planted directly on the unprepared surface. Additionally, *Haloxylon* and *Tamarix* were planted in the following variants:

- Removal of salt crust
- Sand heaps around the stem base of seedlings and saplings
- Plantings in furrows

Most of the seedlings died (less than 1% rate of establishment). Very few *Haloxylon* saplings stayed alive. But these few individuals became parallel to the slowly decreasing salinity in soil the main centres for a subsequent secondary colonization. In the furrow plantations the rate of establishment was comparatively good at 10–14%. After 8 years the plants reached a height of about 129–143 cm, with a diameter of 130 m × 167 cm. Between and in the furrows the secondary colonization was considerable. The young offspring were about 2–4 years old. On average there were three to four offspring per 10 m in the furrows. Between the furrows about five to ten individuals per square metre were counted. Plantings in furrows exhibited first a rather low rate of establishment, but later a considerable and successful secondary colonization could be observed.

After 3 years the rate of establishment of *Halocnemum strobilaceum* was 0–59.8% for spring plantings and 0–28.1% for autumn plantings. *Tamarix* died off.

Experimental Plot 6

This plot is between plots 3 and 5. The soils are moderately saline (Table 15.1). There is a sandy layer between 0 and 10 cm, below that down to 2 m there is loam and in lower parts there is increasing clay. The groundwater is at 1.9 m.

During the planting time in 1989, the soil surface was free of any vegetation; it was densely covered by shells. In the following 4 years, seedlings of *Haloxylon aphyllum* und *Halocnemum strobilaceum* were planted in autumn and in spring directly in the unprepared substrate in long rows of 100-m length. In each row the distance between seedlings was between 1 and 1.6 m. The rows were separated by 10, 20 and 30 m for *Haloxylon aphyllum*, and 5, 10, and 20 m for *Halocnemum strobilaceum*.

After 3 years the rate of establishment of *Haloxylon* was 3.0–42.0% for spring plantings and 8.0–21.4% for autumn plantings, and that of *Halocnemum strobilaceum* was 23.8–76.0% for spring plantings and 0–20.5% for autumn plantings. *Tamarix* had a very low establishment rate (2–4%).

Experimental Plot 17

This plot is 8.5 km west of plot 16, about 51 km west of the former coastline. The soil is a degraded loamy-clayey coastal solonchak (Table 15.1). The soil profile is sandy only from 0 to 10 cm, and below that it is loam with clay. The groundwater level was at 2.1 m.

In 1989 this plot was devoid of any vegetation, except for very few dead *Atriplex pratovii* remnants. Between 1990 and 1992, in autumn and spring seedlings of *Haloxylon aphyllum, Tamarix* and *Halocnemum strobilaceum* were directly planted in the unprepared soil.

After 3 years the rate of establishment of *Haloxylon* was 0–18.2% for spring plantings and 0–3.0% for autumn plantings, and that of *Halocnemum strobilaceum* was 0–45.6% for spring plantings and 0–26% for autumn plantings. *Tamarix* had a very low establishment rate (0–3%).

15.2.1.4 Discussion

The rate of establishment and the growth of *Haloxylon aphyllum* is very variable. On plots 3, 12, 14 and 16, with low salinities in the upper horizon, the establishment was successful (Fig. 15.3). On plot 5 establishment happened only with improved substrate conditions, e.g. in furrows. On plot 1, where the salinity was high, almost no success with saxaul was recorded. Instead, a community with *Halocnemum strobilaceum* started to develop.

Plantings in furrows is preferential for establishment. *Haloxylon aphyllum* exhibits a higher rate of establishment for spring plantings, especially on sandy and sandy-loamy substrates (Figs. 15.4 and 15.5).

Fig. 15.3 *Haloxylon aphyllum* plantings (12 years old) on the Kaskakulan transect (Photo: Wucherer, September 2003)

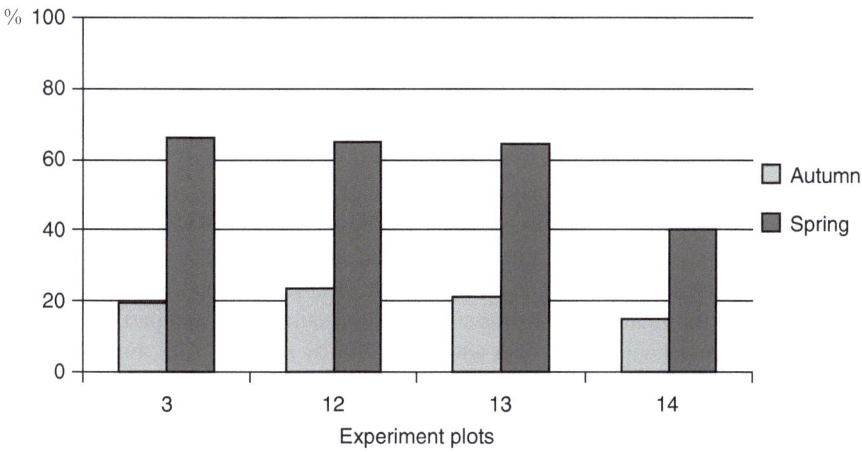

Fig. 15.4 Establishment rate of *Haloxylon aphyllum* on sand soils from autumn and spring plantings

Despite the rather good rate of establishment in spring plantations, the variability of *Haloxylon* is still very high even within plots:

- On sandy soils, 1.3–64.8%
- On sand-covered loamy-clayey soils, 0–68.4%
- On loamy-clayey soils, 0–40.6%

For *Halocnemum strobilaceum* this variability is very similar in and between plots and from year to year:

- On sandy soils, 0–69%
- On sand-covered loamy-clayey soils, 0–75.6%
- On loamy-clayey soils, 0–76.0%

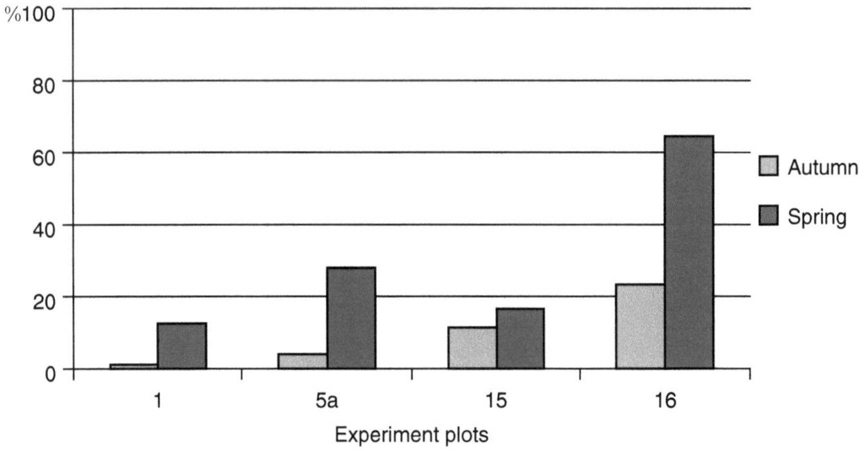

Fig. 15.5 Establishment rate of *Haloxylon aphyllum* on coastal solonchak with a layer of sand in the topsoil from autumn and spring plantings

This also indicates very high variability from year to year according to climatic conditions.

The removal of the salt crust and accumulation of sand around the seedlings exhibits only a marginal positive effect. Seedings with *Haloxylon* are, however, distinctly more successful in mixtures of seed and sand (1:5). Also, granulated seeds can be used. Seeding on a snow cover in spring is especially successful.

In general, the establishment was successful on sandy soils and partly on loamy coastal solonchaks with sand cover and a low salinity of the topsoil. Establishment on the loamy coastal solonchaks occurred only with improvement of the soil surface, such as in the furrows. Springtime is more effective for the accomplishment of the plantings: the establishment is clearly higher. Precipitation variability plays a key role in the establishment of species and in the sustainability of plantings, which is very difficult to achieve.

Haloxylon aphyllum is the most promising culture species for phytomelioration on the desiccated seafloor. Experiments have shown that the plantings were clearly more effective than sowing. In all cases, even only a few surviving individuals play a decisive role in the formation of a future vegetation cover within only a few years.

15.2.2 Growth Rates

The growth rates of *Haloxylon aphyllum*, *Halocnemum strobilaceum* and *Tamarix* on various stands were investigated during the planting experiments (Table 15.2). *Haloxylon aphyllum* bushes reached a maximum height of 2.56 m after 10 years on sandy sites, and a diameter of 2.20 m × 3.20 m. The vitality on loamy-clayey

Table 15.2 Growth rates of planted species

Planted species	Age (years)	Biometrical parameters (cm)		
		Height	Diameter of bushes	
			Along the rows	Perpendicular to the rows
Sandy soils				
Haloxylon aphyllum	3	94.7 ± 5.4	81.1 ± 5.2	79.4 ± 5.8
	5	129.8 ± 4.8	104.7 ± 6.4	115.2 ± 5.6
	6	142.6 ± 3.2	123.2 ± 5.6	151.5 ± 6.3
	7	195.9 ± 5.0	170.0 ± 10.1	205.8 ± 7.5
	8	239.4 ± 6.7	254.2 ± 10.9	287.5 ± 9.4
	10	255.8 ± 8.8	222.8 ± 7.5	322.6 ± 12.7
Halocnemum strobilaceum	5	33.4 ± 2.5	88.2 ± 3.6	91.5 ± 2.9
	6	34.9 ± 1.3	94.4 ± 2.6	98.5 ± 3.2
	7	41.1 ± 1.5	112.3 ± 5.6	119.8 ± 4.6
	8	47.6 ± 2.3	143.8 ± 3.9	151.8 ± 4.7
	10	46.0 ± 0.9	139.4 ± 3.4	157.3 ± 3.1
Tamarix sp.	5	48.8 ± 14.8	57.5 ± 10.3	52.5 ± 10.3
	6	74.5 ± 5.9	71.8 ± 4.5	76.4 ± 5.7
	7	122.5 ± 31.7	161.3 ± 43.7	151.3 ± 36.1
Loamy-clayey soils, with sand cover				
Haloxylon aphyllum	4	126.4 ± 6.9	103.6 ± 7.1	110.1 ± 6.4
	5	154.9 ± 1.4	129.1 ± 3.7	143.6 ± 4.8
	6	166.2 ± 5.2	160.8 ± 7.3	164.4 ± 7.7
	7	165.6 ± 8.4	154.9 ± 10.7	161.3 ± 11.2
Halocnemum strobilaceum	4	32.4 ± 2.9	99.7 ± 3.1	98.1 ± 3.1
	5	44.4 ± 1.6	111.0 ± 3.2	114.0 ± 2.8
	6	45.4 ± 1.9	105.5 ± 3.6	125.0 ± 3.6
	7	33.8 ± 3.1	88.8 ± 8.1	108.4 ± 6.9
Tamarix sp.	5	70.6 ± 4.3	78.8 ± 4.8	83.9 ± 4.9
	6	81.1 ± 4.9	101.3 ± 5.9	104.8 ± 4.7
	7	93.9 ± 8.1	98.9 ± 9.1	99.4 ± 8.8
Loamy-clayey soils				
Haloxylon aphyllum	4	96.9 ± 7.9	84.6 ± 6.8	85.0 ± 6.9
	5	103.3 ± 6.4	93.4 ± 7.8	95.6 ± 7.5
	6	157.4 ± 4.1	134.1 ± 3.9	127.0 ± 4.7
Halocnemum strobilaceum	4	23.3 ± 1.4	61.1 ± 4.7	68.2 ± 4.9
	5	27.4 ± 3.8	76.5 ± 2.2	77.9 ± 2.3
	6	31.0 ± 1.2	103.4 ± 3.0	95.9 ± 2.8
	7	41.3 ± 3.8	125.3 ± 4.1	119.3 ± 4.1
Tamarix sp.	6	72.2 ± 18.3	70.7 ± 9.6	62.9 ± 8.9
	8	100.9 ± 5.6	113.0 ± 7.8	117.8 ± 7.6

soils with blown sand cover was highest after 6 years on sand-covered coastal solonchaks with 1.66-m height and 1.60 m × 1.64 m diameter (Fig. 15.6, Table 15.2).

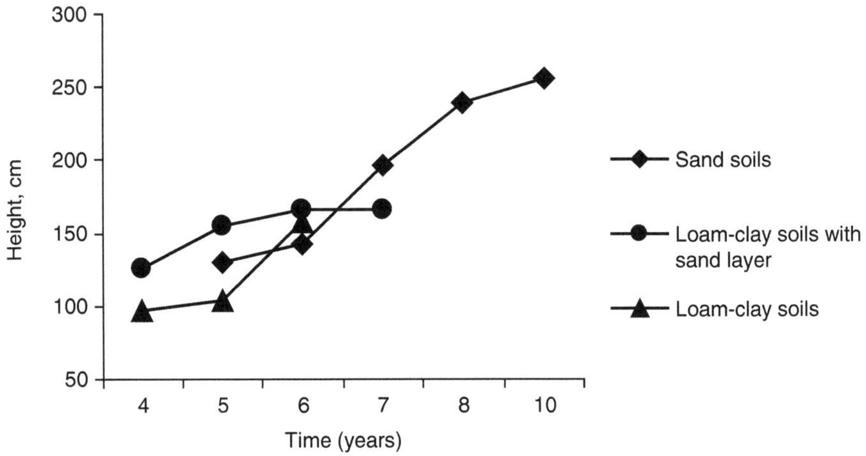

Fig. 15.6 Growing rate of *Haloxylon aphyllum* on different soil types

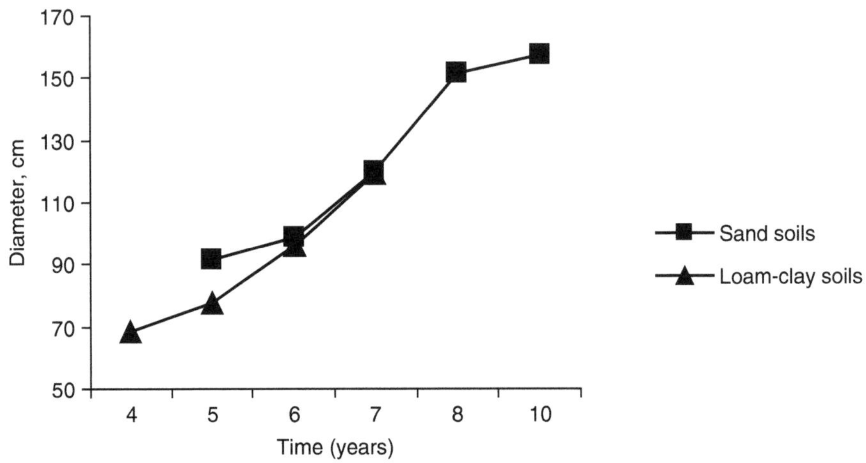

Fig. 15.7 Growing rate of *Halocnemum strobilaceum* on different soil types, indicated by the diameter of bushes

Halocnemum strobilaceum has a maximal vitality of 8–10 years old on sandy soils (height 48 cm, the diameter reached 1.44 m × 1.52 m). The vitality is the same on sandy and on loamy-clayey sites with blown sand cover in 6–7-year-old plants (Figs. 15.7, 15.26).

On heavy soils and with higher salinities, the vitality of *Haloxylon* is reduced. In *Halocnemum strobilaceum* the growth rate is comparable with the growth rates for the other two habitats.

15.2.3 Technology for the Stabilization of Barchans

Fixation and stabilization of barchan dunes was done with distinct methods of mechanical protection. The Forest Institute in Tashkent suggests using mechanical protection and then plantings of *Atriplex* and *Tamarix* with a distance of 4–5 m between the rows. Our investigations have shown that even only 3-m distance between rows is too much for a sufficient fixation of barchans. The optimal distance between rows was checked using reed bundles of 10–12-cm diameter and plastic covers with a width of 30–35 cm. In summer six experimental plots were chosen on six barchans.

15.2.3.1 Barchans (Habitat 4: Plots 7, 8–10, 19–21)

Experimental Plot 7

This plot is 5 km west of plot 5. The soil is covered with small barchan dunes. The soil profile exhibits sand from 0- to 70-cm depth and is slightly saline (Table 15.1). Below 70 cm there is loam and clay mixed until 2-m depth.

The plot area was free of vegetation in 1989. Within the following 3 years, seedlings of *Haloxylon aphyllum* and *Tamarix* species were planted in spring and autumn on one barchan surface, and seeds of *Haloxylon aphyllum* were sown. The length of the barchan was 42 m, the width was 20 m and the maximal height was 106 cm. The surfaces of barchan sands were covered by bundles of reeds in rows separated by 3 m.

Experimental Plots 8–10

These plots were similar to plot 7, bare of vegetation with isolated barchan dunes. The plantings were performed in the same manner as in plot 7. Fixation of the barchan surface, however, was different. In plot 8, reed bundles were placed not horizontally but vertically up to 30 cm; in plot 9 a plastic cover was used, fixed about 20 cm above the surface, in 15 rows. The same was done for plot 10, with 20 rows, each 35 m long, and with 3 m between rows.

Experimental Plots 19–21

All three plots are south of plot 5. The soil is covered by a rather dense pattern of barchans, with low to moderate salinity (Table 15.1).

In 1989 this plot was devoid of any vegetation. In 1991 one barchan in each plot was covered with reed bundles in rows separated by 1, 2 and 3 m. In 1992 *Haloxylon aphyllum* and *Tamarix ramosissima* were planted along these rows.

15.2.3.2 Discussion

The distances between rows were 1.0, 2.0 and 3.0 m. The average size of one barchan was about 0.12 ha, and the whole area fixed was 0.15 ha. This fixation experiment used 2,700 m of reed bundles and 2,700 m of plastic ribbons. Half of the barchans was covered in spring and the other half was covered in autumn for the relevant plantings. Along the protective rows on the barchans, seedlings of *Haloxylon, Tamarix, Eremosparton, Astragalus* and *Calligonum* were planted in spring and autumn. Additionally seedings with *Haloxylon, Calligonum* and *Nitraria* were done.

The seedings were not successful. The saplings of *Astragalus, Eremosparton* and *Calligonum* did not establish themselves. The rate of establishment of *Haloxylon* remained rather low (Table 15.3) at 2.5–7.6% for 3-year-old plants (height 0.55–0.87 m) and 1.2–7.5% for 7-year-old plants (height 1.87–2.12 m) of

Table 15.3 Growth rates of *Haloxylon* and *Tamarix* on barchans

Planted species	Age (years)	Biometrical parameter (cm)			
		Height (cm)	Diameter of plants		
			Along the rows	Perpendicular to the rows	Establishment rate (%)
Distance between reed bundles 1.0 m					
Haloxylon aphyllum	3	50.5 ± 2.9	–	–	2.5
	5	155.7 ± 7.1	164.3 ± 11.6	147.1 ± 8.6	1.2
	7	192.3 ± 7.2	222.0 ± 15.7	215.0 ± 12.4	1.2
Tamarix sp.	3	65.0 ± 8.8	–	–	1.5
	5	122.1 ± 14.1	122.2 ± 5.4	133.6 ± 6.3	1.2
	7	158.3 ± 10.7	196.7 ± 14.1	202.8 ± 16.4	1.2
Distance between reed bundles 2.0 m					
Haloxylon aphyllum	3	65.0 ± 8.8	–	–	7.0
	5	158.6 ± 19.1	179.3 ± 21.5	170.7 ± 17.2	2.0
	7	187.1 ± 10.1	232.1 ± 13.6	222.9 ± 13.8	2.0
Tamarix sp.	3	45.5 ± 9.5	–	–	1.5
	5	135.0 ± 21.0	147.5 ± 25.0	172.5 ± 23.0	1.0
	7	124.2 ± 11.7	128.3 ± 24.0	132.5 ± 22.6	0.5
Distance between reed bundles 3.0 m					
Haloxylon aphyllum	3	60.1 ± 9.1	–	–	–
	5	180.8 ± 4.4	144.3 ± 4.2	157.0 ± 3.7	7.6
	7	197.8 ± 4.3	184.4 ± 5.5	197.6 ± 5.4	4.7
Tamarix sp.	3	50.5 ± 8.4	–	–	7.5
	5	93.1 ± 5.9	100.9 ± 4.8	111.1 ± 6.1	3.2
	7	100.9 ± 5.6	92.8 ± 3.7	103.7 ± 6.2	3.2
Distance between plastic covers 3.0 m					
Haloxylon aphyllum	3	87.4 ± 6.6	–	–	5.6
	6	212.4 ± 8.3	178.9 ± 7.4	200.5 ± 8.8	2.7
Tamarix	3	42.3 ± 2.4	–	–	3.4
	6	122.5 ± 4.2	137.5 ± 14.4	132.5 ± 11.8	3.0

Haloxylon aphyllum. In *Tamarix* the height of 3-year-old plants was 0.42–0.65 m, and that of 7-year-old plants was 1.22–1.58 m.

The distance between rows did not have any influence on establishment or growth in *Haloxylon*. In *Tamarix* the vitality was lower in the rows separated by 3 m. Any mechanical fixation on barchan dunes creates more favourable conditions for the germination of diaspores from adjacent plants. Additionally, sedimentation of dust leads to better soil conditions after 7 years, which can be seen by the extraordinary germination rate of 28,000 seedlings per hectare, mainly along the protective rows. *Haloxylon* starts flowering at the age of 3–4 years; additionally, with immigrating therophytes the plant cover can then reach up to 50–60%. In general, on the barchan plots the primary establishment rate was relatively low (below 20%), but the secondary colonization was very effective.

15.3 Combating Desertification and Rehabilitation of the Salt Deserts on the Desiccated Seafloor

15.3.1 Reforestation Potential of Sandy and Salty Sites and Habitat Conditions

Phytoreclamation activity was conducted in the framework of the BMBF-GTZ/CCD project in 2002–2005. The Department of Ecology (University of Bielefeld in Germany) cooperated with the Institute of Botany and Phytointroduction (Almaty), the Institute of Agroecology (Kyzyl-Orda), the Institute of Forest Economy (Kokchetav), the forest firm Syr-Tabigaty (Kyzyl-Orda), and Barsa-Kelmes Nature Reserve (Aralsk). The following tasks were envisaged:

- To identify the habitat diversity in relation to reclamation
- To plant on large areas with different techniques
- To test the phytoreclamation properties of *Haloxylon aphyllum* in all its ecological amplitude from the sandy soils and barchans to the crusty-puffy coastal solonchaks
- To develop recommendations for phytoreclamation on large areas of the dry seafloor

The sandy areas of the desiccated seafloor define the older surfaces which became dry in the 1960s and 1970s, mainly between the deltas of the Syr Darya and the Amu Darya and around the former islands of Barsa-Kelmes and Vozrozhdeniya. Fine-grit sands with a grain size of 0.1–0.5 mm are characteristic of the whole coastal region. Medium-grained sands are only found in the estuary areas of the Syr Darya and the Amu Darya. The rock-forming material is mainly quartz. The amount of shell limestone in the sand ranges widely from 30% to 70%. The sand depositions at the eastern coast lie at 46–53 (rarely at 43) m asl, above sea level (asl), at the northern coast at 50–53 (rarely at 48) m asl and at the southeastern

Fig. 15.8 Salt desert on the dried-out lake bed of the Aral Sea around the former island of Barsa-Kelmes (Photo: Wucherer, June 2004)

coast at 36–53 m asl. The sandy depositions have been dry for 30–50 years. Plantings on sandy substrates are seen as a good prospect in the whole Central Asian region.

A much greater challenge is the salty substrates. The dry seafloor area of the 1980s, 1990s and even partly of the 1970s (about 70% of the surface area) is salt desert (Fig. 15.8). Here, the surface is covered with a salt crust. These salt surfaces are difficult to approach. The surface of the dry seafloor of the 1990s can exceptionally be approached smoothly in the area of the Bayan transect between the islands of Barsa-Kelmes and Kokaral. It is a mosaic of salt desert and sparse therophytic plant communities. These therophytic communities are dynamic, they are not stable and are no guarantee for surface protection. The salt deserts are very variable, and they represent a plurality of soil types: marsh solonchaks, sandy solonchaks, crusty-puffy coastal solonchaks, loamy coastal solonchaks with sand cover, takyr-like solonchaks, degraded solonchaks, etc. (Ishankulov and Wucherer 1984; Yair 2001). The soil type, horizonation and grain size of the lake deposits are crucial for salinity balance and the phytomeliorative properties of the soils. The conditions of reforestation are very hard and normally need improvement of soils. The following recommendations for phytoreclamation from Yair (2001) were accepted:

1. The sedimentary history of the Aral Sea area, characterized by frequent fluctuations of sea level, resulted in a complicated mosaic of environmental and edaphic conditions. This initial puzzle was affected later on by sand deposits in some parts of the desiccated sea bottom. This evolution, coupled with the high sensitivity of desert plants to slight spatial and temporal changes in salinity and water regime, determines the guidelines for a realistic phytoreclamation management policy (Yair 2001). The following principles are hereby indispensable:
   - Reclamation measures cannot be based on a single principle. They should be closely adapted to the local environmental conditions. In other terms, the recovery of the natural vegetation, and the prospects for reclamation, are not

only a function of time since sea recession, but rather a function of the local sedimentary sequence.

- The prospects for reclamation are highly conditioned by particle size and thickness of the sedimentary units, which determine soil salinity and soil water regime.
- Sandy areas, characterized by a low salinity, good infiltration, good water preservation and low groundwater levels, represent the most favourable environment for the natural recovery of the vegetal cover as well as for phytoreclamation. At the other end of the spectrum we find the solonchak sites characterized by a very high salinity, limited infiltration and shallow saline groundwater. These areas are devoid of any vegetation and represent the main source of airborne dust and salt transported towards cultivated lands and villages. They represent the most problematic areas for reclamation, additionally being the main cause for the deterioration of adjoining arable lands.

2. The results of the experiments presented can serve as the basis for reclamation policy in the area. They lead to the conclusion that solonchak areas are very resilient to a change in their edaphic conditions. Planting without changing the environmental conditions is not an appropriate way to improve the situation. Reclamation of such areas must therefore be based on the alteration of their hydrological regime, if not all over the surface, at least locally.
3. Sand deposition on arable lands is very often regarded as a major cause of land degradation and desertification. This approach is not relevant in the case of sand deposition over silty-clayey, saline, compacted and hydrophobic areas, such as solonchaks, underlain by shallow saline groundwater. Sand deposition over such surfaces has only a beneficial effect. As stated earlier, a topsoil sandy layer has five beneficial effects: it improves infiltration depth, lowers the level of the saline groundwater, increases salt leaching, reduces capillary rise and increases water storage by the creation of a perched freshwater lens on top of an impervious saline substratum. Any reclamation policy should therefore attempt to artificially increase sand cover on top of saline solonchak surfaces.

According to the conditions of afforestation, we basically distinguish the following types of phytomelioration sites on the dry seafloor (Wucherer and Breckle 2005):

(A) Sand deserts

(A1) Sandy soils

(a) Sandy nonsaline soils, soil profile sandy of 2-m depth or more
(b) Sandy saline soils, soil profile sandy of 1–2-m depth

(A2) Barchan or dune fields

(a) A dune height of approximately 1 m
(b) A dune height of 1–3 m (rarely more)
(c) Sand dunes higher than 3 m (only the dunes along of the former coastline)

(B) Salt deserts

(B1) Coastal degraded solonchaks with a layer of sand in the topsoil and with a maximum of salinity in a certain soil depth (often from 20 to 100 cm)

(a) Loamy coastal degraded solonchaks with sand cover, soil profile mostly loamy-clayey, surface layer of sand
(b) Loamy coastal degraded solonchaks, soil profile with a sandy horizon down to 1 m

(B2) Crusty-puffy coastal solonchaks, soil profile loamy-clayey with or without sand cover of 1-m depth (with rather high salinity)

(a) Coastal solonchaks with maximal salinity on the soil surface
(b) Coastal solonchaks with maximal salinity at a certain soil depth (often 20–40 cm).

(B3) Marshy solonchaks

15.3.2 Methodical Approach

The planting plots were selected by Wucherer and Zhamantikov. The technology for the plantings was developed by Wucherer, Breckle and Kaverin.

The substrate surface was prepared and plantings were conducted in November 2002 and March 2004. Several plots (Fig. 15.9, Table 15.4,) were used for plantings south of the former island of Kushzhitmes (Fig. 15.1, area 1). It comprised an area of 248.1 ha of salt deserts with different salinity and texture on the dry seafloor.

The methods were:

1. Planting with use of a commercial tree planter, depth of tillage approximately 20–25 cm (238.6 ha);
2. Manual planting into previously prepared furrows for sand accumulation and into pits filled by sand (9.5 ha)

The plantings were arranged as follows:

- Double-row strips (with a separation of 5 m) from 200- to 1,000-m length with a distance of 30–35 m between the strips.
- One-row strips of 200 m length with a distance of 20 m between the strips.
- Plantings in pits filled with sand.
- Different locations on sandy and loamy soils with various salinities were chosen.

The evaluation of the plantings was done 1 and 2 years later (Fig. 15.10).

Seeds and saplings of local plants (*Haloxylon aphyllum, Tamarix laxa, Halocnemum strobilaceum*) were planted in November 2002 to March 2004. These species have different ecological adaptations to salinity (Chap. 12).

Haloxylon aphyllum (saxaul) is a stem-succulent, shrubby halophyte, which can grow up to 4–5 m, with a wide Irano-Turanian distribution. This species is

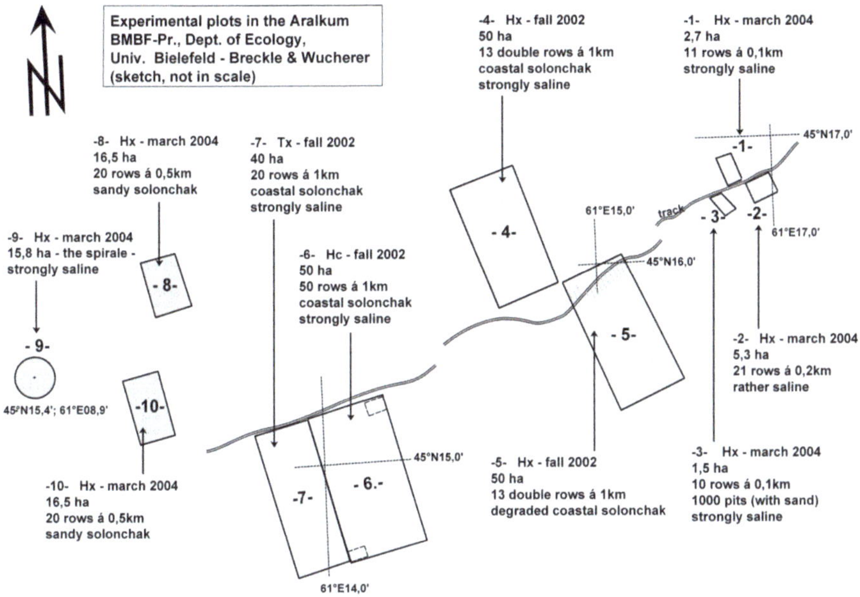

Fig. 15.9 Distribution of experimental plots in the surroundings of the former island of Kushzhitmes at the eastern coast of the Aral Sea

Table 15.4 Scheme of plantings on coastal solonchaks

Phytomeliorant species	Total planting area (ha)	Area of coastal solonchaks (ha)	Area of coastal solonchaks with sand layers (ha)
Haloxylon aphyllum	158.3	59.5	98.6
Halocnemum strobilaceum	50	50	–
Tamarix hispida	40	40	–
Total area	248.1	149.5	98.6

Fig. 15.10 German–Kazakh expedition in the surroundings of the former island of Kushzhitmes at the eastern coast of the Aral Sea (Photo: Wucherer, June 2004)

Table 15.5 Vitality parameters of *Haloxylon aphyllum* under different groundwater levels

Groundwater table (m)	2.5–5.0	5.0–7.0	>7	Very deep
Height (m)	4–7	2.5–4.0	1.5–2.5	1–1.5
Dry matter of wood (t ha^{-1})	30–50	10–25	5–10	<5

After Leontyev 1954

characterized by a broad ecological amplitude. It is normally found on moderate to slightly saline soils, and also on sand dunes and even on hamada rocky sites. Vitality strongly depends on groundwater availability; it produces a huge and deep root system (Table 15.5). Reproduction of saxaul is normally by seeds, but the cuttings from thin plants branches can regenerate vegetatively. The germination rate is very high and often up to 80–90%. The rate of establishment is much lower. Under natural conditions, only in very favourable years with enough soil humidity from winter rain and snow does a conspicuous rejuvenation take place, which is only every 3–5 years. The lifetime of saxaul is about 30–50 years but individuals of more than 50 years up to 100 years of age have been observed (Repetek desert research station in Turkmenistan).

Tamarix hispida is a recreting halophyte, a shrub with an Irano-Turanian distribution. Reproduction is by seeds as well as by suckers. Growth height under optimal conditions can reach 3 m, but normally it is 1–2 m. On saline sites *Tamarix hispida* needs a good groundwater level (1–2 m) for survival. The lifetime is only 10–15 years. On recent assimilating shoots of 30–40-cm length numerous scalelike leaves are formed; their length is 2–9 mm. These scaly leaves exhibit numerous salt glands (about 150–240 per square millimetre) on both surfaces. Recretion of salt is mainly by the salt glands as crystaline NaCl, but also by shedding leavy scales and branches. In July and August about 30–50% of the young branches can be lost depending on the drought situation.

Halocnemum strobilaceum is a stem-succulent euhalophyte, a dwarf shrub with a huge distribution from the Mediterranean to China. The size of the shrubs rarely exceeds 70 cm; the diameter can reach more than 2 m. Reproduction is by seeds and vegetative. The lifetime is about 10–12 years. Optimal development is observed with a groundwater depth of about 1–2 m. Vitality is less with a groundwater depth below 3 m, but vegetative sprouting is enhanced. This flat shrub often forms small mounds (chokolaks) by accumulation of dust and sand, forming a nepkha desert. Seed germination only occurs on wet solonchaks. *Halocnemum strobilaceum* is a halophyte which exhibits one of the highest salinity tolerances; accumulation in the small scaly succulent leaves can reach 50% of dry matter.

15.3.3 *Haloxylon aphyllum* Plantings on Different Sites

15.3.3.1 Plantings on Sandy Soils with Low Salinity (A1a Habitat)

The rate of infiltration of water in the sand is very high; capillary movement is very limited. This leads quickly to eluviation of the salts and destruction of the salt crust

(Yair 2001). The bigger the portion of sand cover in the soil profile, the faster leaching starts. These soils are almost unexceptionally spread over the dry surface of the 1960s. The plantings by the Institute of Forestry of Tashkent at the Aral Sea's southern coast on these surfaces, which were carried in the early 1990s, were successful. In the course of 13–15 years after the desiccation, the soil profile was desalinized up to 2-m depth. The causes were the sandy soils and the low saltwater concentration of the former lake water (10–15 g/l) at the beginning of the Aral Sea's desiccation.

The soil is low in humus (0.33% in the crust, 0.16% in lower horizons) and low in total nitrogen (0.011–0.024%). Carbonate is maximal in the lower horizons (4.9%; 1.52% on the surface). The pH is about 8.0–8.2. CEC is very low (3–6 meq/ 100 g) with prevailing calcium. The salinity in the soil profile varies from 0.36% to 0.58%. The salinity is due to sulphate and chloride.

The ecological conditions are very good for the natural germination and establishment of psammophytes (sand plants) (Wucherer and Breckle 2001, 2005; see Chap. 10). Within 10–20 years after desiccation, a persistent vegetation had developed on these sites. Blowing out of sand and salt dust from these substrates is very low.

15.3.3.2 Plantings on the Barchan and Dune Fields (A2a,b Habitat)

The lack of diaspores on the sandy sites of the desiccated seafloor of the 1970s and 1980s leads to a very slow primary succession (see Chap. 10). The huge open areas exhibit strong deflation and quick creation of dune fields. Along the eastern coast and the former islands of Uzun-Kair, Kyzylbai, Akpasty and Kosaral and in the Akpetki Archipelago the dune fields are widespread. The dune fields are oriented towards the south, southwest and southeast. The barchan fields between the former islands and the eastern coast are located on the lee side, and they became slowly and sparsely covered with vegetation (Fig. 15.11). The barchan fields on the western side of the islands are usually without vegetation, they spread to the west and cover

Fig. 15.11 Vegetation cover of the dune fields with *Haloxylon aphyllum* (Photo: Wucherer, September 2003)

Fig. 15.12 Plantings of *Haloxylon aphyllum* in the dune region on the dry seafloor of the 1970s (Photo: Wucherer, September 2004)

the primary loamy coastal solonchaks and thus diminish the salt-dust asset of the primary lake bottom.

Small-scale plantings in the dune region at the eastern coast (Fig. 15.1, area 1) were successful (Fig. 15.12). The establishment rate of *Haloxylon aphyllum* was above 80%.

At the former southern coast, extensive plantings on the dune fields were conducted successfully by the GTZ project (Hüfler and Novitskiy 2001).

15.3.3.3 Plantings on Poorly to Moderately Saline Sandy Soils (A1b Habitat)

These soils are sandy usually up to 1-m depth, in the upper soil desalinized and in the middle part of the soil profile poorly to moderately saline (Fig. 15.9, plot 10, 16.5 ha, 20 rows in 0.5 km). Such soil conditions are beneficial for the settlement and establishment of different plant species as well as for reforestations. Naturally, mixed plant communities are formed on these habitats with psammophytes, halophytes and tugai species. But vast dune fields and sand deserts are created because of limited seed input. At such a location, successful planting was carried out in March 2004. In June 2004 the establishment rate of *Haloxylon aphyllum* on average was over 90% (Figs. 15.13 and 15.14). By stabilization of the groundwater table there is a possibility to develop long-term plant communities.

15.3.3.4 Plantings on Coastal Solonchaks with a Sand Layer (B1a)

This habitat indicates a loamy-clayey soil profile, which has a sand layer on the soil surface. The sandy overlay can be of natural origin or secondarily formed by aeolian activities. Such a sandy layer eases the germination and establishment conditions for plants. This habitat is present at the spreading border of the sandy sedimentation deposits on the dry seafloor and at the edge of the newly forming dune fields and salt deserts. The soils offer good germination conditions for species

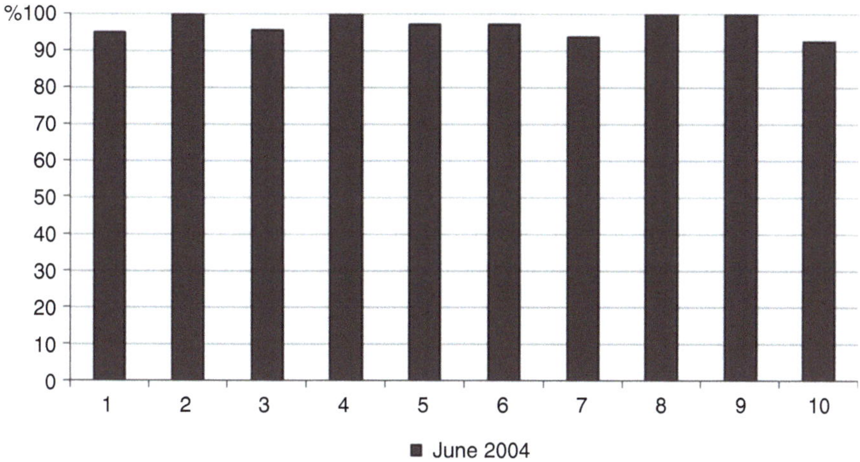

Fig. 15.13 Establishment rate of *Haloxylon aphyllum* on poorly to moderately salinized sand soils, ten strips, 100 m long (Plantings in March 2004 evaluation in June 2004)

Fig. 15.14 Plantings of *Haloxylon aphyllum* on poorly to moderately salinized sand soils (Plantings March 2004; record, June 2004) (Photo: Wucherer)

which tolerate moderate salinity, such as *Haloxylon aphyllum* and *Nitraria schoberi*. This location is not favourable for psammophytes (real sand plants such as *Calligonum*). The spontaneous vegetation is some annual saltworts and *Senecio noeanus, Strigosella circinnata, Atriplex pratovii* and *Salsola nitraria* with up to 15–20% coverage.

The soil is low in humus (0.62% in the crust, 0.25% in the lower horizons), and is low in total nitrogen (0.017–0.045%). Soil carbonate is up to 6%, 3% in the upper horizons. The pH is 8.0–8.2. CEC is rather high in clay horizons (26.5 meq/100 g), but is low in the sandy surface horizon (6.3 meq/100 g). Calcium prevails, but there is considerable magnesium in lower horizons, whereas the amount of sodium is rather low (2%, indicating no solonetzic structure). The soil profile is slightly saline down to 23 cm (0.18–0.93%). The lower horizon has a salt content of 2.13%. The salinity is due to chloride and sulphate.

The plantings on the coastal solonchaks with a thick sandy layer (50 cm, 20 rows, 0.5 km long) from March 2004 were successful (Fig. 15.9; plot 8, 16.5 ha). The establishment rate varied between 80 and 100% and did not change in autumn (Figs. 15.15 and 15.16).

Another habitat for the plantings of *Haloxylon* (plot 5, 50 ha, Fig. 15.9) is a loamy coastal solonchak with sandy cover of 30 cm on topsoil (autumn plantings). The 0–30-cm horizon is sandy and loamy sand. The topsoil to about 15-cm depth is dry, but from 15- to 30-cm depth is moist. In the horizons of 30–50 cm, thin layers

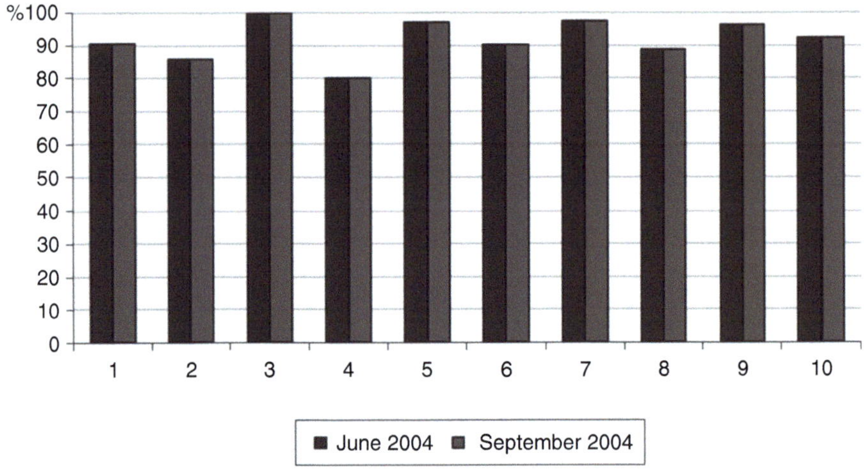

Fig. 15.15 Establishment rate of *Haloxylon aphyllum* on coastal solonchaks with thick sand layer, ten strips, 100 m long (Plantings in March 2004, record in June and September 2004)

Fig. 15.16 Plantings of *Haloxylon aphyllum* on coastal solonchaks with thick sand layer (Plantings – March 2004) (Photo: Wucherer, September 2004)

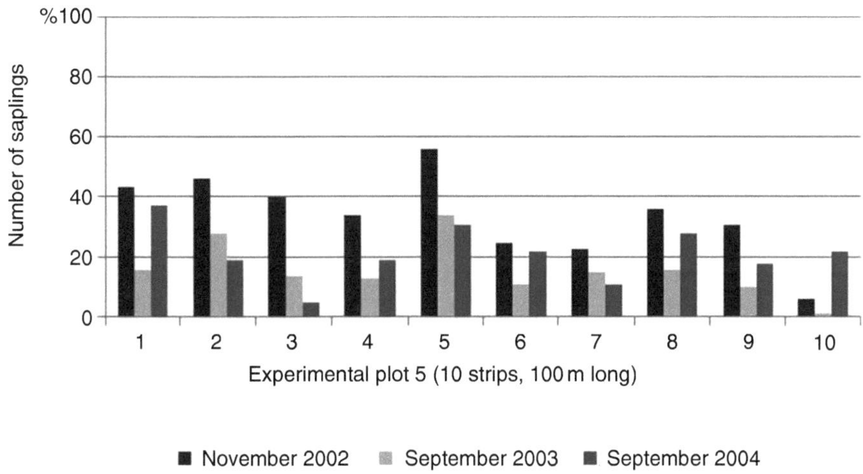

Fig. 15.17 Number of established saplings and seedlings of *Haloxylon aphyllum* on loamy coastal solonchaks with a sand layer, ten strips, 100 m long (Planting in November 2002, record in September 2004)

of sand and loam alternate. Beneath 50 cm, the soil is composed of loam and clay to a depth of 150 cm. The establishment rate of *Haloxylon aphyllum* in the first year (2003) was moderate and variable and ranged from 10 to 35 saplings on 100-m-long strips (Fig. 15.17). An increase in the number of saplings in 2004 was observed, most probably caused by further seeds from 2003 that did not show any morphological differences from the saplings in autumn 2004. The survival rate of saxaul saplings at the end of first year after planting was 37.4%. After 2 years, 26.2% of saplings had survived; about 300 seedlings per hectare were calculated in second year. Finally the planting on plot 5 was successful.

15.3.3.5 Plantings on Crusty-Puffy Coastal Solonchaks (B2b)

These soils are deeply salinized and are the main source of the salt-dust output from the Aral Sea floor. They are spread on the dry surface of the more recent seafloor desiccation, including the dry seafloor of the 1990s, of the 1980s and partly that of the 1970s. A coastal solonchak with a *Suaeda acuminata* plant community was chosen for the plantings (plot 6, with an area of 50 ha, Fig. 15.9) The annual halophytes appear either directly after the desiccation of the seafloor as water-level plant communities on the marshy solonchaks or on the coastal solonchaks in high-rainfall years. *Suaeda acuminata* is an indicator of a high salinity rate in the soil profile. Other annual saltworts (*Climacoptera aralensis* and *Salsola nitraria*) increase the coverage to 15–20%, rarely to 40–45%.

Exceptionally, here the topsoil (up to 4 cm) and the horizon of 105–130 cm are composed of loamy sand and coarse sand. Normally, the whole soil profile is composed of loam and clay. According to our classification, this site is a loamy coastal solonchak, the soil profile of which shows a sandy horizon within 1-m depth. But the thin sandy layer on the surface is insignificant for establishment.

The soil is low in humus (0.70% in the crust, 0.62% in the upper horizons, 0.21% in the lower horizons), an is low in total nitrogen (0.017–0.052%). Soil carbonate is up to 5.41%, and 3.38–2.87% in the upper horizons. The pH is 8.0–8.1. CEC is moderate (16.5 meq/100 g) in the surface horizon, but drops (7.2–14.0 meq/100 g) in the 1–35-cm horizon. Calcium prevails, but sodium is present at 9.8% (weakly solonetzic structure). The soil surface is saline (1.87%), an the profile is slightly saline down to 11 cm (0.33%). The lower horizons have a salt content of 0.99–1.59%, and are loamy clay. The salinity is due to sulphate and chloride.

The survival rate of saxaul saplings was 12–14%. Seedlings of saxaul produced roots at a moderate rate. A calculation in second year showed that saxaul saplings survived with a rate of less than 10%.

The afforestation conditions for plantings of *Haloxylon* are still unfavourable. The site conditions of this plot for *Haloxylon apyllum* are almost beyond the physiological amplitude.

15.3.3.6 Plantings on Crusty-Puffy Coastal Solonchaks in Furrows (Soil Site B2a,b)

These soils are saline and covered by a salt crust. The plot area was 8.0 ha (Fig. 15.9, plots 1 and 2; 2.7 and 5.3 ha). Plots 1 and 2 differ from each other. The plantings on plot 2 were performed in 11 rows, each 200 m long. The furrows were about 30 cm deep and about 60 cm wide. The soil profile is composed of clay and is strongly saline (B2a). The salinity is due to chloride and sulphate, the salt content is about 2.77% (salt crust 5.48%). The natural vegetation was a loose therophytic plant community with *Climacoptera aralensis,* a few *Atriplex pratovii* and some *Suaeda acuminata*.

In September 2004 the rate of establishment of *Haloxylon* was normally below 10% and the vitality of the saplings was rather bad (Figs. 15.18 and 15.19). This demonstrates, that these sites with loamy-clayey crusty-puffy coastal solonchaks are hardly suitable for the establishment of *Haloxylon aphyllum* even with improvement of the soil conditions, since all lower soil horizons are very clayey (75.6%).

Plantings on plot 1 were performed in 21 rows each 100 m long. The furrows had the same size as those for plot 2. The soil profile is mainly loamy and moderately saline (B2b). This coastal solonchak with a salinity maximum rather deep in the soil profile carried a natural vegetation of loose therophytic aggregations with *Climacoptera aralensis*, *Atriplex pratovii* and *Salsola nitraria*. The site conditions are more moderate, and the rate of establishment is thus higher, 20–80%, and there is a better vitality of the saplings (Figs. 15.20 and 15.21).

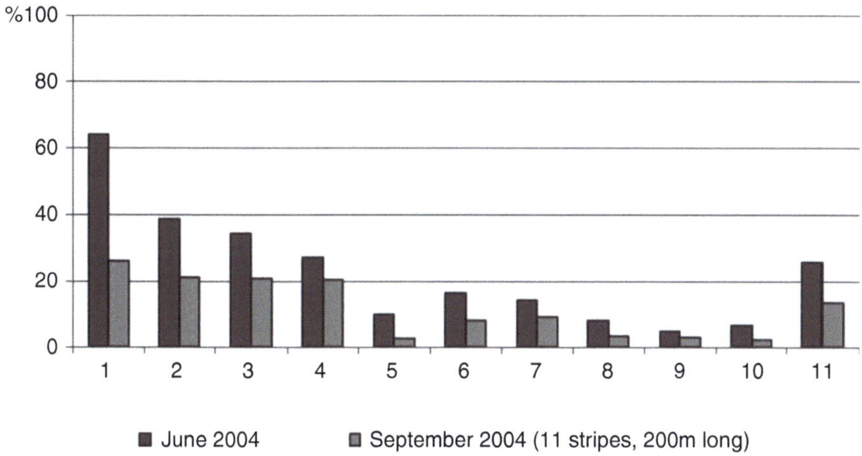

Fig. 15.18 Establishment rate of *Haloxylon aphyllum* on crusty-puffy coastal solonchaks with furrows, 11 strips, 200 m long (Plantings in March 2004, coverage in September 2004)

Fig. 15.19 Plantings of *Haloxylon aphyllum* on crusty-puffy coastal solonchaks (Plantings in furrows March 2004) (Photo: Wucherer, September 2004)

15.3.3.7 Plantings in Pits (Soil Site, Habitat B1b)

In autumn 2002, 1,000 pits were prepared (up to 100-cm depth and 60-cm diameter) on a coastal solonchak (the salinity maximum on the soil surface), and filled with sand. In March 2004, *Haloxylon aphyllum* was planted in the pits. In June 2004, the establishment rate was over 90% and in September it was still over 70% (Figs. 15.22 and 15.23). The improvement of the soil conditions had definitely increased the establishment rate.

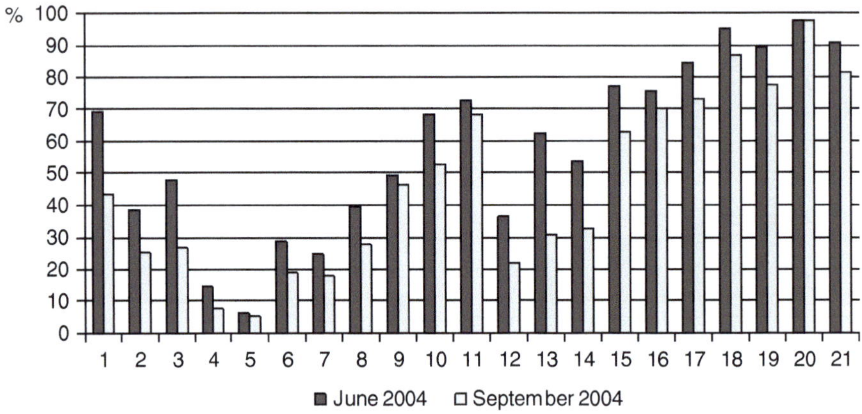

Fig. 15.20 Establishment rate of *Haloxylon aphyllum* on coastal solonchaks with an adjournment of the salinization maximum at a certain soil depth (Plantings in furrows in March 2004, record in June and September 2004)

Fig. 15.21 Plantings of *Haloxylon aphyllum* on coastal solonchaks with an adjournment of the salinization maximum at a certain soil depth (Plantings in 21 furrows in March 2004) (Photo: Wucherer, June 2004)

15.3.3.8 Efficiency of Plantings in Spring and in Autumn

In comparing two equal sites, it is obvious (Fig. 15.24) that the rate of establishment of *Haloxylon aphyllum* in spring plantings is about double that for autumn plantings.

15.3.3.9 Sowing and Germination of *Haloxylon aphyllum*

The seeds of *Haloxylon aphyllum* were sown on three planting plots with *Haloxylon* and *Halocnemum*. The seeds were collected 300 km from the Aral Sea's coast in the area of the village of Karmaktschi. The seeds were sown simultaneously with the

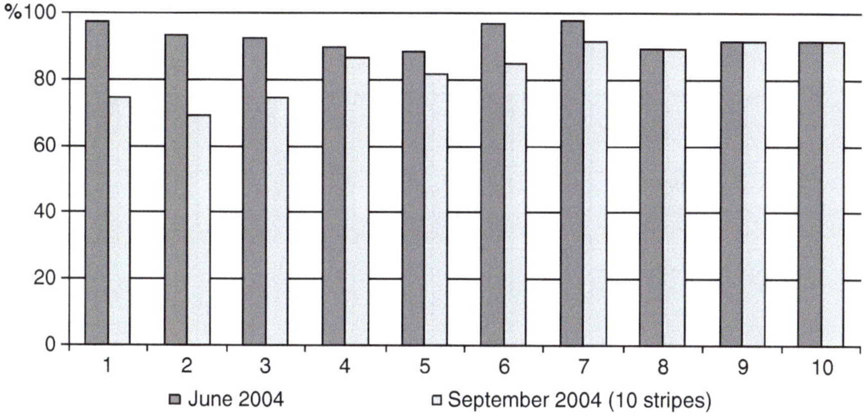

Fig. 15.22 Establishment rate of *Haloxylon aphyllum* on degraded coastal solonchaks (Plantings in pits in March 2004, ten strips, coverage in June and September 2004)

Fig. 15.23 Plantings of *Haloxylon aphyllum* on degraded coastal solonchaks (Plantings in pits, March 2004; coverage, June and September 2004) (Photo: Wucherer, June 2004)

plantings of the seedlings. In the *Haloxylon* plantings on the coastal solonchak only very few seeds germinated. This germination is statistically irrelevant. The germination and establishment of the first year on the loamy coastal solonchaks with sand cover were quite high. The extreme variability of the germination on the loamy coastal solonchak with sand cover is influenced by the prevailing ecological conditions in the topsoil.

15.3.4 *Halocnemum strobilaceum* Plantings

Halocnemum strobilaceum is a dwarf shrub, a euhalophyte withstanding very high salinities. Soils of natural sites with *Halocnemum* are often coastal solonchaks with

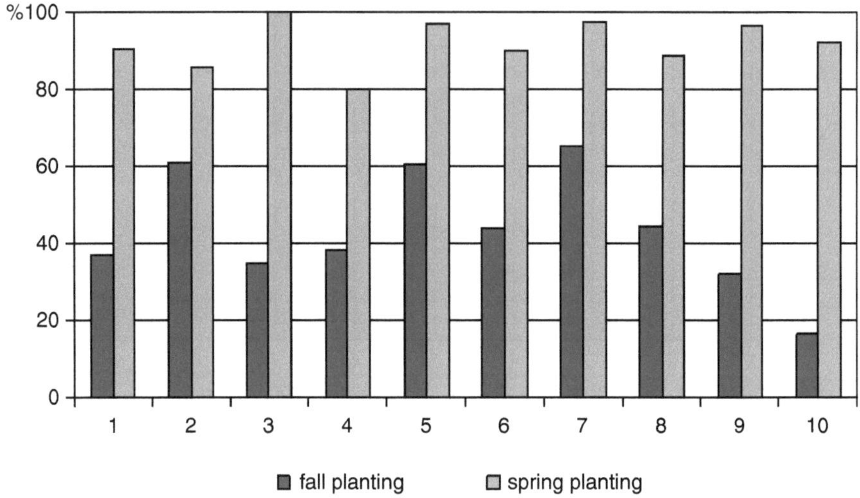

Fig. 15.24 Establishment rate of *Haloxylon aphyllum* on the coastal solonchaks with a sandy layer (autumn plantings in November 2002 and record in September 2003; spring plantings in March 2004 and record in September 2004)

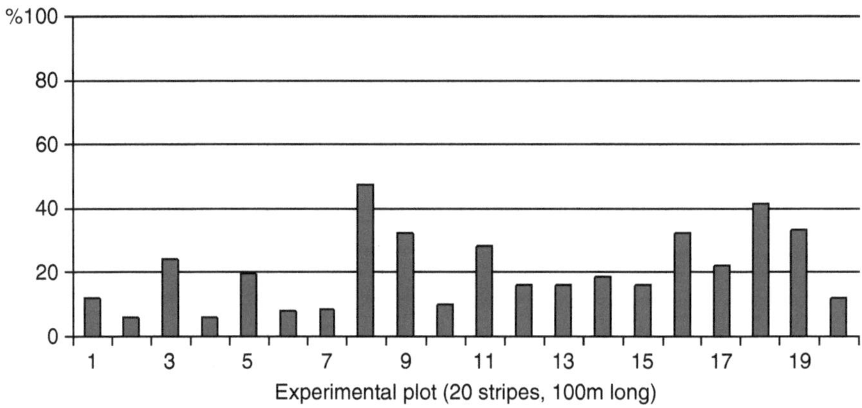

Fig. 15.25 Establishment rate of *Halocnemum strobilaceum,* ten strips (Plantings in November 2002, record in September 2003)

loamy and clayey horizons, often with sandy horizons in middle parts (1 m) of the soil profile (sites B1b). Those sites are obviously the ecological optimum (Fig. 15.26) for *Halocnemum.* Plantings were done on those sites without soil improvements (plot 6, 50 ha, Fig. 15.9). One year after the plantings, the rate of establishment was about 20–40% (Fig. 15.25). But the vitality of the saplings was rather bad, and dropped in the second year to less than 1%. This was mainly due to the dry summer, but was also due to technical difficulties in planting saplings of *Halocnemum.* Roots of *Halocnemum* do not have a strong taproot, and thus are

Fig. 15.26 *Halocnemum strobilaceum* plantings on the Kaskakulan transect (Plantings by V.S. Kaverin, 12 years old) (Photo: Wucherer, June 2003)

often stuck in the dry upper soil during the mechanical planting procedure. The planting success depends very much on the skill of the workers. Plantings by hand were successful (Fig. 15.26).

15.3.5 Tamarix Plantings

Plantings with *Tamarix* were done on plot 7 (Fig. 15.9, 40 ha in 20 rows of 1,000-m length) on strongly saline coastal solonchak. Plantings were done with cuttings, mainly *Tamarix hispida, Tamarix laxa* and *Tamarix elongata*, in autumn 2002. The first evaluation in spring 2003 indicated a severe loss of saplings. Almost none of the saplings were detected (50 seedlings and three saplings per hectare); the very few remnants exhibited severe signs of grazing by rodents, most probably hares. Another test planting with protection against grazing has not been done yet.

15.3.6 Physicochemical Properties of Saline Soils on Planting Plots

Not only grazing but also the specific physicochemical properties of the saline soils (Table 15.6) determine the survival rates of planted material. The evaluation of the physicochemical characteristics of soils was done for most of the experimental plots of the BMBF-GTZ/CCD project by L.A. Dimeyeva and V.N. Permitina. The results allow predictions to be made regarding the results and success of phytomelioration.

Soil samples for the study were collected from the rooted horizons. Chemical analyses of soil samples were done with the support of the Japan Research Association with Kazakhstan (PIE/JRAK).

Table 15.6 Physicochemical characteristics of saline soils in experimental plots

Plot no.	Depth (cm)	Toxic ions (mg eq/100 g soil)	Toxic salts (%)	Degree of salinity	Cl^-/SO_4^{2-} molar ratio	Soil texture
1	0–13	6.19	0.52	High	0.52	Sandy loam
	13–22	29.9	2.43	Very high	1.55	Clay
	22–90	35.5	2.56	Very high	3.18	Clay
1a	0–0.5	201.5	13.7	Very high	5.01	Clay
	0.5–25	57.1	3.9	Very high	2.83	Clay
1b	0–1	68.6	4.9	Very high	2.81	Loam
	1–25	8.63	0.83	Very high	0.84	Clay loam
2	0–0.5	13.2	1.12	Very high	0.76	Sandy loam
	0.5–11	2.01	0.25	Moderate	0.6	Sandy loam
	11–35	9.73	0.86	Very high	1.4	Loam
	35–65	19.5	1.16	Very high	2.71	Clay loam
3	0–0.5	5.36	0.51	High	0.54	Sandy loam
	0.5–8	1.05	0.08	Slight	0.77	Sand
	8–23	2.09	0.31	Moderate	0.23	Sandy loam
	23–80	3.87	1.59	Very high	0.76	Loam
4	0–0.5	3.59	0.29	Moderate	1.67	Sandy loam
	0.5–26	3.55	0.28	Moderate	1.94	Sand
	26–60	3.12	0.31	Moderate	1.5	Sand
	60–70	6.14	0.49	High	1.9	Sand

15.3.6.1 Total Effect of Toxic Ions

Soils of the Aral Sea coast have a significant salt content. Depending on qualitative and quantitative attributes of saline soils, they have different degrees of salinity and unequal toxicity of freely soluble salts. Only the toxic salts Na_2CO_3, $MgCO_3$, $NaHCO_3$, $Mg(HCO_3)_2$, Na_2SO_4, $MgSO_4$, NaCl, $MgCl_2$ and $CaCl_2$ have a hazardous influence on plants. Therefore it is more reasonable to examine the degree of salinity of soils by the presence of toxic ions and salts. Determination of the degree of salinity on the basis of the "total effect of toxic ions" (Vityzev 1973) by calculation defined the toxic ions in aquous extracts. That may not need differentiation of soils by salinity type. The concentration of ions in milliequivalents per 100 g of soil is used for calculation of toxic ions, and the percentage of toxic salts is used as an assessment of the degree of salinity. CO_3^{2-} ions form toxic salts (Na_2CO_3 and $MgCO_3$). HCO_3^- ions form toxic $NaHCO_3$ and $Mg(HCO_3)_2$ salts and nontoxic Ca $(HCO_3)_2$ salts. SO_4^{2-} ions can form toxic (Na_2SO_4, $MgSO_4$) and nontoxic ($CaSO_4$) salts. The same is true of Cl^- ions, the major salt formed being NaCl.

Coastal solonchak soils (plot 1) have considerable amounts of salts throughout the soil profile. Maximum of salts with prevalence of chlorides is found in lower horizons showing fewer desalinization processes. The total effect of toxic ions ranges from 6.19 meq/100 g of soil in the upper horizon to 35.5 meq/100 g of soil in the lower horizon (Table 15.6). The degree of salinity increases with depth. The content of toxic salts amounts to 2.5%. The degree of salinity is very high.

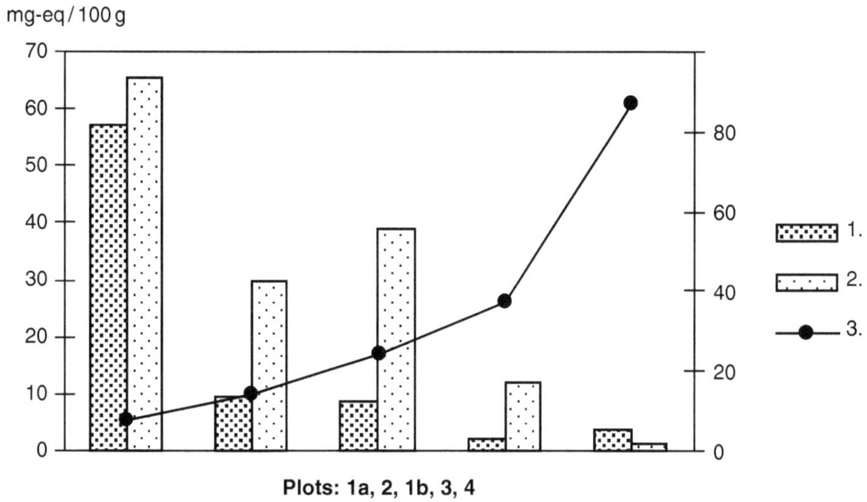

Fig. 15.27 Survival rates of saxaul saplings (*Haloxylon aphyllum*) on saline soils (in root-inhabited horizon) with differing physicochemical site characteristics: *1* total concentration of toxic ions (mg eq/100 g of soil); *2* content of silt and clay (%); *3* survival rate (%) in the first vegetation season

Replacement of saline soils by nonsaline substrates is the most promising method of land reclamation of coastal solonchaks and coastal soils of heavy texture. Addition of sand improves the physicochemical properties of saline soils and increases the survival rate of saxaul saplings up to 91.6%.

There is an important direct dependence of the survival rates of saplings and the total effect of toxic ions together with the texture of saline coastal soils (Fig. 15.27, see also Fig. 15.9, Table 15.6).

15.4 Shelterbelts for Villages

Experiments on rehabilitation of coastal ecosystems were continued by AEON Environmental Foundation/Japan, 2005–2006, and the Japan Fund for Global Environment, 2006–2008, in cooperation with L.A. Dimeyeva. Two tasks were included in the study:

1. Establishment of green shelterbelts around villages
2. Afforestation of the dry seafloor

Two villages were chosen for planting activity with participation of local people. Saplings of *Haloxylon aphyllum* were planted in a plot of 1 ha in a dry bed of Bozkol Bay (close to Kaukey village). The soil conditions are very harsh–highly saline clay. The method of pits with sand was used to increase the survival rate, which was 25% due to reclamation work. A green shelterbelt of 2.5 ha near

Karateren village was established in more favourable soil conditions (medium salt content, loam). The survival rate of saxaul saplings was 60% in the first year and 57% in the second year.

Thus, a modern approach in reclamation of salt deserts and combating desertification in the coastal areas is the creation of small plantations (oases, "green spots"). This will support the natural processes of propagation and creation of seed banks for acceleration of natural succession. On the other hand, green shelterbelts around villages will be created step by step (Dimeyeva 2008).

15.5 Perspectives and Necessity of Phytomelioration on the Desiccated Seafloor

15.5.1 Minicatchment Experimental Sets on the Bayan Transect at the Eastern Coast

These experiments were done within the Bayan transect close to the former eastern coast of the Aral Sea. Experimental plots were chosen on sites of the desiccated seafloor from the 1990s. On those sites only therophytic communities were present; there were no perennial species.

15.5.1.1 Plantings with Minimal and Maximal Technical Approach

In November 1998 on the experimental plots 1- or 2-year old saplings of *Tamarix laxa, Halostachys caspica, Halocnemum strobilaceum* and *Haloxylon aphyllum* were planted. Two variants were tested: planting on the smooth surface and planting in pits about 30 cm deep. About 600 plants were used. The rate of establishment was only about 2%.

The first half in 1999 was extremely dry and almost devoid of precipitation. In any case it is clear that the first year after phytomelioration plantings is very decisive for the success of the plantings. A minimal technical approach is not sufficient for success, especially in rather saline sites with puffy-crusty coastal solonchaks. It remains an open question if it is possible by better technological means to achieve better results even in very dry subsequent years.

In April 2000 again two plantings series were performed. One used furrows about 20 cm deep and 30 cm wide, filled with sand. The other used a modification of the minicatchment procedure by using rather deep pits of about 1-m and 50-cm diameter. One-third of the pit holes were filled by sand, and the upper parts were smoothed to create catchment areas with 10–15° runoff slopes. These slopes covered a length of 1.5–3 m, corresponding to an area of 2.25 m^2 and 9 m^2 for the bigger pits. In total, 64 small and 60 bigger pits were tested.

In the pits cuttings from *Haloxylon aphyllum, Tamarix hispida* and *Tamarix elongata* were planted. In the furrows *Haloxylon aphyllum* and *Halocnemum strobilaceum* were tested. The minicatchment test exhibited rates of establishment for *Haloxylon aphyllum* of 11.4% (planting on the smooth surface) and 22.2% (planting in pits about 30 cm deep), and for *Tamarix hispida* they were 71% (planting on the smooth surface) and 58% (planting in pits about 30 cm deep).

For plantings in furrows about 20 cm deep and 30 cm wide filled with sand the rate of establishment for *Haloxylon aphyllum* was 100% and for *Halocnemum strobilaceum* it was 42%. However, these tests revealed that only a rather laborious method is successful, which only can be used in small-scale plantings. Only the furrow technique is applicable on a larger scale.

15.5.2 Results of the Plantings in the Kushzhitmes Area

Satisfactory and good results were obtained for *Haloxylon*, worse results were obtained for *Halocnemum* and the worst results were obtained for *Tamarix*. The survival rate for *Haloxylon* in conditions of coastal solonchaks of heavy texture was 12–14% at the end of first vegetation period, was under 20% in coastal soils with thin blown sand cover and was up to 76–98% in coastal soils with thick blown sand cover between sand dunes. The establishment rate of *Haloxylon aphyllum* on the barchans was above 80%. The survival rate for *Halocnemum* in conditions of coastal soils with blown sand cover was 14% at the end of first vegetation period; only 1% of saplings survived 1 year later. Germination of *Haloxylon* seeds was higher in coastal soils with sand cover. Survival rates of *Haloxylon* saplings in furrow plots ranged from 10% to 90% on plot 1, which was caused by different soil conditions. Experiments with plantings into pits with sand exhibited a very high survival rate of saplings (above 90%).

The experimental studies resulted in a number of articles and recommendations (Breckle 2003; Breckle and Wucherer 2006, 2007; Dzhamantikov et al. 2003; Ogar et al. 2005a, b; Dimeyeva and Ogar, 2006; Dimeyeva and Permitina 2006; Kaverin et al. 2005; Wucherer 2001; Wucherer et al. 2005; Wucherer and Breckle 2005). Important information on methodology and site selection for phytoreclamation activity can be found in Dzhamantikov et al. (2003).

Green plantations are created by means of sowing seeds and manual or mechanized plantings. Seeding is possible only in years with favourable wet conditions, which cannot be predicted. There were good conditions twice during the 7-year experimental study. Therefore, seeding should constitute 10% of the total amount of forest melioration. Better results could be obtained by seeding in sand-accumulating furrows with sand cover not less than 15–20 cm. Planting should be conducted in early spring after the melting of snow cover or in first half of November.

Results on experimental phytoreclamation in different years and seasons have shown that the most promising time for planting is early spring. The most promising forest culture is *Haloxylon aphyllum*. The most suitable conditions for reforestation in the dry seabed are light textured soils (coastal sandy soils and coastal soils with sand cover). Soils of heavy texture need amelioration, e.g. by sand-accumulating furrows near to sand massifs and replacement of heavy ground by sand.

15.5.3 Phytomelioration Properties of *Haloxylon aphyllum*

The establishment and the growth of *Haloxylon* is dependent on the weather conditions, the hydrological conditions and the season of the plantings. Lake deposits or soils, of which the topsoil is lightly grained, are favourable for the plantings. The establishment was successful on the poorly to moderately saline sandy soils and the loamy coastal solonchaks with sand cover and a low salinity degree of the topsoil. On the crusty-puffy coastal solonchaks, however, the rate of establishment of *Haloxylon* is only up to 10% and the vitality of the young plants is very bad. The plantings on the coastal solonchaks with an improvement of the soil conditions (in furrows and in pits) increase the establishment rate of *Haloxylon*. The establishment rate for spring plantings is higher than that for autumn plantings.

Sowing of the *Haloxylon* seeds (fruits) on the loamy coastal solonchaks with sand cover produced very variable results, but was also effective. Sowing of granulated seeds on the snow cover in spring is more effective compared with sowing on the soil surface without snow. The mechanical treatment of the soil surface increases the establishment rate. Sowing on the crusty-puffy coastal solonchaks even with treatment of the soil surface or the improvement of the topsoil is almost hopeless, even though most sandy and sandy loam coastal soils do not have any morphological features of alkalinity.

However, the establishment of separate individuals on sites with extreme conditions is of crucial relevance for the success of the plantings and spreading of a spontaneous phytomelioration (Fig. 9.23, Chap. 9).

Established plants produce seeds after a few years, often 5–6 years. With an improvement of the site conditions or in wet years, the availability of seed material (from stock plants) can be sufficient to develop thick assets or patches of plant communities. This example shows that even the establishment emanating from very few individuals is an important part in the creation of the future vegetation cover and accordingly in the control of salt-dust blowouts.

On which sites should *Haloxylon* reforestation be organized? In the sand deserts adjacent to the Aralkum no big dune or barchan fields without vegetation are present; they are mostly covered with vegetation. The emergence of local barchan fields is a result of an anthropogenic influence by keeping the sand surface open by overgrazing and trampling adjacent to villages. Normally those sandy sites have a naturally sufficient self-growth potential in this region. If the impact by the people has stopped, a rather fast natural resettlement and spontaneous regeneration of the vegetation starts. This is also due to the sandy sites of the dry seafloor from the 1960s and early 1970s. They do not have to be planted, as they are usually already covered with vegetation.

The younger dune fields of the dry seafloor are usually devoid of vegetation. The salinity of the dune fields, which is rarely linked to the groundwater table, is rather low compared with the salinity of the coastal solonchaks. The salt-dust output of the dune fields is minimal. The creation of dunes favours the transfer and spreading of the sand in all directions (also in the direction of the actual coastline) and favours the covering of the crusty-puffy coastal solonchaks with sand. The leaching procedure is initiated and the salt-dust asset decreases as a consequence of an already thin cover of 3–5 cm sand on the soil surface. Hence, sand is an important natural phytomelioration factor on the dry seafloor of the Aral Sea. The dune fields do not have to be planted, but rather should have contrarily a free development to fulfil their natural phytomeliorative function. The barchan fields at the eastern coast of the Aral Sea on the western side of the former islands (Kaskakulan, Uzun-Kair, Kushzhitmes, Kozzhetpes) are also spreading also in the western direction (to the lake) and cover the primary loamy coastal solonchaks of the Aral Sea floor.

The loamy coastal solonchaks with a sand cover can and must be planted for two reasons:

1. There are often widespread crusty-puffy coastal solonchaks. These saline soils indicate good site conditions for germination on the dry seafloor of the 1970s and 1980s. Plantings can remarkably contribute to minimizing the salt-dust output. The technological effort is not excessively expensive.
2. There are often island-like sand accumulations on the dry seafloor (even on the dry seafloor of the 1990s), where the distribution pattern of the lake deposits is very variable (e.g. at the northern side of the former island of Barsa-Kelmes or the town of Muinak). The plantings on these sand islands can lead to the creation of vegetation islands in the salt desert and this can lead to a second spreading of vegetation.

On the crusty-puffy coastal solonchaks *Haloxylon* plantings can only be successfully performed with an improvement of the soil conditions (e.g. in the furrows or pits, filled with sand). Some seedlings can reach the generative phase, however, rather fast, but a second spreading does not occur, since the seeds of *Haloxylon* do not germinate on coastal solonchaks. Therefore, plantings on crusty-puffy coastal solonchaks are only worth performing at locations where natural desalinization of the topsoil is expected, and where sand islands are present.

15.5.4 Development of New Planting Strategies

15.5.4.1 Change of Areal Priorities

A vast area of the dry seafloor has developed between the eastern coast of the former Aral Sea and the former island of Vozrozhdeniya. A salt swamp and a huge salt desert have been created. The dry surface is about 60,000 km. It will not grow considerably in the coming years. The salt-dust output, however, will increase immensely. The experiments and the plantings of the institutes of forestry of Kokchetav and Tashkent as well as of the GTZ projects in Uzbekistan and of the Department of Ecology of the University of Bielefeld project in Kazakhstan are predominantly limited to the dry surface of the 1960s, 1970s and 1980s. These experiments have provided worthwhile information on the technology of plantings with *Haloxylon aphyllum* and a few other species on special soil types. However, for the plantings emphasis must be placed on the dry surfaces of the 1980s and 1990s as well as on the surface, which recently dried out. These expanded deflation surfaces are the source of the salt-dust output, the salinity of which is usually very high, and chemistry of which is rather variable. Plantings must accordingly be conducted on moderately to highly saline soils and solonchaks.

15.5.4.2 Selection of Reforestation Species

A forest will never be able to grow according to the climatic conditions of this continental desert with temperature extremes between −45°C and 45°C and an average yearly precipitation of about 100 mm. The phytomelioration in the region of the Aral Sea, this means plantings of species which are suitable for the conditions of the climate and the soil, have to take into account the relevant ecological situation. This is a significant limitation, which even under optimal conditions only allows a reconstruction of a low desert bush vegetation, which still features a possibly high cover degree to minimize the open soil surface. The most suitable species for many sites is the bush *Haloxylon aphyllum*; however, this species only thrives on soils of moderate salinity. *Haloxylon* can only be planted on coastal solonchaks with an improvement of soil conditions. Finally, the use of *Haloxylon* is limited for plantings on the dry seafloor of the 1980s and 1990s. The selection must hence be extended to more euhalophytic species, which can withstand more salinity in the soil, such as *Halocnemum strobilaceum.*

In the region there are many other perennial euhalophytes from the genera *Halostachys, Kalidium, Salsola,* and *Suaeda*. However, the technology for planting them is practically unknown and must be developed further.

15.5.4.3 Adjustment to the Planting Technologies and Location Mapping

Experimental studies on phytomelioration of saline soils of different texture have shown that the most important factors affecting the survival rate of saplings are the total effect of toxic ions determined by the degree of salinity and the total amount of toxic salts, and the soil texture of the rooted soil horizons.

The technologies tested (construction of furrows, mechanical enforcement of the topsoil, removing the salt crust, minicatchment procedure, etc.) should also be used for the plantings of euhalophytes, but only with consideration of the specific habitats and the biological properties of the reforestation species. Total soil and vegetation mapping of the dry seafloor has been conducted only roughly. This is an essential requirement for plantings. The planting experiments have shown that the sites within 50 ha were not uniform. Hence, the planting plots should be smaller. The basis for a second settlement wave should be formed by small centres of fruiting old plants. For the expansion to other saline sites on the dry seafloor without vegetation, remote sensing by satellites is essential.

15.5.4.4 Consideration of Other Planting Priorities

Loose sands and migrating dunes are dangerous for humans in the village areas as a consequence of the regional overgrazing and deforestation. Therefore, plantings of loose local sand dunes surrounding villages and residential areas must be encouraged. These wind-protection plantings should also involve bigger tree species, as long as watering is possible, and in pits. According to the previous experiences, those plantings around villages must be performed, secured, treated and protected from fire and grazing in cooperation with the rural population.

15.6 Discussion and Final Conclusions

Experimental studies on phytomelioration of saline soils of different texture have shown that the most important factors affecting the survival rate of saplings are the total effect of toxic ions determined by the degree of salinity and the total amount of toxic salts, and the soil texture of the root-inhabited horizons.

The assortment of plant species for phytomelioration of the dry seafloor with a good prospects is very small. Saxaul (*Haloxylon aphyllum*) is noticeably distinguished among them. It has a wide ecological amplitude and can grow in diverse sites of salinity and soil texture environments. However, the disturbed root system of saplings is affected by salinity and drought stress. Methods of melioration can help to overcome toxic stress. Soils of heavy texture need reclamation, e.g. by establishment of sand-accumulating furrows near sand massifs taking into account the main direction of the prevailing wind and replacement of heavy ground by sand.

Plantings of sarsazan (*Halocnemum strobilaceum*) saplings should be realized in plots with a groundwater table not lower than 1.5 m. Such conditions have usually developed after 10–12 years of the continental regime. This is not always under the influence of the continued desiccation process.

Planting *Tamarix* saplings in the dry seabed of the 1970s is problematic because of this biotope is inhabited by hares, which greatly or totally destroy the plantations.

The improvement of the environmental situation and the living conditions of the people and the struggle against poverty in the region at the Aral Sea is one of the biggest challenges for the struggle against desertification according to the United Nations Conventions to Combat Desertification. It can only be managed by numerous measures by countries concerned with the help of the global community. Here it is indispensable to have constant scientific guidance and to build up and to constitute appropriate monitoring. Planting on saline soils is an effective measure for decreasing the salt-dust output from the dry seafloor and for soil stabilization. The two most important goals of the phytomelioration are attainable. Firstly, there is the protection of the population from sandstorms, which is a local problem. For this, local wind-protection plantings must be established in and around the villages and residential areas with their degraded environment. Secondly, the protection of the population from salt-dust storms, which is the bigger and especially not a local but a regional problem. In addition to this, large-scale plantings must be done directly at the sources of the salt-dust output. The solution of the environment problems is vitally important for the population in the region around the former Aral Sea.

Acknowledgements The experiments and the plantings in the region at the Aral Sea in Kazakhstan were performed by the interdisciplinary research projects "Succession processes on the desiccated sea floor of the Aral Sea and perspectives of land-use" and "Combating desertification and rehabilitation of salt deserts in the Aralkum", funded by the Federal Ministry of Education and Research (BMBF project 0330389) and the latter was also supported by the GTZ/CCD project in the name of the Federal Ministry for Economic Cooperation and Development (BMZ). Funds for students came from DAAD. All these are greatly acknowledged.

References

Agakhanjanz OE, Breckle S-W (1994) Ökologie der Erde, vol 3, 2nd edn, Spezielle Ökologie der gemäßigten und arktischen Zonnen Euro-Nordasiens. Fischer, Stuttgart

Breckle S-W (2003) Rehabilitation of the Aral Sea environment, Kazakhstan. In: Proceedings of the International workshop, Aleppo, May 2002: Combating desertification – rehabilitation of degraded drylands and biosphere reserves. UNESCO-MAB dryland series No 2: pp 47–57

Breckle S-W, Wucherer W (2006) Combatting desertification in the Northern Aral Sea region. In: Gao J et al (eds) Restoration and stability of ecosystems in arid and semi-arid areas. Science Press, Beijing, pp 304–316 (Peking-Symposium China August 2004)

Breckle S-W, Wucherer W (2007) What will be the future of the Aral Sea? In: Lozan JL et al (eds) GLOBAL CHANGE: enough water for all? Wissenschaftliche Auswertungen GEO, Hamburg, pp 142–146

Dimeyeva L (2001) Methods of conservation and restoration of vegetation cover on the Aral Sea coast. In: Breckle S-W, Veste M, Wucherer W (eds): Sustainable Land-Use in Deserts. Springer/Heidelberg, p. 69–73

Dimeyeva L (2007) Primary Successions on the new Aral Seashore. Basic and Applied Dryland Research 1:1–16

Dimeyeva LA (2008) Green spots as a tool to combat desertification in the Aral Sea region. In: Proceedings of international conference on drylands, deserts and desertification, Israel, 14–17 Dec 2008, pp 42–43

Dimeyeva L, Meirman G, Budnikova T et al (2000) Experimental methods of Phytoreclamation in the Aral Sea Coast. In: Proceedings of the international conference on reality and perspectives of sustainable development of the ecosystems in the Aral Sea region, Almaty, 2000, pp 9–11

Dimeyeva L, Ogar N (2006) Rehabilitation of salt deserts in the Aral Sea coast. In: Proceedings of international conference on ecological restoration in East Asia, Osaka, 16–18 June 2006, pp 55

Dimeyeva L, Permitina V (2006) Effect of physical-chemical characteristics of saline soils on phytoamelioration results in the dried seafloor of the Aral Sea. Arid Ecosystems 12–29:82–93
Dzhamantikov Kh, Dzhamantikov E, Daldabayeva G et al (2003) Recommendations on assortment and technology of tree-shrub halophytes culture in the dry seabed of the Aral Sea. Institute of Agroecology, Kyzyl-Orda
Geldyeva GV, Ogar NP, Scorintseva IB et al (2001) Monitoring and modeling of desertification processes in the Syr Dar'ya and Amu Dar'ya River deltas, for GIS. In: Ecological research and monitoring of the Aral sea deltas. A basis for restoration. UNESCO Aral Sea Project. 1997–2000 Final Scientific reports. Book 2, pp 119–153
Hüfler F, Novitskiy S (2001) Wald auf dem ausgetrockneten Boden des Aralsees. BMZ/GTZ Broschüre, Tashkent (German-Russian)
Ishankulov MSh, Wucherer W (1984) Nature complexes of the east coast of the Aral Sea in 1982. Probl Osvo Pustyn 1:52–58 (Russian)
Kaverin VS, Koltunov AA, Solov'ev VA, Salimov AB, Cherevatenko VP (1994) Temporary recommendations on assortiment of species and technology of phytoreclamation in the dry seafloor of the Aral sea (in the limits of Kazakhstan). Institute of Agroforestry, Kokchetav
Kaverin VS, Salimov BA (2000) Perspective ecological rehabilitation of the dry seabed in the dry seabed of the Aral Sea. News Kazakhstan Sci (Almaty) 1:21–23
Kaverin VS, Salimov BA, Shahmatov PF (2005) About necessity of phytomelioration of the dry seabed of the Aral Sea. In: Actual questions of Agroforestry and green planting in Kazakhstan. Almaty, pp 89–96
Kurochkina LYa, Makulbekova GB (1984) To question about phytomelioration of drying shores of the Aral Sea. Probl Desert Dev (Ashgabad) 4:61–71
Kurochkina LYa, Makulbekova GB, Wucherer W, Malaisarova A (1983) Vegetation of the dry seafloor of the Aral Sea. In: State of the water area and the dry seabed of the Aral Sea, Kaz. AN, Alma-Ata, pp.91–128
Leontyev VL (1954) Saksaulovye lesa pustyni Kara-Kum (Saxaul forests of the desert Karakum). AN SSSR, M.-L. [monograph]
Makulbekova GB, Wucherer W (1990) Ispol'zovaniye rastitel'nosti osushennogo dna Aral'skogo morya (use of vegetation of the dry seafloor of the Aral Sea). In: Kurochkina LYa, Shabanova LV (eds) Kompleksnaya kharakteristika pastbishch pustynnoi zony Kazachstana. Nauka, Alma-Ata, pp 221–223
Meesa F, Singer A (2006) Surface crusts on soils/sediments of the southern Aral Sea basin, Uzbekistan. Geoderma 136(1–2):152–159
Meirman G, Dzhamantikov Kh, Kaverin VS (2001a) Phytomelioration on new landscapes of the dried Aral Sea bottom/Ecological research and monitoring of the Aral sea deltas. A basis for restoration. UNESCO Aral Sea Project. 1997–2000 Final Scientific reports. Book 2, pp 203–211
Meirman G, Dimeyeva LA, Dzhamantikov Kh, Wucherer W, Breckle S-W (2001b) Seeding experiments on the dry Aral Sea floor for phytomelioration. In: Breckle S-W, Veste M, Wucherer W (eds) Sustainable land use in deserts. Springer, Berlin, pp 318–322
Novitskiy ZB (1997) Nauchnye osnovy zashitnogo lesorasvedeniya na osushennom dne Aralskogo morya (Research basic of the phytomelioration on the dry sea floor of the Aral Sea). Avtor. doctoral dissertation, Tashkent (in Russian)
Ogar N, Bizhanova G, Dimeyeva L, Permitina V (2005a) Phytoreclamation of salt deserts in the Aral Sea coast. Rep Nat Acad Sci Rep (Kazakhstan) 1:89–93
Ogar N, Kaverin V, Wucherer W (2005b) Experimental works on phytoreclamation of a dry seabed of the Aral Sea. Actual questions of Agroforestry and green planting in Kazakhstan. Academy of Science, Almaty, pp. 157–162
Vityzev VG (1973) Manual on soil-meliorative basing of reclamation construction and mapping. Gyprovodhoz, Moscow
Wucherer W, Breckle SW (2001) Vegetation dynamics on the dry sea floor of the Aral Sea. In: Breckle SW, Veste M & Wucherer W (eds.): Sustainable Land-Use in Deserts. Springer/ Heidelberg, pp 52–68

Wucherer W, Breckle S-W (2005) Desertifikationsbekämpfung und Sanierung der Salzwüsten am Aralsee. Sukzession und Phytomelioration, Naturschutz und nachhaltige Entwicklung. Bielefelder Ökologische Beiträge 19, pp 93

Wucherer W, Breckle S-W, Kaverin VS, Zhamantikov Kh, Ogar N (2005) Phytomeliorative Eigenschaften von *Haloxylon aphyllum* und perspektiven der anpflanzungen in der region am Aralsee. In: Veste M, Wucherer W, Homeier J (eds) Ökologische Forschung im globalen Kontext. Cuvillier, Göttingen, pp 109–127

Yair A (2001) Sedimentary environments in the desiccated Aral Sea floor: vegetation recovery and prospects for reclamation. In: Breckle S-W, Veste M, Wucherer W (eds) Sustainable land use in deserts. Springer, Berlin, pp 310–317

Chapter 16
Phytomelioration in the Southern Aralkum

Z.B. Novitskiy

16.1 Introduction

The problem concerning the desiccation of the Aral Sea is among those which are internationally urgent to solve and some crucial tasks have to be solved immediately. The consequences of the Aralkum syndrome in its current stage have greatly impacted the whole Central Asian region. Mainly Uzbekistan and Kazakhstan suffer from the prevailing situation of frequent salt-dust storms as the states bordering the remnants of the Aral Sea.

We have compared some of the catastrophes which happened globally during the last 25 years and found that the Aral ecological catastrophe is so pernicious and huge on temporal and spatial scales that it could really be compared with the Chernobyl tragedy. Nuclear material has a certain half-life decay period, but the existing ecological situation in the Aral region has no time limits; it is worsening from year to year. The desiccation is still continuing, releasing more and more saline soils from water (see Chap. 2). The total salt accumulation in the air by dust affects the people's health and causes numerous and rising not yet studied diseases in the wider Aral region. It is often said that the air in the Aral region can contain components harmful to the health of humans and animals in a concentration 10 times or even 100 times higher (on the desiccated seafloor) than the international norms allow.

To minimize the overall effects of dust storms, an afforestation project in the Uzbekistan part of the desiccated seafloor of the Aral Sea was launched. Mainly sandy and moderately saline sites were chosen to test large-scale plantings and their effects on particle deflation. The only way to minimize dust storms is by phytomelioration as in the north (Chap. 15).

S.-W. Breckle et al. (eds.), *Aralkum - a Man-Made Desert: The Desiccated Floor of the Aral Sea (Central Asia)*, Ecological Studies 218,
DOI 10.1007/978-3-642-21117-1_16, © Springer-Verlag Berlin Heidelberg 2012

Fig. 16.1 Desiccated seafloor of the Aral Sea between the Akpetki islands close to the former southeastern coast of the Aral Sea

16.2 The Present Critical Situation

Currently, the desiccated seafloor area comprises several million hectares, from which annually up to 75 million tons of salt, dust and sand are blown out into the atmosphere (see Chap. 5). Taking into account sedimentation and accumulation of the atmospheric dust, we cannot exclude the mountains where the Central Asian river springs are located. As salt from the Aral Sea was found even in Arctic ice, it is easy imaginable what would happen to the region's population in the near future. The development of the Aralkum has already led to a degradation of two million hectares of former croplands, and the crop capacity on the other lands is harshly reduced by increasing salinity. If nothing is undertaken, there is the danger of not only an ecological but also a physical starvation in the Aral Sea region. The deflation processes from the dry seafloor develop very rapidly and the blowing out forms moving sands (see Chap. 5) in the form of spits and dunes (Fig. 16.1).

The saline dust is composed of very small particles; they are very light, and thus they are blown high into the atmosphere (see Chaps. 5–7), become mixed with clouds and travel long distances of hundreds of kilometres, after which they fall as salty rain or snow. On the desiccated seafloor of the Aral Sea, snow in winter or spring may be somewhat salty.

With wind velocities of 8–10 m/s, all kinds of fine substrate particles from the dry seafloor are prone to strong wind erosion. Fines of sand, dust and silt are primarily blown out. Their percentage decreases in the remaining substrate significantly; concentration of coarse and mean sand particles distinctly increases.

16.3 Aims of the Project

Nowadays, the normal way of living in the Aral region is harsh. The ecological problems need an urgent solution. Several huge projects have been discussed. One way would be to find means to provide water in such an amount that the former sea level of the Aral Sea will be reached. This is utopian. It would need to refuse all irrigation farming in the area. This is impossible, since the population has increased more than fourfold in the last 50 years (Talanova 1997).

The second way is to accept the present situation and to adapt and to improve the regional ecological and economic conditions. As pointed out in Chap. 15, phytomelioration is one of the important tasks now and in the future. The situation can be improved by "afforestation" of the dry seafloor with desert tree or shrub plants such as *Haloxylon aphyllum*, *Salsola richteri*, C*alligonum caput-medusae* and a few others.

Such plantations of wood on the desiccated seafloor of the Aral Sea perform the following functions:

- Roots of the created plantations of shrubs and trees reduce salt, dust and sand blowout. They decrease deflation processes. Dust storms on the long run may be kept down.
- Plantations improve soil conditions by slowly accumulating some humus.
- Part of the plantations could be transformed to pastures. Fodder cropping may be possible to some extent and under strict grazing management systems with plants such as chogon (*Aellenia subaphylla*), teresken (*Ceratoides papposa*), keyreuk (*Salsola orientalis*) and some others.
- Where good water sources are available (there are a few springs), even plantations as shelterbelts and small-scale farming may be possible.
- Any plantations absorb carbon dioxide and evolve oxygen.

Long-term research by the Uzbek Scientific and Research Forestry Institution has demonstrated that the problem of forest tracts as a climate-regulating factor is important for the Central Asian region, where the share of forest plantations is estimated to be only 4.7%. In Uzbekistan, a country characterized by a very small share and irregular spread of forest tracts throughout the whole territory (the share of forest plantations is 5.1%), as well as the contrasting climatic conditions, the role of protective forest plantations or woods would be to enable maintenance of a better ecological environment.

Only a wide-scale phytomelioration of the dry seafloor can mitigate ecological tension in the southern Aral region and create more favourable living conditions and a sound environment for the local population.

16.4 The Recent History of Phytomeliorative Plantations

During the 1980s, when everybody became aware of the Aral Sea problem and its threat, nobody had experience of how to curb it. All were hoping to use water, i.e. how to find water reserves to rehabilitate the sea level. This was impossible. The threat and health problems faced by the people in Central Asia are increasing year by year. Scientists from the Sector for Forest Melioration of Deserts, Central Asian Scientific and Research Forestry Institution (the current Uzbek Scientific and Research Forestry Institution) pooled their expertise (on desert afforestation over more than 50 years) and decided to tackle the problem of melioration of the desiccated seafloor of the Aral Sea with desert plants.

In 1981–1985 a scientific project on the background and examination of experimental sectors with the forest melioration methods for reclamation of the dry seafloor of the Aral Seawas developed.

In 1986–1990 a scientific project on development and introduction of desert afforestation on the desiccated seafloor of the Aral Sea started to prevent negative ecological consequences and to create pastures with desert plants.

In 1991–1996 a scientific project on development of protective forest plantations on the seafloor sediments suitable for melioration was conducted.

In 1997–2000 a scientific project on development of methods and technologies for improved phytomelioration with wood plantings on the desiccated saline soils on the seafloor of the Aral Sea (Hüfler and Novitskiy 2001).

Besides the major projects mentioned above, the Forestry Institution was developing other projects on methods and technologies for fixation of moving sands, and for methods for fixation and melioration of fine sand dune sediments on the desiccated seafloor at the southern Aral Sea coast. Presently, scientists are working on a project to develop methods for accelerated overgrowing and vegetation cover of the desiccated seafloor of the Aral Sea by means of assisting the renewal of natural seeds.

Nobody had investigated phytomelioration of the desiccated seafloor before the Forestry Institution started in 1981. The desiccated seafloor exhibits diverse substrates with differing particle size, light or heavy, with all kinds of salinity and free of any humus accumulation. Such sterile soils “in nature” are subject to a slowly ongoing primary succession (Chap. 10). Phytomelioration needs scientific achievements according to the differentiation of soil types and the level of its reclamation. By this, we identified three categories of substrates:

1. The first one covers soil types for which reclamation is possible only after their preliminary treatment with different methods (subsandy, sandy, subclay and other soil types).
2. The second covers soil types for which reclamation is possible only after preliminary relief fixation by sand-retaining protections (moving sands).
3. The third category covers soil types which are presently unfit for wood plantings (alkali soil).

Phytomelioration is the most important and realistic way (Chap. 15) to change the environment in Central Asia.

16.5 Melioration Project

16.5.1 Necessary Preconditions

Desertification in the Aral Sea region (Chap. 18) is a product of complicated interactions between the social and economic systems (diseases, poverty, starvation, unsustainable economy) and natural factors (drought, water erosion, soil salinization, desiccation of the sea and severe soil deflation, as well as degradation of vegetation, arable land and reduced crop productivity). These components can easily interrelate and accelerate the desertification processes (Giese 1997; Müller et al. 2006).

Both the former USSR and foreign countries have accumulated quite a rich expertise on melioration of sandy regions, but not for solonchak soils of the seafloor. Research results have shown that plantations can improve the agrarian production and its environment by means of a better hydroclimatic regime and improvement of fodder crop capacities, of soil protection against wind erosion, of drought mollification and abatement of hot winds, of involvement in biological accomplishment of ecosystems (including the attraction of birds and wild animals) and of a rational utilization of nature. Thus, people's living standard could be improved (see Chap. 18).

The following maps created by scientists from the Forestry Institution for the whole of the area covered by the desiccated seafloor of the Aral Sea in Uzbekistan are very important for the melioration practices:

- Map of the location of desiccated seafloor sediments
- Map of the classification of sediments according to their salinity
- Map of the projected desiccated seafloor surface cover with plants
- Map of the desiccated seafloor zones for phytomelioration with woods
- Map of the classification of the dry seafloor for suitability for afforestations

The maps enable one to assess the real situation on the desiccated seafloor of the Aral Sea and to identify the areas with a primary need for phytomelioration and reclamation, as well as to select correctly technologies and to assort wood plants for every separate sediment type.

On the dry seafloor of the Aral Sea, a protective wood or shrub planting based on one common scheme is not possible because of the obvious variety of natural environment and soils, in particular. Therefore, an edaphic typology network of conditions was developed to adapt plantations on the desiccated seafloor of the Aral Sea to a distinct grain size of the substrate, to the levels and types of soil salinities and to groundwater depth. Phytomelioration is now using this edaphic typology network of conditions for the plantation sites. Accordingly, it enabled a plant assortment to be selected for every sediment type (Novitskiy 1997). On the desiccated seafloor of the Aral Sea the following wood plants are used: black saxaul (*Haloxylon aphyllum*), Richter's saltwort (*Salsola richteri*) and Medusa's head kandym (*Calligonum caput-medusae*). For pastures and fodder crops the following plants are used: grey teresken (*Ceratoides papposa*), keyreuk (*Salsola orientalis*), chogon (*Aellenia subaphylla*), izen (*Kochia prostrata*), boyalich (*Salsola arbuscula*) and selin (*Aristida karelinii*).

16.5.2 Effects of Wind Conditions and Planting Results

Wind and the type of vegetation are equally involved in relief-forming processes on the dry seafloor. Aeolian transformation of dunes into semimoving, fixed or overgrown parts is interrelated with vegetation growth and density. Plants influence the wind regime. This depends on the form, size and density of the single plants and the density of vegetation coverage.

Roots system excavations of 2-year-old black saxauls (*Haloxylon aphyllum*) planted on moving sand dunes showed that the role of plants' underground parts is conspicuous in sand reinforcement. At this age, most parts of active roots are concentrated in the upper 30 cm sand layer and are able to withstand high summer temperatures (on the dry seafloor the sand surface temperatures can reach 60°C or even higher). In the process of plants' living activity and by decay of their dead underground and surface litter, the chemical and granulometric content of the sand changes with time. This provides its compression and enrichment with fine-dispersed material. Mechanical and soil-forming activity of plants leads to a decrease of aeolian processes and their adverse effects.

Nepkha formation by single plants on the desiccated seafloor depends on the type of plant. In general, dune hill size depends on the height, shape and growing stage of the plant, on surface relief and on the effective wind power and direction. Trunk-forming plants such as sandy acacia (*Ammodendron conollyi*) and camel burr (*Alhagi*) do not accumulate sand and thus no hills are formed. But dense plants with a broad shape can accumulate huge sand hills, e.g. *Tamarix* and *Nitraria*.

Forest plantations on the desiccated seafloor should possess a good wind-protecting function. Wind velocity was studied near the surface using anemometers as far as it influences the sand transfer within 0–20 cm (Table 16.1).

Table 16.1 Interrelation between wind velocity (20 cm above ground) and age of shrubs

Age of plants (years)	Wind velocity ($m\ s^{-1}$)	Reduction of wind velocity (%)
Open area (control)	12.7	0
1	10.1	20.5
2	8.3	34.6
3	5.7	55.2
4	3.3	74.0
5	2.6	79.5
6	1.1	91.3
7	0	100
10	0	100

As we have observed, in the sector of 1-year-old plants the wind velocity was reduced by 20.5% and in the sector of 2-year-old plants it was reduced by 34.6%. At the age of 5–6 years, forest plants formed a kind of natural grassy vegetation and saxaul self-seeded crop, which caused a significant reduction of the wind velocity and thus deflation.

In another massive plantation, not only reduction of wind velocity but also salt and dust transfer was investigated. The control sandy zone without plants exhibited a wind velocity of 15 m/s. Salt and dust transfer after transmission of the various plantation lines was measured with dust collectors. The results are shown in Table 16.2.

Table 16.2 Reduction of wind velocity and dust transport by wood plantation lines of saxaul (*Haloxylon aphyllum*). The distance between lines of plantations was 10 m, the age of the woods was 4 years and the height was about 2.5 m

Plantation line	Wind velocity ($m\ s^{-1}$)	Reduction of wind velocity (%)	Dust transfer reduction (%)
Sandy zone without plants (control)	15	0	0
Line 1	9	40	37
Line 2	7	53	64
Line 3	5	67	85
Line 4	4	73	94
Line 5	2	87	≈100
Line 6	2	87	≈100
Line 7	1	99	≈100

Fig. 16.2 Seedlings of new *Haloxylon aphyllum* taken from 10-year-old mother plants of *Haloxylon aphyllum* near the former Gulf of Muinak

After 4–5 years, when the saxaul plants start fruiting and diaspores are disseminated with the wind throughout the whole area to other open flats, spreading of young seedlings and saplings starts and additional parts of the desiccated seafloor will be conquered in a natural way, as it is to be seen on older parts of the desiccated seafloor (Fig. 16.2).

On the open dry seafloor the wind often blows from the sea to the mainland, sometimes parallel to the recent shoreline. Newly formed sandy hummocks are often ephemeral and not constant – they disappear or change their shape if the wind direction changes or a plant dies. For this reason, sandy hills on the desiccated seafloor change their orientation, often reflecting the last strong wind direction.

Accumulation of sand by bushes and fixation of sand by leaves, litter, salt and dust is a kind of a conservation process (Fig. 16.3). Conserving abilities of plants depend on the accumulating ability as well as on the quantity of litter produced by the plants and accumulated around the base of the hillocks. Table 16.3 gives an impression of the amount of sand accumulated by various bushes.

For all plantations it is necessary to know the threshold value of vegetation coverage where deflation of soil is significantly reduced. With a coverage of 70%, deflation is almost completely stopped. But also a much lower coverage percentage is good, since wind speed is lowered, and thus much less strong winds (above 5 m/s and only these are responsible for strong deflation) occur, especially in older plantations (Figs. 16.4–16.6). As expected, the quantity of blown sand is in direct proportion to the increasing wind velocity (Table 16.4). The most sand is blown out from a dune

Fig. 16.3 Experimental wood plantings of saxaul (*Haloxylon aphyllum*) in lines

Table 16.3 Sand accumulation by various plant bushes

Plant	Age of plant (years)	Height of plant (cm)	Circumference of plant (cm)	Amount of accumulated sand (m^3)	Cover of stems (%)
Haloxylon aphyllum	7	265	320	10.6	28
Salsola richteri	Not known	220	260	5.6	23
Calligonum caput-medusae	Not known	110	210	6.9	55

Fig. 16.4 Twelve-year-old plantings of saxaul (*Haloxylon aphyllum*) in lines

Fig. 16.5 Massive forest plantings of saxaul (*Haloxylon aphyllum*) (age 10 years)

Fig. 16.6 Saxaul (*Haloxylon aphyllum*, age 6 years) planted on moving sands on the desiccated bottom of the Akpetki Archipelago, in 2005

without a plant cover and the least is blown out from an area protected by wood plantations. The amount of soil matter moved by the wind at the surface (saltation) depends not only on the wind velocity, but also on soil structure, on humidity, on litter

Table 16.4 Interrelation between the quantity of blown sand (g m^{-2}) and wind velocity (m s^{-1}) in 100 h in the summer–autumn period

Location for sand-blowing measurements	Wind velocity (m s^{-1})		
	2–5	6–10	11–16
	Amount of blown sand (g m^{-2})		
Dune not covered with plants	226	294	326
Dune with 1 year old plantation	196	265	289
Sandy valley with 30% grass coverage	156	214	254
Middle zone of 9-year-old wood plantation	38	85	109

Table 16.5 Sand transfer distance (percentage of transported sand mass) by increasing wind velocities on the dry seafloor

Wind velocity (m s^{-1})	Distance from the sand surface source (cm)			
	0–5	6–10	11–15	16–30
0–3	63	28	6	3
4–7	56	30	9	5
8–11	48	34	11	7
12–15	35	29	26	10

Table 16.6 Chemical characteristics and amount of elements of sand blown out from 1 m^2 in 100 h with an average wind velocity of 11–16 m/s. Cl salinization was 0.078% and dry residuum was 0.75%, equivalent to 2.16 g/m^2/100 h

Locality	Dry residuum (g)	Amount (g/m^2/100 h)				
		Na	Cl	S	Ca	Mg
Dune with 1-year-old plantation	2.16	0.01	0.22	0.87	0.26	0.17
Middle zone of 9 year old wood plantation	0.81	0.005	0.08	0.33	0.09	0.06

of plants and on surface roughness. This was studied by sticking paper strips on the desiccated seafloor of the Aral Sea (Table 16.5). About 85–90% of the whole amount of sand transferred by wind is in the narrow layer very close to the surface (0–10 cm). Sand grains are pulled up higher in the air and to rather great distances only with strong winds (12–15 m s^{-1}, see Table 16.5). More than 300 days/year, the wind regime in the Aral region is in the category of low and medium wind velocity. This means that most of the sand and wind flow transfer takes place very close to the land surface and even relatively low plants can successfully perform filter functions by increasing surface roughness. Thus, most plants directly accumulate sand and salt dust in their lee side.

Desalinization of the soil of the desiccated seafloor is a slow process. Whether it is accelerated by plantings is not exactly known. Soils under 12-year-old saxaul plants had a reduction within 12 years from an original 0.35% Cl^- salinization of the 0–10 cm horizon (with 2.13% dry residuum) to 0.02% (0.17% dry residuum). A similar tenfold reduction of salinity was observed at the 10–20 cm horizon. Thus, it is important to know not only the quantitative amount of blown sand, but also its chemical structure. Therefore, samples of blown sand were analysed by the agro-chemical laboratory at the Forestry Institution. The results are shown in Table 16.6.

It has been demonstrated that plantations reduce not only the amount of blown-out soil but also significantly reduce blowing out of chemical elements, however here with rather low salinity.

If the results are extrapolated to 1 ha, the amount of fixed sand and dust is about 200 t. For the whole plantations made on the desiccated seafloor hitherto, six million tons of fixed sand, salt and dust can be estimated. And thanks to plantings, currently 630,000 t of dust and 130,000 t of NaCl are prevented from being blown out from the desiccated seafloor of the Aral Sea.

16.5.3 Projects of Plantings

Naturally, all woody, herbaceous and grassy plants play a soil-protective and antideflation role. However, the desiccated seafloor displays a big variety of saline soils, from completely saline to separate saline spots and to varying sandy layers. Undoubtedly, it is not advisable to start plantations by the method of dense sowings or seedlings. The application of differential methods according to the purpose and soil conditions is needed. Wood plantations for melioration were divided into the following groups taking into account the types of soils and their salinity: patchy (isolated spots, "nidus" type), local and massive plantations on moving sands, for melioration fodder, for pasture protection and for other purposes.

16.5.3.1 Massive Wood Plantings

Massive forest planting in principle uses the suitable bigger woods (desert shrubs, trees) of *Haloxylon aphyllum* and *Salsola richteri*. They were planted in lines with a distance of 8–10 m between them (Figs. 16.4–16.6). During the fourth to fifth years after the plants started fruiting, the space between the lines was filled with plenty of natural rejuvenation (four to six saplings per square metre). The result is a two-storey vegetation, which is already a good shelter for wild animals and a pasture for cattle. The project has created 2,450 ha of such plantations at the southeastern Aral Sea floor.

16.5.3.2 Local Wood Planting

Alkali soils on the dry seafloor are almost devoid of plants for a long time, except for ephemeral ones. Plantings on this soil directly is not possible. But at the border of small alkali soil spots or at their periphery from their wind-percussive side, local plantings are advised. The easiest method is to plant two strips with three lines each, with a distance of 10 m between the strips and 3 m between the lines.

Again, during the 6–7 years a two-storey vegetation spot caused by rejuvenation between the strips developed and a 22-m-wide plantation formed. A slow desalinization was observed. In those plantations, deflation of salt dust has almost stopped and has been reduced by 58% at a distance of 50 m from the plantation. Local plantations can promote the self-covering with wood and grassy plants in their vicinity. By this principle, the project has created 2,100 ha of wood plantations and has minimized salt and dust blowout and stopped deflation on those localized alkali soils.

16.5.3.3 Patchy Wood Planting

There are huge areas of solonchaks on the desiccated seafloor and these will even increase in the future (Chap. 2). Massive or even local plantings do not seem to be not possible. With the nidus (nest) principle, however, small islands can be found, where small nests (patchy spots) for planting saxaul (*Haloxylon aphyllum*) and saltwort (*Salsola richteri*) can be created (Fig. 16.7). Our experiments show that during the fifth year after planting, new seeds are produced and distributed by the wind for a distance of up to 1,000 m. This provides a rich self-seed crop on other suitable spots, slowly providing a coverage of wood and bushes, which protects soils and reduces deflation processes. This area under the self-seed crop increases during favourable years. By the patchy wood planting principle, 3,800 ha of forest plantations has been created.

16.5.3.4 Pasture Protection and Melioration Fodder Wood Planting

Melioration on the desiccated seafloor of the Aral Sea enables one not only to fix and improve the soil but further to use the created plant coverage as pastures. On suitable sites we located pasture-protecting wood strips composed of *Haloxylon aphyllum* and *Salsola richteri* 90 cm from each other with three lines at a distance of

Fig. 16.7 Phytomelioration by the planting of saxaul (*Haloxylon aphyllum*) in spots, in areas where degradation is pretty high (age 5 years)

3 m from each other. Among the pasture-protecting wood strips we sowed one line of melioration and fodder strips composed of chogon (*Aellenia subaphylla* = *Halothamnus subaphyllus*), teresken (*Ceratoides papposa*), boyalich (*Salsola arbuscula*) and keyreuk (*Salsola orientalis*) with a distance of 20 m between the strips. We found that those fodder crops can produce seeds during the first year, which then are disseminated within the whole distance among the strips forming preliminary pastures. This principle was applied on 2,400 ha.

16.5.3.5 Wood Plantings on Moving Sands

The area of moving sands in Uzbekistan comprises 1.8 million hectares, and 0.8 million hectares is located on the desiccated seafloor of the Aral Sea. For decades, sands in Central Asia were very harmful to railway lines, channels, gas mainlines, agricultural territories, etc., as in some other desert regions (Müller et al. 2006; Gao et al. 2007). We found that without preliminary fixation of the dune relief wood plantings on moving sands are not successful. For fixation of sands we used different chemical materials and their modifiers: slate pitch, sulphite and spirit distillers, bitumen and poly(vinyl acetate) emulsions, raw oil, latex, gossypol slate, polyethylene layers and others. All the named materials were tested and permitted for use in the desert zone by the Chief Sanitary Office of Uzbekistan. But those chemicals cannot be widely used because of their high cost and the lack of technologies for automatic spraying. Therefore, we have developed a new and suitable method for fixing mobile sands with clay material.

The water content in soil under the protections was 21.4% higher than in the control sites. This may be important on the desiccated seafloor and may positively influence plant root growth.

16.5.4 Project Sectors

Before 1991 forest melioration works on the desiccated seafloor of the Aral Sea were financed centrally from the national budget and scientific research works by the Soviet State Committee on Science and Technology. After the dissolution of the Soviet Union all the work was financed by the local budget, which did not allow a widening of the scope for works on the desiccated seafloor. However, the Aral Sea problem became worse. Social organizations actively joined the attempts to solve it—they tried to find financing sources but sometimes were not successful. Lately, the world community has not ignored the people—one example is the Uzbek–German project of GTZ, which started on the desiccated seafloor in September 2000 (Hüfler and Novitskiy 2001). It is based on the scientific developments of the Uzbek Scientific and Research Forestry Institution (see earlier). This Uzbek–German cooperation, backed by the Uzbek Government, is a joint project on phytomelioration of the desiccated seafloor of the Aral Sea to improve the environmental situation of Aral region and to create thousands of hectares of plantations.

Before the project started, a complex scientific expedition was sent to the desiccated seafloor with the aim of choosing the sites for implementation of the project. The Akpetki Archipelago linking the desiccated seafloor of two states – Uzbekistan and Kazakhstan – was chosen as a place to conduct the works. Five types of seabed sediments were chosen for wood plantings:

1. Sector 1 – slightly sandy valleys
2. Sector 2 – subsandy valleys
3. Sector 3 – sands with small and medium-sized dunes
4. Sector 4 – sandy valleys with dunes starting to form and areas with big dunes
5. Sector 5 – sands with small hills slightly covered with grass and sands with slight dunes not covered with plants and easily deflating

These sectors represent most of the typical older desiccated seafloor sites (see Chap. 2) and are located in various typological zones. For the phytomelioration works of the project, the Government of the Republic of Karakalpakstan selected an area on the desiccated seafloor comprising 80,000 ha. The organization and structure of the project included two teams consisting of 30 people each, provided with a complete set of working uniforms, hot meals three times per day, wagons with sleeping facilities, mobile ambulance points and telephone communication from the workers' camp to Chimbay and Nukus. A temporary creative team of scientists and constructors joined the project. They developed for some cases modernized machines and teached mechanics required for the phytomelioration project on the desiccated seafloor. The project is provided with various ground techniques and seed stations for determining seed quality and capacity. Keynotes for phytomelioration on the desiccated seafloor of the Aral Sea were prepared by the GTZ project (Hüfler and Novitskiy 2001), and disseminated among all workers, being the major document in the phytomelioration process.

The Uzbek–German project follows not only the aim of planting woods, serious attention is also paid to seed quality and plant resistance against hazards and diseases. The whole technological cycle is included in the project. It starts from preparation of fruits/seeds to growing seedlings in the nurseries before their planting on the desiccated seafloor (Fig. 16.8). Black saxaul (*Haloxylon aphyllum*) in many parts of Central Asia exhibits certain diseases (floury dew, hallisas, galls) which negatively influence seed set and growth. Therefore, fruits/seeds were prepared in seed sectors founded in 1982–1987 by the Forestry Institution scientists. A natural variety of black saxaul resistant to hazards and diseases, especially the floury dew disease, was selected from Bukhara province. Every year seeds (fruits) are bought there, up to an amount of 10,000 kg. Thousands of hectares of plantations are created by the project from these seeds bought in Bukhara province. Therefore, we now have a constant forestry and seed sector and it takes care of healthy plants with high capacity, from which the healthy fruits are taken annually. To improve the crop and to increase the amount of fruit, the project treats the plants with ground sulphur (25 kg per hectare) twice a year to protect them from the floury

Fig. 16.8 Forestation nursery of black saxaul (*Haloxylon aphyllum*) on the desiccated seafloor of the Aral Sea, with ten million seedlings in 2006

dew disease. The first treatment takes place in May and the second one takes place 15 days later. By this, the infections of plants are reduced by 85% and this enables healthy fruits to be collected in November to seed them in the nurseries.

Forest plantations can be created from seeds and saplings (see Chap. 15). Unfortunately, no good results are achievable by seedings. Good results depend, first of all, on fruit quality and on the natural environment. During the first years of the project, seeds and seedlings were planted. Better results were achieved with the seedlings. Therefore, the project now pays great attention to nurseries. Each year nurseries were founded in the field with an area of about 50 ha. The output of standard seedlings is about 250,000–300,000 per hectare per year.

For every sector the project developed a different technology for phytomelioration.

For the first and second sectors (slightly sandy and subsandy valleys) furrowing machines were used. In 2003 we started using furrowing equipment (as in northern parts, see Chap. 15) to make soil furrows of 25-cm depth. This enhances water accumulation from winter and spring precipitation and leads to the formation of a small water reservoir along the furrow, and the furrow is slowly covered with sand (Chap. 15). Seedlings were planted 30–35 cm deep into the ground and the main root was put into this slot with water and fed with it in spite of the absence of water in the upper ground layers. Our research showed that soil humidity in the lower layers is 10–20% higher in the sector where this furrowing equipment was used, and the rooting of seedlings was 15–20% higher. This technology was developed in the GTZ project framework and showed itself as the most suitable one. In these sectors the distance between the lines was 10 m and woods were planted massively.

In the third and fourth sectors, where moving sands prevail, before the plantings the relief had to be fixed with rush protections. The distance between the lines was 7 m. Also, in the fifth sector (easily deflating small sand hills, partly vegetated), we stabilized the relief with rush protections and with seedlings of black saxaul (*Haloxylon aphyllum*, Fig. 16.9), Richter's saltwort (*Salsola richteri*) and Medusa's

Fig. 16.9 Young seedlings of saxaul (*Haloxylon aphyllum*) on sandy plains of the desiccated sea bottom of the Akpetki Archipelago, 2005

Fig. 16.10 Moving sand will be fixed through seedings from *Calligonum*, later by *Haloxylon aphyllum* and *Salsola richteri*

head kandym (*Calligonum caput-medusae*) planted along them (Figs. 16.10 and 16.11). The distance between the lines was normally 3–5 m.

In 2000–2004, the Uzbek–German GTZ project created 24,805 ha of wood plantations (in 2000 2,000 ha; in 2001 6,050 ha; in 2002 6,945 ha; in 2003 6,010 ha; in 2004 3,800 ha). On about 70% of the area a mixture of *Haloxylon aphyllum* and *Salsola richteri* was planted, on about 20% of the area *Calligonum caput-medusae* was planted and on about 10% of the area fodder crops such as teresken (*Ceratoides papposa*), chogon (*Halothamnus subaphyllus*) and boyalich (*Salsola arbuscula*) were planted. In spring 2005 selin (*Aristida karelinii*) seeds were sown on big areas with the aim of sand fixation (Fig. 16.12).

The selected five sectors on which the work was done represent 90.4% of the sediment types of the older seafloor of the Aral Sea (sector 1, 19.2%; sector 2, 17.9%; sector 3, 12.2%; sector 4, 18.3%; sector 5, 22.8%; but see also Chaps. 3, 5, 12). Together with the plantation project on the desiccated seafloor, the project conducted scientific work, in particular:

Fig. 16.11 Moving sand will be fixed through *Calligonum caput-medusae* (age 3 years)

Fig. 16.12 *Aristida karelinii* and *Stipagrostis* plants are fixing moving sand and thus forming new dunes

- Study of the environment for wood plantings to obtain information about the major typology developed
- Develop the methods for accelerated plant cover development by promoting the process of natural seed and fruit production and renewal by germination
- Compose a scheme for phytomelioration and reclamation of various sediment types by the GTZ project within the next few years
- Determine the order of their reclamation and type of wood plantings depending on the purpose of the plantations

The works conducted were part of an extensive program of the Uzbek–German project. They reflect some of the scientific and research parts of the Aralkum problem.

Having made huge scientific developments and possessing experience on all types of sediments within the last 25 years, the GTZ project is now able to solve the most complicated practical problems in any part of the desiccated seafloor. For the next 10 years phytomelioration on the desiccated seafloor should be directed to minimize the harmful impact of northerly winds in the Akpetki Archipelago region. By natural revegetation, the ecological situation in the Aralkum will change too slow. Therefore, plantations are also needed in other parts of the desiccated seafloor (see Chap. 15). This is possible by increasing capacities, the ecological situation may improve, additional pastures will be formed and new opportunities for cattle breeding and beekeeping may arise. Thus, plantations also contribute to poverty reduction (Chap. 18).

In 2005 the Uzbek–German project became interregional and started its work in the Republic of Kazakhstan. The department of Natural Resources and natural usage regulation of the Kyzyl-Orda province chose a part of the dry seafloor (10,000 ha) for phytomelioration by a new Uzbek–Kazakh–German GTZ project.

Concerning the current situation of the desiccated seafloor of the Aral Sea (based on the aerophotography material) it became clear that the Aralkum has a size of about 60,000 km^2 (in 2007) and is still increasing by further desiccation of the South Aral Sea (Chap. 2). For plantations to reduce wind velocity to less than 4.5–5.0 m s^{-1}, a distance of not more than 5 m between plantation strips would be needed. According to our calculations, about 2,750 km^2 would thus need urgent phytomelioration. This would also contribute to removal of a small portion of CO_2 from the atmosphere. Table 16.7 indicates the amount of CO_2 assimilated by black saxaul (*Haloxylon aphyllum*) and Richter's saltwort (*Salsola richteri*) in plantations of different age. These figures directly depend on the assimilating shoot mass of the plants. While getting older, saxaul and saltwort reduce the green mass of assimilation sprouts, as well as oxygen extraction and carbon dioxide absorption correspondingly. The maximum is shown for 4-year-old plantations.

For the future Kazakhstan phytomelioration project of the GTZ, eight melioration regions were chosen. Their soil sediment types can be categorized: completely suitable (66.2%), averagely suitable (12.2%), conditionally suitable (11.6%) and not suitable (10.0%). Suitable types of seafloor sediments mostly contain soils with a light granulometric structure and with chloride content of only 0.01–0.08%. The groundwater depth is mostly 2–4 m. These soil types, in particular, are suitable for phytomelioration. The younger seafloor of the Aral Sea with heavy solonchaks is almost not suitable for phytomelioration with saxaul, or can be regarded as a great

Table 16.7 Amount of absorbed CO_2 (kg ha^{-1}) for saxaul and saltwort for different ages

Species	Age (years)					
	2	3	4	5	6	7
Haloxylon aphyllum	558	812	1,158	935	855	678
Salsola richteri	1,041	747	1,548	1,327	1,217	1,098

challenge (Chap. 15). It is a big and provoking challenge of the future to find technical means and adapted species, e.g. *Halocnemum strobilaceum*, for phytomelioration of those saline soils (Breckle and Wucherer 2006, 2007; Wucherer et al. 2005a, b).

16.6 Conclusions

The Aralkum is source of severe salt-dust storms. Phytomelioration in Kazakhstan (Chap. 15) as well as in Uzbekistan (Chaps. 16 and 17) has started on the desiccated seafloor. The desiccated seafloor of the Aral Sea should be granted a better public status and more political attention. Those who live and work there should be enabled to take care of this fragile but dynamic new mosaic of ecosystems (Chap. 18).

Long-term experience in phytomelioration by various small and large-scale projects should be exchanged. Rehabilitation projects, including wood plantings, should be executed by specialized enterprises with experienced staff possessing corresponding labour, material and technical resources, and sophisticated suitable sources with a scientific basis. Cooperation with specialized groups, e.g. for remote sensing, is needed (Chap. 7) in all those projects.

The desiccated seafloor is partly an unexplored territory; therefore, the following matters should be theoretically developed (see also Chaps. 10 and 13):

- Impact of same-aged forest or wood plantations on the natural environment
- Ideal model of forest landscape applicable to various types of seabed sediments
- Possibilities of beekeeping and wild cattle and poultry breeding in the plantations, as well as other useful applications

Research should be conducted constantly and continuously. Monitoring research should include:

- Observing model phytomelioration plantations and their dynamics (e.g. growth, reproduction) depending on sediment types and salinity
- Observing changes of flora (see Chaps 8–10) and fauna (see Chap. 11) in the new landscapes and open parts of the dry seafloor

One of the most important studies will be to look for the enhanced succession by natural fruiting of the planted species and invasion properties of adjacent open seafloor spaces. This might also enable us to find an even better, simpler and more reliable way to phytomeliorate the desiccated seafloor by means of its own dynamic of seed renewal and invasions, enhancing also soil rehabilitation.

Further development of phytomelioration on the desiccated seafloor is not appropriate without theoretical development and fundamental research. The scientific and technological progress achieved in different spheres of knowledge can provide us with new sophisticated research methods and scientific and technological capacities. These are system analysis, physical and mathematical modelling, biotechnologies and genetic engineering, remote sensing and land monitoring. Work on the desiccated seafloor should be conducted in a complex and diversified way to create plantations with a high productive capacity.

Acknowledgement The German Agency for Technological Cooperation (GTZ) and the Ministry for Economic Cooperation and Development (BMZ) financed the German–Uzbek project on afforestation on the dry seafloor of the Aral Sea, which is greatly acknowledged.

References

Breckle S-W, Wucherer W (2006) Combatting desertification in the Northern Aral Sea region. In: GAO J et al (eds) Restoration and stability of ecosystems in arid and semi-arid areas. Science Press, Beijing, pp 304–316, Peking-Symposium China Herbst 2004

Breckle S-W, Wucherer W (2007) What will be the future of the Aral Sea? In: Lozan JL et al (eds) GLOBAL CHANGE: enough water for all? Wissenschaftliche Auswertungen GEO, Hamburg, pp 142–146

Gao J, Nishio T, Ichizen N, Veste M, Breckle S-W (2007) Desertification and rehabilitation in China. Japan. J Arid Land Studies 17(3):101–112

Giese E (1997) Die ökologische Krise der Aralseeregion. Geogr Rdschau 49:293–299

Hüfler F, Novitskiy, SB (2001) Wald auf dem ausgetrockneten Boden des Aralsees. BMZ/GTZ Broschüre, Taschkent (German-Russian)

Müller JV, Veste M, Wucherer W, Breckle S-W (2006) Desertifikation und ihre Bekämpfung – eine Herausforderung an die Wissenschaft. Naturwissenschaftliche Rundschau 59(11): 585–593

Novitskiy, SB (1997) Nauchnye osnovy zashitnogo lesorasvedeniya na osushennom dne Aralskogo morya (Research basic of the phytomelioration on the dry sea floor of the Aral Sea). (in Usbekistan)

Talanova O (ed) (1997) Aral – yesterday and now. Photoalbum. MFSA, Almaty

Wucherer W, Breckle S-W, Kaverin VS et al (2005a) Phytomeliorative Eigenschaften von *Haloxylon aphyllum* und Perspektiven der Anpflanzungen in der Region am Aralsee. In: Veste M, Wucherer W, Homeier J (eds) Ökologische Forschung im globalen Kontext. Festschrift zum 65. Geburtstag von Prof. Dr. S.-W. Breckle. Cuvillier, Göttingen, pp 109–128

Wucherer W, Veste M, Herrera Bonilla O, Breckle S-W (2005b) Halophytes as useful tools for rehabilitation of degraded lands and soil protection. Proceedings of the first international forum on ecological construction of the Western Beijing, Beijing: 87-94 (English); 169-175 (Chinese)

Chapter 17
Phytomelioration of Solonchaks in the Uzbekistan Pre-Aral Region Under Recent Climate Change

Zh.V. Kuzmina and S.Ye. Treshkin

17.1 Introduction

In 2001 the Aral Sea crisis developed in such a way that the remaining water surface of the Large Aral Sea (South Aral Sea) split into a western and an eastern part, where the salt content continued to increase drastically and in April 2009 reached 177 and 243 g/L, respectively. On 13 June 2009, the analytical laboratory of Johnson Cosmic Center (NASA 2009) documented for the first time the complete drying of the shallow eastern part of the Large Aral Sea. The desiccated sea bottom was completely transformed into an area covered by solonchaks (a huge salt swamp). Owing to drying of the sea bottom, dust storms (see Chaps. 5 and 7) became more prominent and resulted in salt removal from the bare newly formed solonchaks with puffy salt crusts (Glazovskii 1990; Rubanov and Bogdanova 1987; Semenov et al. 2006).

The Aral Sea crisis started because regulation of river flows and extension of areas under irrigation caused not only the development of solonchaks on the dry sea bottom, but also a major part of solonchaks occurred within the irrigation regions because of flooding and desiccation of the "living" floodplains as a result of regulations of the river flows (Akzhigitova 1982; Kuz'mina and Treshkin 1997; Novikova et al. 1998; Breckle et al. 2001). For this reason, soil salinization is by far the main factor responsible for land degradation in the southern parts of the Pre-Aral region.

The present study aimed at growing stable halophytic plant communities on solonchaks of marine and floodplain origin – on strongly and moderately saline soils of the Aral Sea dry bottom and floodplains of the Amu Darya delta under recent climatic conditions.

The objectives were as follows:

1. Creation of test areas to study phytomelioration under irrigation by freshwaters
2. Study of microclimatic conditions by original devices to record temperature and moisture in plantations on solonchaks

S.-W. Breckle et al. (eds.), *Aralkum - a Man-Made Desert: The Desiccated Floor of the Aral Sea (Central Asia)*, Ecological Studies 218,
DOI 10.1007/978-3-642-21117-1_17,

3. Study of salinization dynamics in solonchaks of marine and floodplain origin
4. Use of different experiment variants for growing of halophytes
5. Monitoring of the growth and development of halophytic plant species
6. Analysis of climate changes by processing of statistical data (daily and those averaged for several years) obtained by meteorological stations

According to the scheme of vegetation dynamics in the Amu Darya delta (Fig. 17.1) and the position of plant communities composed of black saxaul (*Haloxylon aphyllum*), cherkez (*Salsola richteri*) and teresken (*Ceratoides papposa*) in the scheme of the succession dynamics of tugai vegetation, the field experiments were conducted jointly with specialists from the Institute of Bioecology in the Karakalpak Department of Academy of Sciences in Uzbekistan with the aim of planting and seeding the above-mentioned plant species to rehabilitate the vegetation cover in the irreversibly changed territory of the Amu Darya delta and on bare solonchaks of the desiccated Aral Sea bottom. Thus, these results can be compared with those from the north (Chap. 15).

In this chapter special attention is paid to artificial formation of pasture vegetation in test areas situated 5 km north of Muinak on the desiccated seafloor that became dry between 1985 and 1990. This is the former Rybatsky Bay now covered by bare clay-loamy solonchaks with salt accumulation in the first 1-m layer (the weighted average salinization is 1.5–2%).

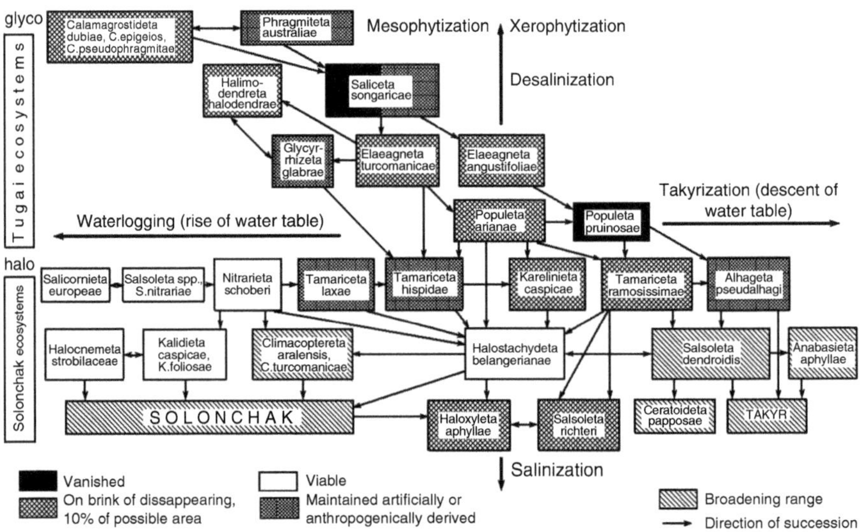

Fig. 17.1 Model of the dynamics of tugai and solonchak vegetation (formations according to Russian nomenclature) in the Amu Darya delta, e.g., indicating shrinking or broadening of the range area and gradients of ecological factors and processes

17.2 Analysis of Recent Climate Changes in the Pre-Aral Region

Our study of solonchak phytomelioration started at the end of 2002 in the period of increasing precipitation and changeable weather in the spring, which is not typical for the southern Pre-Aral region.

Following the climatic zonation accepted in countries of the former USSR, the territory of the southern Pre-Aral region, where our experiments were performed, is referred to as the subtropical zone (43–44°N), particularly to central deserts in terms of botany and geography (Table 17.1; Beresneva 2006; Rachkovskaya et al. 2003). This means that the desert zone is divided into semideserts or northern, central and southern deserts. However, the characteristics of the geothermal regime indicated in the above-mentioned zonation (Beresneva 2006) do not correspond to the recent climatic conditions in the southern Pre-Aral region.

In recent years, the sum of annual precipitation has increased and has reached 160–300 mm (240 mm in 2002, 314 mm in 2003, 164 mm in 2005; Table 17.2), thus considerably exceeding the rate of long-term precipitation (Kouzmina et al. 2006), but see also Chap. 4. The spring started unusually early; the vegetation period was absent in spring because of drought; late frosts occurred in the warm period – at the end of April and in early May (Table 17.3). In view of this, our experiments with young plants and seedings revealed extremely unstable survival rates in different years, as shown in Table 17.4.

Table 17.1 Climatic mode of steppes and deserts of Kazakhstan and Central Asia (Beresneva 2006, p. 258)

Geographical zones	Temperature (°C)			Period with	Precipitation (mm)			Index		
	$t_{January}$	t_{July}	t_{year}	$t > 5°C$	Rn_{year}	Rn_{V-X}	Rn_h	I_d	I_c	N
Dry steppes	−15.2	22.8	4.1	180	300	145	70	3.7	23	48
Northern deserts	−13.3	24.5	6.3	195	265	110	62	5.0	22	40
Central deserts	−10.5	25.6	8.4	215	190	75	40	6.0	22	39
Southern deserts (conditional subtropics)	−2.5	29.2	13.6	250	160	35	25	6.5	45	22
Southern deserts (true subtropics)	2.8	28.8	15.9	300	185	25	16	7.5	0	15

Rn_h is precipitation in a hard state (snow, soft hail, snow pellets, hail)

I_d is the radiation index of dryness (Budyko 1956, 1974; Grigoriev and Budyko 1959), $I_d = R/L\ Rn_{year}$, where R is the annual radiation balance of the Earth's surface (MJ/cm), L is the latent heat of vaporization (2.45 MJ kg^{-1}) and Rn_{year} is the annual amount of precipitation (mm)

I_c is the coefficient climate continentalization (Revyakin 1981), $I_k = t < 0/Rn_h$, where Rn_h is the precipitation in a hard state (snow, soft hail, snow pellets, hail) and $t < 0$ is the sum of air temperatures for a period with negative temperatures (below 0°C)

N the coefficient of irregularity (seasonal prevalence) of atmospheric precipitation (Beresneva 2006), shows the ratio of summer and annual precipitation, $N = (Rn_{V-X}/Rn_{year})$ 100%, where Rn_{V-X} is precipitation in May to October and Rn_{year} is the annual amount of precipitation (mm)

Table 17.2 Comparison of the sums of atmospheric precipitation for recent years (2002–2008) with the precipitation norms for different time periods (Chimbay station): 1937–1965 – before the Aral crisis; 1965–2002 – full period of the Aral crisis; 1980–2002 – active phase of the Aral crisis

Years and periods	Precipitation in millimetres and as a percentage of the norm for the period												
	Months												Year
	1	2	3	4	5	6	7	8	9	10	11	12	
2002 (mm)	15.3	53.4	18.3	33.8	42.7	55.1	4.6	11.9	0.7	0.7	0.4	2.9	239.8
2003 (mm)	5.9	17.8	35.0	21.3	72.2	31.2	1.9	0.3	0	15.9	45.1	69.0	315.6
2004 (mm)	2.4	36.7	0.0	64.6	33.5	3.8	3.9	0	0.8	2.8	7.0	8.1	163.6
2005 (mm)	20.8	0.3	9.4	3.9	12.1	4.4	2.5	2.3	0	2.4	10.1	22.6	90.8
2006 (mm)	16.4	12.2	6.8	9.7	3.1	2.6	4.9	0	2.5	17.2	38.3	3.6	117.3
2007 (mm)	4.1	7.1	34	20.4	3.9	4	3.1	0	0	2	6	14.7	99.3
2008 (mm)	5.7	2.8	0.6	10.5	19.2	0.3	20.6	0	5.6	10.3	0.7	16.3	92.6
1937–1965 (mm)	8.4	11.4	13.6	13.2	10.3	4.2	3.5	2.7	3.1	7.7	5.5	7.1	90.8
1965–2002 (mm)	12.2	10.4	19.8	19.4	16.6	5.5	2.5	2.9	3.5	8.7	11.3	12.6	125.3
1980–2002 (mm)	12.3	9.9	21.5	16.7	19.6	6.4	2.3	3.8	3.5	7.5	12.6	12.4	130.6
2002(%) from 1937 to 1965	182	467	134	255	415	1,301	130	435	22	9	7	41	264
2003(%) from 1937 to 1965	70	156	257	161	702	737	54	11	0	206	820	972	348
2004(%) from 1937 to 1965	29	322	0	489	325	90	111	0	26	36	127	114	180
2005(%) from 1937 to 1965	248	3	69	30	117	105	71	85	0	31	184	318	100
2006(%) from 1937 to 1965	196	145	50	73	30	62	140	0	81	223	675	51	128
2007(%) from 1937 to 1965	49	85	250	155	38	95	89	0	0	26	109	206	109
2008(%) from 1937 to 1965	68	25	4	80	186	7	589	0	181	130	13	230	102
2002(%) from 1965 to 2002	126	515	92	174	257	1,003	186	405	20	8	4	23	191
2003(%) from 1965 to 2002	48	172	177	110	434	568	76	10	0	183	398	548	252
2004(%) from 1965 to 2002	20	353	0	333	202	69	156	0	23	32	62	64	131
2005(%) from 1965 to 2002	170	3	47	20	73	80	100	79	0	28	89	179	72
2006(%) from 1965 to 2002	134	117	34	50	19	47	196	0	71	198	339	29	94
2007(%) from 1965 to 2002	34	68	172	105	23	73	124	0	0	23	53	117	79
2008(%) from 1965 to 2002	47	27	3	54	116	5	824	0	160	115	6	129	74
2002(%) from 1980 to 2002	125	539	85	203	218	867	202	315	20	9	3	23	184
2003(%) from 1980 to 2002	48	180	163	128	369	491	83	8	0	212	358	556	242
2004(%) from 1980 to 2002	20	371	0	387	171	59	170	0	23	37	56	65	125
2005(%) from 1980 to 2002	169	3	44	23	62	69	109	61	0	32	80	182	70
2006(%) from 1980 to 2002	133	123	32	58	16	41	213	0	71	229	304	29	90
2007(%) from 1980 to 2002	33	72	158	122	20	63	135	0	0	27	48	119	76
2008(%) from 1980 to 2002	46	28	3	63	98	5	896	0	160	133	6	131	71

Table 17.3 Monthly mean (average) temperatures of air in southern Priaralye from December 2004 to April 2009 in comparison with the norm for different decades

Meteorological stations	Period analysed	Temperature type	Month				
			12	1	2	3	4
Chimbay	2004–2005	Monthly mean temperatures	−2.1	−4.1	−5.6	8.5	15.4
	2005–2006		0.2	**−13.6**	0.6	7.5	15.2
	2006–2007		−2.4	−4.0	−0.2	4.5	15.5
	2007–2008		−3.3	**−15.0**	−3.7	**12.0**	16.4
	2008–2009		−4.2	−5.2	0.7	8.5	**11.1**
	Different decades from 1936 to1990	The norm					
		Maximum	−1.4	−3.3	−2.9	3.9	18.5
		Minimum	−5.7	−7.7	−5.5	0.9	11.7
		Average	(−3.4)	(−5.5)	(−4.0)	(2.8)	(12.5)
Nukus	2004–2005	Monthly mean temperatures	−1.4	−3.9	−6.1	9.4	15.9
	2005–2006		0.7	**−13.5**	1.2	8.4	15.6
	2006–2007		−2.0	1.1	0.5	4.5	15.8
	2007–2008		−2.5	**−15.6**	−3.7	**14.4**	17.5
	2008–2009		−3.8	−4.1	1.5	9.3	**11.8**
	Different decades from 1936 to 1990	The norm					
		Maximum	2.5	−3.0	1.5	5.2	19.2
		Minimum	−2.7	−4.3	−3.8	3.1	14.5
		Average	(−0.4)	(−3.7)	(−1.7)	(4.1)	(16.3)

Owing to the difficulties which occurred in the first few years of our experiments and owing to the weather fluctuation during the last 10 years, it became necessary to study the climate trends in the region with the aim of correcting the time of and technology used in field experiments on phytomelioration of solonchaks. A comprehensive analysis allowed us to study the daily weather trends using data obtained from eight meteorological stations included in the World Meteorological Organization (Turgay, Irgiz, Aral Sea, Chimbai, Turkestan, Tamdy, Chardzhou, and Samarkand; Fig. 17.2). The latter provided information on such data for the long period of time: since the time of their foundation until 2005–2009 (Table 17.5). For general climatic conditions, see also Chap. 4.

To estimate adequately trends of precipitation and air temperature, the amplitudes of their changes were calculated by the following ratio – the *index of changes in trend values* of precipitation or temperature for the long-term period to the *index of the fluctuation amplitude of their factual (measured) values* (Table 17.5).

The changes in the distribution of precipitation and temperature proved to be identical in the southern Pre-Aral region and neighbouring territories of Central Asia. They reveal an increase in the hot summer and warm period and a rise in temperature and annual atmospheric moisture (Figs. 17.3 and 17.4) at the expense of 6 cold months as well as the winter and autumn period (Kuz'mina 2007; Kuz'mina and Treshkin 2009).

Table 17.4 The basic characteristics of climatic conditions [temperature (*T*), atmospheric relative humidity (*RH*), rainfall (*R*)] and survival rate of plantings of *Haloxylon aphyllum* in various years of experiments on solonchak sites of the dry bottom of the Aral Sea

Periods of the analysis, (seasons of planting)	Survival rate (%)		Source of the data	*R* (mm) (precipitation)	*T* (°C)					RH (%)		
	With watering	Without watering			Absolute maximum	Absolute minimum	Average (mean)	Average (minimum)	Average (maximum)	Absolute maximum	Absolute minimum	Average (mean)
1 November 2002 to 31 October 2003 (spring)	9–14	46–59	Chimbay (WMO)	203.6	42.1	−21.1	11.8	−8.0	28.1	100	20	58.4
1 November 2002 to 31 October 2003			Kungrad (WMO)	162.2	40.0	−24.1	11.4	−7.8	27.3	100	15	57.8
1 November 2002 to 31 October 2003			Chimbay (UzGM)	204.8	41.8	−20.3	11.5	−8.0	26.9	–***	–	61.1
1 November 2002 to 31 October 2003			Muinak (UzGM)	180	41.4	−24.0	10.7	−9.2	28.7	–	–	59.9
1 November 2003 to 31 October 2004 (winter)	39	32–43	Chimbay (WMO)	225.4	39.0	−12.3	13.2	−1.0	26.4	100	13	59.1
1 November 2003 to 31 October 2004			Kungrad (WMO)	140.9	40.0	−13.8	13.1	−1.1	26.7	100	10	55.8
1 November 2003 to 31 October 2004			Chimbay (KKGM)	261.8	39.7	−17.5	13.1	−1.2	26.6	–	–	63.6
1 November 2003 to 31 October 2004			Muinak (UzGM)	144.3	41.0	−17.6	12.8	−2.0	27.1	–	–	56.3
1 November 2005 to 31 October 2006 (spring)	0	0.5	Chimbay (WMO)	–	42.4	−26.2	12.6	−13.5	27.9	100	16	60.2
1 November 2005 to 31 October 2006			Chimbay (UzGM)	108.1	43.3	−26.9	12.4	−13.6	28.0	–	–	61.1
1 November 2005 to 31 October 2006			Muinak (UzGM)	66.5	43.0	−32.2	13.4	−15.1	29.0	–	–	–
1 November 2005 to 31 October 2006			Kungrad (KKGM)	92.7	43.1	−28.4	12.2	−14.1	28.0	100	13	57.3

1 November 2006 to 31 October 2007 (spring)	0.5	2.5	Chimbay (UzGM)	120.5	42.2	−20.7	12.9	−2.3	28.1	–	–	–
1 November 2006 to 31 October 2007			Muinak (UzGM)	84.3	42.0	−25.7	12.7	−3.3	28.8	–	–	–
1 November 2006 to 31 October 2007			Chimbay (KKGM)	115.2	42.2	−20.7	12.5	−4.0	28.2	100	5	59.3
1 November 2006 to 31 October 2007			Kungrad (KKGM)	75.0	42.5	−21.2	12.7	−0.5	28.0	100	10	58.6
1 November 2007 to 31 October 2008 (spring)	0.1	3.5	Chimbay (KKGM)	96.3	44.5	−26.6	12.2	−15.0	29.0	100	11	53.1
1 November 2007 to 31 October 2008			Kungrad (KKGM)	78.8	45.1	−28.7	12.0	−14.8	29.1	100	5	–
2 May 2008 to 2 May 2009 (spring)	0.1	3.5	Aral Sea bottom (Hydrochron)		45.9	−21.6	12.5	−7.4	29.9	100	5.2	53.4
2 May 2008 to 2 May 2009			Muinak (Hydrochron)		44.6	−19.0	13.5	−6.3	30.9	100	5.3	51.2
2 May 2008 to 2 May 2009			Chimbay (WMO)	–	43.6	−20.0	12.8	−5.2	28.9	100	11	54.9
2 May 2008 to 2 May 2009			Kungrad (WMO)	–	44.5	−17.7	12.5	−5.6	29.1	100	5	55.5

WMO World Meteorological Organization, *UzGM* Hydrometeorological Centre of Uzbekistan (Tashkent), *KKGM* the Hydrometeorological Centre of Karakalpakstan (Nukus)

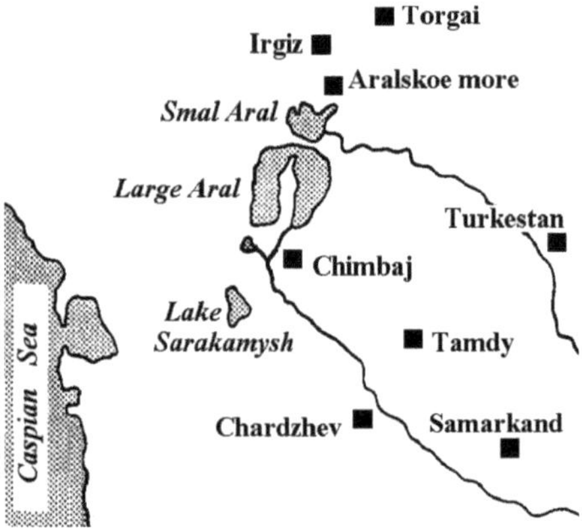

Fig. 17.2 Location of meteorological stations included in the World Meteorological Organization. *Small Aral* North Aral Sea, *Large Aral* South Aral Sea

Having analysed trends and characteristics of climate change, we thought it reasonable to estimate Pedya's aridity index (Ped' 1975). This includes the temperature and precipitation rate and allows one to compare the trends of different meteorological stations and seasons of the year. Pedya's index (Eq. 17.1) was calculated for the period 1961–1990:

$$I_{\text{Pedya}\,1} = \frac{\Delta t_i}{\sigma_{\text{t}}} - \frac{\Delta p_i}{\sigma_{\text{p}}}, \tag{17.1}$$

where Δt and Δp are anomalies of the trend in the mean air temperature and precipitation (the anomalies are deviations from the mean actual values for 1961–1990), and σ_{t} and σ_{p} are average quadratic deviations of the mean air temperature and precipitation. Moreover, the values of this index were estimated as averaged using the data from every meteorological station for the whole period of our observations. For this purpose, Pedya's modified index of aridity (Eq. 17.2) was calculated:

$$I_{\text{Pedya}\,2} = \frac{\Delta T_i}{\sigma_{\text{t}}} - \frac{\Delta P_i}{\sigma_{\text{p}}}, \tag{17.2}$$

where ΔT and ΔP are deviations from the average level of mean air temperature and the sum of precipitation for the entire period of the field experiments. These indices allowed us judge the conditions of water and heat supply and determine the values with different signs. Positive index values corresponded to dry periods with a higher thermal regime, and negative values correspond to wet periods characterized by increasing cold weather.

A conjugated analysis of Pedya's aridity indices (Eqs. 17.1 and 17.2) showed a slight tendency of changing climatic conditions in the Pre-Aral region and adjacent territories (Table 17.5). It became slightly wetter in the annual cycle at the expense of 6 cold months and winter in particular, the air temperature increased, but the amount of precipitation decreased during the 6 warm months including the summer

Table 17.5 Analysis of trends of long-term changes in the precipitation and air temperature within the semidesert and desert zones of Central Asia

Station	Periods			Precipitation (mm)						Pedya index		Air temperature (°C)																	
												Mean						Absolute minimum						Absolute maximum					
	Seasons, months		Years	*Y*	T	*r*	mn	Δ	%	$I_{Pedya\ 1}$	$I_{Pedya\ 2}$	*Y*	T	*r*	mn	Δ	%	*Y*	T	*r*	min	Δ	%	*Y*	T	*r*	max	Δ	%
Aral Sea (Kazakhstan) N 35746	Year	1–12	1906–2005	100	+	**0.41**	118.1	**62.2**	**25.2**	**0.4**	**−0.2**	93	+	**0.46**	7.7	**2.1**	**35.0**												
	Warm	4–9		100	+	**0.19**	55.9	**22.2**	**11.0**	**0.5**	**−0.2**	93	+	**0.47**	20.0	**1.8**	**27.1**	91	0 (+)	0.11	−15.9	–	–	85	+	**0.26**	46.9	**3.5**	**31.2**
	Cold	10–3		100	+	**0.48**	62.4	**40.0**	**28.7**	**0.4**	**−0.1**	93	+	**0.34**	−4.7	**2.4**	**27.1**	94	+	**0.30**	−37.9	**3.9**	**22.8**	82	+	**0.27**	32.6	**3.8**	**22.0**
	Spring	3–5		98	+	**0.27**	36.8	**21.9**	**19.7**	**0.4**	**0.0**	97	+	**0.33**	8.3	**2.4**	**22.4**	89	+	**0.33**	−36.1	**7.2**	**23.2**	85	+	**0.22**	39.9	**2.4**	**15.6**
	Summer	6–8		99	0 (+)	0.10	28.5	–	–	**1.3**	**1.6**	95	+	**0.34**	25.1	**1.2**	**23.5**	92	+	0.17	3.6	–	–	84	+	**0.26**	46.9	**3.5**	**31.2**
	Autumn	9–11		98	0 (+)	0.13	31.0	–	–	**1.0**	**1.6**	96	+	**0.40**	7.9	**2.2**	**19.6**	90	+	**0.19**	−31.6	**4.0**	**15.2**	80	+	**0.41**	41.0	**4.1**	**30.4**
	Winter	12–2		99	+	**0.46**	26.4	**18.4**	**28.7**	**0.1**	**−0.2**	96	+	**0.20**	−10.6	**2.0**	**11.1**	94	+	**0.31**	−37.9	**4.0**	**22.8**	83	+	**0.34**	17.8	**4.1**	**19.7**
Chimbaj (Uzbekistan) N 38262	Year	1–12	1937–2009	72	+	**0.36**	113.4	**57.3**	**21.3**	**0.5**	**0.0**	73	+	**0.70**	11.1	**2.6**	**55.1**												
	Warm	4–9		72	+	**0.27**	46.4	**27.6**	**19.5**	**0.1**	**0.1**	73	+	**0.78**	21.4	**2.5**	**59.1**	73	+	**0.40**	−8.8	**3.6**	**28.5**	72	+	**0.48**	43.9	**2.3**	**34.0**
	Cold	10–3		72	+	**0.32**	67.0	**29.7**	**17.2**	**0.9**	**0.1**	73	+	**0.48**	0.7	**2.8**	**35.9**	73	+	**0.24**	−33.7	**3.9**	**19.6**	72	0 (−)	0.02	34.0	–	–
	Spring	3–5		72	+	**0.29**	49.2	**29.5**	**22.6**	**0.6**	**0.2**	73	+	**0.51**	12.1	**2.9**	**34.0**	70	+	**0.27**	−22.6	**4.2**	**25.4**	72	0 (+)	0.13	41.1	–	–
	Summer	6–8		72	+	0.15	11.0	–	–	**1.5**	**1.6**	73	+	**0.76**	25.6	**2.7**	**60.3**	73	+	**0.33**	2.6	**2.5**	**21.3**	72	+	**0.50**	43.9	**2.7**	**40.8**
	Autumn	9–11		72	+	**0.22**	20.6	**10.7**	**16.7**	**0.6**	**0.1**	73	+	**0.57**	10.3	**2.8**	**40.5**	72	+	**0.22**	−25.5	**3.5**	**17.6**	72	+	**0.22**	39.1	**1.5**	14.8
	Winter	12–2		72	+	**0.20**	32.6	**11.8**	**13.6**	**1.2**	**0.1**	73	+	**0.28**	−4.1	**2.2**	**22.2**	71	+	**0.23**	−33.7	**3.8**	**19.2**	72	0 (+)	0.06	24.4	–	–

Reliable values are shown in *bold*

Y number of years; ***T*** direction of a trend – increase (+), reduction (−); *r* correlation coefficient of the linear trend and curves showing actual values of precipitation and temperature for many years (mean, absolute minimum and maximum); *mn*, *min*, *max* actual mean, absolute minimal and absolute maximal temperatures (and precipitation) for many years; **Δ** the change index of trend values (precipitation and temperature) obtained for the period under consideration calculated by trend, $\Delta = |F(t_n) - F(t_1)|$; **%** the change amplitude as a percentage (according to the trend) – relative change in the actual mean values of precipitation and temperature calculated as the ratio of the change index of trend values to the index of the fluctuation amplitude of the actual values of this parameter for the long-term period: $K_{change} = \frac{|F(t_n)-F(t_1)|}{|t_{max}-t_{min}|} \times 100\%$, where $F(t_1)$ and $F(t_n)$ are the initial and final values of the linear trend, and t_{max} and t_{min} are maximum and minimum actual (measured) values of this parameter for many years (relative change in actual average values of precipitation and temperature,%); $I_{\mathbf{Pedya\ 1}}$ index of aridity (Eq. 17.1); $I_{\mathbf{Pedya\ 2}}$ the modified index of aridity (Eq. 17.2)

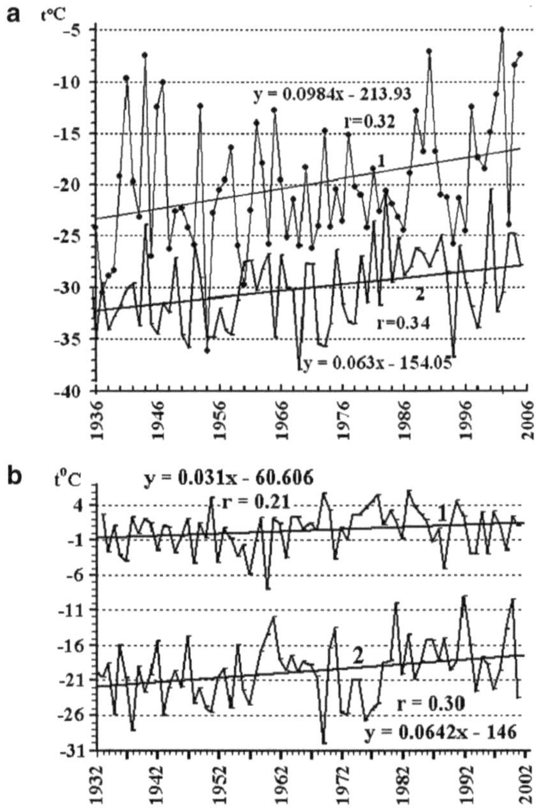

Fig. 17.3 Trends of increasing absolute minima of air temperatures: (**a**) in seasons of the year (*1* spring, *2* winter) according to data from the Aral Sea meteorological station (1936–2005) in the northern Pre-Aral region; (**b**) in half-yearly periods (1 the warm half of the year – months 4–9 months; *2* the cold half of the year-months 1–3 and 10–12) according to the data from the Tamdy meteorological station (1932–2001) in the southern Pre-Aral region

and autumn. In the region of our experiments, the precipitation became higher particularly in the winter and spring – from March–April to the end of May–June, which is not typical for this southern region.

17.3 Microclimatic Conditions for Experimental Plantations According to Data Obtained by Device Recorders

Automatic device recorders for measuring the temperature and relative moisture were used to analyse climatic data of the air and soils in the test areas of our experiments (iBDL Thermohygrograph DS1923-F5, Thermochron DS19221-F5, Hygrochron DS1923-F5). To compare the data obtained with those from meteorological stations (Chimbai, Kungrad), one of the above-mentioned devices was installed under the shade of awning in the bare area at a height of 1.8 m near Muinak. The other device was set in a control variant of our experiments (without watering) under the shade of saxaul at a height of 0.8 m. The remaining nine recorders were in different soil horizons of the test areas. Thus, in spring of 2008

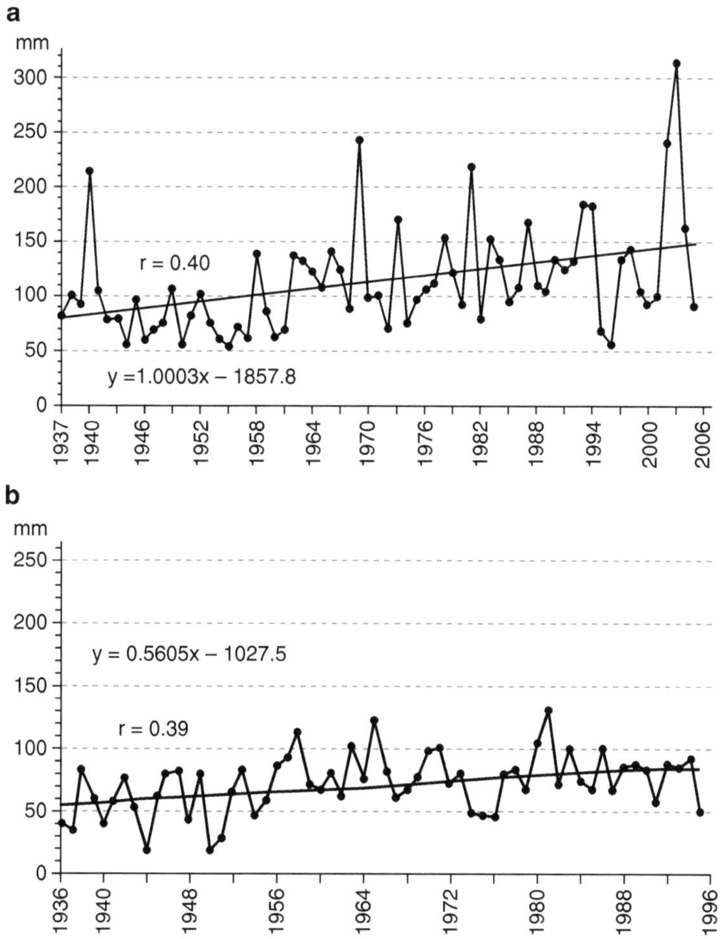

Fig. 17.4 Trends of increasing sum of precipitation (**a**) for a year according to data from Chimbai the meteorological station (1937–2005) in the southern Pre-Aral region and (**b**) for 6 cold months (months 4–9) according to data from the Irgiz meteorological station (1936–1995) in the northern Pre-Aral region. Equations of trends and their correlation coefficients are presented

it was possible to evaluate air and soil temperatures and moisture in the area covered by plantations and to organize the monitoring of microclimatic conditions for solonchakous biotopes.

On the basis of a comprehensive analysis of trends in long-term climatic characteristics since the early twentieth century up to 2002–2005 (Kuz'mina 2007; Kuz'mina and Treshkin 2007, 2009) as well as the climatic data obtained for the last 6 years (from 1 November 2002 to 2 October 2008) and the data recorded by devices (from 2 May 2009 to 10 October 2009), it seemed reasonable to draw a careful conclusion about rapid climate changes taking place in the

Pre-Aral region. The pronounced climatic fluctuation of the geothermal regime is observed from year to year, and is especially notable in the cold period. The following trends (which need much longer observations) exist:

- The annual climate warming is especially in the spring–summer–autumn period (from May to October).
- The periodic (in a year) sharply expressed fluctuations of temperature (maximal and average monthly) are mainly in the winter from very low ($t_{av} = -15.1°C$, $t_{max} = -32.2°C$) to high ($t_{av} = -1.1°C$).
- The pronounced (every 2–3 years) fluctuation of precipitation had a maximum of 240–314 mm in 2002–2003 and minima of 90 and 89 mm in 2005 and 2007.

It was possible to notice the following climate changes during the vegetation period from 2 May to 2 October 2008 in test areas on the dry bottom of the Aral Sea by comparing the climatic data from meteorological stations (Kungrad and Chimbai) with those obtained by recorders in the field:

- Increase in the average daily air temperature (by 1.3°C and 0.9°C)
- Increase in the maximum temperature (average maximum by 1°C and absolute maximum by 1°C and 1.4°C)
- A higher frequency in recording the maximum (by 8–12 times) and minimum (below 10°C; by 4.5 times) air temperatures
- Increase in the fluctuation amplitude of average daily temperature by 3.1–3.3°C
- Increase in the fluctuation amplitude of average daily air moisture by 19–20%

Thus, it was revealed that the data on air temperature and relative moisture obtained in local meteorological stations (Chimbai and Kungrad) do not reflect the real climatic conditions inherent to test areas of our experiments on the Aral Sea dry bottom. In fact, they appeared more continental.

The data from recorders in test areas covered by vegetation (*Haloxylon aphyllum*) and without it (in Muinak) served as evidence that the vegetation on the Aral Sea dry bottom is conducive to a decrease of the average daily air temperature by 0.9–1°C as compared with that in the bare area (Fig. 17.5). The data obtained helped us analyse the distribution of soil temperature in different horizons of marine and floodplain solonchaks both under solonchakous vegetation and without it. One should conclude that the soil temperature even at a depth of 150 cm changes under the influence of shade by vegetation on the soil surface (Fig. 17.6). There is a fall in temperature in horizons of solonchaks only during the vegetation period. In the 6 cold months the temperature in soil horizons remains identical in solonchaks both covered by vegetation and without it (Fig. 17.6). The difference in soil temperature during the 6 warm months is fixed in two upper (35–40 and 60–65 cm) and lower (95–100 cm and 150 cm) horizons. However, solonchaks of northern Karakalpakstan, as referred to the subzone of northern deserts (Beresneva 2006), revealed an almost identical difference in soil temperature for the vegetation period in the area covered by halophytic plants and in the area without them. On average it is 1°C in both the upper (35–40 cm) and the lower (95–100 cm) soil horizons (Fig. 17.6). At the same time, in solonchaks of southern Karakalpakstan, in

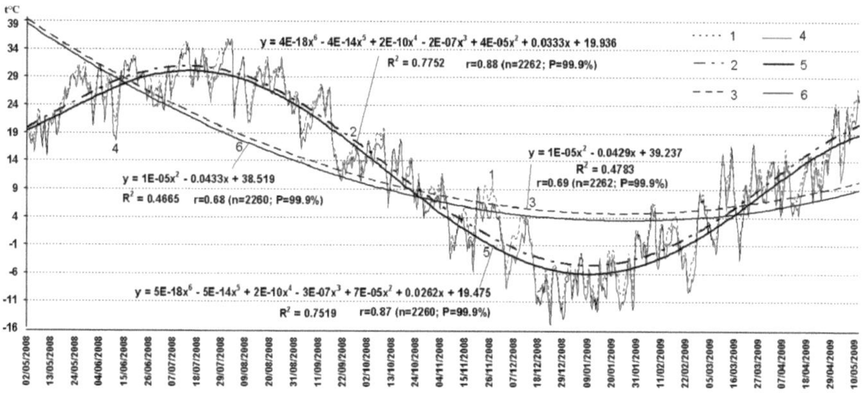

Fig. 17.5 Comparison of curves of daily averaging trend line (six-period moving average, curves *1* and *4*) and polynomial trend lines [two-period (curves *3* and 6) and six-period (curves *2* and *5*)] of air temperatures (°C) on plots without vegetation (curves *1*, *2*, *3*) and on a key site with vegetation *Haloxylon aphyllum* (curves 4, 5, 6) on the dried bottom of the Aral Sea in Muinak district of Karakalpakstan in the period from 2 May 2008 to 10 May 2009 (averaging trend line for the data from daily measurements every 4 h)

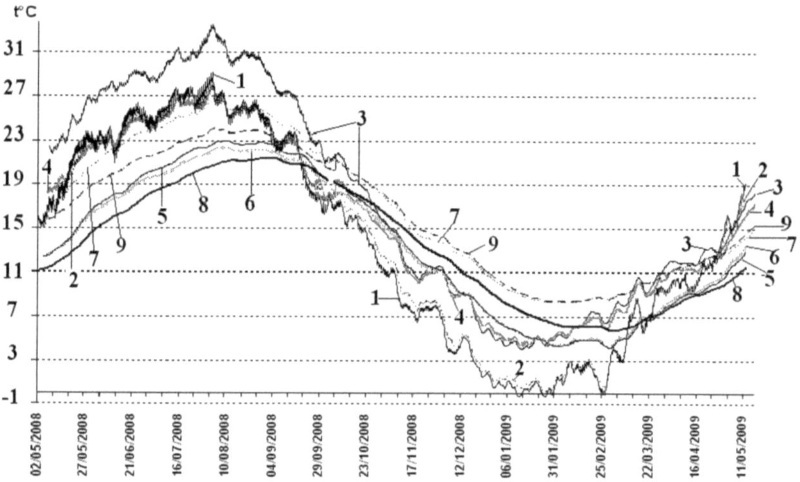

Fig. 17.6 Comparison of curves for soil temperature (°C) in different horizons (35–40 cm, 60–64 cm, 95–100 cm and 150 cm) according to recorder data in the soil marine (on the dried bottom of the Aral Sea in Muinak district of Karakalpakstan; curves *1*, *2*, *5*, *6*, *8*) and floodplain (dried floodplain of the Amu Darya in Biruni district of Karakalpakstan; curves *3*, *4*, *7*, *9*) solonchaks from 2 May 2008 to 10 May 2009: *1* in the 35 cm horizon on bare solonchaks (without vegetation) in Muinak district, *2* in the 40 cm horizon under saxaul in Muinak district, 3 in the 60 cm horizon on bare solonchaks (without vegetation) in Biruni district, *4* in the 64 cm horizon under shrubs in Biruni district, *5* in the 95 cm horizon under saxaul in Muinak district, *6* in the 100 cm horizon on bare solonchaks (without vegetation) in Muinak district, *7* in the 150 cm horizon on bare solonchaks (without vegetation) in Biruni district, *8* in the 150 cm horizon under saxaul in Muinak district, *9* in the 150 cm horizon under shrubs in Biruni district

Biruni, as referred to the subzone of southern deserts, the difference in the temperature of the soil horizons seemed higher both in areas shaded by vegetation and in bare areas. It was up to 5–6°C in upper (60–65 cm) and 2–2.5°C in lower (to a depth of 150 cm) soil horizons.

17.4 Location of Test Areas

Our experimental investigations were conducted in different regions of Karakalpakstan:

1. The test area of 4 ha was created on automorphic solonchaks of the Aral Sea dry bottom, 5 km north of Muinak with a groundwater depth of 6–8 m during the annual cycle. In this area the experiments were performed with the aim at seeding black saxaul (*Haloxylon aphyllum*), teresken (*Ceratoides papposa*) and annual grasses as well as planting seedlings of saxaul (*Haloxylon aphyllum*), cherkez (*Salsola richteri*), kandym (*Calligonum caput-medusae*) and tamarisk (*Tamarix hispida*).
2. The other test area (0.5 ha) was on meadow solonchaks, alluvial clay-loamy and hydromorphic slightly and moderately saline soils with the groundwater level at a depth varying from 2.0 to 4.5 m during the year. This area is near the Glavmyaso channel on lands of the “Aral” collective farm 15 km south of the first test area. This test area was used for seeding of black saxaul (*Haloxylon aphyllum*), teresken (*Ceratoides papposa*) and annual species (*Kochia iranica* and *Climacoptera lanata*).

17.5 Planting Technology

In the course of field investigations on automorphic solonchaks which started in 2002 the different plantation techniques (harrowing, furrow planting every 1 m and every 2 m in the row) were tested. A variety of halophytic plant species (shrubs, perennial and annual grasses) were studied as was the different salt content of the irrigation water (not more than 2.5 g/l). Furrow planting of halophytic shrubs seemed most favourable. Harrowing caused abundant appearance of annual alkali species (*Salsola* spp., *Suaeda* spp.) that oppressed the planted young plants and led to the formation of a salt crust at the soil surface.

The soil was subjected to deep ploughing combined with furrowing. Every six furrows had an open area (6 m wide) for passage of agricultural machines. The space between rows was 2 m and the ditches were 30–45 cm deep for accumulation of atmospheric precipitation, prevention of the salt crust at the soil surface and for irrigation in the summer. The seedlings of halophytic shrubs were planted into the furrows every 2 m and every 1 m.

17.6 Watering

In 2003–2008 the watering of test areas on the Aral Sea dried bottom was implemented 2–3 times during the vegetation period, as a rule in mid-May and/or early June or July. An amount of 1,300–1,670 m^3/ha irrigation water with a salt content of 2.5–3.5 g/l (chloride–sulphate and manganese–calcium type of salinization) was used. The irrigation water was provided by a fire engine (800 l/min) from a collector located 300 m away. However, the water did not remain the space between the rows, being infiltrated down through suffusional cracks and leaving dry the upper 0.3–0.5 m soil layer. This is explained by the drying character of the territory and the parent materials. Hence, the watering of furrows was possible only at the expense of capillary rise from lower soil horizons.

17.7 The Main Results from Plantations

The long-term experiments and monitoring of plantations on saline soils showed that the black saxaul (*Haloxylon aphyllum*), cherkez (*Salsola richteri*) and partially teresken (*Ceratoides papposa*) seemed the most promising for rehabilitation of marine and floodplain solonchaks. These halophytic shrub species on automorphic solonchaks have specific features in their growth and development as compared with those grown on sandy slightly saline soils with the groundwater level at a depth of less than 4 m. Halophytic shrubs on automorphic solonchaks require watering (2–3 times a year) in small amounts (from 100 to 1,800 m^3/ha water but not more than 3,000 m^3/ha water) in the first 2 years after planting. The plant height is affected by such watering to a lesser extent in comparison with that in controls (Fig. 17.7); however, the survival of saxaul is increased by 2 times and that of cherkez by 3 times as compared with controls (Fig. 17.8).

The minimal amount their seedlings grow is identically bad under watering and without watering. That is why the medium-height and tall planting material (more than 10 cm in height) should be recommended for plantations on solonchaks (see also Chap. 15). On the whole, the results obtained from 6 years of monitoring of the black saxaul and cherkez growth and development on solonchaks proved to be rather similar under watering and without it. In May 2009 in the test areas with watering the maximal, middle and minimal heights of black saxaul at the age of 6.5 years were 289 cm, 189.4 cm and 75 cm, respectively, whereas in test areas without watering they were 250 cm, 170.1 cm and 95 cm, respectively. The maximal, middle and minimal heights of cherkez were 188 cm, 89.9 cm and 15 cm in the case of watering and were 120 cm, 50.1 cm and 13 cm in controls (without watering). It is evident that the growth of cherkez is affected by watering to a more considerable extent as compared with that of saxaul. The middle height of cherkez increased by 1.8 times in the case of watering, whereas the watering did not affect the growth of black saxaul. Both halophytic shrubs started their growth on

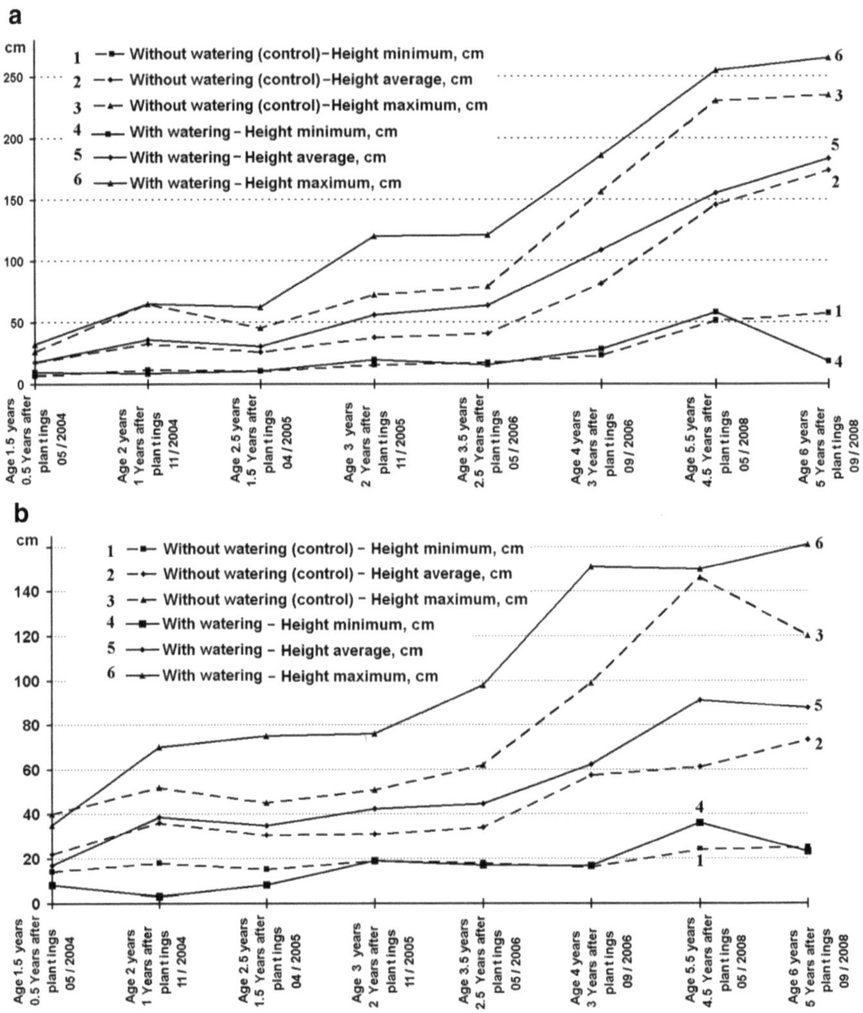

Fig. 17.7 The growth of plantings of saxaul (*Haloxylon aphyllum*) plants (**a**) and cherkez (*Salsola richteri*) plants (**b**) on solonchaks of the dried bottom of Aral Sea

solonchaks almost 2 times later (Fig. 17.7); on sandy soils they were 2.5–3 m in height, whereas the middle height of black saxaul at the age of 6.5 years was only 170–190 cm. The maximum increase in the height of halophytic shrubs took place during the warm period – from April to November – and the heights of cherkez were 2 times lower than those of saxaul. Having grown on solonchaks, these shrubs had their first fruiting rather late, in September 2008 for the black saxaul at a age of 6.5 years and for teresken at an age of 4.5 years.

The analysis of the height of saxaul planted every 1 m and every 2 m in furrows under watering and without it showed quite different results (Fig. 17.9). The plants

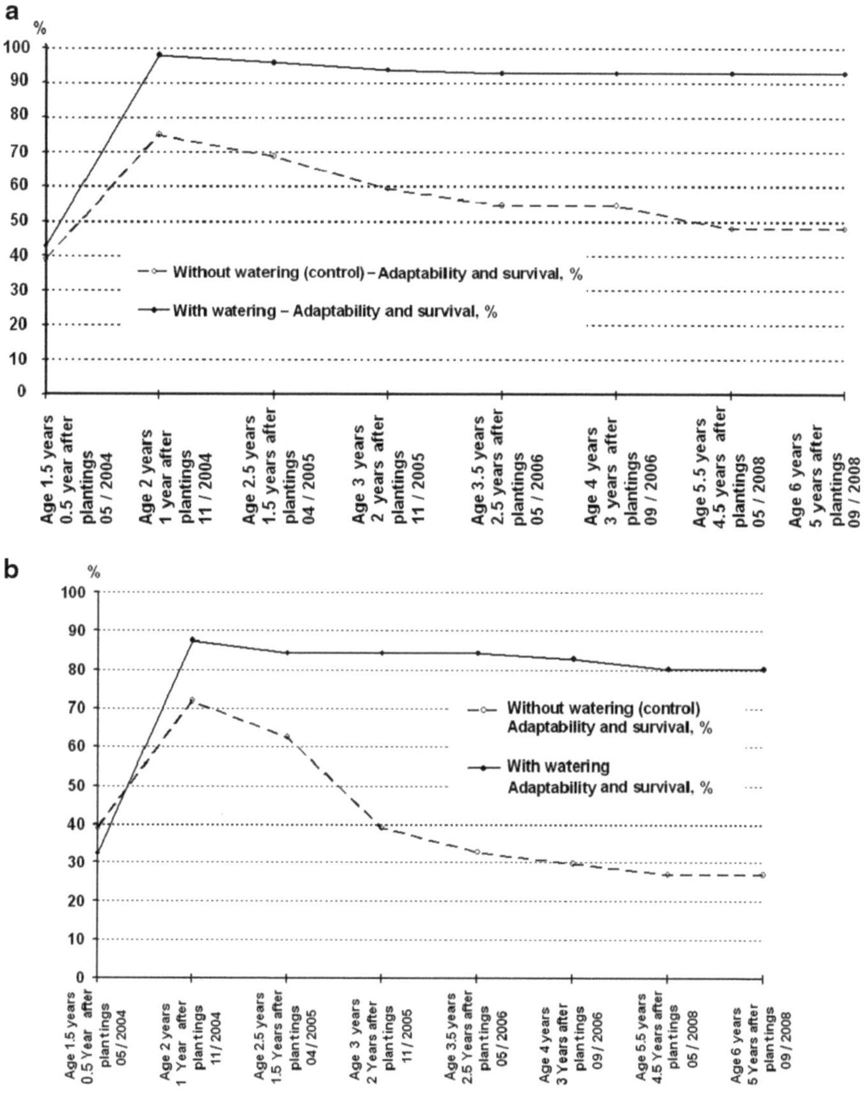

Fig. 17.8 Survival and adaptability (%) of (**a**) saxaul (*Haloxylon aphyllum*) and (**b**) cherkez (*Salsola richteri*) on solonchaks of the Aral Sea dried bottom. The plant adaptability was estimated with dependence on survival taken as 100% in the first half a year after planting

at the age of 6.5 years planted every 1 m seemed to be very high with a small diameter of their crown (Fig. 17.9, curves 1 and 2). At the same time, the saxaul planted every 2 m revealed a larger crown diameter, being low in height (Fig. 17.9, curves 4 and 3). Thus, the indices of the height of saxaul planted every 1 m seemed to 20 cm greater than those of the crown, whereas the indices of the crown diameter of shrubs planted in the furrows every 2 m seemed greater by 12 cm than those of

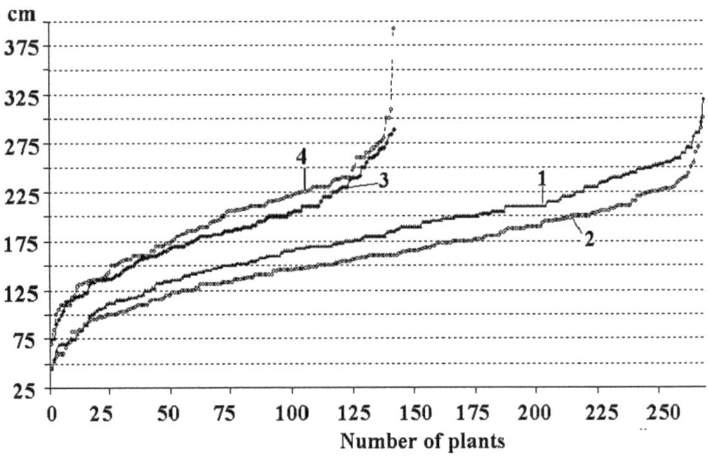

Fig. 17.9 Distribution of the height (curves *1* and *3*) and crown diameter (curves *2* and *4*) of black saxaul (*Haloxylon aphyllum*) at the age of 6.5 years (in May 2009) planted in furrows every 1 m (curves *1* and *2*) and every 2 m (curves *3* and *4*)

their height. On the whole, the height and crown diameter of shrubs planted every 2 m displayed an increase of 10–33 cm and 35–90 cm, respectively, as compared with shrubs planted every 1 m.

The distribution of crown diameters of 6.5-year-old saxaul showed a close correlation with the height of these shrubs planted every 1 m and every 2 m under watering ($r = 0.68$[1] and $r = 0.65$ for $\alpha = 0.001$) and without it ($r = 0.78$ and $r = 0.65$ for $\alpha = 0.001$) (Fig. 17.10). As is evident, in the areas covered by dense saxaul plantation (every 1 m), the crown diameter is highly dependent on the plant height because of intraspecific competition of plants.

It has been established that the sparse plantation of black saxaul should be recognized as the most favourable for solonchaks of the Aral Sea dried bottom. The denser the saxaul plantation (every 1 m), the lower the plant height and crown diameter. The height exceeds the crown diameter and the plants shoot up and become very high. In contrast, in the case of sparse planting (every 2 m), the plants develop with a sprawling crown.

Under natural conditions the plantations of black saxaul (*Haloxylon aphyllum*) on solonchaks of the Aral Sea dried bottom suffer from water deficit caused by a decreasing amount of precipitation to a lesser extent than those of cherkez (*Salsola richteri*). The saxaul at the age of 2.5–4 years reveals an increase in height as compared with cherkez. The latter survives for 2 times less in controls (without watering) in comparison with black saxaul.

When the watering was discontinued, the adaptability of teresken (*Ceratoides papposa*) decreased in all the experiment variants. But in test areas with watering it

[1] r is the correlation coefficient between the crown diameter and the height of saxaul.

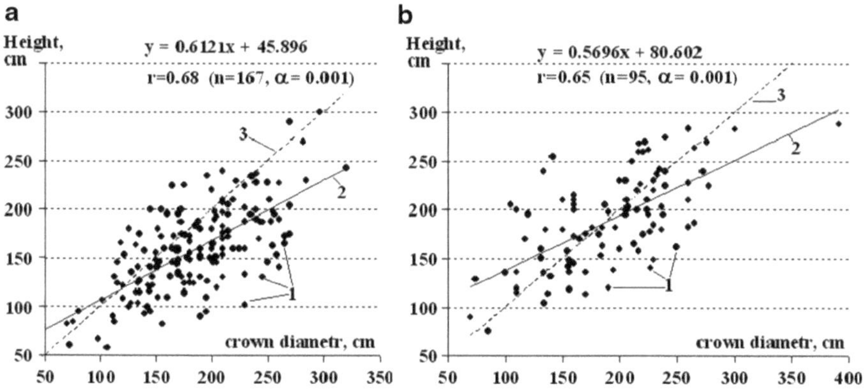

Fig. 17.10 Dependence of the crown diameter on the plant height in communities of saxaul (*Haloxylon aphyllum*) at the age of 6.5 years (in May 2009) in test areas under watering for the first 2 years. Furrow plantation on solonchaks of the Aral Sea dried bottom (**a**) every 1 m and (**b**) every 2 m. *1* the points of actual selection, *2* linear trend of selection, *3* straight line directly proportional to $y = x$ dependence (for comparison of trend deviation)

remained 2 times higher and the growth of these plants was 1.5 times better under watering.

It is worth emphasizing that the watering of plantations on solonchaks implemented during the spring–summer period improves the soil moisture both in the root 40–110 cm layer and down to a depth of 200–300 cm. However, it does not affect the 0–20 cm topsoil.

The furrow planting of *Haloxylon aphyllum* and *Salsola richteri* seedlings and the seeding of *Certoides papposa* should be considered as advisable on bare solonchaks of the Aral Sea dried bottom with minimum watering or without it. In years with favourable weather conditions, the survival of these shrubs reaches 32–59%, but is only 0.5–3.5% under unfavourable climate conditions (see also Chap. 15). However, they have the best growth and development on the nonsaline soil under irrigation by freshwaters. In the first 1.5 years after seeding, the adaptability was 70–80% on slightly saline soils (0.3%) but because of intraspecific competition about 50% of plants remained in the composition of this community every half a year. Thus, 3 years later, the adaptability of black saxaul on nonsaline alluvial soils in Muinak was only 8–10%. According to the extremely ecotopic conditions the phytomelioration of solonchaks on the Aral Sea dried bottom should be performed using the seedlings at the age of 1 year.

Our study showed that saxaul and cherkez seedlings best survive on solonchaks of the Aral Sea dried bottom only under favourable weather and climatic conditions. The amount of precipitation should be 140–220 mm, and there should be no cold winters (the mean temperature in the coldest month −1 to −2°C), no hot summers (the mean temperature in the warmest month 26–27°C) and no frosts in the spring period.

The forecasted effect of climate change on the water resources in Central Asia is a decrease of the amount of river water by 20–40% (Alamanov et al. 2006). It will be conducive to increase the further drying of recent floodplain territories and intensify the formation of solonchaks. As a result, the tugai vegetation could be completely lost (Kuz'mina and Treshkin 2001).

To maintain tugai and solonchakous types of vegetation under climate change, new approaches are urgently required to stabilize these ecosystems when the floods are completely absent. It is impossible to lose the tugai vegetation, and the formation of solonchaks at the site of former tugai alluvial slightly or moderately saline soils cannot be tolerated. But now owing to the rise of air temperature, these soils are quickly transformed into solonchaks, the rehabilitation of which can be implemented only by additional leaching operations. Diversion of lands for nonagricultural purposes should be reduced. Moreover, the valuable and productive tugai ecosystems can be lost irreversibly. To prevent partially the tugai loss in the drying floodplain territories, it is necessary to plant some more saxaul, cherkez and teresken in the degrading tugai communities of halophytic shrubs. This is the artificial formation of more stable mixed halophytic–tugai ecosystems as analogues of relict tugai halophytic ecotones, the latter still being encountered in the old delta of the Ili River (Novikova 1987) or in oases of the Gobi in Mongolia (Pankova et al. 1996). In 2005–2006 similar plantations were tested on dried floodplain soils in the territory of Nukus forestry.

We should mention several regularities identified in the experiments on automorphic solonchaks in Muinak district:

- The experiments serve as evidence that the salt-resistant shrub species *Salsola richteri* and *Haloxylon aphyllum* should be planted as seedlings at the age of 1 year on typical and strongly saline solonchaks. Their survival and adaptability in the first 2 years after planting are 10–15 times higher as compared with those of seedings on moderately and strongly saline soils.
- It has been established that the shrubs *Haloxylon aphyllum* and *Salsola richteri* growing on solonchaks of the Aral Sea dried bottom seem very sensitive to changes in moisture conditions. Their maximum growth is observed under artificial watering as well as owing to increase in atmospheric precipitation.
- In the case of lacking irrigation, the black saxaul (*Halophylon aphyllum*) is more resistant than *Salsola richteri*. It endures the water deficit due to a fall in precipitation and has increased height at the age of 2.5–3.5 years in test areas with watering and under natural conditions as compared with *Salsola richteri*.
- The height and survival of *Haloxylon aphyllum* and *Salsola richteri* on solonchaks are greater under watering. Thus, the recreation or formation of the soil cover (pastures and elements of tugai vegetation) on solonchaks should be accompanied by infrequent watering to maintain the moisture conditions within the upper 0–100 cm soil layer.
- Our attempts to plant seedlings of *Calligonum caput-medusae* on solonchaks of the Aral Sea dried bottom (in April 2003 and in December 2003) failed.

These shrubs displayed the minimum of survival (0.03–6.1%) under watering and without it.

- The results obtained in our experiments showed that *Tamarix hispida* is also unsuitable for the formation of a stable vegetation cover on solonchaks. This is in accordance with the results in the north (Chap. 15)
- The distribution of plants at the ages of 0.5, 1.5 and 2 years according to their height independent of species and growth conditions is described by one type of curve – a polynominal distribution of the sixth order.
- Such regularities identified in our experiments may be used as a basis for improving the methods of growing salt-resistant plant species on solonchaks within the arid and subarid zones of Central Asia. Rehabilitation of solonchaks may be considered as the first step in reconstruction of initial typical ecosystems (both woods and shrubs).

17.8 Main Conclusions

1. In view of climate change and the rather varying weather conditions in the Pre-Aral region and of increasing aridity and continentality of microclimatic conditions in field as compared with the data from meteorological stations, the phytomelioration of solonchaks should be implemented only in periods of forecasted favourable weather conditions.
2. The vegetation-shading bare solonchaks in the southern part of the Pre-Aral region favourably affects the microclimatic conditions of biotopes both in their terrestrial and their subsoil parts: the air temperatures and soil temperatures to a depth of 100–150 cm become lower during the vegetation period.
3. Without artificial (constant) moisture in the upper soil horizons, the phytomelioration of automorphic solonchaks is impossible under the recent climatic conditions (with trends of increasing temperature and decreasing precipitation in the summer–autumn period).
4. For artificial formation of the vegetation cover on automorphic moderately and strongly saline solonchaks of marine and floodplain origin, a few halophytic shrubs seem the most suitable: black saxaul, cherkez (seedlings) and partially teresken (seeding). For better survival of halophytic shrubs on automorphic solonchaks, it is necessary to implement watering (2–3 times at the beginning of the vegetation period) with slightly saline waters (to 2 g l^{-1}) during the first 2 years after their planting.
5. It seems reasonable to reconstruct the tugai ecosystems, which perished because of man-made regulation of the river flows of the Amu Darya and the Syr Darya, into halophytic productive pastures on desiccated delta sites. Shrubs such as saxaul, cherkez and teresken are recommended for planting to improve the plant communities under degradation. This may also be reasonable if the forecasts of a decreased river flow of the Amu Darya and the Syr Darya in view of changing

climatic conditions and mainly of the growing deficit of water resources in Central Asia come to reality.

6. The mixed halophytic–tugai ecosystems formed in such a way (turanga–saxaul, tamarisk–black saxaul and others) will be conducive to maintaining the productivity of ecosystems and the main elements of tugai flora to be human-modified analogues of relict tugai halophytic ecotones. In the case of an increase of the natural moisture conditions or an improvement of the water management situation, they can soon be restored or rehabilitated into typical tugai ecosystems.

References

Akzhigitova NI (1982) Halophytic vegetation of Central Asia and it indicator properties (*Galophitnaya rastitelnost v Sredney Asii i ejo indikazionnye svoystva*). FAN UzSSR, Tashkent (Russian)

Alamanov SK, Lelevkin VM, Podresov OA, Podresov OA (2006) Change of a climate and water problems in Central Asia (*Izmenenie klimata i vodnye problemy v Zentralnoy Asii*). Moscow-Bishkek. (Russian)

Beresneva IA (2006) The climate of arid zones (*Klimaty aridnoy zony*). Biological resources and natural conditions of Mongolia. Works of joint Russian-Mongolian complex biological expedition (*Biologicheskie resursy i prirodnye usloviya Mongolii. Trudy sovmestnoy Rossiysko-Mongolskoy komplexnoy biologicheskoy expedizii*). XLVI. M.: the Science (*Nauka*). (Russian)

Breckle S-W, Wucherer W, Agachanjanz O, Geldyev BV (2001) The Aral Sea crisis region. In: Breckle S-W, Veste M, Wucherer W (eds) Sustainable land Use in deserts. Springer, Berlin, pp 27–37

Budyko MI (1956) Heat balance of the Earth surface (*Teplpvoy balans zemnoy poverhnosti*). Hydrometeoizdat, Leningrad (Russian)

Budyko MI (1974) Change of climate (*Izmenenie klimata*). Hydrometeoizdat, Leningrad (Russian)

Glazovskii NF (1990) Aral crisis (*Aral'skii krizis*). The Science (Nauka), Moscow, Russian

Grigoriev AA, Budyko MI (1959) Classification of climates USSR (*Klassiphikaziya klimatov SSSR*). Informations of academy of sciences USSR. A geographical series. (*Izvestiya akademii nauk SSSR. Seriya geographicheskaya.*) Moscow. No 3. pp 3–18 (Russian)

Kouzmina JV, Treshkin SY, Mamutov NK (2006) The results of experimental of natural vegetation formation on saline soils on the dried bottom of the Aral sea. Arid Ecosystems (Aridnye Ekosistemy) Moscow 12(29):27–39 (Russian, resume English)

Kuz'mina ZhV (2007) Analysis of changes of long-term meteorological characteristics and their influence on dynamics of ecosystems (*Analiz izmemeniy mnogoletnikh meteorologischeskikh kharakteristik i ikh vozdeystvie na dinamiku ecosystem*). News of high schools. The North-Caucasian region. Sciences about Earth. Natural sciences. Special issue (*Izvestiya vusov. Severo-Kavkazskiy region. Nauki o zemle. Estesvennye nauki. Spezvypusk*). No 6. pp 73–78 (Russian)

Kuz'mina ZhV, Treshkin SE (2001) Modern state of flora and vegetation of reserve "Baday-Tugay" in connection with change of a hydrological regime (*Sovremennoe sostoyanie phlory i rastitelnosti zapovednika "Baday-Tugay" v svyazi s izmeneniem gidrologischeskogo rezhima*). Botanical Journal (Botanischeskiy zhurnal) Sankt-Petersburg 86(1):73–84, Russian

Kuz'mina ZhV, Treshkin SE (2007) Recultivation of solonchaks of drying bottom land of Aral Sea in variating climatic conditions (*Rekultivaziya solonchakov obsokhshego dna Aralskogo moray v izmenyauschikhsya klimaticheskikh usloviyakh*). The Biological recultivation and monitoring of disturbed lands (*Biologischeskaya rekultivaziya i monitoring narushennykh zemel*). Publishing house of the Ural University, Ekaterinburg, pp 391–408 (Russian)

Kuz'mina ZhV, Treshkin SE (2009) Formation of vegetation on Solonchaks of the dried Aral Sea Bed under changing climate conditions. Russian Agricultural Sciences 35(1):37–41 (English)

Kuz'mina ZhV, Treshkin SE (1997) Soil salinization and dynamics of tugai vegetation in the Southeastern Caspian Sea region and in the Aral Sea coastal region. Eurasian Soil Science 30 (6):642–649, Translated from Pochvovedenie. Moscow. No 6. pp 726–734 (English)

NASA (2009) Image Science and Analysis Laboratory, NASA-Johnson Space Center. "The Gateway to Astronaut Photography of Earth". http://eol.jsc.nasa.gov/scripts/sseop/clickmap.pl?hashnumber=7871&fieldlist=GEON,FEAT,LAT,LON,PDATE&options=none (8 November 2009 02:31:58)

Novikova SS (1987) Vegetation of dry river bed of an ancient delta of river Ili. Problem of deserts development (Problemy osvoeniya pustyn) Aschgabad 3:21–28 (Russian and English)

Novikova NM, Kust GS, Kuzmina JV, et al (1998) Contemporary plant and soil cover changes in the Amu-Dar'ya and Syr-Dar'ya river deltas. In: Ecological research and monitoring of the Aral Sea deltas. UNESCO Aral Sea Project. Book 1,UNESCO, Paris, pp 55–80

Pankova EI, Kuz'mina ZhV, Treshkin SE (1996) The nature of the southern Gobi oases as related to reclamation of the tugai vegetation. Arid Ecosystems (Aridnye Ekosistemy) Moscow 2(2–3):131–144 (Russian and English)

Ped' DA (1975) On exponents of drought and excessive moisture (*O pokasatelyakh zasukhi i izbytochnogo uvlazhneniya*). Proceedings of hydrometeorological centre of the USSR (*Trudy Hydrometzentra SSSR*). Vol 156. pp 19–39 (Russian)

Rachkovskaya EI, Volkova EA, Khramtsov VN (eds) (2003) Botanical geography of Kazakhstan and Middle Asia (desert region). St Petersburg. (Russian and English)

Revyakin VS (1981) Natural ice of Altay-Sayan mountain area. (*Prirodnye ldy Altae-Sayanskoy gornoy oblasti.*). Hydrometeoizdat, Leningrad

Rubanov IV, Bogdanova NM (1987) Quantitative evaluation of salt deflation on drained bottom of the Aral sea. Problem of desert development (Problemy Osvoeniya Pustyn) Aschgabad 3:9–16 (Russian and English)

Semenov OE, Shapov AI, Galaeva OS, Idrisova VP (2006) Wind deflation and sand and salt deposition form the drained part of the Aral sea bottom. Arid Ecosystems (Aridnye Ekosistemy) Moscow 12(29):47–58 (Russian, resume English)

Chapter 18
The Aralkum Situation Under Climate Change Related to Its Broader Regional Context

G. Winckler, E. Kleinn, and S-W. Breckle

18.1 The Background of Desertification and Land Degradation in the Central Asian Region

The ecological and social situation in the Aralkum area is only understandable in its regional context. Dry areas (arid, semiarid and subhumid zones) represent nearly 80% of the five land-locked countries Kazakhstan, Kyrgyzstan, Tajikistan, Turkmenistan and Uzbekistan, and Afghanistan as another Central Asian country should not be ignored. The ecological conditions of the region (geography, climate and natural resources) are highly diverse. As in many other dry regions of the world, problems related to desertification and degradation of natural resources are increasing in Central Asia. Although the situation of the Aralkum is particular, many of its background features are symptomatic for the general land use situation in most of the Central Asian region (Bot et al. 2000; Zonn et al. 2009).

18.1.1 The Underlying Problem: The Dimension and Consequences of Land Degradation

In eastern Central Asia the Tien Shan and Pamir ranges represent some of the world's largest glaciers and freshwater reserves. Western Central Asia consists of arid lowland plains. Most of the area (68%) is occupied by sparsely vegetated deserts (including the Karakum and the Kyzylkum), semideserts and steppes. Annual precipitation is generally much less than 400 mm, with variable and uncertain rainfall and periodic droughts. Around 65% of the total land area is dedicated to pastures and rangelands. Forests cover approximately 4%. Roughly one third of the agricultural land is irrigated, with the remaining land supporting rain-fed agriculture (Fig. 18.1).

Historically, the fragile dryland ecosystem of Central Asia – arid and semiarid areas of very low natural productivity – has long sustained small populations of

S.-W. Breckle et al. (eds.), *Aralkum - a Man-Made Desert: The Desiccated Floor of the Aral Sea (Central Asia)*, Ecological Studies 218, DOI 10.1007/978-3-642-21117-1_18,

Table 18.1 Percentage of land with soil degradation

Country	Salinity	Erosion
Kazakhstan	7.91	2.87
Kyrgyzstan	0.50	28.2
Tajikistan	4.89	25.9
Turkmenistan	14.9	1.43
Uzbekistan	14.1	2.92

Based on Bot et al. (2000)

nomadic pastoralists and, in more favourable areas, subsistence farmers. Under the agricultural production system of the Soviet era, large-scale collective and state farms controlled 95% of agricultural land and produced most of the commercially marketed output. Since their independence, Central Asian countries have continued to design agricultural practices for maximum production, with limited regard for their sustainability. In lowland areas agricultural production largely depends on irrigation (Fig. 18.1). Water use has been badly managed and resulted in waterlogging and increasing salinity (Table 18.1), rendering an increasing portion of land useless or with little remaining productivity. Poor irrigation management has resulted in falling river levels, with less available water downstream. In upland areas, unsustainable crop and animal production has caused erosion and reduced fertility of cropland and pastures; areas under forests and their productivity have decreased.

After independence, land reforms focused mainly on the "decollectivization" of the large state-owned enterprises and privatization of holdings. In the 1990s, the pressure on pastures diminished somewhat owing to migration of the rural population to urban areas. In the meantime, however, people have been returning to rural areas, many of them lacking basic knowledge of sustainable agriculture and pasture management. Reduced mobility in pasture management leads to serious degradation close to human settlements caused by heavy overgrazing.

Presently, cereals (mainly wheat) are grown on 40% of the arable area of Central Asian countries. Cotton production was the main target of irrigation infrastructure development during the Soviet period and cotton is today still the dominant cash crop. Income inequality between the agricultural and nonagricultural population is growing in most of the countries because of low economic efficiency of agricultural production, low capital investments in agriculture and slow institutional and infrastructural development (IFPRI 2008).

During the Soviet era, the rural population used to be supplied with coal for household energy. Now most people depend on wood, shrubs and animal dung for cooking and heating, which considerably accelerates desertification in Central Asian countries.

What is desertification in this respect? Desertification describes a process in which desert-like conditions develop as a result of degradation processes, most notably because soil productivity declines significantly. Worst affected are regions with arid, semiarid and dry climatic conditions, which are highly sensitive to overuse by people and animals, but also to climate change. The outcome is that soils become eroded and salinized (Table 18.1), they lose their capacity to store

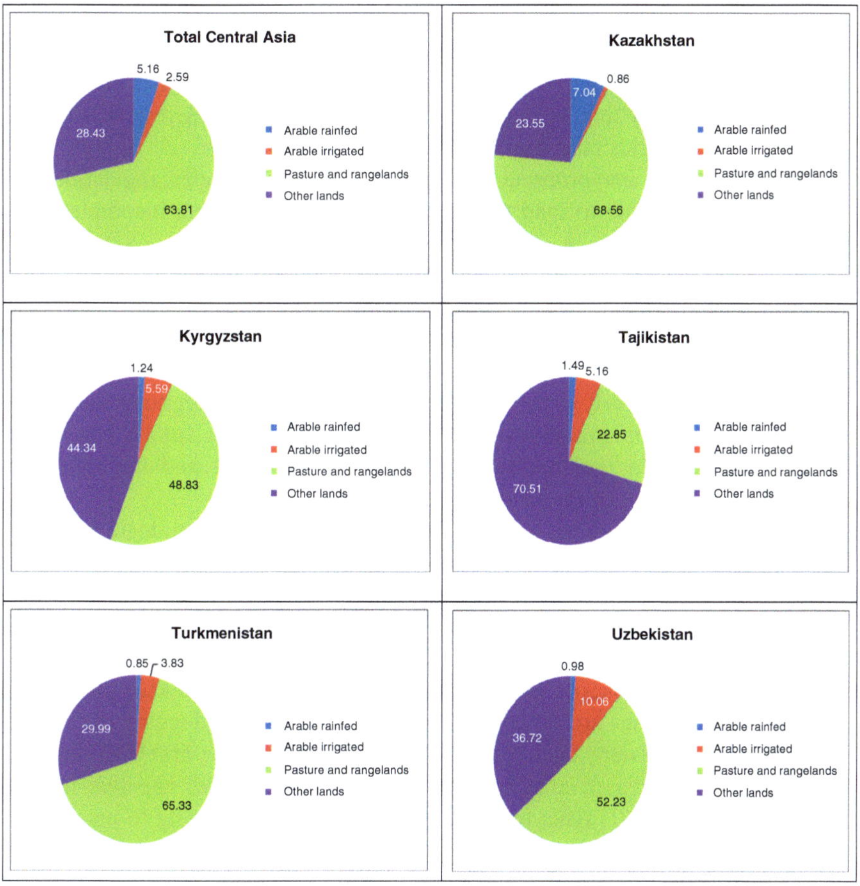

Fig. 18.1 Land use in Central Asia, indicating the very small portions of arable rain-fed and irrigated land and the much bigger portions of pastures and rangelands as well as of unproductive and very arid lands (other lands). (Source IFPRI-ICARDA 2008 based on FAO 2006)

water, groundwater levels fall and the remaining plant cover is further reduced or disappears altogether. In short, the land becomes infertile and degraded (GTZ UNCCD Project 2007a).

Land degradation is a serious economic, social and environmental problem in the transition economies of the Central Asian countries of Kazakhstan, Kyrgyzstan, Tajikistan, Turkmenistan, and Uzbekistan. A reduced productivity of land resources and weakening services derived from natural systems directly affects the livelihood of the rural population. Across the region, agricultural yields are reported to have declined by 20–30% over the last decade. Annual losses of agricultural production from salinization alone are estimated at $2 billion (ADB 2006). The abuse and overexploitation of the natural resource base is the main cause of land degradation, largely attributable to inappropriate and unsustainable agricultural practices, overgrazing, deforestation and forest degradation. By remote sensing techniques, the degree of desertification of various land use categories is nowadays easily detectable (Bai et al. 2008).

What are the root causes of land degradation? Today, Central Asia is one of the regions of the world that is most severely threatened by desertification. In all Central Asian countries land degradation can be traced to three sets of issues.

The first includes inappropriate agricultural and water policies, regulations and incentives. These relate to land tenure insecurity resulting from incomplete land reforms, crop-growing such as the promotion of irrigated cotton production with inadequate drainage and continued subsidies for irrigation, market restrictions, and limited access to production factors.

Second is the inability of governments to adopt and implement sustainable land management because of a relatively weak capacity of central and local governments as well as private or civil society to formulate and monitor integrated sustainable land use development initiatives, and the lack of incentives for land managers (farmers, pastoralists, communities) to invest in protecting and improving the long-term productivity of the land.

Third, there is a whole set of environmental problems – forest degradation and deforestation, water pollution and natural resource management and biodiversity conservation in general. In part, these result from environmental laws across the Central Asian countries that suffer from inadequacy, inconsistency and weak implementation (CACILM 2006).

The principal forms of land degradation currently experienced across the Central Asian countries are (a) erosion, salinization and water logging (b) deteriorating fertility of pasture land (c) decrease in fertility of the arable drylands of the steppes (d) decreased area and productivity of forests (e) negative impacts of mining operations (f) risks from landslides and flooding owing to poor watershed management (g) reduced ecological stability of desert, mountain, wetland and riparian ecosystems and (h) inadequate assessment and monitoring of land degradation.

Thus, the main cause of the degradation of agricultural land is ruthless overexploitation of natural resources by humans. But in doing so, humans themselves become the primary victims. Since the dissolution of the Soviet republics, the rural populations have become more desperate to sustain their livelihoods, thus accelerating the degradation of the unique dryland and mountain ecosystems. The sustainable livelihoods of millions of farmers and pastoralists in this region are threatened by reduced productivity and services of ecosystems.

Land degradation processes are also causing significant social costs. Mainly, the increased salinity and presence of agricultural chemicals in drinking water supplies may be having negative health effects on humans and livestock, further affecting productivity. In some downstream areas, groundwater salt levels exceed by far the maximum limit of 1 g/L recommended by the World Health Organization (IFPRI 2008). High levels of dioxins have been found in the breast milk and blood of pregnant women and recent mothers in Karakalpakstan (Médicines sans Frontières 2003). Another study found that DNA from inhabitants of this region showed evidence of genetic damage among some inhabitants, leading to large increases in cancer rates (Blua 2004). Environmental pollution and child health is strongly correlated (Jensen et al. 1997).

One of the most renowned cases of the consequences of subsidizing production at the expense of the environment is the reduction of the volume of the Aral Sea due to diversion of irrigation water from the Syr Darya and the Amu Darya. The sea has lost 90% of its original volume, and the level of salinity in the sea has increased tremendously. This has caused an increase in poverty and subsequent migration from the region owing to the breakdown of the local fishing industry. Dust storms (see Chaps. 5 and 7) have become a widely recognized source of air pollution. They carry particulates from millions of hectares of openly exposed salt- and pesticide-contaminated dust on the dry seabed hundreds of kilometres away (CACILM 2006, see Chaps. 5 and 7). High rates of respiratory and tuberculosis infections in the surrounding Karakalpakstan region, particularly among children, can be observed (Médicines sans Frontières 2003).

Water consumption per capita in Central Asia is among the highest in the world, with the rate of per capita water consumption in Uzbekistan being ten times the rate in Jordan, whereas Turkmenistan's rate of consumption is more than twice Uzbekistan's rate. The Central Asian countries are dry countries that depend on the snow and glaciers of the high mountains for their water needs. The glaciers of Central Asia are the only main source of freshwater in the region, and whereas two thirds of the water is generated in the mountains, two thirds is currently consumed downstream. Therefore, the economies of the Central Asian states are highly interconnected by water discharge of the Syr Darya and the Amu Darya and other transboundary rivers. During the Soviet Union era, regional institutions regulated the use of water resources on an interstate basis, contributing significantly to the stability in the region. However, in the post-Soviet era the region is facing a number of challenges:

- A common framework on water and energy management does not exist.
- The Central Asian states do not have binding agreements for regional water management. The mandates and responsibilities of regional institutions for water management need to be reformed (Fang et al. 2005).
- The river basin approach is still in its infancy in Central Asian countries. Especially for transboundary rivers, joint water management would be required (Fang et al. 2005). However, river basin management plans are not yet been developed owing to lacking qualifications of the administrative structures. Consequently, projects or measures with impacts for the whole basin (based on environmental and social impact assessments) have not yet been developed
- The impacts of climate change on water availability are not yet really predictable

The cropland area under irrigation has almost doubled from 4.5 million hectares in 1960 to nearly 8 million hectares today. With the projected scenarios of climate change in Central Asia, the crop demand for water will grow because of rising of temperatures. On the other hand, the availability of water in the Syr Darya may decrease by up to 30% and in the Amu Darya by up to 40%, leading to increased drought stress and lower crop yields.

Excessive irrigation and poorly designed and maintained irrigation and drainage systems have already led to severe land and water degradation in Central Asia (Saiko and Zonn 2000). Only 50–70% of the irrigation water used reaches crops because of water losses in irrigation pipes, with 10–25% of the water lost in the interfarm irrigation network and 20–30% lost in the intrafarm network (Bekturova and Romanova 2007). In irrigated cropland areas, salinity and waterlogging are the major problems undermining agricultural productivity. Between 40% and 60% of irrigated croplands in Central Asia are salt-affected and/or waterlogged (Qadir et al. 2009). In some regions the share of waterlogged lands is as high as 92%. Salinization is particularly acute in the downstream areas closest to the Aral Sea. In Turkmenistan, nearly all of the irrigated area is affected by salinity. Even the availability of drinking water is a problem; new technologies could help (Khaydarov and Khaydarov 2007; Larchet et al. 2002).

The impacts of waterlogging and salinization include decreased agricultural productivity and adverse health and environmental effects. Salinity slows the growth of plants by inhibiting their ability to absorb water (Chap. 12). For cotton, the Central Asian Scientific Research Institute for Irrigation estimates yield losses of 20–30% on slightly salinized land, 40–60% on moderately salinized land and up to 80% or more on severely salinized land (Table 18.2) (Bucknall et al. 2003).

Rangelands represent 77–95% of the agricultural area in the five Central Asian countries, the world's largest contiguous area of grazed land (Fig. 18.1). Many rangeland areas in the region are poorly managed, leading to poor feed, land degradation, loss of plant biodiversity, and desertification.

Central Asian rangelands are expected to be affected by climate change. Significant decreases in forage biomass production are to be expected, except in mountain rangelands, where the rise in temperatures could positively influence pasture productivity. Rangeland degradation in Central Asia is not only threatening the livelihoods of local populations, but can also have global implications in affecting climate change since it will result in significant releases of CO_2 and other trace gases to the atmosphere, with possible effects on the global climate.

The Central Asian Countries Initiative for Land Management (CACILM) National Programming Frameworks for the Central Asian countries provide a comprehensive list of causes of rangeland degradation, including poor pasture management, overgrazing, cutting of shrubs, lack of maintenance of rangeland infrastructure, difficult conditions for mobile grazing, insufficient introduction of

Table 18.2 Salinization of irrigated areas in Central Asia

Country/region	Irrigated area affected by salinization (%)
Kyrgyzstan	11.5
Tajikistan	16.0
Kazakhstan	33.0
Turkmenistan	95.9
Uzbekistan	50.1
Central Asia	47.5

From Bucknall et al. (2003)

Table 18.3 FAO estimates of land area degrading and population affected

Country	Area degrading (%)	Population affected (%)
Kazakhstan	17.9	13.3
Kyrgyzstan	11.7	12.7
Tajikistan	5.9	2.4
Turkmenistan	0.3	0.3
Uzbekistan	1.3	2.2
Central Asia	13.2	6.0

Based on Bai et al. (2008)

rangeland rotation, unclear regulations about leasing, limited awareness of rangeland degradation issues and approaches, weak policy and institutional capacity to manage rangeland sustainably, increased demand for other land uses such as agriculture, industry and infrastructure development, and insufficient supply of water.

The above list shows not only the complexity of the problems related to rangeland management in Central Asia, but also the huge financial dimension of investments needed if the situation should be reversed (Table 18.3).

18.1.2 All Countries in Central Asia Face Similar Challenges

All Central Asian countries are still undergoing transitions following the dissolution of the Soviet Union. They are still in the process of developing the appropriate capacity to secure sustainable land use. The present agricultural and water policies, regulations and incentives (with respect to land tenure, decisions on crop choices and marketing, access to factors of production, etc.) have to a large extent been inherited from the past and are often inappropriate for current conditions. In all countries inconsistent environmental laws and regulations contribute to the agricultural problems.

Land degradation is spreading without regard for national borders; common water resources are being overexploited. The reasons for desertification are similar in all Central Asian countries. More than 80% of the Central Asian countries (over four million square kilometres) consists of dry zones. Frequently, common land use strategies are inappropriate for the prevailing ecological conditions. In addition, water resources are unevenly distributed – Kyrgyzstan and Tajikistan hold about two thirds of the total reserves, whereas Uzbekistan, Turkmenistan and Kazakhstan are more or less dependent on these. Conflict seems inevitable.

Large areas of the Central Asian countries are already been severely degraded for a number of reasons: decades of centrally planned economy during the Soviet era; state-dictated sedentarism; legal and institutional uncertainties; poverty resulting from high unemployment following independence. The fact that all Central Asian countries are now suffering from the consequences of overexploitation of their natural resources is aggravated by the effects of population growth and social biases in policies and in the allocation of public goods.

In all Central Asian countries, land degradation processes lead to negative socioeconomic problems such as increasing poverty and negative population shifts, reduced food security and life expectancy, increases in healthcare costs, social, economic and political instability, and deterioration of environmental quality (CACILM 2006). In all Central Asian countries, the paradigm of "production at all costs" continues to be expressed through numerous government decrees maintaining two essential barriers for sustainable land use:

1. A low level of consciousness on the part of decision makers, who are still not aware of the costs of land degradation in real terms.
2. The countries have yet to develop the capacity to develop comprehensive action programmes that will lead to the necessary policy reforms in each country, including the mainstreaming and simplification of diverse and often conflicting policies. This is aggravated by weak capacity in key ministries and agencies at central government level together with an unclear definition of tasks, structures, commitments and duties, procedures, and human resource deployment.

In the future these problems will be accentuated by the effects of climate change.

18.2 Climate Change and Land Use Development in the Central Asian Region

18.2.1 Climate Change Scenarios and Predictions

The assessment report "Climate Change 2007" of the Intergovernmental Panel on Climate Change (IPCC) states that climate change is already under way. Even if immediate mitigation action is taken, the consequences of climate change are no longer reversible.

The IPCC also concludes that Central Asia is one of the regions of the world that is most threatened by climate change. According to the IPCC's forecasts, in Central Asia the temperature will rise, precipitation will be reduced and, in the long term, melting glaciers will threaten water supply in the region. Experts are already warning that no region in the world outside North Africa will be more badly hit by water shortages due to climate change than Central Asia. Between 1957 and 1980, glaciers in Central Asia shrank by 19%, and the glaciers surrounding Lake Issyk-Kul in Kyrgyzstan shrank by about 8%. The small lower glaciers may completely disappear in the middle of the twenty-first century if melting continues at the same pace.

According to the IPCC 2007 Assessment Report, the Central Asian region in detail is likely to warm on average by 3.7 K by 2100 if CO_2 levels double; this is well above the global mean prediction of 2.8 K. This phenomenon can be explained according to Giese and Moßig (2004) by the anthropogenic influence – mainly diversion of the water resources on a very large scale in whole of Middle Asia, and specifically by the desiccation of the Aral Sea and the formation of huge dust clouds in many parts of the Aralkum (see Chap. 4). This hypothesis looks reasonable, but to clarify the situation many more data are needed from large-scale studies in the area and special research with remote sensing techniques.

Winter precipitation is likely to increase and summer precipitation is likely to decrease in Central Asia. All models predict temperature increases, ranging between 3 K and 6 K. Most models predict an increase of precipitation in Kyrgyzstan, Kazakhstan and Uzbekistan and a decrease in Turkmenistan.

Emissions per capita range from around 1 t CO_2 per person per year in Tajikistan and Kyrgyzstan to 13.3 t per person per year in Kazakhstan. In global terms, Central Asian countries are low emitters but they are high carbon intensity economies, meaning that emissions are high per dollar of gross domestic product. In most republics, energy consumption and industrial processes make up between 80% and 90% of all emissions, the rest coming from agriculture and waste. In the agricultural sector, most emissions are from enteric fermentation and manure storage; Tajikistan and Uzbekistan show significant N_2O emissions from fertilizer application. Since 1990 greenhouse gas emissions have decreased in Kyrgyzstan, Kazakhstan and Tajikistan and increased in Turkmenistan and Uzbekistan. In some countries a trend towards lower carbon sequestration can be observed. This trend is due to cutting of trees and shrubs for fuel and ploughing up of pasture land to grow subsistence crops. In Kazakhstan, on the other hand, abandoned cropland has led to an increase in carbon removal from the atmosphere by fallow vegetation.

18.2.2 Potential Consequences of Climate Change in Central Asia

As a result of the predicted climate change, climate zones will shift and hence agricultural land, pasture and forest will very likely move northwards by up to 400 km. A new hyperarid zone will be created by this process.

In the short term, runoff peaks will increase and occur earlier in the spring as a result of glacial melting and warmer springtime, whereas the summer runoff will be lower than at present. This presents a serious problem for low-level Central Asian countries such as Uzbekistan, Turkmenistan and part of Kazakhstan for which runoff is critical for irrigation during the spring period. Over the long term, total annual runoff is predicted to decrease in most basins, particularly those fed predominantly by snowmelt and small glaciers. The Amu Darya and Syr Darya basins may see runoff decreases of up to 40% and 28%, respectively, if CO_2 levels double. Under worst-case scenarios, Kazakhstan basins may experience losses of between 30% and 70%.

On the one hand, an increase of CO_2 in the atmosphere may benefit crop growth provided that soil nutrients are not limiting; longer growing seasons may improve the possibility to grow certain types of vegetable at higher altitudes or more northern latitudes. On the other hand, extreme prolonged periods of high temperatures in the south will have negative effects on cotton production even if irrigation is maintained. Negative effects on animal and pasture productivity will have to be expected when temperatures increase.

In the event of failure to adapt land use to these changes, degradation will increase sharply, particularly in the semiarid and arid areas of Central Asian countries. The IPCC assumes an overall loss of productivity of 40–90% in the dry zones of Central Asian countries. In the long term the availability of water resources will decline, although in the medium term river runoff in summer will increase owing to increasing melting of glaciers. However, some regions of Central Asian countries might benefit from climate changes, at least in the short term.

Forests in the mountainous regions, for example in Kyrgyzstan, could spread and forest density could increase owing to the higher temperatures and improved water availability. Higher yields are also expected in arable farming. In Kazakhstan, on the other hand, freshwater resources will decrease by 20–30% in the longer term, and grain yields and pasture productivity are expected to fall by up to 90%. Expectations are similar for Turkmenistan.

In Tajikistan, summer pastures are expected to benefit from climate changes, but winter pastures, on the other hand, will become degraded. In the medium term, a reduction in glacier ice volume of up to 30% is expected, with a likely significant reduction in water runoff from the rivers in the longer term. This could have catastrophic consequences for irrigation agriculture, water supply and hydropower generation (ADB 2006).

In the Aralkum itself, the migration and dispersal of desert shrubs and the formation of semidesert pastures, at least on the areas of the desiccated seafloors from the 1970s and 1980s, can be expected.

18.2.3 Priority to Stabilize Agricultural Production

Against this background, combating desertification and promoting sustainable land use takes on even greater significance, because climate change, desertification processes and land use are intimately interconnected. It is certain that the predicted climate change will lead to a decline in the productivity of agricultural land, particularly if current patterns of intensive and inappropriate land use continue. The consequences of climate change – water shortage, decline in soil fertility, loss of harvests and incomes, and scarcity of food and drinking water – will affect a region where rural poverty is already widely spread. The social consequences, such as loss of food security, conflicts over water and migration to more fertile areas, are presently impossible to predict but seem to be inevitable.

Since the demands for wheat and rice are projected to increase, and with wheat at the upper limits of its temperature threshold because of global warming, planning for investment in productivity will be required to enable an increase in domestic production on a limited and deteriorating land base. The World Development

Report 2008 of the World Bank recommends a future concentration of the agricultural production in Central Asian countries. It will be inevitable to modernize the agricultural production systems in Central Asian countries in order to be compatible with the international and increasingly globalized markets. The protection and sustainable use of their natural resources is an essential prerequisite for the stabilization and modernization of the agricultural production systems in Central Asian countries.

18.3 International Treaties Governing Land Management, Biodiversity and Climate Change in Central Asia

The following international conventions and regimes are relevant for the management of natural resources in Central Asia, and are also relevant for the special situation of the dry seafloor of the former Aral Sea:

- The United Nations Convention to Combat Desertification (UNCCD)
- The United Nations Convention on Biological Diversity (UNCBD) together with its principles for the sustainable use of biodiversity (Declaration of Addis Ababa)
- The United Nations Framework Convention on Climate Change (UNFCCC)
- The Convention on Access to Information, Public Participation in Decision-Making and Access to Justice in Environmental Matters (Aarhus Convention)
- The Convention on Migratory Species, and the Convention on International Trade in Endangered Species of Wild Fauna and Flora
- The Framework Convention on Environmental Protection for Sustainable Development in Central Asia (Ashgabat Convention)

All Central Asian countries have ratified the UNCBD, UNCCD and UNFCCC, as well as the Aarhus Convention, except Uzbekistan.

Especially, the adoption of the UNCCD in 1996 meant that for the first time there was an instrument enabling Central Asian countries to find common solutions to the problem of desertification, which had worsened dramatically following the demise of the Soviet Union. National action programmes (NAPs) provided a strategic instrument for combating desertification at national level; a subregional action programme (SRAP) to combat desertification for Central Asia covered the regional and transborder issues related to desertification (Winckler and Wucherer 2003).

Within the framework of the SRAP, the following priority lines of subregional cooperation were chosen:

- Monitoring and assessment of desertification processes and environmental impact assessment in the region
- Improving water use in agriculture; combating erosion, soil salinization and salt-marsh formation

- Development of forest amelioration and forest resource management in both the plains and the mountains; watershed management
- Range management
- Nature protection and biodiversity conservation; ecological tourism development
- Establishment of early warning system for drought and drought mitigation

Government funding for these NAPs remained uncertain, however, and the necessary investment for the implementation of both NAPs and the SRAP did not materialize as envisaged. For this reason, the Central Asian countries and a number of their international partners decided to create the CACILM, as a multi country partnership framework, and the CACILM programme.

CACILM represents the first ever attempt by a group of donors to support a whole region to implement the UNCCD and undertake concrete measures to combat desertification and improve land management. The Global Environment Facility (GEF) as the financial mechanism of UNCCD has therefore made Central Asia a priority region and, as a result, CACILM could become a worldwide model. Most of the measures will be conducted at the country level. Some activities, for example training and education or information exchange, are planned for the regional level.

CACILM pursues a "bottom-up" approach to planning. The rural population and government institutions and nongovernmental organizations are encouraged to develop their own structures and strategies. In the countries themselves, implementation is to be managed by an interdisciplinary working group with the participation of civil society and development organizations operating in the country. For the overall CACILM process, a steering committee has been established through which implementation of the initiative is managed by country representatives together with the group of donors (CACILM 2006).

Significant progress has been made in achieving expected outputs during the inception phase, which started in 2006. Some national programme areas and at the same time multi country activities (activities spanning the Central Asian countries) were formulated:

- Knowledge management and information dissemination (Fig. 18.2)
- Capacity building for mainstreaming sustainable land management and ensuring that it is integrated into planning and management
- Development of a sustainable land management information system
- Sustainable land management research (Cai et al. 2003)

It is expected that over the 10-year CACILM programme period, the goal of restoring, maintaining and increasing the productivity of land and improved economic and social well-being will be attained, for example through:

- Increases in productivity of lands and pasturelands, decreases in the amount of waterlogged and salinized land and increases in forest cover
- Increase in the size of biodiversity conservation areas
- Improved capacity of institutions in Central Asian countries to adopt integrated land use planning and management (Fig. 18.3)

Fig. 18.2 The abandoned terskent biological research station north of Saksaulsk (47°N17,2′ – 60°E 59,5′) should be urgently restaured for capacity building (photo: Breckle)

Fig. 18.3 Saving old scientific books from wet cupboards from the abandoned Terskent biological station by Liliya Dimeyeva and Walter Wucherer (photo: Breckle)

18.4 The Role of Science in Combating Desertification and Improving Land Management in Central Asia

18.4.1 Scientific Potential of the Area

Scientific structures dealing with the desert and desertification problems originated in Central Asia in the 1960–1970s. First of all, the Desert Institute in Ashgabad, Turkmenistan (now the National Institute of Deserts, Flora and Fauna, Ashgabat), in the former USSR became the centre of scientific research in the sphere of desert problems in the Soviet Union. It has held scientific meetings, organized international training courses and published a monthly periodical (*Problems of Desert Reclamation*). The situation changed in the 1990s with the independence of the Central Asian countries. The scientific research funding was drastically reduced, and many skilled specialists left their institutes. Nevertheless, the still existing scientific potential in the region provides a basis for carrying on the necessary research on desertification problems.

In Central Asia there are several field research stations where fundamental and applied research was conducted, such as the Terskent research station in the Aral basin zonal deserts (see Figs. 18.2 and 18.3), which is now deteriorating, the Kyzylkum station in Uzbekistan, the Nebit-Dag and Karryh-Kul stations in Turkmenistan, the Kazalinsk ecological station in the Kyzyl-Orda region of Kazakhstan and the Muinak ecological station in Karakalpakstan (Uzbekistan). Today the condition of all these stations is very poor, and their future is unclear. There is no doubt that these field stations can only continue to serve as strong points in the future regional arid research network if they are adequately supported.

The proposed CACILM research programme takes into consideration the requirement for broad stakeholder involvement. It will also take into account the numerous ongoing research programmes whose results are often difficult to obtain and to adapt to practical use. The proposed research programme will strongly support scientific capacity building, especially fostering the training of young scientists.

Fig. 18.4 Vegetable cultivation as an alternative source of income in Nokhur, Turkmenistan (photo: Kleinn)

Fig. 18.5 After deforestation, the forest near Barvoz is slowly recovering as a result of forest management (photo: Kleinn)

A number of gaps concerning knowledge of land degradation and sustainable land management practices have been identified which will become priority areas for further research:

- Improved agronomic, soil and water conservation practices (Fig. 18.4). Especially, the rational use of water to minimize risks of salinization and at the same time to save water is a big challenge (Abdullaev 2008, 2009; Martius et al. 2009).
- Diversification of cropping and livestock production systems for increased income generation.
- Germplasm collection and improvement particularly for stress tolerance to drought, salinity and extreme temperatures.
- Improved water-use efficiency.
- Better rangeland and livestock management (Gintzburger et al. 2003).
- Improved forest and woodlot management (Fig. 18.5) (Khamzina et al. 2008).
- Adequate methods of land degradation assessment and monitoring (Figs. 18.6 and 18.7).

Fig. 18.6 Aralsk, the former harbour basin is full of rubbish and garbage, saline soils and a source of contamination and health problems (photo: Breckle)

Fig. 18.7 Sand dunes cross the village of Bugun and destroy house walls by sand blasting (photo: Breckle)

Fig. 18.8 Barsa-Kelmes Nature Reserve: the old map, shown at the nature reserve's administrative site in Aralsk, indicating the former situation (photo: Breckle)

Fig. 18.9 Barsa-Kelmes Nature Reserve. New posters and boards indicating prohibition of hunting and penalties are prepared by the nature reserve's administration. Hunting saiga antelopes will cost 183,800 tenges, hunting a gazelle will cost 367,600 tenges and uprooting a saxaul shrub will cost 3,400 tenges as a penalty. Hunting a wild ass is even more expensive, it will be punished with 1,838,000 tenges (1€ is equivalent to about 190 tenges). It is not known how many prosecutions there have been (photo: Breckle)

18.4.2 Barsa-Kelmes: An Example for a Strategic Change in Nature Protection in the Aralkum Region

There is only one nature reserve in the Greater Aral Sea area: Barsa-Kelmes (see Chap. 14, Sect. 14.2). It was founded in 1939 and its area amounts to about 17,000 ha

(Fig. 18.8). Until 1998 Barsa-Kelmes was an island in the Aral Sea. Between 1996 and 1998, as a consequence of the falling water table of the Aral Sea, a "bridge" was formed between the island of Barsa-Kelmes and the mainland at the eastern coast.

Because of its natural isolation, Barsa-Kelmes Nature Reserve used to be one of the best protected nature reserves in Central Asia (Wucherer and Breckle 2005; Breckle 2011). Shortly after Barsa-Kelmes became accessible via the mainland in 1998, hunting activities started, particularly for antelopes and wild boars (Fig. 18.9).

In the meantime, wildlife (mainly antelopes and wild boars) have spread to a larger, freely accessible space between the former island of Barsa-Kelmes and the eastern coast. This change as well as economic factors compelled the relevant

Fig. 18.10 Hot spring at Akbasti with 49°C water, day and night (photo: Breckle)

Fig. 18.11 The restored mosque at Turkestan, an impressive building and park, famous for tourist visits (photo: Breckle)

Fig. 18.12 Harvesting teresken in the Pamir Mountains (photo: Kleinn)

Fig. 18.13 Dung from cows, dried in front of a farm house and stable, north of Aralsk, used as the main energy source for cooking and heating (photo: Breckle)

government services to abandon the strict approach of nature reserve and develop a buffer zone model instead (see Chap. 14).

The example of Barsa-Kelmes Nature Reserve demonstrates how changing the system of nature protection can link biodiversity conservation and economic interests of the local population in the Aralkum region. In future, ecotourism could be developed since there are several hot springs in the area (Fig. 18.10) as well as famous historical sites (Fig. 18.11).

18.4.3 Desertification, Water Supply and Energy Supply

In 2001 the GTZ-CCD project group, together with the individual Central Asian countries, initiated pilot projects in each Central Asian country, which later became part of a regional network for exchange of experience and information on participatory approaches as well as technical and structural solutions for sustainable land management and combating desertification. In the meantime, most of them have developed into more substantial subprojects.

Fig. 18.14 Preliminary pontoon bridge over the Syr Darya (photo: Breckle)

Fig. 18.15 Fishing boats on the North Aral Sea, near Bugun, indicate the recovery of fish populations after the construction of the dam. The main fish population is European plaice (*Pleuronectes platessa*), a commercially important flatfish introduced by a Danish consortium (photo: Wucherer)

Two of these subprojects in Tajikistan and Turkmenistan will be briefly mentioned. Both tackle the Aralkum basin in its wider sense. In Tajikistan the project dealt with "Combating poverty on the roof of the world: Desertification driven by poverty and energy crisis" (CACILM 2006; GTZ CCD Project 2008). The riparian forests indigenous to the river valleys and the teresken bushes (*Ceratoides papposa*)—a plant not only required for yak rearing—in the mountainous regions are severely threatened, as the energy crisis has meant that local people are using them intensively as fuel (Fig. 18.12) for heating and cooking (Ahmadov et al. 2006); thus, the demand for this slow-growing bush is high. By using this plant as fuel, people are burning their livestock's fodder and their only natural defence against erosion. This specific situation is called "teresken syndrome" and is paralleled in Bolivia by the "tola syndrome" (Breckle and Wucherer 2006). Often dung from cows is collected, dried and used as an energy source for cooking and heating (Fig. 18.13).

Fig. 18.16 The fruit market at Aralsk (photo: Breckle)

A pilot project in Turkmenistan delivered the basis and starting point for a UNDP/GEF/GTZ cooperation project which started in 2006 and which is part of the CACILM implementation process in Central Asia. The project aims at creating a partnership for sustainable land management between the Government of Turkmenistan, local government structures, communities and the civil society.

On the ground, investments by the project will focus on three ecologically different regions in the country: the mountain area of Kopet-Dag; the sandy desert region of the Karakum and the area of intensive irrigated agriculture of Mary. In these regions, the project will enhance sustainable pasture management practices, identify and support alternative sources of income for the population and install technical measures to prevent further damage through wind and water erosion as well as salinization processes (GTZ CCD Project 2007b). Thus, it places strong importance on the water dependence of the Karakum Canal on the Amu Darya as the main source for water in Turkmenistan and improved water management.

Environmental and economic river basin assessments will be made which will include water demands for irrigation and drinking water and aspects related to safety of hydrotechnical facilities and dams, including bridges (Fig. 18.14).

In most subprojects value chain promotion is a key component of the programme. The value chains are set so as to generate explicit environmental benefits, for example, by applying principles of sustainable resource use and management in agriculture, by introducing integrated pest management – including the use of so-called beneficial organisms – and by adopting product quality standards. The value chain promotion focuses on developing the potentials of the following subsectors: melons and sugar melons, greenhouse products, fodder crops, livestock, aquaculture and fisheries (Fig. 18.15 and 18.16). Furthermore, the programme promotes energy efficiency as well as decentralized renewable energy technologies.

18.5 Main Challenges for Sustainable Land Management in the Central Asian Region

18.5.1 Three Big Basic Challenges

Natural resources – forests, pasture, arable land and water – form the basis for subsistence of most people living in the Central Asian countries. Poverty, however, often drives them to put even greater strain on natural resources that are already overexploited. That means that environmental and social problems have to be tackled simultaneously. There are three main challenges lying ahead for the Central Asian countries:

1. The first crucial challenge consists of breaking the vicious circle of poverty and desertification. Only if people learn to use natural resources sustainably and if their living conditions improve at the same time can this vicious circle be broken. What is needed first and foremost is the active participation of the resource users in all field activities aiming at sustainable land use. Only if people in Central Asia are able to continue to derive benefits from the use of their land will they be motivated to use their land resources sustainably. That means that activities aiming at sustainable land use have to take technical and socioeconomic aspects into consideration at the same time.
2. Adapting land use to the declining productivity of agricultural and pastoral land induced by desertification and climate change is another main challenge. To ward off the increasing risk of degradation, it is vital to make land use more flexible, for example by planting new types of crop, making pasture use more mobile, introducing other types of livestock and protection measures against erosion
3. Central Asia is a region with scarce water resources that are used intensively for hydropower and irrigation, often with conflicting claims for use. The economies of the five independent Central Asian countries and Afghanistan are highly interconnected by water discharge of the Syr Darya and Amu Darya and other transboundary rivers. The inhabitants along the lower reaches of the rivers in Central Asia need substantial quantities of water for their irrigated agriculture. The inhabitants of the upper reaches, however, are aiming to use hydropower more intensively as a source of energy. One of the crucial challenges therefore is sustainable and equitable water management in the region, in order to avoid conflicts over water issues, and integration of water issues with energy and climate issues.

The essential preconditions to meet these challenges are enhanced policy coherence through mainstreaming of sustainable land management principles into national policies and legislation; country ownership of sustainable land management; improved interaction between state agencies and land users through human resource development and capacity building (MCB, 2009); strengthening policy, legislative and institutional frameworks in each country; operational land management information systems; development of sustainable land management research programmes; learning and knowledge sharing; and sufficient financial resources for on-the-ground project investments.

Measures already under way, such as the implementation of the regional initiative CACILM and the numerous projects in the Central Asian countries themselves, form a good basis for an adaptation strategy of this sort. It will be important, however, that they provide the people directly affected by the consequences of climate change with competencies for combating desertification and managing increasingly scarce resources more flexibly.

18.5.2 Adaptation Strategies

Major adaptation strategies should try to make land use more flexible and less vulnerable to more extreme weather conditions:

- Interventions in the water management sector are necessary to minimize evaporation, reduce losses from the irrigation network, reduce water consumption through modern irrigation techniques, introduce community water management schemes and regional quotas, etc. The introduction of water-saving technologies could be a measure for adaptation to climate change in the region. For example, introduction of water-saving technologies in cotton cultivation in Turkmenistan, allowing the use of 7,000 m^3 of water instead of the usual 12,000 m^3, could allow savings of 20% of the country's current water use. However, effective extension and training approaches are needed to disseminate these technologies. Beyond this, some technologies require investments in new types of equipment or materials that farmers lack access to, and that are expensive (IFPRI/ICARDA 2009)
- In agriculture, proposals include mulching, use of cover crops, crop switching and introduction of drought-resistant varieties. There is a good basis of a great variety of crops which are cultivated in a traditional way, which serve for the local markets (Figs. 18.4 and 18.16) and which have great potential for improved and sustainable yields.
- Pasture management proposals include control of animal numbers and more mobile pasture, for example through rotation and introduction of better adapted types of livestock. Possible options for addressing rangeland degradation could be, especially for the Aralkum, controlled grazing, rotational grazing, water harvesting, establishing protected area, reseeding rangelands, planting salt-tolerant and drought-tolerant species, fertilizer application, water harvesting, and the creation of an enabling institutional and policy environment
- Intensive research on climate change and vulnerability is needed; monitoring and communication systems have to be developed to establish early warning systems for drought and other extreme climatic events in the whole Aral Sea basin and in the Aralkum region.

- Public awareness campaigns are also necessary to win public support for climate change measures.

The adaptation strategies have to integrate politically challenging issues such as land reform, pasture reform and other measures to facilitate livestock mobility, and the consequences of demographic development.

18.5.2.1 Water-Saving Technologies

Many water and land management technologies and practices are available that can substantially improve the efficiency of irrigation water use in Central Asia. Among these are drip irrigation, sprinklers or microsprinklers, cutback and alternate furrow irrigation, raised bed planting, field levelling, contour furrows, microfurrows, mulching, plastic chutes for conveying water, reuse of drainage water for irrigation, and planting of more and more-water-efficient crops. Numerous experiments have been conducted with such technologies in Central Asia, with most showing very promising results. In almost all cases, substantial savings in water use were achieved (as much as 60% reduction with drip or sprinkler irrigation), and the efficiency of water use (production per unit of water) was increased substantially. In most cases, crop yields were also increased, indicating that there were generally no trade-offs between yields and water use efficiency for the technologies tested.

Irrespective of the prevailing image and economic difficulties at present, the regional vision shows that the water resources of the Aral Sea basin as a whole are sufficient to provide adequate nutrition and use for a population twice the size of the present one. Water supply and sanitation can be improved without excessive cost, with the inherent substantial reduction of child mortality. Furthermore, enough cash crops could be produced to sustain diversified economic growth and still save enough water for a healthy environment. The main condition is that agricultural productivity per cubic metre of water must increase considerably along with the careful selection of cropping patterns and varieties (UNESCO 2000; Breckle and Wucherer 2007; Breckle and Küppers 2007).

In Turkmenistan, Uzbekistan and Kazakhstan the major potential for reducing emissions of greenhouse gases lies in power generation. Consequently, most future investments are planned here. There is little potential for the development of renewable energy, except in Kazakhstan, where wind energy and hydropower investments are planned. In Tajikistan and Kyrgyzstan major investments will be undertaken in the field of hydropower, in order to produce more electricity for the national market and for export. Particularly in Tajikistan investment in hydropower will ease the energy crisis, which is a source of land degradation and loss of vegetation. Other proposals include:

- Production of biogas from manure as a method of reducing emissions, and particularly rational use of industrial fertilizer to reduce N_2O emissions
- Capturing of methane emissions from waste for fuel through recycling schemes or waste-disposal plants

- Reforestation schemes to increase sinks
- Legislative measures to regulate greenhouse gas emissions such as taxes, subsidies and emission permits

18.5.2.2 Nature Protection and Conservation as an Element of Sustainable Land Use and Combating Desertification

Biodiversity conservation is particularly important for Central Asia. Long-term biodiversity conservation, however, implies an effective balance between protection goals and the interests of the local population. Protection of biodiversity will only be accepted by members of the local population if their economic interests are taken into account. Therefore, creating alternative sources of income through sustainable use of natural resources and biodiversity is of particular importance.

Effective biodiversity protection can only be implemented by simultaneous changes at the national, regional and local levels. Here the following aspects have to be taken into account:

- Participation of all interest groups in decision-making processes
- Decentralization of decision making and participation of the local population
- Development of adequate incentives for the user groups and ensuring their user rights, among others by creating an appropriate legal framework and through integration into land use planning
- Coordination with other sectors (agriculture, tourism, nature conservation, etc.); improvement of transboundary cooperation

18.6 Conclusions

The situation in the Aralkum region is basically a consequence of problems and complex processes having their seeds outside the region. They crucially influenced the development of the Aralkum region in the past. One of the most important influencing factors is the national and regional management of water resources and its association with aspects of energy supply as well as agricultural and pastoral production. The impact of climate change has to be added.

However, as pointed out, the complex problems of the Central Asian region will continue to have a decisive influence on the ecological and socioeconomic situation in the Aralkum and determine its economic future. In other words, the situation in the Aralkum will only have a chance of a positive evolution if solutions for the structural problems which threaten the future of the Central Asian region as a whole can be found.

Research results from the last decade in the Aralkum are distinctly important because they identify concrete possibilities and limits for adjusting to ecological change at the local level. A choice of possible actions on various levels in the region will arise from these results. Many other small bricks are necessary to provide an overall improvement of the living conditions of the local people, and to switch local and regional administrations to active projects in the whole Aralkum region to find new ways to adapt and to take responsibility for positive actions, new perspectives and even use of advantages of the new situation in the Aralkum.

References

Abdullaev I (2008) Socio-technical aspects of water management in CA: emerging water governance issues at the grass root level. In: Varis O, Rahaman M (eds) Central Asian Water: water and development publications. Helsinki University of Technology, Helsinki, pp 89–103

Abdullaev I, Manthritilake H, Kazbekov J (2009) Water and geopolitics in Central Asia. In: Arsel M, Spoor M (eds) Water, environmental security and sustainable rural development: conflict and cooperation in Central Eurasia, Routledge ISS studies in rural livelihoods. Routledge, New York, pp 125–144

ADB (March 2006) Multi Country Partnership Support (Project Document)

Ahmadov KM, Breckle SW, Breckle U (2006) Effects of grazing on biodiversity, productivity and soil erosion of alpine pastures in Tajik mountains. In: Spehn E, Körner C, Lieberman M (eds) Land-use changes and mountain biodiversity. Taylor & Francis, Boca Raton, pp 253–261

Bai ZG, Dent DL, Olsson L, Schaepman ME (2008) Global assessment of land degradation and improvement. 1. Identification by remote sensing. Report 2008/01. Wageningen, Netherlands, ISRIC – World Soil Information

Bekturova G, Romanova N (eds) (2007) Traditional land management knowledge in Central Asia: resource pack. S-print, Almaty

Blua A (2004) Aral Sea catastrophe leaving its mark on DNA of local populations. EurasiaNet 7

Bot A, Nachtergaele F, Young A (2000) Land resource potential and constraints at regional and country levels. World Soil Resources Report No. 90. Rome, FAO

Breckle SW (2011) Der verlorene Aralsee – aber 60 000 km^2 Neuland! In: Anhuf D, Fickert T, Grüninger F (eds) Ökozonen im Wandel. Passauer Kontaktstudium Geographie 11: 91–100

Breckle S-W, Küppers W (2007) Water conserving plants for agriculture in arid and semi-arid areas. In: Lozan Jl et al. (eds) Global Change: Enough water for all? Wiss. Auswertungen/GEO Hamburg, p.310–314 (3rd ed. 2011, p. 518–528 in German)

Breckle SW, Küppers M (2011) Wassersparende Nutzpflanzen in ariden und semi-ariden gebieten. In: Lozan JL, Graßl H, Hupfer P, Menzel L, Schönwiese Ch-D (eds) Warnsignal Klima: Genug Wasser für alle? Wissenschaftliche Auswertungen, in Kooperation mit GEO, 3rd edn (in press)

Breckle S-W, Wucherer W (2007) What will be the future of the Aral Sea? In: Lozan Jl et al. (eds) Global Change: Enough water for all ? Wiss. Auswertungen/GEO Hamburg, p.142–146 (3rd ed. 2011, p. 210–219 in German)

Bucknall J, Klytchnikova I, Lampietti J, Lundell M, Scatasta M, Thurman M (2003) Irrigation in Central Asia: social, economic and environmental considerations. Europae and Central Asia Region. Environmentally and Socially Sustainable Development Division. World Bank, Washington (English, Russian). Also available at http://www.worldbank.org/eca/environment, (accessed July, 19th, 2011)

CACILM (May 2006) Central Asian Countries Initiative for Land Management (CACILM brochure)

Cai X, McKinney DC, Rosegrant MW (2003) Sustainability analysis for irrigation water management in the Aral Sea region. Agric Syst 76:1043–1066

Fang L, Wang LZ, Hipel KW (2005) Securing fair water allocation in the Aral Sea Basin. Systems and human science – for safety, security, and dependability. Pp.159–170. Elsevier/Amsterdam

Giese E, Moßig I (2004) Klimawandel in Zentralasien. ZEU Discussion Papers No. 17

Gintzburger G, Toderich KN, Mardonov BK, Mahmudov MM (2003) Rangelands of the arid and semi-arid zones in Uzbekistan. CIRAD-ICARDA, Montpellier/France; Syria
GTZ CCD Project (2007a) Acting locally – cooperating regionally: combating desertification in Central Asia
GTZ CCD Project (2007b) Capacity building and on-the-ground investments for sustainable land management in Turkmenistan (A project leaflet)
GTZ CCD Project (2008) Climate change and land degradation in Central Asia: scenarios, strategies and funding opportunities
IFPRI (2008) Annual report. Supported by the GGIAR. (http://www.ifpri.org), accessed july, 19th, 2011)
IFPRI/ICARDA (2009) Economic analysis of sustainable land management options in Central Asia (Second Progress Report)
Jensen S, Mazhitova Z, Zetterström R (1997) Environmental pollution and child health in the Aral Sea region in Kazakhstan. Sci Total Environ 206:187–193
Khaydarov RA, Khaydarov RR (2007) Solar powered direct osmosis desalination. Desalination 217:225–232
Khamzina A, Lamers JPA, Vlek PLG (2008) Tree establishment under deficit irrigation on degraded agricultural land in the lower Amu Darya River region, Aral Sea Basin. For Ecol Manage 255:168–178
Larchet C, Eigenberger G, Tskhay A, Tastanov K, Nikonenko V (2002) Application of electro-membrane technology for providing drinking water for the population of the Aral region. Desalination 149:383–387
Martius C, Froebrich J, Nuppenau EA et al (2009) Water resource management for improving environmental security and rural livelihoods in the irrigated Amu Darya lowlands. In: Brauch HG, Oswald SÚ, Grin J (eds) Facing global environmental change: environmental, human, energy, food, health and water security concepts. Hexagon series on human and environmental security and peace, vol 4. Springer, Berlin, pp 749–762
MCB (2009) The Multi-Country Capacity Building (MCB, 2009); strengthening Project (Kazakhstan, Kyrgyz Republic, Tajikistan, Turkmenistan and Uzbekistan) UNDP Project Document
Médicines sans Frontières (2003) Karakalpakstan: a population in danger. Public relations, Tashkent
Qadir M, Noble AD, Qureshi AS, Gupta R, Yuldashev T, Karimov A (2009) Salt-induced land and water degradation in the Aral Sea basin: A challenge to sustainable agriculture in Central Asia. Natural Resources Forum 33(2):134–149
Saiko TA, Zonn IS (2000) Irrigation expansion and dynamics of desertification in the Circum-Aral region of Central Asia. Appl Geogr 20:349–367
UNESCO (2000) Water related vision for the Aral sea basin for the year 2025. UNESCO, Paris
Winckler G, Wucherer W (2003) Desertification – an ecological and social-economical problem in Central Asia. In: Rachkovskaya EI et al. (eds) Botanical geography of Kazakhstan and Middle Asia. GTZ-CCD-Project, St Petersburg, pp 348–351
World Development Report (2008) Agriculture for development. The World Bank, Washington, DC
Wucherer W, Breckle SW (2005) Desertifikationsbekämpfung und Sanierung der Salzwüsten am Aralsee. Sukzession und Phytomelioration, Naturschutz und nachhaltige Entwicklung. Bielefelder Ökologische Beiträge 19:1–93
Zonn I, Glantz M, Kostianoy A, Kosarev A (2009) The Aral Sea encyclopedia. Springer, Berlin

Part D
Conclusions

Chapter 19
The Aralkum, a Man-Made Desert on the Desiccated Floor of the Aral Sea (Central Asia): Final Conclusions and Comments

S-W. Breckle and W. Wucherer

The Aral Sea no longer exists. Within about 50 years a huge new desert has developed: the Aralkum. Public awareness of this area has grown considerably in the last few years.

Two research projects of the Department of Ecology at the University of Bielefeld, financed by BMBF (German Federal Ministry of Education, Research and Technology), were only two of several attempts to reveal the dynamic situation. The first was to study "Succession processes on the desiccated sea floor of the Aral Sea and perspectives of land-use" (1998–2001), the second was to study "Combating desertification and rehabilitation of salt deserts in the Aralkum" (2002–2005). Some of the open scientific questions are answered in chapters in this book. Both projects revealed interesting and new scientific data and results. They were also able to contribute to better socioeconomic conditions in some villages, and to increase the participatory involvement of stakeholders by capacity building. The cooperation with scientific institutions in Kazakhstan and Uzbekistan, and with the relevant political administrations, gave good opportunities for information exchange and increased mutual understanding in future applications of phytomelioration measures and nature conservation topics. For both, knowledge of the general situation is a precondition.

The abiotic dynamics in the area was demonstrated by Chaps. 2–4. On 13 June 2009, the analytical laboratory of Johnson Cosmic Center (NASA 2009) fixed for the first time the complete drying of the shallow eastern part of the Aral Sea (see also Fig. 2.6b). Today, the Aral Sea consists of several remnant water bodies – the Small Aral Sea (North Aral Sea) in the territory of Kazakhstan and the Large Aral Sea (West Aral Sea or Southwest Aral Sea) as a deepwater part of the former sea in Uzbekistan and Kazakhstan. A small remnant basin is the Tschebas basin in the northwest. Within about 50 years a huge new desert has developed. This desert is caused by human activities; thus, it is an artificial desert, but all the ongoing processes follow natural laws and are most interesting for science. The area of the dry seafloor of the Aral Sea, called the Aralkum (Agachanjanz and Breckle 1993; Breckle and Agachanjanz 1994; Breckle et al. 2001; Breckle and Wucherer 2011), is still increasing and is currently about 60,000 km. The present huge salt desert area

S.-W. Breckle et al. (eds.), *Aralkum - a Man-Made Desert: The Desiccated Floor of the Aral Sea (Central Asia)*, Ecological Studies 218,
DOI 10.1007/978-3-642-21117-1_19,

of the desiccated seafloor is now comparable with the Iranian salt desert or the Great Basin salt desert in Utah (USA). But the fundamental difference is the timescale. The latter deserts were formed by a geological process, whereas the Aralkum has been formed within a few decades, meaning a much faster speed of development.

The dry seafloor of the Aral Sea is a new terrestrial surface. It has developed to form a new geographical object, a new desert, that has a strong environmental impact on the surroundings of the Aral basin. The Aralkum is located within the Asiatic desert belt. In the Aralkum area, two climatic subdistricts can be distinguished – the Aralsk subdistrict in the north and the Muinak subdistrict in the south. The limit between these two districts is defined. The northern part of the Aralkum is part of the Kazakh-Dzungarian or northern Turanian deserts, and the southern part is part of the southern Turanian deserts (see Chap. 4). The climatic conditions of the area are governed by global trends. The global trend of rising temperatures is also to be seen in the area of the Aralkum. The annual mean temperatures as well as most of the monthly mean temperatures of almost all stations around the Aral Sea have risen since 1960 by about 1–2K. The warming trend was also indicated by Kuzmina (2007), who gave smoothed functions for the seasons for the Aralsk meteorological station, which indicate a rise in temperature of about 2–3 K between 1937 and 2002. The same is true for the hinterland stations (Kazalinsk, Kyzyl-Orda, Chilik-Rabat, Kuruk). The annual precipitation seems to be rising too, but is very dynamic and thus not yet statistically significant.

In the Aralkum, about 20% (about 12,000 km^2) is sand and sandy-loamy deserts, mainly the first parts of the seafloor which became dry in the 1960s and 1970s. The majority is salt desert (70%, about 42,000 km^2), and this situation is rather different from that in other deserts, where sometimes no salt desert may exist (e.g., in the Negev, the Sonoran Desert or the Chihuahua; Table 1.1). The remaining area (10%) is wetlands and remnants of tugai scrub, as well as transformed landscapes that are under the influence of the Amu Darya and the Syr Darya (see Chap. 1). The Aralkum is a new continental area that is governed by geohalomorphic as well as by aeolian processes. The vast-plain open-sea bottom is not totally plain; it exhibits old submarine channels, hills, ridges, sands splits and tombolos, which, are now overformed. The desiccated soil sediments along the coast and the forms of the primary relief play an important role in the formation and development of the present landscape structure in the Aralkum. The north and the west of the Aralkum are limited by the high structural denudation plateaus of Turgay and Ustyurt, in the east by the sand deserts Priaralski Karakum and Kyzylkum and in the south by the Amu Darya delta. The elevation amplitude of the Aralkum is now between 25 m above sea level (asl) and 58 m asl, including the old and new Aral terraces (see Chap. 3).

The biotic dynamics, the flora and vegetation cover, and the already rich fauna of the Aralkum were discussed in Chaps. 8–11. The rapidly changing conditions and the fast successional processes are fascinating and give a new picture of the not at all constant species pattern (Chaps. 10, 12 and 13). The flora of the Aralkum consists of about 370 species, belonging to 44 families and 178 genera (Chap. 8). The flora of the Aralkum is rather young and has formed since 1960 as a typical immigration and desert flora. Formation of the Aralkum flora in ancient and modern times occurred and is occurring simultaneously with the development of landscapes, vegetation and soils. It is certainly a flora where species have to be adapted to the rather severe ecological conditions of drought and heat in summer, to frost in winter and to moving sand conditions and to varying salinity, as well as to inundation and anaerobic conditions of solonchak soils.

The Aralkum region is a biodiversity centre of halophytes in Middle Asia and Central Asia (Wucherer et al. 2001). Plants have developed various mechanisms to cope with salinity (Chap. 12). Often morphological structures are typical for distinct adaptation strategies. Especially halosucculence of stems or leaves, or both, is very common in halophytes strongly adapted to salinity (Breckle 2002). The halophytic species, nevertheless, are on the other hand indicators of the degree of salinity on their site, and thus can be used to monitor salinity. A novel list of indicator values for salinity is presented. All species of the Aralkum flora are listed in Table 12.9 with their ecological salinity indicator value (*S* value), their halophytic strategy type and their life form. This list may serve as an easy way to check the degree of salinity at adjacent agricultural plots.

The Aral Sea depression is part of the temperate continental desert and semidesert belt. Accordingly, the vegetation of the Aralkum is a desert vegetation (Chap. 9). It is in various stages of development. The main vegetation types of the Aralkum are halophytic, psammophytic, tugai and salt-meadow communities. The salt and sand deserts and accordingly the halophytic and psammophytic vegetation types dominate. The typical pattern of landscapes, vegetation and soils is striated. The southern and southeastern coasts of the former Aral Sea including the dry seafloor belong to the southern Turanian phytogeographical area, and the western, northern and northeastern coasts belong to the northern Turanian area. The dry seafloor of the Aral Sea is a new surface, where terrestrial plants (including seed banks) and animals have not existed before. It is now actively populated by organisms. The formation of plant communities, soils, a new groundwater level, aquifers and all components and processes of ecosystems is occurring more or less simultaneously (Wucherer and Breckle 2001). It is a typical primary succession (Chap. 10). The succession on the dry seafloor has continued for the last 50 years. The distribution and dynamics of the vegetation and ecosystems were surveyed along transects. On average, the succession on loamy stands can be described by two to four stages, and that on sandy soils can be described by three to five stages. The existence of a distinct stage is a consequence of the ecological conditions and stability, and thus might range between 2 and 30 years. The geological, geomorphological, climatic, edaphic and aeolian factors controlled the vegetation development in the Aralkum in the first few years after the desiccation. But in the late phase, mechanisms of facilitation and tolerance have played a crucial role, especially by the psammophyte succession.

The fauna of the Aralkum (Chap. 11) has been studied only partly. Although many faunistic data have been collected in the Aral Sea basin over the past decades, there is no systematic monitoring of the faunal settlement of the Aralkum. But lists of mammals of the Kazakhstan part of Aral Sea region, the migratory breeding bird species and rare winter visitors, the resident breeding bird species, passage visitors (birds), vagrant birds, reptiles collected around the Aral Sea in the 2002–2004 and the taxonomical diversity of insect orders and other groups have been produced. The shoreline of the Aral Sea in Kazakhstan belongs mainly to two zoogeographical zones (in relation to mammals) – northern Aralo-Caspian deserts and the Kyzylkum. The southern border of the first zone runs along the state border between Kazakhstan and Uzbekistan (Karakalpakstan) and along the Syr Darya. The Kyzylkum zone comprises the Kyzylkum from the Syr Darya to the state border with Uzbekistan, and the Uzbekistan coast. This conclusion confirms the climatic (Chap. 4) and botanical-geographical (Chap. 9) border between the northern and southern Aralkum.

The desiccated sea bottom was completely transformed into an area covered by sands and solonchaks. There are many plans for tackling and combating the disastrous situation caused, for example, by the dangerous salt-dust storms. The salt-dust storms are particularly endangering human health as well as productivity of agricultural areas by enhancing salinization in adjacent agricultural regions. This is a prominent problem in the Aralkum area as a whole. Semenov has developed a physical model to evaluate the amount of aerosols transported from the desiccated seafloor of the Aral Sea (Chap. 5). This is a regional problem. Monitoring of land cover condition and analysing land cover change in the Aralkum is of great importance, since the ecological situation is still very dynamic and a large part of the landscape in the Aralkum is unstable (Chap. 6). The results demonstrate that MODIS time series classification is a valuable tool to produce accurate landscape classification, landscape change maps and statistics for large areas. For the southern Aralkum, the results of the postclassification change detection revealed that (a) between 2000 and 2008, no significant vegetation cover emerged on the former seabed in the study area; (b) the shrubland and reed classes show high interdependence; and (c) the potential source area for dust and salt storms in the study area has increased. The results of analysis of satellite data for the period 2005–2008 (Chap. 7) show significant differences in the frequency and magnitude of dust storms from year to year. A monotonic increase in the annual activity of the transport processes is observed. It is obvious that the process of removal of salt-dust aerosol particles from the dry bottom of Aral Sea is the result of progressive drying of the former seabed and the formation of a growing area with an unstable surface not fixed by vegetation.

This problem could be solved by huge plantings of heat- and frost-resistant as well as drought- and salt-resistant woody plant species to minimize wind speed and deflation processes on the soil surface of the dry seafloor. Huge experimental plantings (Chaps. 15–17) with plots up to 250 ha have shown that only very few species are suitable for this purpose: *Haloxylon aphyllum* and *Halocnemum strobilaceum*. Some other species still need to be tested accordingly (*Halostachys caspica* and *Tamarix*, *Ammodendron*, *Suaeda* and *Salsola* species). Furthermore, recent evaluation of various experimental sets revealed that special techniques to plant saplings have to be applied and have to be adjusted to the relevant soil situation.

The problems of the local sandstorms (Chaps. 5 and 15), threatening villages and infrastructure, must be solved by local wind shelter programmes and rational grazing management, involving activities of all inhabitants.

The problems of nature conservation in the Aralkum region are described in Chap. 14. The territory of the nature reserve in the north was expanded in 2006 by almost 10 times (160,826 ha), including territories of the dry seafloor. The area now consists of two cluster areas: (1) the former island of Barsa-Kelmes with the surrounding dry seafloor; (2) the former islands of Kaskakulan and Uzun-Kair with the surrounding dry seafloor. These are the main habitats of onager (*kulan*; *Equus hemionus onager*) and Persian gazelle (*jairan*; *Gazella subgutturosa*), to where they migrated after the area became dry. The flora and vegetation represent typical combinations of plants and vegetation types for the region as well as unique trends of the successional development on the desiccated seafloor of the Aral Sea.

Chapter 18 presents a synthesis on the complexity of the social and economic situation in the post-Soviet states as well as an approach to the applications of scientific results to combat desertification also regarding the issue of the sensitivity of the area to climate change. An important issue will certainly be the rebuilding of efficient agriculture with water-saving techniques and plant crops for advanced irrigation (Breckle and Küppers 2007; Breckle and Wucherer 2007).

To conclude, the Aralkum is characterized by a very high percentage of salt desert plains (Table 1.1). Those sites only can be invaded by halophytes (Chap. 12) from the adjacent deserts (Kyzylkum and Karakum), which themselves are centres of biodiversity in Chenopodiaceae and other halophytic and also psammophytic plant families. Both play an important role in ecosystem development. The dry seafloor is the largest area worldwide where a primary succession is taking place. Knowledge of vegetation dynamics in the Aralkum, which is a mosaic of sand and salt desert ecosystems, is important for understanding of the ecosystem dynamics in the whole Central Asian area. The desiccation of the Aral Sea and the formation of the Aralkum are without doubt important milestones in the formation and evolution of new Turanian geoecosystems and bioecosystems and for the Middle Asian region as a whole. Future studies may reveal if there is a steady state of a dynamic equilibrium of water balance as well as of ecosystems has been reached. Most probably, the development of species patterns adapted to the harsh environmental conditions will be very dynamic.

The importance of this huge new desert with the remnant water bodies for the Central Asian states may be seen as an area for gas and oil mining, but also as an area where nature may have a chance to develop rich semidesert and desert ecosystems as basic parts of a protected nature reserve or a national park (Chap. 14).

Within this volume some important data and basic scientific knowledge are brought together in an interdisciplinary and international approach by not only scientists from the fields of ecology and geography, but also social scientists and economists. To improve the regional situation, cooperation of nature conservation organizations, developmental and health agencies and other main stakeholders in the area with administrations, decision makers and regional and national politicians is an urgent need (Chap. 18).

The Central Asian states involved are strongly advised to fulfil their mutual treaties (Chap. 18) as well as to use international help and funds for joint development of adapted water and energy supplies for the people living adjacent to the

Aralkum. The Aralkum is a focal point for combating desertification on many levels. Joint programmes for social and economic development of the area with promotion of education and capacity building of the people provide a great chance for a better future by minimizing the negative effects of the irreversible desiccation of the Aral Sea and by accepting the new desert Aralkum as an opportunity for the next generations.

References

Agachanjanz OE, Breckle SW (1993) Umweltsituation in der ehemaligen Sowjetunion.– Naturwiss. Rdschau 47:99–106

Breckle SW (2002) Salinity, halophytes and salt affected natural ecosystems. In: Läuchli A, Lüttge U (eds) Salinity: environment – plants – molecules. Kluwer, Dordrecht, pp 53–77

Breckle SW, Agachanjanz OE (1994) Spezielle Ökologie der gemäßigten und arktischen Zonen Euro-Nordasiens, Zonobiom VI-IX, vol 3, 2nd edn, Ökologie der Erde. Fischer, Stuttgart

Breckle S-W, Küppers W (2007) Water conserving plants for agriculture in arid and semi-arid areas. In: Lozan Jl et al. (eds) Global Change. Enough water for all? Wiss. Auswertungen/GEO Hamburg, p.310–314 (3rd ed. 2011, p. 518–528 in German)

Breckle S-W, Wucherer W (2007) What will be the future of the Aral Sea? In: Lozan Jl et al (eds) Global Change. Enough water for all? Wiss. Auswertungen/GEO Hamburg, p.142–146 (3rd ed. 2011, p. 210–219 in German)

Breckle SW, Wucherer W, Agachanjanz O, Geldyev BV (2001) The Aral Sea crisis region. In: Breckle SW, Veste M, Wucherer W (eds) Sustainable land use in deserts. Springer, Berlin, pp 27–37

Kuzmina JV (2007) Long-term dynamics according to meteorological characteristics in the Prearalie Region. Izvest vizs ujebnich zawedenii pp. 73–78 (in Russian)

NASA (2009) Image Science and Analysis Laboratory, NASA-Johnson Space Center. The Gateway to Astronaut Photography of Earth. http://eol.jsc.nasa.gov/scripts/sseop/clickmap.pl?hashnumber=7871&fieldlist=GEON,FEAT,LAT,LON,PDATE&options=none. Accessed 8 Nov 2009 02:31:58

Wucherer W, Breckle SW (2001) Vegetation dynamics on the dry sea floor of the Aral Sea. In: Breckle SW, Veste M, Wucherer W (eds) Sustainable land use in deserts. Springer, Heidelberg, pp 52–68

Wucherer W, Breckle SW, Dimeyeva L (2001) Flora of the dry sea floor of the Aral Sea. In: Breckle SW, Veste M, Wucherer W (eds) Sustainable land-use in deserts. Springer, Heidelberg, pp 38–51

Subject Index

S.-W. Breckle et al. (eds.), *Aralkum - a Man-Made Desert: The Desiccated Floor of the Aral Sea (Central Asia)*, Ecological Studies 218,
DOI 10.1007/978-3-642-21117-1, © Springer-Verlag Berlin Heidelberg 2012

T

Taxonomic Index

S.-W. Breckle et al. (eds.), *Aralkum - a Man-Made Desert: The Desiccated Floor of the Aral Sea (Central Asia)*, Ecological Studies 218, DOI 10.1007/978-3-642-21117-1, © Springer-Verlag Berlin Heidelberg 2012

MIX
Papier aus verantwortungsvollen Quellen
Paper from responsible sources
FSC® C105338

If you have any concerns about our products,
you can contact us on
ProductSafety@springernature.com

In case Publisher is established outside the EU,
the EU authorized representative is:
Springer Nature Customer Service Center GmbH
Europaplatz 3, 69115 Heidelberg, Germany

Printed by Libri Plureos GmbH
in Hamburg, Germany